Enantioselective Synthesis, Enantiomeric Separations and Chiral Recognition

Enantioselective Synthesis, Enantiomeric Separations and Chiral Recognition

Special Issue Editors

Maria Elizabeth Tiritan
Madalena Pinto
Carla Sofia Garcia Fernandes

MDPI • Basel • Beijing • Wuhan • Barcelona • Belgrade • Manchester • Tokyo • Cluj • Tianjin

MDPI

Special Issue Editors

Maria Elizabeth Tiritan
Cooperativa de Ensino Superior
Politécnico e Universitário (CESPU)
Universidade do Porto
Portugal

Madalena Pinto
Universidade do Porto
Portugal

Carla Sofia Garcia Fernandes
Universidade do Porto
Portugal

Editorial Office
MDPI
St. Alban-Anlage 66
4052 Basel, Switzerland

This is a reprint of articles from the Special Issue published online in the open access journal *Molecules* (ISSN 1420-3049) (available at: https://www.mdpi.com/journal/molecules/special_issues/Chiral_separation_recognition).

For citation purposes, cite each article independently as indicated on the article page online and as indicated below:

LastName, A.A.; LastName, B.B.; LastName, C.C. Article Title. *Journal Name* **Year**, *Article Number*, Page Range.

ISBN 978-3-03936-238-7 (Hbk)
ISBN 978-3-03936-239-4 (PDF)

Cover image courtesy of Maria Elizabeth Tiritan.

Contents

About the Special Issue Editors . . . vii

Preface to "Enantioselective Synthesis, Enantiomeric Separations and Chiral Recognition" . . ix

Maria Elizabeth Tiritan, Madalena Pinto and Carla Fernandes
Enantioselective Synthesis, Enantiomeric Separations and Chiral Recognition
Reprinted from: *Molecules* **2020**, *25*, 1713, doi:10.3390/molecules25071713 . . . 1

Diana Ibrahim and Ashraf Ghanem
On the Enantioselective HPLC Separation Ability of Sub-2 µm Columns: Chiralpak® IG-U and ID-U
Reprinted from: *Molecules* **2019**, *24*, 1287, doi:10.3390/molecules24071287 . . . 7

Ali Fouad, Montaser Sh. A. Shaykoon, Samy M. Ibrahim, Sobhy M. El-Adl and Ashraf Ghanem
Colistin Sulfate Chiral Stationary Phase for the Enantioselective Separation of Pharmaceuticals Using Organic Polymer Monolithic Capillary Chromatography
Reprinted from: *Molecules* **2019**, *24*, 833, doi:10.3390/molecules24050833 . . . 19

Anamarija Knežević, Jurica Novak and Vladimir Vinković
New Brush-Type Chiral Stationary Phases for Enantioseparation of Pharmaceutical Drugs
Reprinted from: *Molecules* **2019**, *24*, 823, doi:10.3390/molecules24040823 . . . 35

Lie-Ding Shiau
Chiral Separation of the Phenylglycinol Enantiomers by Stripping Crystallization
Reprinted from: *Molecules* **2018**, *23*, 2901, doi:10.3390/molecules23112901 . . . 51

Andreea Elena Bodoki, Bogdan-Cezar Iacob, Laura Elena Gliga, Simona Luminita Oprean, David A. Spivak, Nicholas A. Gariano and Ede Bodoki
Improved Enantioselectivity for Atenolol Employing Pivot Based Molecular Imprinting
Reprinted from: *Molecules* **2018**, *23*, 1875, doi:10.3390/molecules23081875 . . . 63

Joana Teixeira, Maria Elizabeth Tiritan, Madalena M. M. Pinto and Carla Fernandes
Chiral Stationary Phases for Liquid Chromatography: Recent Developments
Reprinted from: *Molecules* **2019**, *24*, 865, doi:10.3390/molecules24050865 . . . 83

Yangfeng Peng, Cai Feng, Sohrab Rohani and Quan He
Improved Resolution of 4-Chloromandelic Acid and the Effect of Chlorine Interactions Using (*R*)-(+)-Benzyl-1-Phenylethylamine as a Resolving Agent
Reprinted from: *Molecules* **2018**, *23*, 3354, doi:10.3390/molecules23123354 . . . 121

Fei Xiong, Bei-Bei Yang, Jie Zhang and Li Li
Enantioseparation, Stereochemical Assignment and Chiral Recognition Mechanism of Sulfoxide-Containing Drugs
Reprinted from: *Molecules* **2018**, *23*, 2680, doi:10.3390/molecules23102680 . . . 133

Yuichi Uwai
Enantioselective Drug Recognition by Drug Transporters
Reprinted from: *Molecules* **2018**, *23*, 3062, doi:10.3390/molecules23123062 . . . 145

Ankur Gogoi, Nirmal Mazumder, Surajit Konwer, Harsh Ranawat, Nai-Tzu Chen and Guan-Yu Zhuo
Enantiomeric Recognition and Separation by Chiral Nanoparticles
Reprinted from: *Molecules* **2019**, *24*, 1007, doi:10.3390/molecules24061007 . . . . **159**

Giovanna Brancatelli, Enrico Dalcanale, Roberta Pinalli and Silvano Geremia
Probing the Structural Determinants of Amino Acid Recognition: X-Ray Studies of Crystalline Ditopic Host-Guest Complexes of the Positively Charged Amino Acids, Arg, Lys, and His with a Cavitand Molecule
Reprinted from: *Molecules* **2018**, *23*, 3368, doi:10.3390/molecules23123368 . . . . **191**

Allan Ribeiro da Silva, Deborah Araujo dos Santos, Marcio Weber Paixão and Arlene Gonçalves Corrêa
Stereoselective Multicomponent Reactions in the Synthesis or Transformations of Epoxides and Aziridines
Reprinted from: *Molecules* **2019**, *24*, 630, doi:10.3390/molecules24030630 . . . . **203**

Farid Chebrouk, Khodir Madani, Brahim Cherfaoui, Leila Boukenna, Mónica Válega, Ricardo F. Mendes, Filipe A. A. Paz, Khaldoun Bachari, Oualid Talhi and Artur M. S. Silva
Hemi-Synthesis of Chiral Imine, Benzimidazole and Benzodiazepines from Essential Oil of *Ammodaucus leucotrichus* subsp. *leucotrichus*
Reprinted from: *Molecules* **2019**, *24*, 975, doi:10.3390/molecules24050975 . . . . **227**

Solida Long, Diana I. S. P. Resende, Anake Kijjoa, Artur M. S. Silva, Ricardo Fernandes, Cristina P. R. Xavier, M. Helena Vasconcelos, Emília Sousa and Madalena M. M. Pinto
Synthesis of New Proteomimetic Quinazolinone Alkaloids and Evaluation of Their Neuroprotective and Antitumor Effects
Reprinted from: *Molecules* **2019**, *24*, 534, doi:10.3390/molecules24030534 . . . . **237**

Jinsong Xuan and Yingang Feng
Enantiomeric Tartaric Acid Production Using *cis*-Epoxysuccinate Hydrolase: History and Perspectives
Reprinted from: *Molecules* **2019**, *24*, 903, doi:10.3390/molecules24050903 . . . . **255**

Carla Fernandes, Maria Letícia Carraro, João Ribeiro, Joana Araújo, Maria Elizabeth Tiritan and Madalena M. M. Pinto
Synthetic Chiral Derivatives of Xanthones: Biological Activities and Enantioselectivity Studies
Reprinted from: *Molecules* **2019**, *24*, 791, doi:10.3390/molecules24040791 . . . . **267**

Anna Poryvai, Terezia Vojtylová-Jurkovičová, Michal Šmahel, Natalie Kolderová, Petra Tomášková, David Sýkora and Michal Kohout
Determination of Optical Purity of Lactic Acid-Based Chiral Liquid Crystals and Corresponding Building Blocks by Chiral High-Performance Liquid Chromatography and Supercritical Fluid Chromatography
Reprinted from: *Molecules* **2019**, *24*, 1099, doi:10.3390/molecules24061099 . . . . **303**

Emilija Petronijevic and Concita Sibilia
Enhanced Near-Field Chirality in Periodic Arrays of Si Nanowires for Chiral Sensing
Reprinted from: *Molecules* **2019**, *24*, 853, doi:10.3390/molecules24050853 . . . . **315**

Bang-Jin Wang, Ai-Hong Duan, Jun-Hui Zhang, Sheng-Ming Xie, Qiu-E Cao and Li-Ming Yuan
An Enantioselective Potentiometric Sensor for 2-Amino-1-Butanol Based on Chiral Porous Organic Cage CC3-R
Reprinted from: *Molecules* **2019**, *24*, 420, doi:10.3390/molecules24030420 . . . . **325**

About the Special Issue Editors

Maria Elizabeth Tiritan has a degree in chemistry and PhD in organic chemistry. Currently, she is Assistant Professor at the Faculty of Pharmacy of University of Porto and Team Leader of the Drug Research line at IINFACTS. Her current interests in research include exploiting new enantioselective analytical methods to quantify and identify metabolites in the biodegradation of chiral drugs in environmental matrices; synthesis of new chiral compounds for diverse biological activities, including potential antitumor and antimicrobial agents; as well as structure–activity–properties relationships. She is also involved in the design of new chiral selectors for enantiomeric separation by liquid chromatography and membranes. https://orcid.org/0000-0003-3320-730X; https://sigarra.up.pt/ffup/pt/func_geral.formview?p_codigo=481977; https://iinfacts.cespu.pt/.

Madalena Pinto is a retired Full Professor from the Faculty of Pharmacy (FFUP) and Researcher and Team Leader of the Group of Natural Products and Medicinal Chemistry at the Interdisciplinary Center for Marine and Environmental Research (CIIMAR) University of Porto, Portugal. Her research areas focus on medicinal chemistry, involving the synthesis and molecular modification of pharmacologically active compounds based on natural models of plant and marine sources, as well as the development of substances with antifouling activity. She has a special interest in chirality, namely in synthesis and bioenantioselectivity studies of chiral bioactive compounds, as well as studies of structure–activity–properties relationships; molecular recognition of artificial receptors as new chiral stationary phases for liquid chromatography is another area of interest. ORCID: http://orcid.org/0000-0002-4676-1409; CIIMAR: https://www2.ciimar.up.pt/team.php?id=226; https://www2.ciimar.up.pt/research.php?team=27; Linkedin: linkedin.com/in/madalena-pinto-14334836; Webpage: http://madalenapinto.com/.

Carla Sofia Garcia Fernandes is a researcher at the Research Center CIIMAR and Assistant Professor of Organic and Medicinal Chemistry in the Faculty of Pharmacy of the University of Porto, Portugal. She obtained her B.S. in pharmaceutical sciences at the University of Porto in 2002; M.S. in quality control - scientific area in drug substances and medicinal plants, from the same university in 2006; and her PhD in pharmaceutical and medicinal chemistry from the same university in 2012. Current research interests include design of chiral compounds with potential biological activity, development of synthetic methodologies for obtaining bioactive chiral compounds, studies of enantioselectivity and structure–activity relationships, synthesis of chiral stationary phases, analytical and preparative enantiomeric separation of diverse racemic mixtures by liquid chromatography and membranes, and chiral recognition mechanisms. ORCID: 0000-0003-0940-9163 URL https://sigarra.up.pt/ffup/pt/func_geral.formview?p_codigo=424646.

Preface to "Enantioselective Synthesis, Enantiomeric Separations and Chiral Recognition"

The importance of producing chiral compounds in enantiomerically pure form is well recognized by academics and industries. Currently, the demand for efficient methodologies to produce chiral compounds with a high degree of enantiomeric purity requires continuous advances in enantioselective synthesis, chiral analyses, preparative enantioseparation, as well as chiral recognition studies.

This book includes both fundamental studies and applications in a multidisciplinary research field that considered commercial chiral compounds with industrial applications, bioactive compounds and pharmaceuticals, and new compounds with promising biological activities.

Nineteen papers published in the Special Issue entitled "Enantioselective Synthesis, Enantiomeric Separations and Chiral Recognition" in *Molecules* are gathered in this edition. The recent developments and innovative approaches in enantiomeric separation, both on the analytical and preparative scale, and in enantioselective synthesis are presented. Many different aspects of chiral recognition, including chiral sensors, recognition in biological systems, and in analytical methods are described..

The editors acknowledge all authors that contributed the papers included in this book.

Ai-Hong Duan; Ali Fouad; Allan Ribeiro da Silva; Anake Kijjoa; Anamarija Knežević; Andreea Elena Bodoki; Ankur Gogoi; Anna Poryvai; Arlene Gonçalves Corrêa; Artur M. S. Silva; Artur M. S. Silva; Ashraf Ghanem; Ashraf Ghanem; Bang-Jin Wang; Bei-Bei Yang; Bogdan-Cezar Iacob; Brahim Cherfaoui; Cai Feng; Carla Fernandes; Concita Sibilia; Cristina P. R. Xavier; David A. Spivak; David Sýkora; Deborah Araujo dos Santos; Diana I. S. P. Resende; Diana Ibrahim; Ede Bodoki; Emília Sousa; Emilija Petronijevic; Enrico Dalcanale; Farid Chebrouk; Fei Xiong; Filipe A. A. Paz; Giovanna Brancatelli; Guan-Yu Zhuo; Harsh Ranawat; Jie Zhang; Jinsong Xuan; Joana Araújo; Joana Teixeira; João Ribeiro; Jun-Hui Zhang; Jurica Novak; Khaldoun Bachari; Khodir Madani; Laura Elena Gliga; Leila Boukenna; Li Li; Lie-Ding Shiau; Li-Ming Yuan M. Helena Vasconcelos; Madalena M. M. Pinto; Marcio Weber Paixão; Maria Elizabeth Tiritan; Maria Letícia Carraro; Michal Kohout; Michal Šmahel; Mónica Válega; Montaser Sh. A. Shaykoon; Nai-Tzu Chen; Natalie Kolderov; Nicholas A. Gariano; Nirmal Mazumder; Oualid Talhi; Petra Tomáškov; Qiu-E Cao; Quan He; Ricardo F. Mendes; Ricardo Fernandes; Roberta Pinalli; Samy M. Ibrahim; Sheng-Ming Xie; Silvano Geremia; Simona Luminita Oprean; Sobhy M. El-Adl; Sohrab Rohani; Solida Long; Surajit Konwer; Terezia Vojtylová-Jurkovičov; Vladimir Vinković; Yangfeng Peng; Yingang Feng; Yuichi Uwai.

Maria Elizabeth Tiritan, Madalena Pinto, Carla Sofia Garcia Fernandes

Special Issue Editors

molecules

MDPI

Editorial

Enantioselective Synthesis, Enantiomeric Separations and Chiral Recognition

Maria Elizabeth Tiritan [1,2,3,*], Madalena Pinto [2,3] and Carla Fernandes [2,3]

1 CESPU, Instituto de Investigação e Formação Avançada em Ciências e Tecnologias da Saúde (IINFACTS), Rua Central de Gandra, 1317, 4585-116 Gandra PRD, Portugal
2 Laboratório de Química Orgânica e Farmacêutica, Departamento de Ciências Químicas, Faculdade de Farmácia da Universidade do Porto, Rua de Jorge Viterbo Ferreira, 228, 4050-313 Porto, Portugal; madalena@ff.up.pt (M.P.); cfernandes@ff.up.pt (C.F.)
3 Centro Interdisciplinar de Investigação Marinha e Ambiental (CIIMAR), Edifício do Terminal de Cruzeiros do Porto de Leixões, Av. General Norton de Matos s/n, 4050-208 Matosinhos, Portugal
* Correspondence: elizabeth.tiritan@iucs.cespu.pt or beth@ff.up.pt; Tel.: +351-913499293

Received: 1 April 2020; Accepted: 2 April 2020; Published: 8 April 2020

Chirality is a geometric property associated with the asymmetry of tridimensional features that accompanies our daily life at macroscopic as well as microscopic molecular levels. Chirality is a hallmark of many natural small molecules, and it is intrinsically associated with chiral building blocks as D-sugars and L-amino acids, intervening in chemical procedures of living cells, for example, as enzymes and receptors constituent proteins. Interestingly, free D-amino acids, which are naturally occurring, are important biomarkers with diagnostic value that demonstrate the importance of chiral analyses [1]. Nevertheless, the importance of chirality is recognized across many related areas as witnessed in wide-ranging fields such as chemistry, physics, biochemistry, material science, pharmacology, and many others (Figure 1).

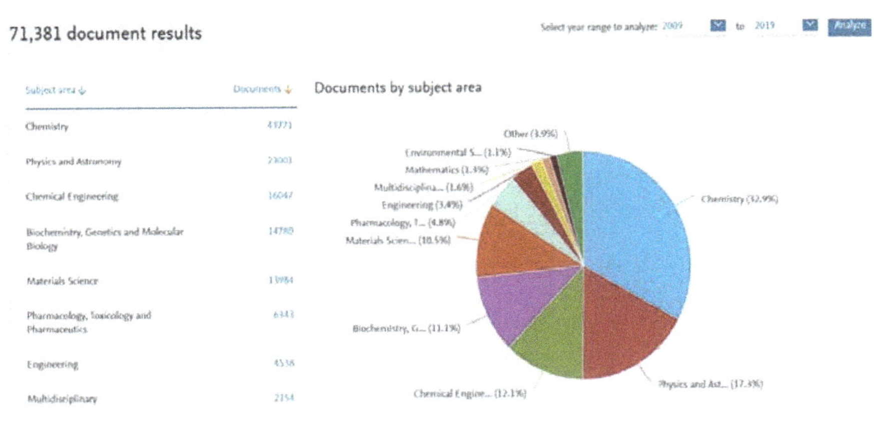

Figure 1. Results analysis for Scopus query "chiral" in titles, keywords, or the abstract section of articles between 2009 and 2019.

Though chirality has a major position in chemistry, compared with other fields, due to the importance of chiral compounds in their pure enantiomeric form, there is a need for the development of analytic methods capable of controlling the enantiomeric ratio, and to understand the behavior of chiral compounds in biological systems and in other matrices in which chirality is also present. Currently, there is a very high demand for efficient methodologies to obtain chiral bioactive compounds with a high degree of enantiomeric purity, which boosts the continuous advances in enantioselective synthesis, chiral analyses, preparative enantioseparation, as well as in chiral recognition studies. The number of publications with chirality as a subject has increased in the last decade and disclosed considerable growth in the last year, demonstrating the importance of the research in this field (Figure 2).

Figure 2. Results analysis for Scopus query "chiral" in titles, keywords, or abstract sections of articles between 2009 and 2019 (Limited to Chemistry).

These demands are related to drug discovery and development, safety in medication, food and environmental quality, materials for fine chemical industry such as chiral building blocks, among others. To meet these needs, it is essential that the international scientific community must work intensively to ensure effective production and quality of analyses of chiral compounds for a diversity of applications. For this reason, Molecules recognized the need to propose the Special Issue "Enantioselective Synthesis, Enantiomeric Separations, and Chiral Recognition". This Special Issue is aimed at offering an opportunity to all the contributors to make their results and techniques more visible, and to present the most recent findings. This Special Issue has received remarkably positive feedback, with many contributions submitted by numerous geographically diverse scientists, resulting in a collection of 19 publications, including six exhaustive review articles [2–7], and thirteen original articles [8–20]. Among the contributing authors, we can find countries of origin such as Algeria, Australia, Brazil, Canada, China, Croatia, Czech Republic, Egypt, India, Italy, Japan, Portugal, Romania, Russia, and Taiwan.

The published articles include findings related to the analytical chiral stationary phases (CSPs) for liquid chromatography (LC), currently the better choice for chiral quality control and determination of enantiomeric ratios. Faster, more efficient, and sensitive methods are urgently needed for chiral analysis, and can be achieved within small particle sizes (sub-2 µm) of the chromatographic support. The ability of the recently commercialized sub-2 µm CSP with different substituents for the fast enantioseparation of a set of drugs was demonstrated in an original article [8]. New selectors for CSPs are always required to show the response in the continuous progress of chiral analyses, and there is a need for better and low cost CSPs. In this context a new brush-Type CSP for LC was reported for enantioseparation of several drugs including nonsteroidal anti-inflammatory drugs and 3-hydroxybenzodiazepine [10]; and a new colistin sulfate CSP for nano-LC reported enantioseparation for α- and β-blockers, anti-inflammatory, antifungal, norepinephrine-dopamine reuptake inhibitors, catecholamines, sedative-hypnotic, antihistaminic, anticancer, and antiarrhythmic drugs [9]. Additionally, an exhaustive review concerning recent developments in CSPs for LC includes many different types of selectors, showing that it continues to be a field of research with great importance [2].

Methodologies regarding innovation in the preparative scale were also comprised in this Special Issue. For example, one article presents the purification of *R*-phenylglycinol from the phenylglycinol enantiomers by stripping crystallization, a new separation technology, which combines melt crystallization and vaporization to produce a crystalline product due to the three-phase transformation [11]. The classical preparative scale approach through diastereomeric salts formation, widely used in the pharmaceutical industry, is also presented with the resolution of 4-chloromandelic acid using the (*R*)-(+)-benzyl-1-phenylethylamine; with diastereomeric salts exhibiting significant differences in solubility and in thermodynamic properties. These differences originate from the distinct supramolecular interactions in the crystal lattice of the pair of diastereomeric salts. In addition to well-recognized hydrogen-bonding, CH/π interactions and aromatic group packing, halogen involved interactions, such as Cl ... Cl and Cl/π were observed as significant contributions to the chiral discrimination [13].

The approach to achieve bioactive enantiomers by enantioselective synthesis is reported in two original publications and two reviews. One article reports the syntheses of a small library of proteomimetic quinazolinone-derived compounds and investigates their action on neurodegenerative disorders as well as the search of their potential as tumor cell growth inhibitors, giving evidence for the influence of stereochemistry on the bioactivity of diverse derivatives. Here, the enantiomeric ratio was determined by a chiral LC [17]. In another original article, the hemi-synthesis of chiral imine, benzimidazole, and benzodiazepine structures is reported by the condensation of (*S*)-(−)-perillaldehyde, the major phytochemical of the *Ammodaucus leucotrichus subsp. leucotrichus* essential oil, with different amine derivatives of 2,3-diaminomaleonitrile, *o*-phenylenediamine, and 3-[(2-aminoaryl)amino]dimedone. The chiral analyses confirm the formation of unique enantiomers and diastereomeric mixtures [16]. Small ring heterocycles, such as epoxides and aziridines, present in several natural products, are frequently involved as highly versatile building blocks in the synthesis of numerous bioactive products and pharmaceuticals. Multicomponent reactions (MCRs) have been explored in the synthesis and ring opening of these heterocyclic units. An exhaustive review about the recent advances in MCRs discuss the synthesis and applications of epoxides and aziridines to prepare other heterocycles, emphasizing the stereoselectivity of the reactions [7]. Synthesis of chiral derivatives of xanthones, an important class of bioactive compounds, as well the enantioselectivity in their biological activities, was also exhaustively revised [3].

Industrial production by biocatalyse using the *cis*-epoxysuccinic acid hydrolases (CESHs) was summarized, as well the perspective on the future research and applications of CESH in enantiomeric tartaric acid production [6].

Additional work concerning chiral recognition are also included in this Special Issue, such as stereochemistry assignment and chiral recognition mechanisms of sulfoxide-containing drugs [14], the structural determination of the crystal structures of three complexes between the Tiiii cavitand as host and positively charged amino acids (Arg, Lys, and His) as guests [15]; a revision concerning enantioselective drug recognition by transporters [4], and another article about enantiomeric recognition and separation by chiral nanoparticles [5]. Molecular imprinting technology is a well-established tool for the synthesis of highly selective biomimetic molecular recognition platforms. One article reports the improvement in chiral selectivity of the important β-blocker atenolol by the addition of a metal pivot *versus* the traditional molecular imprinted polymer formulation [12].

Finally, original works related to special materials as chiral liquid crystals and components for chiral sensing are presented. For the proper function of liquid crystals-based devices, not only chemical but also optical purity of materials is strongly desirable, since any impurity could be detrimental to the self-assembly of the molecules. One article demonstrated that LC with UV detection and supercritical fluid chromatography with UV and mass spectrometry detection enables full control over the chemical and enantioselectivity of the synthesis of a novel type of lactic acid-based chiral liquid crystals and the corresponding chiral building blocks [18]. Regarding chiral sensing, one article reports a path to enhanced near-field optical chirality, by means of symmetric Si nanowires arrays, which support leaky waveguide modes that enhance the near-field optical chirality of circularly polarized excitation in the shorter wavelength part of the visible spectrum, which is of interest for many chiral molecules [19]. Another article reports an enantioselective potentiometric sensor composed of a polyvinyl chloride membrane electrode modified with CC3-R porous organic cages material, used for the recognition of enantiomers of 2-amino-1-butanol [20].

This Special Issue is accessible thought the following link: https://www.mdpi.com/journal/molecules/special_issues/Chiral_separation_recognition.

As Guest Editors for this Special Issue, we would like to thank all the authors and co-authors for their contributions and all the reviewers for their effort in the careful and rapid evaluation of the manuscripts. Last but not least, we would like to appreciate the hard work done by the editorial office of the Molecules journal, as well as their kind assistance in preparing this Special Issue.

Funding: This work was supported by the Strategic Funding UID/Multi/04423/2019 through national funds provided by FCT—Foundation for Science and Technology and European Regional Development Fund (ERDF), through the COMPETE—Programa Operacional Factores de Competitividade (POFC) program in the framework of the program PT2020; Project No. POCI-01-0145-FEDER-028736, co-financed by COMPETE 2020, under the PORTUGAL 2020 Partnership Agreement, through the European Regional Development Fund (ERDF), and the project CHIRALBIOACTIVE-PI-3RL-IINFACTS-2019.

Conflicts of Interest: The authors declare no conflict of interest.

References

1. Kimura, R.; Tsujimura, H.; Tsuchiya, M.; Soga, S.; Ota, N.; Tanaka, A.; Kim, H. Development of a cognitive function marker based on D-amino acid proportions using new chiral tandem LC-MS/MS systems. *Sci. Rep.* **2020**, *10*, 1–12. [CrossRef] [PubMed]
2. Teixeira, J.; Tiritan, M.E.; Pinto, M.; Fernandes, C. Chiral Stationary Phases for Liquid Chromatography: Recent Developments. *Molecules* **2019**, *24*, 865. [CrossRef] [PubMed]
3. Fernandes, C.; Carraro, M.L.; Ribeiro, J.; Araújo, J.; Tiritan, M.E.; Pinto, M. Synthetic Chiral Derivatives of Xanthones: Biological Activities and Enantioselectivity Studies. *Molecules* **2019**, *24*, 791. [CrossRef] [PubMed]
4. Uwai, Y. Enantioselective Drug Recognition by Drug Transporters. *Molecules* **2018**, *23*, 3062. [CrossRef] [PubMed]
5. Gogoi, A.; Mazumder, N.; Konwer, S.; Ranawat, H.; Chen, N.-T.; Zhuo, G.-Y. Enantiomeric Recognition and Separation by Chiral Nanoparticles. *Molecules* **2019**, *24*, 1007. [CrossRef] [PubMed]
6. Xuan, J.; Feng, Y. Enantiomeric Tartaric Acid Production Using cis-Epoxysuccinate Hydrolase: History and Perspectives. *Molecules* **2019**, *24*, 903. [CrossRef] [PubMed]
7. Da Silva, A.R.; Dos Santos, D.A.; Paixao, M.W.; Corrêa, A.G. Stereoselective Multicomponent Reactions in the Synthesis or Transformations of Epoxides and Aziridines. *Molecules* **2019**, *24*, 630. [CrossRef] [PubMed]
8. Ibrahim, D.; Ghanem, A. On the Enantioselective HPLC Separation Ability of Sub-2 μm Columns: Chiralpak®IG-U and ID-U. *Molecules* **2019**, *24*, 1287. [CrossRef] [PubMed]
9. Fouad, A.; Shaykoon, M.; Ibrahim, S.M.; El-Adl, S.M.; Ghanem, A. Colistin Sulfate Chiral Stationary Phase for the Enantioselective Separation of Pharmaceuticals Using Organic Polymer Monolithic Capillary Chromatography. *Molecules* **2019**, *24*, 833. [CrossRef] [PubMed]
10. Knežević, A.; Novak, J.; Vinković, V. New Brush-Type Chiral Stationary Phases for Enantioseparation of Pharmaceutical Drugs. *Molecules* **2019**, *24*, 823. [CrossRef] [PubMed]
11. Shiau, L.-D. Chiral Separation of the Phenylglycinol Enantiomers by Stripping Crystallization. *Molecules* **2018**, *23*, 2901. [CrossRef] [PubMed]
12. Bodoki, A.E.; Iacob, B.-C.; Gliga, L.E.; Oprean, S.L.; Spivak, D.A.; Gariano, N.A.; Bodoki, E. Improved Enantioselectivity for Atenolol Employing Pivot Based Molecular Imprinting. *Molecules* **2018**, *23*, 1875. [CrossRef] [PubMed]
13. Peng, Y.; Feng, C.; Rohani, S.; He, Q.S. Improved Resolution of 4-Chloromandelic Acid and the Effect of Chlorine Interactions Using (R)-(+)-Benzyl-1-Phenylethylamine as a Resolving Agent. *Molecules* **2018**, *23*, 3354. [CrossRef] [PubMed]

14. Xiong, F.; Yang, B.-B.; Zhang, J.; Li, L. Enantioseparation, Stereochemical Assignment and Chiral Recognition Mechanism of Sulfoxide-Containing Drugs. *Molecules* **2018**, *23*, 2680. [CrossRef] [PubMed]
15. Brancatelli, G.; Dalcanale, E.; Pinalli, R.; Geremia, S. Probing the Structural Determinants of Amino Acid Recognition: X-Ray Studies of Crystalline Ditopic Host-Guest Complexes of the Positively Charged Amino Acids, Arg, Lys, and His with a Cavitand Molecule. *Molecules* **2018**, *23*, 3368. [CrossRef] [PubMed]
16. Chebrouk, F.; Madani, K.; Cherfaoui, B.; Boukenna, L.; Valega, M.; Mendes, R.F.; Paz, F.A.A.; Bachari, K.; Talhi, O.; Silva, A.M.S. Hemi-Synthesis of Chiral Imine, Benzimidazole and Benzodiazepines from Essential Oil of Ammodaucus leucotrichus subsp. leucotrichus. *Molecules* **2019**, *24*, 975. [CrossRef] [PubMed]
17. Long, S.; Resende, D.I.S.P.; Kijjoa, A.; Silva, A.M.S.; Fernandes, R.; Xavier, C.P.R.; Vasconcelos, M.H.; Sousa, M.E.; Pinto, M. Synthesis of New Proteomimetic Quinazolinone Alkaloids and Evaluation of Their Neuroprotective and Antitumor Effects. *Molecules* **2019**, *24*, 534. [CrossRef] [PubMed]
18. Poryvai, A.; Vojtylová-Jurkovičová, T.; Šmahel, M.; Kolderová, N.; Tomášková, P.; Sýkora, D.; Kohout, M. Determination of Optical Purity of Lactic Acid-Based Chiral Liquid Crystals and Corresponding Building Blocks by Chiral High-Performance Liquid Chromatography and Supercritical Fluid Chromatography. *Molecules* **2019**, *24*, 1099. [CrossRef] [PubMed]
19. Petronijevic, E.; Sibilia, C. Enhanced Near-Field Chirality in Periodic Arrays of Si Nanowires for Chiral Sensing. *Molecules* **2019**, *24*, 853. [CrossRef] [PubMed]
20. Wang, B.-J.; Duan, A.-H.; Zhang, J.-H.; Xie, S.-M.; Cao, Q.-E.; Yuan, L.-M. An Enantioselective Potentiometric Sensor for 2-Amino-1-Butanol Based on Chiral Porous Organic Cage CC3-R. *Molecules* **2019**, *24*, 420. [CrossRef] [PubMed]

molecules

MDPI

Article

On the Enantioselective HPLC Separation Ability of Sub-2 μm Columns: Chiralpak® IG-U and ID-U

Diana Ibrahim and Ashraf Ghanem *,†

Chirality Program, Faculty of Science and Technology, University of Canberra, ACT 2601, Australia; diana.ibrahim@canberra.edu.au

* Correspondence: ashraf.ghanem@canberra.edu.au; Tel.: +61-(02)-6201-2089 or 0429319993; Fax: +61-(02)-6201-2328

† Dedicated to Prof. Frantisek Svec on the occasion of his 75th birthday.

Received: 4 February 2019; Accepted: 29 March 2019; Published: 2 April 2019

Abstract: Silica with a particle size of 3–5 μm has been widely used as selector backbone material in 10–25 cm HPLC chiral columns. Yet, with the availability of 1.6 μm particles, shorter, high-efficiency columns practical for minute chiral separations are possible to fabricate. Herein, we investigate the use of two recently commercialized sub-2 μm columns with different substituents. Thus, Chiralpak® IG-U and ID-U were used in HPLC for the fast enantioseparation of a set of drugs. Chiralpak® IG-U [amylose tris (3-chloro-5-methylphenylcarbamate)] has two substituents on the phenyl ring, namely, a withdrawing chlorine group in the third position and a donating group in the fifth position. Chiralpak® ID-U [amylose tris (3-chlorophenylcarbamate)] has only one substituent on the phenyl ring, namely a withdrawing chlorine group. Their applications in three liquid chromatography modes, namely, normal phase, polar organic mode, and reversed phase, were demonstrated. Both columns have similar column parameters (50 mm length, 3 mm internal diameter, and 1.6 μm particle size) with the chiral stationary phase as the only variable. Improved chromatographic enantioresolution was obtained with Chiralpak® ID-U. Amino acids partially separated were reported for the first time under an amylose-based sub-2-micron column.

Keywords: Chiralpak® ID-U; Chiralpak® IG-U; mobile phase modifiers; polar organic and reversed phase modes; sub-2 μm particles

1. Introduction

In nature and chemical systems, enantiomeric distinction and chiral recognition are fundamental occurrences [1]. This phenomenon has had a profound impact on a plethora of scientific fields, though the pharmaceutical industry significantly drives developments in chirotechnologies to cater to the demands of drug discovery [2,3]. There is no option when it comes to chiral considerations; all enantiomers must be tested in isolation of each other before being introduced to the market [3]. As a result, high performance liquid chromatography (HPLC) has emerged as the workhorse for racemate resolution [4]. HPLC enantiomer separation using chiral stationary phases (CSPs) is known to be one of the most convenient and versatile methods for the separation of chiral drugs [4].

In the last few decades, numerous CSPs have been developed and become commercially available [5,6]. CSPs filled in conventional columns of 4.0–4.6 mm internal diameter (i.d.) are the most widely used for analytical scale enantioseparation for industrial applications [5,6]. Nonetheless, conventional chiral columns are expensive; they consume large volumes of hazardous solvents and have long analysis times, and due to the dimensions of these large columns they are of limited throughput [6]. One of the possible solutions to enhance the speed of the analysis is to use columns filled with a CSP of smaller particles (sub-2 μm) and hence a smaller theoretical plates height [7].

Sub-2 µm totally porous particles can be used to speed up analysis without loss in efficiency, as the optimal flow rate is inversely proportional to particle diameter [8]. The main limitation of using totally porous particles is the induction of high back pressure across the column induced by the friction of the mobile phase percolating through the particles generating heat, which hinders their usage within conventional HPLC systems [9]. Studies suggest that small i.d. columns can be used to minimize the frictional heating effect since heat dissipation is faster within such a narrow-bore column compared to conventional 4.6 mm i.d. columns [10]. Narrow-bore columns have a lower internal volume (2.1 mm i.d.) than the standard HPLC columns and thus achieve fast analysis [10,11]. They operate at lower flow rates (0.1–0.5 mL/min) with much reduced peak volumes, resulting in reduced mobile phase consumption and increased sensitivity [11,12].

Mobile phases can be modified to achieve higher enantioselective separation of racemates via improvement of complementary interactions between functional groups on the chiral selector and the analyte structure [13]. Pirkle and Welch have studied modifier effects on chiral selectivity and found that the influence of the mobile phase modifier was dependent upon the analyte structure [13–15]. Tambute and co-workers have also examined the use of modifiers and concluded that selectivity in their system depends on the steric hindrance of the alcohol modifier [14–16]. Researchers believe that the mobile-phase modifiers not only compete for chiral bonding sites with chiral solutes but can also alter the steric environment of the chiral grooves on the CSP by binding to the achiral sites at or close to the groove [13,17]. Enantioselective resolution is mainly due to the overall combination of all types of bonding [18]. Thus, not only the steric but also the substitutes of a certain chiral compound and the CSP should be taken into consideration to elucidate chiral recognition mechanisms [19].

Here we evaluate and compare the enantiorecognition abilities of two amylose-based sub-2 µm CSPs towards 28 compounds, as they differ in the substituents on the phenyl ring. Recently commercialized Chiralpak® IG-U [amylose tris (3-chloro-5-methylphenylcarbamate)] possesses an extra donating methyl group in the fifth position compared to the prototype Chiralpak® ID-U [amylose tris (3-chlorophenylcarbamate)]. This investigation was performed using an operational instrument at an HPLC system pressure of 500 bar at which frictional heating is not very significant. Hence, thermal gradients inside the column were not expected to affect the efficiency.

2. Experimental

2.1. Instrumentation

The mobile phase for the HPLC was filtered through a Millipore membrane filter (0.2 µm) and degassed before use. The HPLC system consisted of a Waters binary pump, Model 1525, (Milford, MA, USA), equipped with a dual wavelength absorbance detector, Model 2487, an autosampler, Model 717 plus, and an optical rotation detector (JM Science Inc., Grand Island, NY, USA) operating at room temperature. The UV-detector was set at 254 nm. Chiralpak® IG-U and ID-U (50 mm column length, 3.0 mm i.d, and 1.6 µm silica gel) were supplied by Daicel (Tokyo, Japan).

2.2. Chemicals and Reagents

All compounds and solvents (HPLC grade) were purchased from Sigma-Aldrich (St. Louis, MO, USA). The choice of compounds was arbitrary and guided by preliminary investigations. The compounds were, namely: beta-blockers (propranolol and atenolol), alpha-blockers (naftopidil), anti-inflammatory compounds (carprofen, naproxen, flurbiprofen, ketoprofen, and indoprofen), anticancers (ifosfamide), sedative hypnotics (aminoglutethimide), antiarrhythmic drugs (tocainide), norepinephrine-dopamine reuptake inhibitors (nomifensine), catecholamines (normetanephrine and epinephrine), antihistamines (chlorpheniramine), flavonoids (flavanone and 6-hydroxyflavanone), miscellaneous (1-acenaphthenol, 1-indanol, 4-hydroxy-3-methoxymandelic acid, propafenone HCL, cizolirtine, and 1-phenyl-2,2,2-trifluoroethanol), amino acids (glutamic acid, tyrosine, and phenylalanine) and antifungals (miconazole and sulconazole).

2.3. Procedures

Mobile phases were filtered through a membrane Sartorius Minisart RC 15 0.2 μm pore size filter (Goettingen, Germany), further used for analysis without dilution, and degassed before use. The chromatographic measurements were performed at a flow rate of 0.5 mL/min at a temperature of 25 °C. All measurements were performed in triplicate with an injection volume of 1 μL. Stock solutions of samples were prepared at a concentration of 1 mg/mL using HPLC-grade 2-propanol as a solvent.

3. Results and Discussion

The potential of the sub-2 μm CSPs to separate the racemic compounds listed above under normal-phase, reversed-phase, and polar organic solvents have been investigated. The influence of the mobile phase composition on the separation (α), resolution (Rs), and retention time (RT) of enantiomers has been examined using (1) non-polar solvents (n-alkanes) containing a polar alcohol modifier, namely, ethanol (EtOH), 2-propanol (2-PrOH), and *n*-butanol (*n*-BuOH), and (2) polar solvents, namely, methyl tert-butyl ether (MtBE), acetonitrile (ACN), 1,4-dioxane, and dichloromethane (DCM). The CSP structural differences under different mobile phase conditions are reflected in some selected chromatograms shown in Figures 1–7.

3.1. Enantioselectivity under Non-Polar Solvents Containing an Alcohol Polar Modifier

The initial mobile phase composition of *n*-hexane/alcohol modifier (90/10, *v/v*) was prepared. Out of the three alcohol modifiers tested, *n*-BuOH showed the lowest enantioselectivity in both tested CSPs, namely, Chiralpak® IG-U and ID-U. This might be due to the difference in the steric bulkiness around the hydroxyl moiety contained in the mobile phase modifier [15–18]. Conversely, EtOH afforded better enantioselectivity for both CSPs. Upon replacement of EtOH with bulkier *n*-BuOH, the competition for hydrogen-bonding sites on these CSPs becomes weaker. This might be due to the fact that lower alcohols such as EtOH are unlike bulkier alcohols and could diffuse more easily into well-defined grooves of the CSP. Thus, more stable diastereomeric complexes with the enantiomers could be formed, consequently resulting in higher Rs and α value [15,17–21]. Of particular interest is that ifosfamide and glutamic acid were only separated under *n*-hexane/EtOH on Chiralpak® ID-U.

In a few cases, such as with 4-hydroxy-3-methoxymandelic acid, 1-acenaphthenol, 1-indanol, and propafenone HCL, the use of 2-PrOH as an alcohol modifier afforded superior Rs and α on Chiralpak® ID-U. By contrast, these compounds expressed the best Rs and α using EtOH on Chiralpak® IG-U. For example, 4-hydroxy-3-methoxymandelic acid expressed a superior Rs of 2.71 and α of 2.12 on Chiralpak® ID-U (Figure 1A) under *n*-hexane/2-PrOH (90/10, *v/v*) compared to Rs 1.63 and α 1.77 under *n*-hexane/EtOH. Chiralpak® IG-U expressed the best Rs 8.74 and α 3.86 under *n*-hexane/EtOH compared to Rs 0.75 and α 1.08 under *n*-hexane/2-PrOH (Figure 1B). In particular, 1-phenyl-2,2,2-trifluoroethanol with Rs 2.38 and α 3.90, cizolirtine with Rs 5.27 and α 3.39, and naftopidil with Rs 1.75 and α 1.95 were only successfully separated under *n*-hexane/EtOH (90/10, *v/v*) using Chiralpak® IG-U.

The results indicate that the different structural features of the CSP, combined with the incorporation of the alcoholic modifiers of different sizes/shapes, ultimately results in a different stereo environment of the chiral cavities in the CSP, yielding different chiral selectivities [21–25].

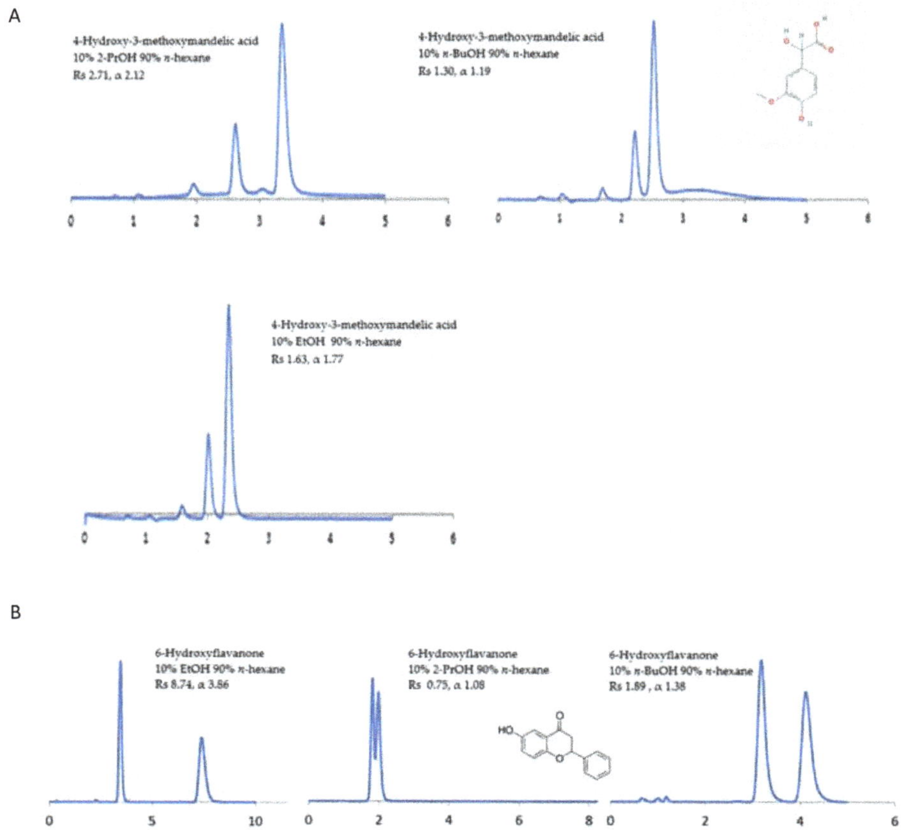

Figure 1. The effect of different alcohol modifiers: 2-propanol (2-PrOH), ethanol (EtOH), and *n*-butanol (*n*-BuOH) on enantioselectivity under two sub-2-micron chiral stationary phases. (**A**) The effect of different alcohol modifiers on 4-hydroxy-3-methoxymandelic acid using Chiralpak® ID-U. (**B**) The effect of different alcohol modifiers on 6-hydroxyflavanone using Chiralpak® IG-U.

Previous studies have showed improvements in selectivity with *n*-heptane over *n*-hexane [26,27]. Therefore, in the current study, *n*-hexane was replaced with *n*-heptane. For example, flavonoids (6-hydroxyflavanone and flavanone) using Chiralpak® ID-U showed an enhanced Rs and α under *n*-heptane. As shown in Figure 2, flavanone showed an enhanced Rs 2.14 and α 1.99 under *n*-heptane/*n*-BuOH (90/10, *v/v*) compared to Rs 1.17 and α 1.74 under *n*-hexane/*n*-BuOH (90/10, *v/v*). The effect of different alcohol modifiers used on Chiralpak® IG-U expressed a range of results in the transition between *n*-hexane to *n*-heptane. For example, chlorpheniramine showed an enhanced Rs which increased from 1.74 to 2.33 and an α which increased from 1.55 to 1.97 using *n*-heptane.

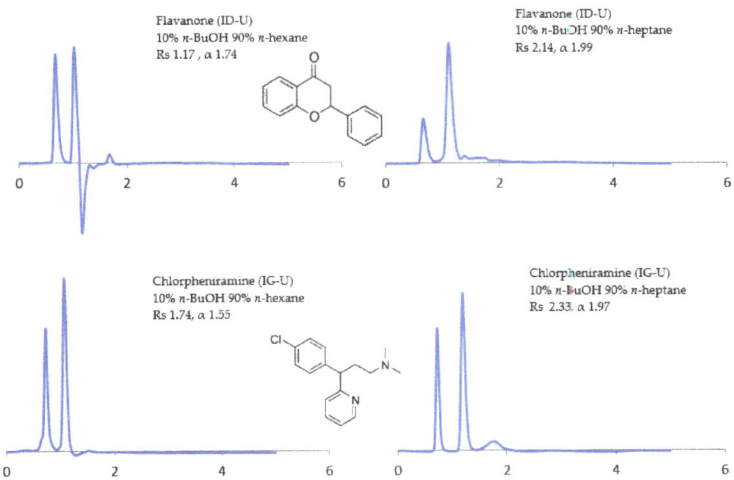

Figure 2. Effect of *n*-hexane versus *n*-heptane on resolution (Rs) and separation factor (α) using Chiralpak® IG-U and ID-U.

3.2. The Effect of Alcohol Modifier Percentage on Enantioselectivity

The composition of the alcohol modifier in the mobile phase was evaluated at 10%–40% *v*. Increasing the composition of the alcohol modifier increases the strength of the mobile phase (the ability of compounds to elute quicker from the column) and hence the RT will consequently be reduced (at the expense of Rs and α, however) [22–25]. For example, 6-hydroxyflavanone achieved baseline separation in 4 min with Rs 3.85 and α 2.89 under 20% EtOH compared to 8 min with Rs 8.74 and α 3.86 with 10% EtOH on Chiralpak® IG-U (Figure 3). These results indicate that alcohol molecules compete with the analytes for achiral and chiral adsorption sites on the CSP. Thus, RT, α and Rs are altered by changes in the concentration of alcohol [22–24].

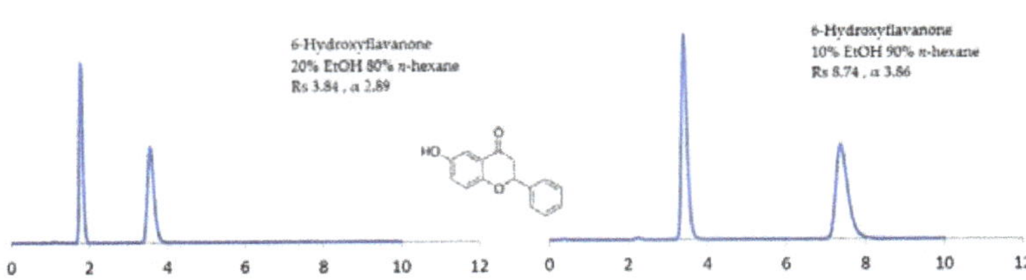

Figure 3. The effect of decreasing the alcohol percentage on chiral selectivity and time taken for the baseline separation of 6-hydroxyflavanone.

3.3. Effect of the Structure of Analytes on Enantiomeric Separation

It is known that the alcohol modifiers used in the normal-phase mode have a profound influence on the chiral selectivity of CSPs. Therefore, gaining structural information regarding the CSPs in contact with mobile phases containing different alcohol modifiers would be of interest. Polar and π-π interactions between the CSP phenyl groups and the functional group of the solute may also play a role in chiral recognition [21,22,28]. It has been hypothesized that with an increase in the mobile phase polarity, the strength of the hydrogen bonds between the analytes and the CSP decreases and the solubility of the analytes in the mobile phase increases [26–29]. Moreover, it is possible that some alcohol molecules are associated with the CSP and cause swelling of the column, which leads to opening of the chiral cavities. Thus, the inclusion interactions of the enantiomers are diminished and RT is decreased [26–29].

3.3.1. β-Blockers

β-blockers are hydroxylamines with functional groups bearing secondary amines or N-isopropyl amines. These drugs also contain aromatic rings with different substituent moieties. The OH and NH groups and an oxygen atom in the model examples of β-blockers studied (propranolol and atenolol) are functional groups which are available to take part in hydrogen bonding with the C=O and NH groups of the CSPs [14,27]. As shown in Figure 4, under 20% 2-PrOH, atenolol has the lowest Rs of 0.75 and α of 1.17. By contrast, propranolol has the largest Rs of 1.00 and α of 1.24. A possible explanation for these results could be that the naphthalene ring of propranolol can form stronger interactions with the CSP [14]. On the other hand, the amide group of atenolol could compete with the groups on the CSP for bonding sites, causing low stereoselective interactions. Furthermore, the CSP-substituted phenyl ring interaction might also be important where the pronounced steric effect could be close to the analyte chiral center, resulting in poor chiral discrimination of atenolol [19]. Both groups adjacent to the chiral centres and the substituent groups on the phenyl rings could contribute to an enhanced separation result [28–32].

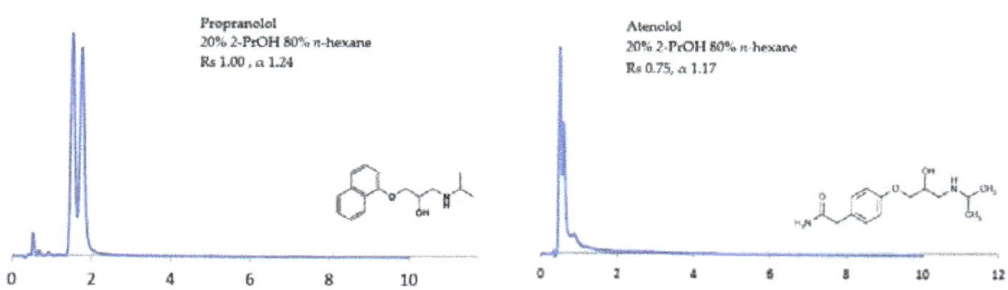

Figure 4. The effect of 20% 2-propanol (2-PrOH) on the stereoselective interactions of β-blockers.

3.3.2. Anti-Inflammatory

Out of the four profens used in this study (flurbiprofen, ibuprofen, naproxen, and ketoprofen), ibuprofen and naproxen achieved the lowest Rs and α values under normal phase conditions (an alkane/alcohol modifier). However, they expressed much higher enantio-separation under reversed phase conditions (100% ACN, *v*), (ACN/H_20, 60/40, *v/v*). On the other hand, flurbiprofen and ketoprofen expressed a higher Rs under normal phase conditions. In particular, Chiralpak® ID-U showed significantly higher enantio-selectivity values for the tested profens. This column was able to partially separate all tested profens while Chiralpak® IG-U was less effective in the chiral separation of ibuprofen and ketoprofen. Contrary to the literature, as shown in Figure 5, the order of increasing enantioselectivity is 2-PrOH < EtOH < *n*-BuOH. It is hypothesized that hydrogen-bonding might be a predominant factor between the solutes and the CSPs [20,33,34].

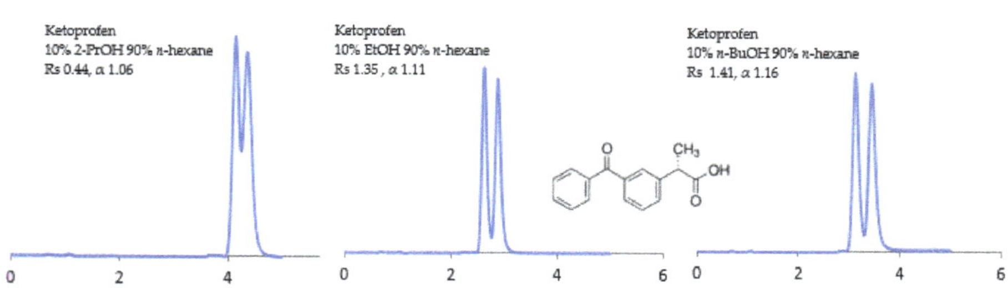

Figure 5. Effect of different alcohol modifiers with *n*-hexane on resolution (Rs) and separation factor (α) of ketoprofen. Ketoprofen expressed an increasingly enhanced Rs and α in the order of 2-propanol (2-PrOH) to ethanol (EtOH) to *n*-butanol (*n*-BuOH).

3.3.3. Amino Acids

Amino acids ($H_2NCHRCOOH$) have three main groups: the carboxyl group, the amino group and a variable (R) group [35,36]. Three model examples have been selected, namely, glutamic acid, tyrosine, and phenylalanine. The analytes used herein form a double hydrogen-bonded complex with the CSP carbamate group. The protonated amino group of the analytes and the carbonyl group of the CSP form hydrogen bonds with the CSP carbonyl and amide groups.

Glutamic acid expressed Rs of 1.71 and α of 1.65 under standard mobile phase composition on Chiralpak® ID-U (Figure 6). This is opposed to the weaker stereoselective results obtained with Chiralpak® IG-U under different standard mobile phase compositions. Glutamic acid is an acidic compound with a hydrogen acceptor atom in its side chain which is negatively charged. It is very polar and can easily engage in ionic bonds through electrostatic attractions [23]. Similarly, tyrosine has both a hydrogen donor and acceptor atoms in its side chain [35–40]. Its hydroxyl group is considered uncharged and can engage in hydrogen bonds [41]. The polarity of glutamic acid and tyrosine could explain the reasons for the unsuccessful separation using [amylose tris (3-chloro-5-methylphenylcarbamate)] or Chiralpak® IG-U, since it exhibits a hydrophobic methyl group.

Conversely, phenylalanine has no hydrogen donor or acceptor atoms in its side chain [38–41], whereas Chiralpak® IG-U has both a methyl and chloro group. This could explain the poor stereoselectivity of Chiralpak® ID-U compared to Chiralpak® IG-U under different mobile phase conditions with the best Rs of 1.83 and α of 1.63 under *n*-hexane/EtOH (80/20, *v/v*) and the lowest Rs of 0.92 and α of 1.40 under MtBE/EtOH (98/2, *v/v*) (Figure 6).

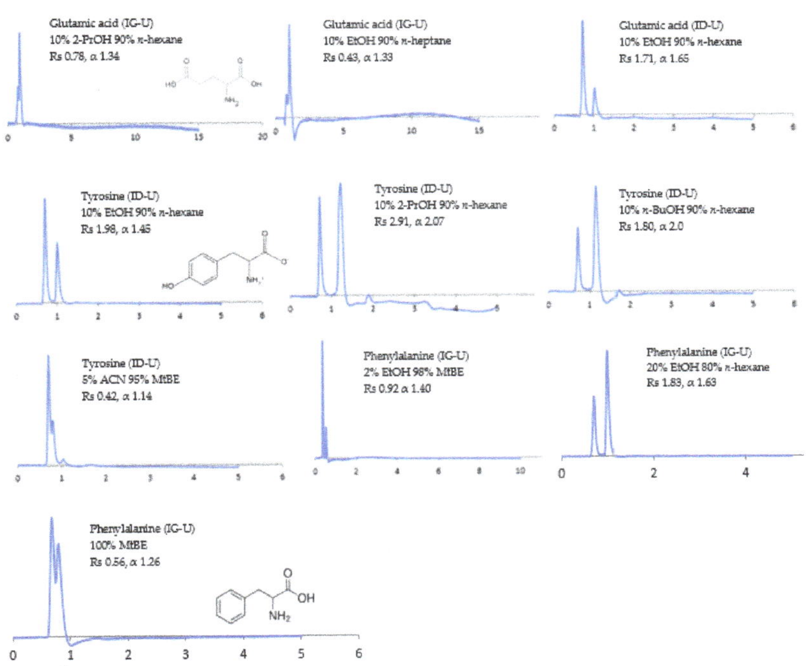

Figure 6. Enantioselectivity of three amino acids under different mobile phase compositions.

3.4. Effect of Polar Solvents on Enantioselectivity

Apart from the standard mobile phase compositions used (alkane/alcohol modifier), the literature reveals that ACN and MtBE, together with the standard solvents, are those with the highest potential in terms of enantioselectivity [42]. Starting with non-standard organic solvents in the mobile phase composition, neat ACN and MtBE (100%, *v*) were investigated as eluents for enantioselective separation.

3.4.1. Acetonitrile

ACN has unique characteristics such as its ability to dissolve a wide range of solutes, low acidity, minimal chemical reactivity, low UV cut-off, and low viscosity. The unique properties of ACN render it the solvent of choice in the separation of pharmaceuticals. However, since ACN is a poor hydrogen bonding solvent, chiral compounds analyzed with large amounts of ACN can form hydrogen bonds with the CSP [23,24,43]. Contrary to our expectations, a large percentage of compounds were separated under neat ACN (100, *v*), though RT was decreased. Of particular interest is that antifungals used in this study were only separated under ACN (100, *v*) using Chiralpak® IG-U, with sulconazole expressing Rs of 1.49 and α of 1.57 and miconazole expressing Rs of 2.00 and α of 1.92.

The addition of water to ACN enhanced Rs at the expense of a longer RT for all tested analytes herein. These results were consistent with a reversed phase mechanism, where the addition of water weakened the mobile phase strength, and RT increased [42]. For example, 6-hydroxyflavanone enhanced Rs from 1.30 to 1.87 and α from 1.24 to 2.57 on Chiralpak® IG-U. On the other hand, the addition of water to ACN decreased Rs from 1.68 to 1.53 and α from 3.70 to 3.02 on Chiralpak® ID-U. Additionally, the use of neat ACN (100, *v*) improved the peak shape on Chiralpak® ID-U (Figure 7).

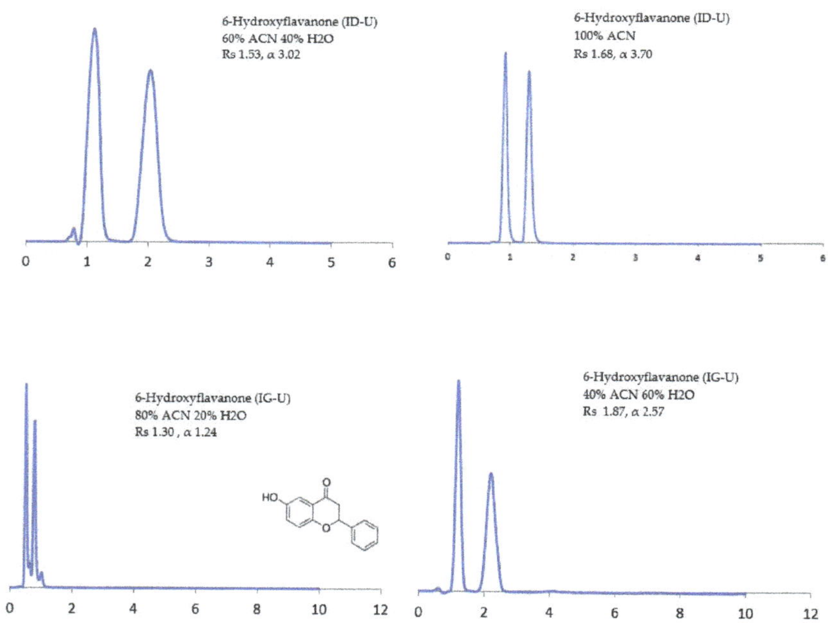

Figure 7. Enantioseparation under organic-aqueous conditions and the effect of water in acetonitrile (ACN) mobile phase on resolution (Rs) and separation factor (α) of 6-hydroxyflavanone.

3.4.2. Methyl tert Butyl Ether (MtBE)

Apart from the alkanes, MtBE has the weakest eluting strength among the solvents investigated in this study. Therefore, it is possible to use it in its pure form. Neat MtBE (100, *v*) showed an enhanced Rs and α under Chiralpak® IG-U for compounds such as nomifensine, normetanephrine, and epinephrine. For example, nomifensine showed Rs of 4.08 and α of 3.86 under MtBE (100, *v*) compared to Rs of 1.78

and α of 2.41 under *n*-hexane/EtOH (80/20, *v/v*). However, it has been proven that neat MtBE (100, *v*) may sometimes not be strong enough for compounds eluted within a reasonable time length and the peak shape is poor: broad peaks with large tailing have been previously observed [42].

Several solvents with higher eluting strength, such as EtOH, ACN, and 1,4-dioxane, can be efficiently used as modifiers in MtBE to improve separations [42]. It should be noted that the modifier providing the best separation results depend on the compound to be resolved [43]. Although the percentage of a modifier is generally low (mostly 2–10% in MtBE), its nature can greatly affect the enantioselectivity of a given compound. For example, the addition of 5% EtOH can reduce RT by half and the peak shape is significantly improved without deteriorating the selectivity [42].

For example, in Chiralpak® IG-U, 6-hydroxyflavanone under MtBE/EtOH, (95/5, *v/v*) resulted in Rs of 0.46 and α of 1.22. The substitution of EtOH with 5% ACN enhanced Rs up to 2.69 and α to 3.69 and resulted in better peak shape. The best Rs and α values were eventually achieved with 10% 1,4-dioxane as a modifier. On the other hand, for the same compound under Chiralpak® ID-U, the lowest Rs and α values were achieved under 10% 1,4-dioxane in MtBE. Five percent EtOH resulted in Rs of 12.12 and α of 4.04. This was further enhanced to Rs of 15.47 and α of 5.90 when ACN was substituted with EtOH. Of particular interest was that compounds such as tocainide, ifosfamide, and amino glutethimide were only separated under MtBE with an organic modifier (2–10%) using Chiralpak® ID-U.

4. Conclusions

In this work, the influence of mobile phase composition on the stereoselectivity of enantiomers was studied on two sub-2 μm columns. Regarding the two non-polar solvents (alkanes) containing a polar alcohol modifier (EtOH, 2-PrOH, and *n*-BuOH), EtOH expressed the best enantioselectivity on the two CSPs. In particular cases, 2-PrOH fit better on Chiralpak® ID-U. For the non-standard solvents (MtBE with organic modifiers), Chiralpak® IG-U expressed the best enantioselectivity using 10% 1,4-dioxane, while 10% 1,4-dioxane was not sufficient on Chiralpak® ID-U. The use of aqueous solutions such as ACN in water enhanced enantioselectivity of all racemates compared to similar separations using neat ACN.

Twenty-seven compounds were baseline/partially separated on Chiralpak® IG-U compared to 22 compounds separated on Chiralpak® ID-U. Chiralpak® IG-U separated compounds that were not separated under any mobile phase composition on Chiralpak® ID-U, namely, cizolirtine, naftopidil, sulconazole, miconazole, 1-phenyl-2,2,2-trifluoroethanol, and phenylalanine. In conclusion, mobile phase composition, the structure of the analytes, and their interaction with the CSP all play a role in enantioselectivity.

Author Contributions: D.I. processed the experimental data. D.I. and A.G. were involved in planning and A.G. supervised the work. D.I. performed the analysis, drafted the manuscript and designed the figures. D.I. performed the chromatogram calculations. D.I. prepared the samples and characterization with HPLC. D.I. and A.G. aided in interpreting the results and worked on the manuscript. All authors discussed the results and commented on the manuscript.

Funding: This research received funding from the Centre for Research on Therapeutic solutions (CResTs) strategic funding.

Conflicts of Interest: The authors declare no conflict of interest.

References

1. Sekhon, S.B. Enantioseparation of chiral drugs—An overview. *Int. J. Pharm. Technol. Res.* **2010**, 2, 1584–1594.
2. Purushoth, P.T.; Shinesudev Kurian, M.; Manjushsa, P.; Minnu, P.S.; Mohamed, H.C.; Shrikumar, S. Pharmaceutical review and its importance of chiral chromatography. *Int. J. Res. Pharm. Chem.* **2016**, *6*, 476–484.
3. Blaser, H.U. Chirality and its implications for the pharmaceutical industry. *Rend. Lincei. Sci. Fis.* **2013**, *24*, 213–216. [CrossRef]

4. Fekete, S.; Veuthey, J.; Guillarme, D. Comparison of the most recent chromatographic approaches applied for fast and high-resolution separations: Theory and Practice. *J. Chromatogr.* **2015**, *1408*, 1–14. [CrossRef]
5. Zhi-Jian, H.; Song, H.; Zhang, Y.W.; Wang, D.C.; Yao, S. Chiral Stationary Phases and their Relationship with Enantiomer Structures in Enantioseparation Research of Analytical Laboratory. *J. Mex. Chem. Soc.* **2014**, *1*, 43–49.
6. Chen, X.; Yamamoto, C.; Okamoto, Y. Polysaccharide derivatives as useful chiral stationary phases in high-performance liquid chromatography. *Pure Appl. Chem.* **2007**, *79*, 1561–1573. [CrossRef]
7. Carr, P.; Stoll, D.; Wang, X. Perspectives on Recent Advances in the Speed of High Performance Liquid Chromatography. *Anal. Chem.* **2011**, *83*, 1890–1900. [CrossRef]
8. Wang, Y.; Ai, F.; Ng, S.C.; Tan, T.T. Sub-2 μm porous silica materials for enhanced separation performance in liquid chromatography. *J. Chromatogr. A* **2012**, *1228*, 99–109. [CrossRef]
9. Nguyen, D.T.; Guillarme, D.; Rudaz, S.; Veuthey, J.L. Fast analysis in liquid chromatography using small particle size and high pressure. *J. Sep. Sci.* **2006**, *29*, 1836–1848. [CrossRef] [PubMed]
10. Fountain, K.J.; Neue, U.D.; Grumbach, E.S.; Diehl, D.M. Effects of extra-column band spreading, liquid chromatography system operating pressure, and column temperature on the performance of sub-2-μm porous particles. *J. Chromatogr. A* **2009**, *1216*, 5797–5988. [CrossRef] [PubMed]
11. Shaaban, H.; Gorecki, T. Green ultra-fast high-performance liquid chromatographic method using a short narrow-bore column packed with fully porous sub-2-micron particles for the simultaneous determination of selected pharmaceuticals as surface water and wastewater. *J. Chromatogr. A* **2011**, *1218*, 252–261. [CrossRef]
12. Gritti, F.; Guiochon, G. The van Deemter equation: Assumptions, limits, and adjustment to modern high performance liquid chromatography. *J. Chromatogr. A* **2013**, *1302*, 1–13. [CrossRef]
13. Sardella, R.; Ianni, F.; Lisanti, A.; Marinozzi, M.; Scorzoni, S.; Natalini, B. The effect of mobile phase composition in the enantioseparation of pharmaceutically relevant compounds with polysaccharide-based stationary phases. *Biomed. Chromatogr.* **2014**, *28*, 159–167. [CrossRef]
14. Stringham, R.W.; Ye, Y.K. Chiral separation of amines by high-performance liquid chromatography using polysaccharide stationary phases and acidic additives. *J. Chromatogr. A* **2006**, *1101*, 86–93. [CrossRef]
15. Blackwell, J.A.; Stringham, R.W.; Xiang, D.; Waltermire, R.E. Empirical relationship between chiral selectivity and mobile phase modifier properties. *J. Chromatogr. A* **1999**, *852*, 383–394. [CrossRef]
16. Scriba, G.K. Chiral recognition in separation science: An update. *J. Chromatogr. A* **2016**, *1467*, 56–78. [CrossRef]
17. Wang, T.; Wenslow, J. Effects of alcohol mobile-phase modifiers on the structure and chiral selectivity of amylose tris(3,5-dimethylphenylcarbamate) chiral stationary phase. *J. Chromatogr. A* **2003**, *1015*, 99–110. [CrossRef]
18. Nguyen, L.A.; He, H.; Pham-Huy, C. Chiral drugs: An overview. *Int. J. Biomed. Sci.* **2006**, *2*, 85–100. [PubMed]
19. Wang, D.; Li, F.; Jiang, Z.; Yu, L.; Guo, X. Chiral Recognition Mechanisms of four β-Blockers by HPLC with Amylose Chiral Stationary Phase. *J. Pharm. Res.* **2014**, *13*, 449–457.
20. Zhang, X.; Li, Z.; Shen, B.; Chen, J.; Xu, X. Enantioseparation of three non-steroidal anti-inflammatory agents on chiral stationary phase by HPLC. *J. Anal. Sci.* **2012**, *2*, 18–23. [CrossRef]
21. Singh, C.; Sharma, C.S.; Kamble, P.R. Amino acid analysis using ion-exchange chromatography: A review. *Int. J. Pharm.* **2013**, 69–75. [CrossRef]
22. Velickovic, T.C.; Ognjenovic, J.; Mihajlovic, L. Separation of amino acids, peptides, and proteins by ion exchange chromatography. *Ion Exc. Tec. II* **2012**, 1–34. [CrossRef]
23. Koga, R.; Miyoshi, Y.; Sato, Y.; Mita, M.; Konno, R.; Lindner, W.; Hamase, K. Enantioselective determination of phenylalanine, tyrosine and 3,4-dihydroxyphenylalanine in the urine of d-amino acid oxidases deficient mice using two-dimensional high-performance liquid chromatography. *Chromatographia* **2016**, *37*, 15–22. [CrossRef]
24. Bhushan, R.; Nagar, H.; Martens, J. Resolution of enantiomers with both achiral phases in chromatography: Conceptual challenge. *RSC Adv.* **2015**, *5*, 316–328. [CrossRef]
25. Batra, S.; Bhushan, R. Determination and separation of enantiomers of atenolol by direct and indirect approaches using liquid chromatography: A review. *Biomed. Chromatogr.* **2017**, *32*, 1. [CrossRef] [PubMed]
26. Yu, J.; Tang, J.; Yuan, X.; Guo, X.; Zhao, L. Evaluation of the chiral recognition properties and the column performances of three chiral stationary phases based on cellulose for the enantioseparation of six dihydropyridines by high-performance liquid chromatography. *Chirality* **2017**, *29*, 147–154. [CrossRef] [PubMed]

27. Wang, T. Solid-state NMR characterization of amylose tris(3,5-dimethylphenylcarbamate) chiral stationary-phase structure as a function of mobile-phase composition. *Anal. Chem.* **2001**, *73*, 4190–4195.
28. Wang, Y.; Ma, C.; Feng, R.; Fu, J.; Sun, Y.; Zhang, Q.; Zhang, Y.; Zhou, Y.; Yang, B.; Zhang, Y.; et al. High performance liquid chromatographic separation of eight drugs collected in Chinese pharmacopoeia 2010 on amylose ramification chiral stationary phase. *Acta Pharm. Sin. B* **2012**, *2*, 527–533. [CrossRef]
29. Younes, A.A.; Galea, C.; Mangelings, D.; Heyden, Y.V. *Normal-Phase and Polar Organic Solvents Chromatography in Analytical Separation Science*, 3rd ed.; Vrije Universiteit: Brussel, Belgium, 2015; pp. 227–244.
30. Zhang, T.; Nguyen, D.; Franco, P. Reversed-phase screening strategies for liquid chromatography on polysaccharide-derived chiral stationary phases. *J. Chromatogr. A* **2010**, *1217*, 1048–1055. [CrossRef]
31. Smith, R.M.; Burr, C.M. Retention prediction of analytes in reversed-phase high-performance liquid chromatography based on molecular structure: VI. Disubstituted aromatic compounds. *J. Chromatogr. A* **1991**, *550*, 335–356. [CrossRef]
32. Atamna, I.Z.; Muschik, G.M.; Issaq, H.J. Effect of alcohol chain length, concentration and polarity on separations in high-performance liquid chromatography using bonded cyclodextrin columns. *J. Chromatogr.* **1990**, *499*, 477–488. [CrossRef]
33. Apaya, R.; Bondi, M.; Price, S. The orientation of N-H..O=C and NH...N hydrogen bonds in biological systems: How good is a point charge as a model for a hydrogen bonding atom? *J. Comput. Aided Mol. Des.* **1997**, *11*, 470–490. [CrossRef]
34. Jie, Z.; Quizheng, D.; Zuzhen, Z.; Fang, S.; Xinyu, L.; Zhenzhong, Z. Enantioseparation of Three Important Intermediates of Tanikolide with Immobilized Cellulose Chiral Stationary Phase. *J. Chromatogr. Sci.* **2015**, *53*, 959–962. [CrossRef] [PubMed]
35. Rebizi, M.N.; Sekkoum, K.; Belboukhari, N.; Cheriti, A.; Aboul-Enein, H.Y. Chiral separation and determination of enantiomeric purity of the pharmaceutical formulation of cefadroxil using coated and immobilized amylose-derived and cellulose-derived chiral stationary phases. *Egypt. Pharm. J.* **2016**, *15*, 88–97.
36. Buddrick, O.; Jones, O.A.H.; Morrison, P.D.; Small, D.M. Heptane as a less toxic option than hexane for the separation of vitamin E from food products using normal phase HPLC. *RSC Adv.* **2013**, *3*, 24063–24068. [CrossRef]
37. Thienpont, A.; Gal, J.; Aeschlimann, C.; Felix, G. Studies on stereoseletive separations of the "azole" antifungal drugs ketocaonazole and itraconazole using HPLC and SFC on silica-based polysaccharides. *Analysis* **1999**, *27*, 713–718.
38. Aboul-Enein, H.Y.; Ali, I. Comparison of the chiral resolution of econazole, miconazole, and sulconazole by HPLC using normal-phase amylose CSPs. *J. Anal. Chem.* **2001**, *370*, 951–955. [CrossRef]
39. Aboul-Enein, H.; Ali, I. Enantiomeric separation of cizolirtine and metabolites on amylose tris (3,4-dimethylphenyl carbamate) chiral stationary phase. *Iran J. Pharm. Res.* **2004**, *59*, 743–746.
40. Kučerováa, G.; Vozka, J.; Kalíková, K.; Geryk, R.; Plecitá, D.; Pajpanova, T.; Tesařováa, E. Enantioselective separation of unusual amino acids by high performance liquid chromatography. *Sep. Purif. Technol.* **2013**, *119*, 123–128. [CrossRef]
41. Tanaka, N.; Goodell, H.; Karger, B.L. The role of organic modifiers on polar group selectivity in reversed-phase liquid chromatography. *J. Chromatogr. A* **1978**, *158*, 233–248. [CrossRef]
42. Matthijs, N.; Maftouh, M.; Heyden, Y.V. Screening approach for chiral separation of pharmaceuticals IV. Polar organic solvent chromatography. *J. Chromatogr. A* **2006**, *1111*, 48–61. [CrossRef] [PubMed]
43. Gyimesi-Forras, K.; Akasaka, K.; Lammerhofer, M.; Maier, N.M.; Fujita, T.; Watanabe, M.; Harada, N.; Linder, W. Enantiomer separation of a powerful chiral auxiliary, 2-methoxy-2-(1-naphthyl) propionic acid by liquid chromatography using chiral anion exchanger-type stationary phases in polar-organic mode; investigation of molecular recognition aspects. *Chirality* **2005**, *17*, S134–S142. [CrossRef] [PubMed]

Sample Availability: Samples of the compounds are not available from the authors.

molecules

MDPI

Article

Colistin Sulfate Chiral Stationary Phase for the Enantioselective Separation of Pharmaceuticals Using Organic Polymer Monolithic Capillary Chromatography †

Ali Fouad [1,2], Montaser Sh. A. Shaykoon [2], Samy M. Ibrahim [3], Sobhy M. El-Adl [3] and Ashraf Ghanem [1,*]

1 Chirality Program, Faculty of Science and Technology, University of Canberra, Canberra, ACT 2601, Australia; alifouad247@gmail.com
2 Pharmaceutical Chemistry Department, Faculty of Pharmacy, Al-Azhar University, Assiut 71524, Egypt; monoceutical@yahoo.com
3 Pharmaceutical Chemistry Department, Faculty of Pharmacy, Zagazig University, Zagazig 44519, Egypt; dr-samy2010@hotmail.com (S.M.I.); sobhyeladl@yahoo.com (S.M.E.-A.)
* Correspondence: ashraf.ghanem@canberra.edu.au; Tel.: +61-(02)-6201-2089 or 0429319993; Fax: +61-(02)-6201-2328

† The article is dedicated to Prof. Frantisek Svec on the occasion of his 75th birthday.

Academic Editors: Maria Elizabeth Tiritan, Madalena Pinto and Carla Sofia Garcia Fernandes
Received: 16 January 2019; Accepted: 21 February 2019; Published: 26 February 2019

Abstract: A new functionalized polymer monolithic capillary with a macrocyclic antibiotic, namely colistin sulfate, as chiral selector was prepared via the copolymerization of binary monomer mixtures consisting of glycidyl methacrylate (GMA) and ethylene glycol dimethacrylate (EGDMA) in porogenic solvents namely 1-propanol and 1,4-butanediol, in the presence of azobisiso-butyronitrile (AIBN) as initiator and colistin sulfate. The prepared capillaries were investigated for the enantioselective nano-LC separation of a group of racemic pharmaceuticals, namely, α- and β-blockers, anti-inflammatory drugs, antifungal drugs, norepinephrine-dopamine reuptake inhibitors, catecholamines, sedative hypnotics, antihistaminics, anticancer drugs, and antiarrhythmic drugs. Acceptable separation was achieved for many drugs using reversed phase chromatographic conditions with no separation achieved under normal phase conditions. Colistin sulfate appears to be useful addition to the available macrocyclic antibiotic chiral phases used in liquid chromatography.

Keywords: colistin sulfate; enantioselective; encapsulation; capillary chromatography; monolith; organic polymer

1. Introduction

Most of the drugs currently in use are worldwide marketed as racemates. Enantiomers can exhibit different activities in biological systems, in particular, their pharmacology, toxicology, pharmacokinetics and metabolism. Therefore, it is important to separate single enantiomers to limit side effects that may arise from unwanted enantiomers [1–3]. To access enantiomerically pure compounds, enantioselective chromatographic techniques have been considered as the most feasible method compared to other more expensive and time-consuming approaches [4–10]. Among these techniques, High Performance Liquid Chromatography (HPLC) is the most widely used technique in enantiomer separation [11–13]. In HPLC, a chiral selector is required to form a Chiral Stationary Phase (CSP), the main driver for the chiral separation. The CSP is normally bound, immobilized adsorbed or otherwise attached to an appropriate support. The enantiomers are then resolved by the formation of temporary diastereomeric complexes between the analyte and the CSP. The stationary phase support plays a very important role in any research investigation in this field [14–16].

Because of their advantages, the use of monoliths as stationary phases for HPLC represents a promising alternative to particle packed columns for CEC, conventional HPLC columns and nano-HPLC capillaries [17–21]. The preparation of organic polymer monolithic stationary phases via surface modification with a suitable precursor followed by the polymerization process results in increased stability of the monolith and affords greater adherence to the confining wall [22–28]. Many CSPs have been previously reported attached to monolithic support for chiral separation, especially in enantioselective capillary chromatography [29–31]. However, only a few were macrocyclic antibiotics [32].

Natural macrocyclic antibiotic materials play a very important role in chiral separation as useful CSPs. In general, the macrocyclic antibiotics most widely used as chiral selectors are vancomycin, vancomycin aglycon, norvancomycin, teicoplanin, and teicoplanin aglycon, ristocetin A, thiostrepton, rifamycin, kanamycin, streptomycin, fradiomycin, eremomycin and avoparcin [33]. The unique features of these chiral selectors include different chiral centers, inclusion cavities, phenyl rings, several hydrogen donor and acceptor sites, sugar moieties, and other groups which are the main drivers for their good chiral recognition abilities in different chromatographic modes. The chiral recognition mechanism in most of these antibiotics chiral selectors relies heavily on complexation, hydrogen bonding, inclusion complex formation, dipole interactions, steric interactions, and anionic and cationic binding. These chiral selectors have been employed for the enantiomeric resolution of a variety of racemates in HPLC, CEC, and CE [34–37]. Furthermore, a few were previously used in preparation of chiral monolithic columns for the enantioselective separation of racemic pharmaceuticals [38]. Colistin sulfate represents a new addition to the macrocyclic antibiotic family enabling its multi-chirality sites and functional groups to provide chromatographic interactions with racemic analytes [38]. Furthermore, the encapsulation of a macrocyclic antibiotic in an organic polymer monolith in capillary HPLC hasn't been previously reported. The ease of in situ preparation in capillaries or narrow channels of microfluidic devices render these ideal stationary phases for microscale separation formats [39].

Here we introduce a new chiral macrocyclic antibiotic, namely colistin sulfate, encapsulated in organic polymer monolithic capillary for the enantioselective nano-liquid chromatographic separation of a set of racemic pharmaceuticals.

2. Results

2.1. Preparation and Characterization of Polymer Monoliths

The use of macrocyclic antibiotics in chiral separation was previously reported in both conventional HPLC and CEC. In particular, macrocyclic antibiotic-based silica monolithic columns were previously studied [32,38,40]. However, the macrocyclic antibiotics were immobilized on the activated monoliths by a tedious reductive amination process [32]. No work was previously reported on the polymer monolithic antibiotic-based CSP in capillary liquid chromatography. Here we report the first use of colistin sulfate as a macrocyclic antibiotic chiral selector entrapped in organic polymer monolith for enantioselective capillary LC (Figure 1). The miscibility and solubility of colistin sulfate was tested in porogenic solvents used in monolith preparation, namely 1,4-butanediol, ethanol and *n*-propanol. When 1,4-butanediol was used in the polymerization mixture, a highly homogeneous solution occurred, however, better solubility was achieved when used in combination with 1-propanol as porogenic solvent.

Colistin sulfate-based polymer monolithic column (C1) was prepared via in situ copolymerization of colistin sulfate with monomers (40%) (GMA (20%) as a functional monomer and EGDMA (20%) as a cross linker) in the presence of a ternary porogenic system composed of 1-propanol (48%), 1,4-butanediol (6%) and chiral selector (6%). The ratio of the monomers to the porogens was fixed at

40:60 w/w, respectively; this was selected to provide columns with a good balance of permeability, surface area and mechanical stability.

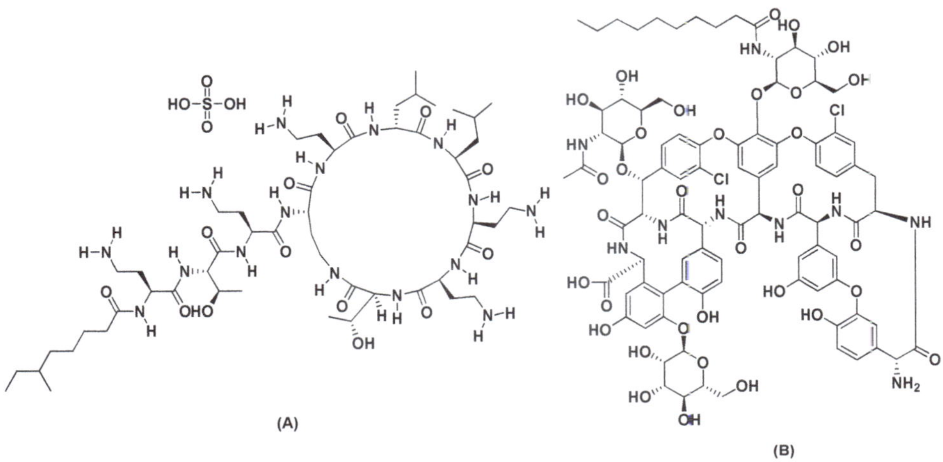

Figure 1. Chemical structure of colistin sulfate (**A**) and teicoplanin (**B**).

Scanning Electron Microscopy (SEM) and Surface Properties of the Monoliths

Scanning electron microscopy (SEM) photos were taken to study the morphology of the prepared monolith. Column C1 showed a porous structure with interconnecting channels allowing the flow of mobile phase with reduced column back-pressure (Figure 2). The textural surface properties of the monolithic columns, including the specific surface area and the pore structure, were previously calculated by our group. The pore size distribution was determined from the adsorption isotherms using the Barrett–Joyner–Halenda (BJH) method. Specific surface area (SBET) was calculated using multi-point adsorption data from a linear segment of the N_2 adsorption isotherms using the Brunauer–Emmett–Teller (BET) theory [41]. The monolithic column previously prepared using similar procedure and demonstrated good enantioseparation exhibited surface area of 28.67 m^2/g, pore size of 169.2 nm and total pore volume of 0.12 cm^3/g.

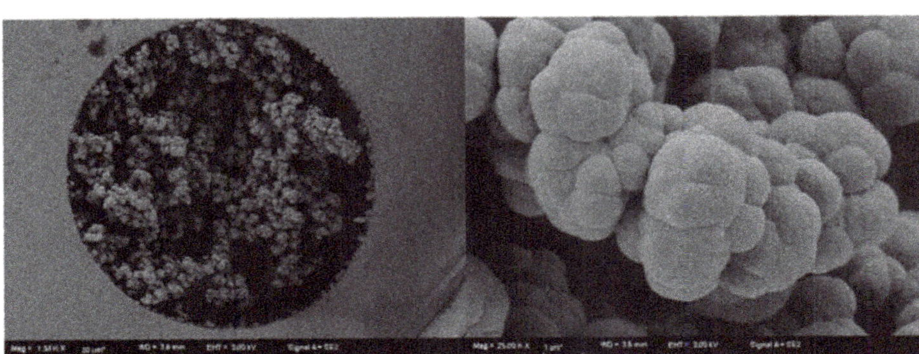

Figure 2. SEM of column C1 at 1200× and 25,000× (left and right, respectively) shows small micro-globules with rough surface.

Elemental analysis was used to determine the nitrogen (1.9 and 2.68 % w/w) and sulfur (0 and 0.4% w/w) content in the C1 column and blank column (G column), respectively. The blank column

(G column) was prepared using the same polymerization mixture with addition of water instead of colistin sulfate. The measured the nitrogen contents were 1.9 and 2.68 % *w/w* and the measured sulfur contents were 0 and 0.4% *w/w* in the C1 column and blank column (G column), respectively. Elemental analysis was conducted to ensure the relevance of the presence of colistin sulfate in the prepared C1 column. These results confirm the presence of the chiral selector in the prepared C1 column.

2.2. Enantioseparation of Different Classes of Pharmaceutical Racemates

The colistin sulfate-based polymer monolithic capillary column was prepared as described above and investigated for the nano-LC enantioseparation of a set of different classes of racemic pharmaceuticals, namely: β-blockers, α-blockers, anti-inflammatory drugs, antifungal drugs, norepinephrine-dopamine reuptake inhibitors, catecholamines, sedative hypnotics, antihistamines, antibacterial drugs, anticancer drugs and antiarrhythmic drugs. Although reversed phase enantio-selective LC examples are limited, macrocyclic antibiotics were previously used in enantioseparation chromatography under reversed phase chromatographic mode [34,36–38,42–45]. The initial mobile phase selected for the enantioseparation separation of racemates **1–37** (Figure 3) was a binary mixture of methanol/water screened from 95:5 to 5:95 *v/v* at 1 mL/min flow rate at fixed UV detection 219 nm with eleven compounds separated (Rs $\geq$ 1) (Table 1). For examples, in MeOH/H_2O 80:20 *v/v*, only ibuprofen (**7**) was separated, while in MeOH/H_2O 40:60, indoprofen (**10**), hexaconazole (**15**) and miconazole (**16**) were separated. In MeOH/H_2O 10:90 *v/v*, aminoglutethimide (**22**), tyrosine (**29**) and *O*-methoxymandelic acid (**34**) were also separated. The addition of an additive, namely triethylamine (TEA) 1% *v/v* in 10:90, resulted in the separation of acebutolol (**4**) normetanephrine (**21**), propafenone (**26**), tyrosine (**29**) and 4-hydroxy-3-methoxymandelic acid (**35**) (Figure 4), while non-acceptable separations were achieved by addition of the acidic additive namely trifluoroacetic acid (TFA). In an attempt to use normal phase namely *n*-hexane/2-propanol mixture ranging from 10–90% (*v/v*) resulted in resolution less than 1. All chromatographic data are summarized in Table 1.

Table 1. Chromatographic data, separation and resolution factors for the significantly resolved compounds, using reversed mobile phases, flow rate: 1 μL/min.

		Column C1 (Colistin Sulfate)				
Phase	**Mobile Phase**	**Drug**	**Rt_1 (min)**	**Rt_2 (min)**	**Separation Factor (α)**	**Resolution (Rs)**
Reversed Phase	Methanol:water 80:20	Ibuprofen (**7**)	23.3	38.9	1.7	1.02
	Methanol:water 40:60	Indoprofen (**10**)	24.3	40.1	1.72	1.6
		Hexaconazole (**15**)	23.8	37.6	1.9	1.2
		Miconazole (**16**)	17.5	23.6	1.4	1.6
	Methanol:water 30:70	Indoprofen (**10**)	23.6	63.4	2.8	1.63
		Miconazole (**16**)	17.6	23.6	1.41	1.6
	Methanol:water 10:90	Aminoglutethimide (**22**)	22.4	33.6	1.5	1
		Tyrosine (**29**)	23.4	35	1.5	1.64
		O-Methoxymandelic acid (**34**)	26.1	40.2	1.52	1.1
	Methanol:water 10:90, 1%TEA	Acebutolol (**4**)	22.8	28.7	1.3	1.14
		Normetanephrine (**21**)	22.3	28.4	1.3	1.3
		Propafenone (**26**)	21.4	30.5	1.3	1.2
		Tyrosine (**29**)	18.1	24.2	1.3	1.7
		4-Hydroxy-3-methoxymandelic acid (**35**)	24.1	31.5	1.3	1.03

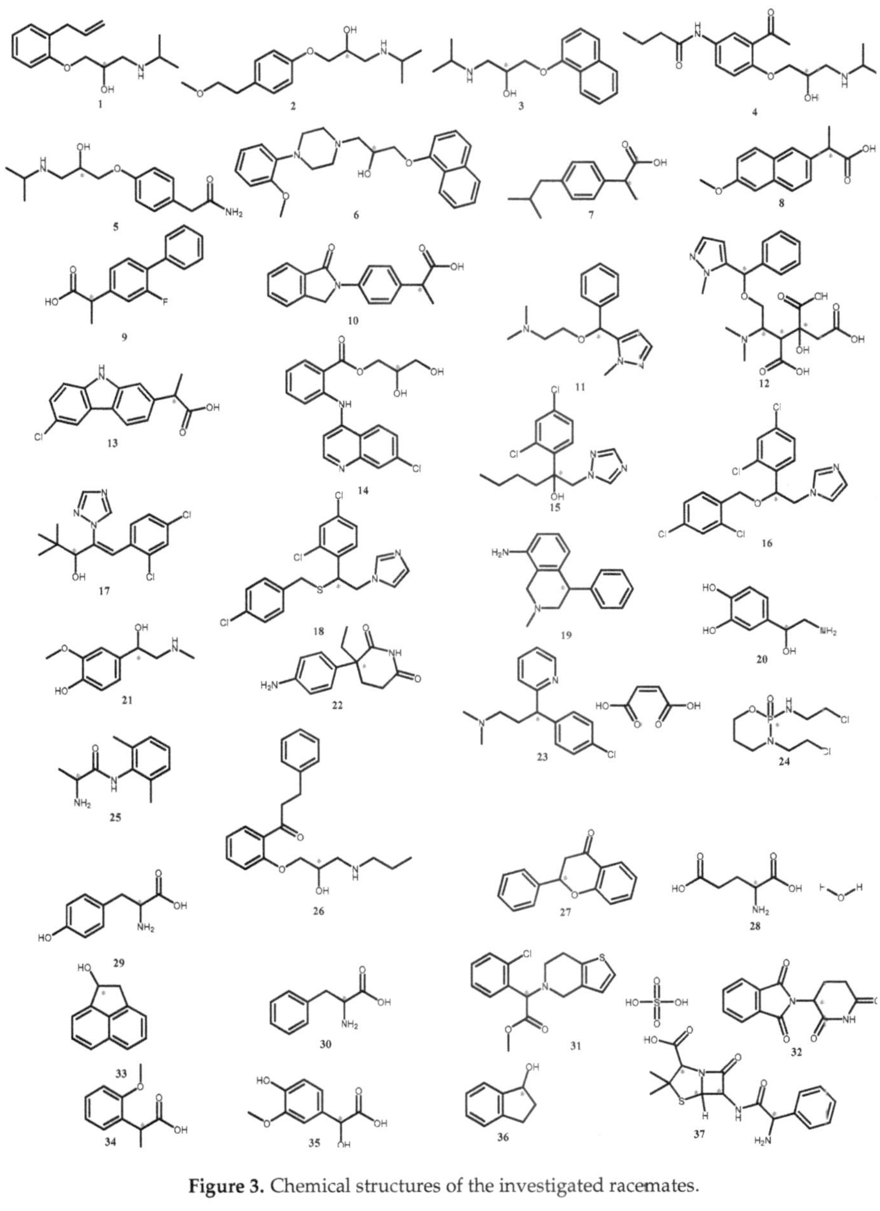

Figure 3. Chemical structures of the investigated racemates.

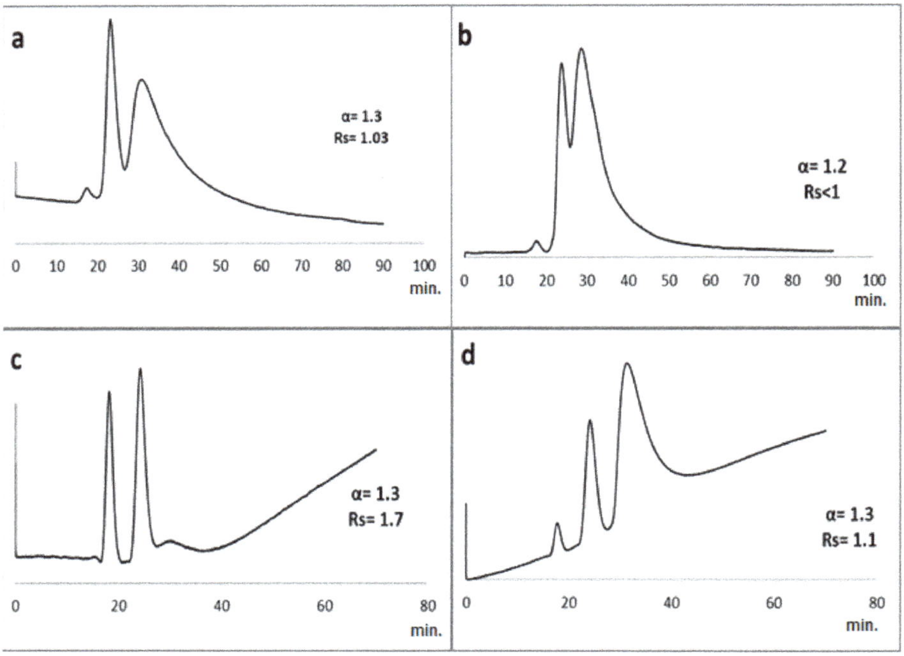

Figure 4. Enantioselective nano-LC separation of; (**a**) racemic 4-hydroxy-3-methoxymandelic acid (**35**); (**b**) phenylalanine (**30**) (mobile phase: methanol/water 40:60 *v*/*v*,); (**c**) tyrosine (**29**) and (**d**) *O*-methoxymandelic acid (**34**) on a C1 capillary column (150 µm ID, 25 cm length). UV: 219 nm, flow rate: 1 µL/min.

Because of the novelty associated with using colistin sulfate as a chiral selector, confirmatory tests were done by injecting the separated enantiomeric drugs using capillary monolithic column without chiral selector (blank column, cf. Figure 1). The injected drugs included tyrosine (**29**), phenylalanine (**30**), *O*-methoxymandelic acid (**34**) and 4-hydroxy-3-methoxymandelic acid (**35**). Only single peaks were obtained under chromatographic conditions similar to those previously used when using capillary columns with colistin sulfate as CSP (C1 column). Furthermore, the single *S*-enantiomer of acebutolol (*S*-acebutolol) was injected on the C1 column (Figure 2). Only a single peak was obtained when used alone and mixed with its racemic mixture, which resulted in a peak with higher intensity, but unfortunately with low resolution. Also it was observed that *S*-acebutolol eluted first in the same retention time as the eluted single peak of single isomer *S*-acebutolol. The results achieved from the injection of the enantiomers on both the blank and C1 column, confirm the presence of the chiral selector in situ the capillary and that it was not washed out or dissolved in the mobile phase. The investigated repeatability of the used C1 column is considered as a proof of stability of the chiral selector contained in the capillary. It was also observed that the chiral separation was mostly achieved at high water content in the mobile phase; although, the chiral selector itself can be dissolved in water. This does not contradict what has been previously reported in literature where similar solvent used for dissolving the chiral selector can be used as mobile phase in the same column [32].

The combination of thin-hair capillary format in capillary HPLC is also beneficial as swapping from existing conventional liquid chromatography LC (mL flow, more solvent) to micro/nano flow LC (less solvent) is beneficial. The expected outcome will be environmentally responsible, cost effective and efficient analytical sample preparation and separation technologies for analytical laboratories throughout the world. Some featured benefits include but not limited to: (1) up to 4× increase in sensitivity; (2) improved turn-around-time with up to 5× faster separations; (3) up to a 95% reduction in mobile phase consumption and (4) improved robustness–less sample introduced into the MS when used in LC/MS and, ultimately, less instrument downtime.

For example, the chiral analysis for one run in conventional HPLC consumes at least 20–30 mL of environmentally unfriendly solvents for 30 min separation. On the other hand, in nano-HPLC, running a similar analysis under reversed phase conditions consumes less than 100 μL of water-based mobile phase. The capillary monolithic column is 10,000 less in internal diameter and operates with one million times less solvent volume than a conventional column. Consequently, this approach is economically efficient where only a single sorbent, namely colistin antibiotic, was used as CSP in capillary HPLC reducing materials/solvents consumption. Taking ibuprofen as an example, it was efficiently separated on the prepared monolithic column (Rs = 1.02) while it was recently enantio-separated (Rs = 1.05) on a mixed sorbents containing eremomycin and bovine serum albumin BSA-based CSP under reversed phase conditions using mobile phase: MeOH:KH_2PO_4 (0.1 M, pH 4.5); 50:50 (*v*/*v*); flow rate: 0.5 mL/min; and fixed UV 220 nm [34]. Another example is the recent use of mixed chiral sorbents based on silica with immobilized macrocyclic antibiotics eremomycin and vancomycin for the enantioselective of β-blockers such as atenolol and amino acids like phenylalanine by conventional HPLC using a mobile phase of MeOH:ACN–TEAA (0.1%, pH 4.5) (95:5, *v*/*v*), and a flow rate of 1 mL/min [37]. Nano-HPLC chromatograms for some of the separated compounds under different ratios of methanol and water are given in Figure 4.

2.3. Column Repeatability

The repeatability is the ability to prepare equally performing capillaries at different time (run to run) and locations (batch to batch). To determine the repeatability of the prepared capillaries, two capillaries were prepared on the same day using the same polymerization mixture to test column-to-column repeatability. Moreover, batch-to-batch repeatability was tested by preparing three different batches at different days using the same polymer mixtures. 4-Hydroxy-3-methoxymandelic acid (**35**) was selected to test the capillaries' performance in terms of repeatability as it was nearly baseline resolved on both columns. Reproducibility of the retention times of both 4-hydroxy-3-methoxymandelic acid (**35**) peaks was satisfactory. In the run-to-run repeatability using one column, the average retention times for the two peaks were 23.5 min (RSD = 1.7%) and 30.6 min (RSD = 1.27%); respectively. In column-to-column repeatability, the average retention times for the two peaks are 23.5 min (RSD = 2.2%) and 30.6 min (RSD = 1.9%); respectively. In batch-to-batch repeatability, the average retention times for peak one and peak two are 22.5 min (RSD = 3.9%) and 31.4 min (RSD = 5.3%); respectively. The retention times and relative standard deviations (RSD) of the retention times ranged between 1.2% and 5.3%. These results suggest that the monolithic capillary columns can be used for reproducible routine analysis. It is worth mentioning that the acceptable %RSD values for intra-batch and inter-batch are 2.5% and 15%; respectively. Furthermore, the column loadability was tested by injecting more than 300 runs on the same column; 4-hydroxy-3-methoxymandelic acid (**35**) was injected in different orders started at run number 160 and ended by run number 307. The same separation was achieved (Figure 5).

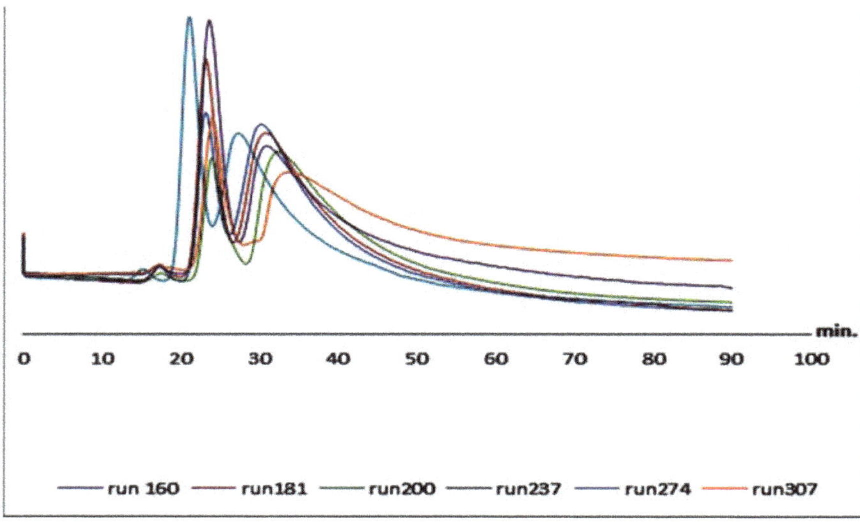

Figure 5. The loadability of the monolithic columns of 4-hydroxy-3-methoxymandelic acid (**35**) started at run no. 160 up to run no. 307, on C1 capillary column (150 µm ID, 25 cm length). mobile phase: methanol/water 40:60 *v*/*v*, UV: 219 nm, flow rate: 1 µL/min.

2.4. *Effect of the Concentration of Chiral Selector*

The optimum concentration of the colistin sulfate in the polymerization mixture was selected after the preparation of three different capillaries with different concentrations of colistin sulfate (10, 20 and 30 mg/mL). The results revealed that 10 mg/mL afforded better separation and resolution while increasing the concentration to 30 mg/mL or more resulted in poor separation and resolution.

3. Discussion

Various macrocyclic antibiotics have been previously synthesized and applied on silica or polymer surfaces as a stationary phase either by immobilization, coating or by covalent bonding [24,28,32–36]. Whilst coating or physical adsorption is considered an suitable method to prepare CSPs, covalent bonding increases the chances for using diverse mobile phases and creates a more robust CSP [46]. It is worth pointing out that most of the CSPs have been prepared via immobilization to bond the chiral selectors to the solid supports. This has resulted in robust and more stable CSP, however, time consuming process offering less coverage of the CS compared to the one pot technique [47]. The way the CSP has been prepared (coating vs. bonding) may affect the chiral recognition mechanism. Thus, bonded-type phase showed a lower chiral recognition power than the coated-type phase.

Schmid et al. have been reporting since 2006 the development of dynamically-coated chiral stationary phases [48] using a macrocyclic antibiotic, namely vancomycin. Few macrocyclic antibiotics were previously used in preparation of chiral monolithic columns for the enantioselective separation of racemic pharmaceuticals [38]. Of interest, in 2010 Schmid et al. [32] published an article describing the preparation of a new chiral stationary phase by dynamic coating of a reversed-phase HPLC monolithic column with vancomycin-derivatives as chiral selector. They then investigated the separation of amino acids using reversed phase chromatographic conditions, namely triethyl-ammonium acetate (TEAA) buffer and methanol. As the underivatized vancomycin is hydrophilic, it could not be adsorbed on the commercial hydrophobic chromolith monolith. Consequently, vancomycin was derivatized to *N*-(2-hydroxydodecyl)-derivative before immobilization on the chromolith. Vancomycin is reversibly adsorbed via a hydrophobic side chain to the reversed-phase material. Similarly, Haroun et al. [49] dynamically coated the macrocyclic antibiotic teicoplanin on RP_{18} and RP_8 columns.

Teicoplanin has a hydrophobic C10 side chain which is attached to the glucopyranosyl group (Figure 1). This side chain was used to immobilize the chiral selector on the hydrophobic reversed phase material. This dynamically coated phase was used for the separation of aromatic amino acids. Similary, in this manuscript, colistine possesses a C9 hydrophobic side chain that can be used for the immobilization on the hydrophobic monolith prepared in this manuscript. It is worth to note that (1) continuous polymers formed from hydrophobic monomers can be used as stationary phase in reversed phase chromatography (RPC) and (2). Solvents used for dissolving the chiral sector can be used as mobile phase (not in excess) on the same column [32].

The chiral recognition of macrocyclic antibiotics used as chiral selectors for the enantio-separation of different compounds is due to the presentation of ionisable acidic or basic functional groups contributing to stereoselectivity, the presence of multiple stereogenic centers, and the presence of both hydrophobic and hydrophilic groups. Therefore, the transient non-covalent diastereomeric complexes with macrocyclic antibiotic are based on both electrostatic interactions and secondary interactions such as hydrophobic, hydrogen bonds, dipole-dipole, π–π interactions, and steric repulsion. Macrocyclic antibiotics have been successfully applied to HPLC and also to CEC for chiral separation of pharmaceutical drugs using stationary phases in the reversed-phase and the normal-phase modes [50–55].

Colistin sulfate has never been used as chiral selector although it possesses many points of interaction which significantly increase its enantiorecognition ability. It is well established that under reversed phase conditions, the formation of inclusion complexes within the cavity is the most predominant mechanism of retention and enantioselectivity. Moreover, the presence of different functional groups creates more points of interaction between the enantiomers and the CSP via π–π bonding, hydrogen bonding, dipole–dipole stacking, etc. which can increase the selectivity towards some analytes. For example, in miconazole (**16**), hydrophobic interactions are the prevailing CSP-analyte interactions, whereas hydrogen bonding seems to be more important in the enantiointeractions between the other analytes and CSPs [32]. Initial testing with mixture of methanol-based mobile phase, enantioselective separation was observed for many analytes with polar groups including acebutolol (**4**), indoprofen (**10**), hexaconazole (**15**), normetanephrine (**21**), aminoglutethimide (**22**), propafenone (**26**), tyrosine (**29**), *O*-methoxymandelic acid (**34**) and 4-hydroxy-3-methoxymandelic acid (**35**). This confirms the importance of solvent polarity in the chiral separation mechanism in terms of the inclusion complex stability. The large retention times observed is due to the very low flow rate used. Higher flow rate may result in high backpressure. Peak tailing of the more retained isomers was overcome by the use of mobile phase additives such as triethanol-amine (TEA), which resulted in improved resolution, although, their negative effect on the lifetime of the capillary columns as well as its potential problems with nano-LC systems (e.g. precipitation in the pumps and valves) [56] can be dominant. No remarkably peak tailing of acebutolol (**4**), atenolol (**5**) and tyrosine (**29**) racemates was observed, ascribed to the existence of the amino groups on the surface of the monolithic matrices. It was also observed that the chiral separation was mostly obtained at high water content of the mobile phase; this indicates that water facilitates the interaction between the CSP and the racemates. We postulate that chiral separation in this study was mainly achieved via the formation of inclusion complexes as discussed previously. The use of normal organic phase resulted in high back pressure and very short life time of the prepared column. Nevertheless, the use of *n*-hexane/2-propanol mobile phase mixture ranging from 10–90% (*v*/*v*) resulted in resolution less than 1.

4. Experimental

4.1. Reagents and Materials

Colistin sulfate (99%), ethylene glycol dimethacrylate (EGDMA, 98%), glycidyl methacrylate (GMA, 98%), 1-propanol (99%), 1,4-butanediol (99%), trifluoroacetic acid (TFA, $\geq$99.5%), sodium hydroxide and hydrochloric acid were purchased from Sigma Aldrich (Milwaukee, WI, USA). Acetone (AR grade) and ethanol (HPLC grade) were purchased from BDH (Kilsyth, Vic., Australia). Methanol (HPLC) grade was purchased from Scharlau (Sentmenat, Spain). All other reagents were of the highest available grade and used as received. The fused-silica capillaries (150 µm internal diameter) were purchased from Polymicro Technologies (Phoenix, AZ, USA). 2,2-Azobis(isobutyronitrile) (AIBN) was obtained from Wako (Osaka, Japan). Water used for dilutions and experiments was purified by a Nano-pure Infinity water system (NJ, USA). The racemic analytes were mostly purchased from Sigma Aldrich.

4.2. Preparation and Characterization of the Monolithic Columns

4.2.1. Activation of the Fused Silica Capillaries

Briefly, the fused silica capillaries were rinsed using a Harvard syringe pump (Harvard Apparatus, Holliston, MA, USA) and a 250 µL gas-tight syringe (Hamilton Company, Reno, NE, USA) with acetone and water 2–3 times each, activated with 0.2 mol/L sodium hydroxide (NaOH) for 6 h confirming the absence of any air bubbles, washed with water 3–4 times till neutral (pH 7), then washed with 0.2 mol/L hydrochloride (HCl) for 12 h, rinsed with water and ethanol 2–3 times each. A 20% (*w*/*w*) solution of 3-(trimethoxysilyl)propyl methacrylate in 95% ethanol adjusted to pH 5 using acetic acid was pumped through the capillaries at a flow rate of 0.25 µL/min for 6 h. The capillary was then washed with acetone one time and dried with a stream of nitrogen for 2 min. then left at room temperature for 24 h.

4.2.2. Preparation of Colistin Sulfate Functionalized Monomer

The short (~25 cm in length) surface modified capillary was filled by Harvard syringe pump with the degassed polymerization mixture at 0.25 µL/min using the syringe pump. Colistin sulfate polymer-based monolithic capillary column was prepared via in situ copolymerization of binary monomer mixtures consisted of GMA (20%) as a monomer and EGDMA (20%) as across linker along with different porogens namely; 1-propanol (48%), 1,4-butanediol (6%), in the presence of 1 wt% AIBN (with respect to monomers) and colistin sulfate (6%) as chiral selector. The blank column (G column) was prepared using the same procedure by addition of water (6%) instead of water. The filled capillaries were then sealed with a septum, placed in 70 °C water bath for 18 h for the polymerization reaction to take place. The unreacted monomers were removed from the monolithic columns by pumping methanol at a flow rate of 100 µL/h for 24 h before being investigated under light microscope to ensure its consistency and homogeneity of the polymerization mixture inside the capillary. The filled capillaries were conditioned with mobile phase for 1–3 days at µL/min (Figure 6). The ratios of the monomers to the porogens were kept 40% and 60%, respectively. The ratios of the porogens were fixed as 48% 1-propanol, 6% 1,4-butanediol and 6% chiral selector, all percentages are *w*/*w*.

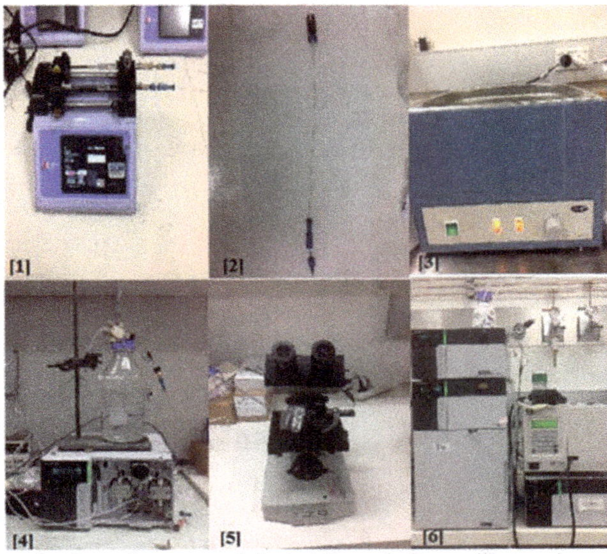

1) The polymerization mixture was injected through the capillary using a Harvard syringe Pump.
2) The capillary ends was Sealed with a rubber septum.
3) Thermal polymerization in a 70°C water bath.
4) The unreacted monomers/porogens was wased with methanol.
5) The capillary for continuous monolith was examined using light microscope.
6) Nano-LC analysis of the continuous monolithic capillaries.

Figure 6. Steps showing the preparation of polymer monolithic capillary columns.

4.2.3. SEM of the Prepared Monoliths

SEM was performed to study the morphology of the prepared capillaries. The capillaries were cut into ~1 cm sections and put perpendicularly on 12.7 mm pin-type aluminum stub using double face epoxy resin tape. SEM was carried out and high resolution images were collected by sputter coating the capillary sections with gold Using ZEISS SIGMA FE-SEMs for High Quality Imaging and Advanced Analytical Microscopy (ZEISS Sigma, Jena, Germany).

4.3. Instrumentation

A nano-liquid chromatographic system consisting of an LC-10AD VP pump (Shimadzu, Kyoto, Japan), injector model 7725i-049 (Rheodyne, Park Court, CA, USA), a UV-Vis detector model MU 701 UV-VIS (GL Science, Tokyo, Japan) and a Shimadzu CDM-20A communications bus module was used. The system flow was split after direct injection. The data was processed by the Shimadzu Lab-Solutions software version 5.54 SP2 (Shimadzu, Kyoto, Japan).

4.4. Standard Solutions and Sample Preparation

Stock solutions of the racemic analytes at concentrations of 1 mg/mL in filtered HPLC grade methanol were prepared. Prior to injection, the stock solutions were further diluted 10× by mobile phase and filtered through Minisart RC 15 0.2 μm pore size filters (Sartorius, Goettingen, Germany). Tested compounds: β-blockers: alprenolol (**1**), metoprolol (**2**), propranolol (**3**), acebutolol (**4**), atenolol (**5**); α-blockers: naftopidil (**6**); anti-inflammatory drugs: ibuprofen (**7**), naproxen (**8**), flurbiprofen (**9**), indoprofen (**10**), cizolirtine (**11**), cizolirtine citrate (**12**), carprofen (**13**), glafenine (**14**); antifungal drugs: hexaconazole (**15**), miconazole (**16**), diniconazole (**17**) sulconazole (**18**); norepinephrine-dopamine reuptake inhibitor: nomifensine (**19**); catecholamines: arterenol (**20**), normetanephrine (**21**); sedative hypnotics: aminoglutethimide (**22**); anti-histamines: chlorpheneramine (**23**); anticancer drugs: ifosfamide (**24**); antiarrhythmic drugs: tocainide (**25**), propafenone (**26**); flavonoids: flavanone (**27**); amino acids: glutamic acid monohydrate (**28**), tyrosine (**29**), phenylalanine (**30**); anti-platelet agents: clopidogrel (**31**); immunomodulatory drugs: thalidomide (**32**); miscellaneous: 1-acenaphthenol (**33**), *O*-methoxymandelic acid (**34**) 4-hydroxy-3-methoxymandelic acid (**35**), 1-indanol (**36**) and ampicillin (**37**). The chemical structures of the investigated racemates are shown in Figure 3.

4.5. HPLC Conditions

The mobile phase consisted of water/methanol (*v*/*v*) for the reversed phase HPLC and *n*-hexane/2-propanol for normal phase HPLC. For all samples, the injected volume was 0.2 μL at room temperature with flow rate 1 μL/min on C1 capillary column (150 μm ID, 25 cm length). Preliminary UV analyses were performed at a wavelength of 219 nm.

5. Conclusions

The macrocyclic antibiotic colistin sulphate has been used for the first time as a chiral selector entrapped in a polymer monolith for enantioselective capillary chromatography. The new capillary column was investigated for the enantioselective separation of a set of racemic drugs. Acceptable separation was achieved for many drugs, including acebutolol (**4**), ibuprofen (**7**), indoprofen (**10**), hexaconazole (**15**), miconazole (**16**), normetanephrine (**21**), aminoglutethimide (**22**), propafenone (**26**), tyrosine (**29**), *O*-methoxymandelic acid (**34**) and 4-hydroxy-3-methoxymandelic acid (**35**) under reversed phase chromatographic conditions, while normal phase conditions did not achieve any acceptable separations. The method provides more economical analysis under environmentally benign reversed phase conditions.

Author Contributions: Conceptualization, M.S.A.S., S.M.I. and S.M.E.-A.; Methodology, A.F.; Software, A.F. and A.G.; Validation, A.F. and A.G.; Formal analysis, A.F. and A.G.; Investigation, A.F. and A.G.; Resources, A.F. and A.G.; Data curation, A.F.; Writing—original draft preparation, A.F.; Writing—review and editing, A.F.; Visualization, A.G.; Supervision, M.S.A.S., S.M.I. and S.M.E.-A.; Project administration, S.M.E.-A. and A.G.; Funding acquisition, M.S.A.S.

Funding: This research was supported, in part, by Al-Azhar University, Egypt and University of Canberra, Australia. Funding was supported by the Egyptian Cultural and Educational Bureau, Minister of Higher Education Egypt (Cultural Affairs and Missions Sector) as Ph.D. joint mission stipend offered to Ali Fouad.

Conflicts of Interest: The authors declare no conflict of interest.

References

1. Lin, G.Q.; Zhang, J.G.; Cheng, J.F. Overview of chirality and chiral drugs. *Chiral Drugs Chem. Biol. Action* **2011**, 14–18.
2. Maier, N.M.; Franco, P.; Lindner, W. Separation of enantiomers: Needs, challenges, perspectives. *J. Chromatogr. A* **2001**, *906*, 3–33. [CrossRef]
3. Lorenz, H.; Seidel-Morgenstern, A. Processes to separate enantiomers. *Angew. Chem. Int. Ed.* **2014**, *53*, 1218–1250. [CrossRef] [PubMed]
4. Ali, I.; Aboul-Enein, H.Y. *Chiral Pollutants: Distribution, Toxicity and Analysis by Chromatography and Capillary Electrophoresis*; John Wiley & Sons: Chichester, UK, 2005.
5. Aboul-Enein, H.; Wainer, I. *The Impact of Stereochemistry on Drug Development and Use*; John Wiley & Sons: New York, NY, USA, 1997; p. 142.
6. Francotte, E.R. Enantioselective chromatography as a powerful alternative for the preparation of drug enantiomers. *J. Chromatogr. A* **2001**, *906*, 379–397. [CrossRef]
7. Gil-Av, E.; Feibush, B. Resolution of enantiomers by gas liquid chromatography with optically active stationary phases. Separation on packed columns. *Tetrahedron Lett.* **1967**, *8*, 3345–3347. [CrossRef]
8. Ward, T.J.; Ward, K.D. Chiral separations: A review of current topics and trends. *Anal. Chem.* **2011**, *84*, 626–635. [CrossRef] [PubMed]

9. Speybrouck, D.; Lipka, E. Preparative supercritical fluid chromatography: A powerful tool for chiral separations. *J. Chromatogr. A* **2016**, *1467*, 33–55. [CrossRef] [PubMed]
10. Fanali, S. Nano-liquid chromatography applied to enantiomers separation. *J. Chromatogr. A* **2017**, *1486*, 20–34. [CrossRef] [PubMed]
11. Lämmerhofer, M. Chiral recognition by enantioselective liquid chromatography: mechanisms and modern chiral stationary phases. *J. Chromatogr. A* **2010**, *1217*, 814–856. [CrossRef] [PubMed]
12. Catani, M.; Ismail, O.H.; Gasparrini, F.; Antonelli, M.; Pasti, L.; Marchetti, N.; Felletti, S.; Cavazzini, A. Recent advancements and future directions of superficially porous chiral stationary phases for ultrafast high-performance enantioseparations. *Analyst* **2017**, *142*, 555–566. [CrossRef] [PubMed]
13. Peluso, P.; Mamane, V.; Aubert, E.; Cossu, S. Recent trends and applications in liquid phase chromatography enantioseparation of atropisomers. *Electrophoresis* **2017**, *38*, 1830–1850. [CrossRef] [PubMed]
14. Glajch, J.L.; Kirkland, J.J.; Köhler, J. Effect of column degradation on the reversed-phase high-performance liquid chromatographic separation of peptides and proteins. *J. Chromatogr. A* **1987**, *384*, 81–90. [CrossRef]
15. Wehrli, A.; Hildenbrand, J.C.; Keller, H.P.; Stampfli, R.; Frei, R.W. Influence of organic bases on the stability and separation properties of reversed-phase chemically bonded silica gels. *J. Chromatogr. A* **1978**, *149*, 199–210. [CrossRef]
16. McNeff, C.; Zigan, L.; Johnson, K.; Carr, P.W.; Wang, A.; Weber-Main, A.M. Analytical advantages of highly stable stationary phases for reversed-phase LC. *LC GC* **2000**, *18*, 514–529.
17. Cabrera, K. Applications of silica-based monolithic HPLC columns. *J. Sep. Sci.* **2004**, *27*, 843–852 [CrossRef] [PubMed]
18. Núñez, O.; Nakanishi, K.; Tanaka, N. Preparation of monolithic silica columns for high-performance liquid chromatography. *J. Chromatogr. A* **2008**, *1191*, 231–252. [CrossRef] [PubMed]
19. Kanatyeva, A.Y.; Kurganov, A.A.; Viktorova, E.N.; Korolev, A.A. Monolithic stationary phases in liquid and gas chromatography. *Russ. Chem. Rev.* **2008**, *77*, 393–400. [CrossRef]
20. Wistuba, D. Chiral silica-based monoliths in chromatography and capillary electrochromatography. *J. Chromatogr. A* **2010**, *1217*, 941–952. [CrossRef] [PubMed]
21. Nakanishi, K.; Soga, N. Phase separation in silica sol-gel system containing polyacrylic acid I. Gel formaation behavior and effect of solvent composition. *J. Non-Cryst. Solids* **1992**, *139*, 1–13. [CrossRef]
22. Turson, M.; Zhou, M.; Jiang, P.; Dong, X. Monolithic poly(ethylhexyl methacrylate-co-ethylene dimethacrylate) column with restricted access layers prepared via reversible addition-fragmentation chain transfer polymerization. *J. Sep. Sci.* **2011**, *34*, 127–134. [CrossRef] [PubMed]
23. Guiochon, G. Monolithic columns in high-performance liquid chromatography. *J. Chromatogr. A* **2007**, *1168*, 101–168. [CrossRef] [PubMed]
24. Gibson, G.T.T.; Mugo, S.M.; Oleschuk, R.D. Surface-mediated effects on porous polymer monolith formation within capillaries. *Polymer* **2008**, *49*, 3084–3090. [CrossRef]
25. Saunders, K.C.; Ghanem, A.; Hon, W.B.; Hilder, E.F.; Haddad, P.R. Separation and sample pre-treatment in bioanalysis using monolithic phases: A review. *Anal. Chim. Acta* **2009**, *652*, 22–31. [CrossRef] [PubMed]
26. Liao, J.-L.; Zhang, R.; Hjertén, S. Continuous beds for standard and micro high-performance liquid chromatography. *J. Chromatogr. A* **1991**, *586*, 21–26. [CrossRef]
27. Peters, E.C.; Petro, M.; Svec, F.; Fréchet, J.M. Molded Rigid Polymer Monoliths as Separation Media for Capillary Electrochromatography. *Anal. Chem.* **1997**, *69*, 3646–3649. [CrossRef] [PubMed]
28. Svec, F. Porous polymer monoliths: amazingly wide variety of techniques enabling their preparation. *J. Chromatogr. A* **2010**, *1217*, 902–924. [CrossRef] [PubMed]
29. Fouad, A.; Marzouk, A.A.; Ibrahim, S.M.; Sobhy, M.; Ghanem, A. Functionalized polymer monoliths with carbamylated amylose for the enantioselective reversed phase nano-liquid chromatographic separation of a set of racemic pharmaceuticals. *J. Chromatogr. A* **2017**, *1515*, 91–95. [CrossRef] [PubMed]
30. Ahmed, M.; Ghanem, A. Chiral β-cyclodextrin functionalized polymer monolith for the direct enantioselective reversed phase nano liquid chromatographic separation of racemic pharmaceuticals. *J. Chromatogr. A* **2014**, *1345*, 115–127. [CrossRef] [PubMed]
31. Ahmed, M.; Yajadda, M.M.A.; Han, Z.J.; Su, D.; Wang, G.; Ostrikov, K.K.; Ghanem, A. Single-walled carbon nanotube-based polymer monoliths for the enantioselective nano-liquid chromatographic separation of racemic pharmaceuticals. *J. Chromatogr. A* **2014**, *1360*, 100–109. [CrossRef] [PubMed]

32. Pittler, E.; Schmid, M.G. Enantioseparation of dansyl amino acids by HPLC on a monolithic column dynamically coated with a vancomycin derivative. *Biomed. Chromatogr.* **2010**, *24*, 1213–1219. [CrossRef] [PubMed]
33. Armstrong, D.W.; Nair, U.B. Capillary electrophoretic enantioseparations using macrocyclic antibiotics as chiral selectors. *Electrophoresis* **1997**, *18*, 2331–2342. [CrossRef] [PubMed]
34. Fedorova, I.A.; Shapovalova, E.N.; Shpigun, O.A.; Staroverov, S.M. Bovine serum albumin adsorbed on eremomycin and grafted on silica as new mixed-binary chiral sorbent for improved enantioseparation of drugs. *J. Food Drug Anal.* **2016**, *24*, 848–854. [CrossRef] [PubMed]
35. Ilisz, I.; Grecsó, N.; Forró, E.; Fülöp, F.; Armstrong, D.W.; Péter, A. High-performance liquid chromatographic separation of paclitaxel intermediate phenylisoserine derivatives on macrocyclic glycopeptide and cyclofructan-based chiral stationary phases. *J. Pharm. Biomed. Anal.* **2015**, *114*, 312–320. [CrossRef] [PubMed]
36. Hroboňová, K.; Lehotay, J.; Čižmárik, J. HPLC Enantioseparation of Phenylcarbamic Acid Derivatives by Using Macrocyclic Chiral Stationary Phases. *Nova Biotechnol. et Chim.* **2016**, *15*, 12–22. [CrossRef]
37. Fedorova, I.; Shapovalova, E.; Shpigun, O. Separation of β-blocker and amino acid enantiomers on a mixed chiral sorbent modified with macrocyclic antibiotics eremomycin and vancomycin. *J. Anal. Chem.* **2017**, *72*, 76–82. [CrossRef]
38. Hsieh, M.L.; Chau, L.K.; Hon, Y.S. Single-step approach for fabrication of vancomycin-bonded silica monolith as chiral stationary phase. *J. Chromatogr. A* **2014**, *1358*, 208–216. [CrossRef] [PubMed]
39. Qin, F.; Xie, C.; Yu, Z.; Kong, L.; Ye, M.; Zou, H. Monolithic enantiomer-selective stationary phases for capillary electrochromatography. *J. Sep. Sci.* **2006**, *29*, 1332–1343. [CrossRef] [PubMed]
40. Claude Guillaume, Y.; André, C. Fast enantioseparation by HPLC on a modified carbon nanotube monolithic stationary phase with a pyrenyl aminoglycoside derivative. *Talanta* **2013**, *115*, 418–421. [CrossRef] [PubMed]
41. Ghanem, A.; Adly, F.G.; Sokerik, Y.; Antwi, N.Y.; Shenashen, M.A.; El-Safty, S.A. Trimethyl-β-cyclodextrin-encapsulated monolithic capillary columns: Preparation, characterization and chiral nano-LC application. *Talanta* **2017**, *169*, 239–248. [CrossRef] [PubMed]
42. Hroboňová, K.; Deáková, Z.; Moravčík, J.; Lehotay, J.; Armstrong, D.W.; Lomenová, A. Separation Of Methionine Enantiomers By Using Teicoplanin And Cyclofructan Columns. *Nova Biotechnol. et Chim.* **2015**, *14*, 1–11. [CrossRef]
43. Maia, A.S.; Castro, P.M.; Tiritan, M.E. Integrated liquid chromatography method in enantioselective studies: Biodegradation of ofloxacin by an activated sludge consortium. *J. Chromatogr. B* **2016**, *1029*, 174–183. [CrossRef] [PubMed]
44. Fanali, C.; Fanali, S.; Chankvetadze, B. HPLC Separation of Enantiomers of Some Flavanone Derivatives Using Polysaccharide-Based Chiral Selectors Covalently Immobilized on Silica. *Chromatographia* **2016**, *79*, 119–124. [CrossRef]
45. Harvanová, M.; Gondová, T. New enantioselective LC method development and validation for the assay of modafinil. *J. Pharm. Biomed. Anal.* **2017**, *138*, 267–271. [CrossRef] [PubMed]
46. Chankvetadze, B.; Kubota, T.; Ikai, T.; Yamamoto, C.; Kamigaito, M.; Tanaka, N.; Nakanishi, K.; Okamoto, Y. High-performance liquid chromatographic enantioseparations on capillary columns containing crosslinked polysaccharide phenylcarbamate derivatives attached to monolithic silica. *J. Sep. Sci.* **2006**, *29*, 1988–1995. [CrossRef] [PubMed]
47. Zhang, Z.; Wu, M.; Wu, R.A.; Dong, J.; Ou, J.; Zou, H. Preparation of perphenylcarbamoylated β-cyclodextrin-silica hybrid monolithic column with "one-pot" approach for enantioseparation by capillary liquid chromatography. *Anal. Chem.* **2011**, *83*, 3616–3622. [CrossRef] [PubMed]
48. Schmid, M.G.; Schreiner, K.; Reisinger, D.; Gübitz, G. Fast chiral separation by ligand-exchange HPLC using a dynamically coated monolithic column. *J. Sep. Sci.* **2006**, *29*, 1470–1475. [CrossRef] [PubMed]
49. Haroun, M.; Ravelet, C.; Grosset, C.; Ravel, A.; Villet, A.; Peyrin, E. Reversal of the enantiomeric elution order of some aromatic amino acids using reversed-phase chromatographic supports coated with the teicoplanin chiral selector. *Talanta* **2006**, *68*, 1032–1036. [CrossRef] [PubMed]
50. Dermaux, A.; Lynen, F.; Sandra, P. Chiral capillary electrochromatography on a vancomycin stationary phase. *J. Sep. Sci.* **1998**, *21*, 575–576. [CrossRef]

51. Karlsson, C.; Karlsson, L.; Armstrong, D.W.; Owens, P.K. Evaluation of a vancomycin chiral stationary phase in capillary electrochromatography using polar organic and reversed-phase modes. *Anal. Chem.* **2000**, *72*, 4394–4401. [CrossRef] [PubMed]
52. Desiderio, C.; Aturki, Z.; Fanali, S. Use of vancomycin silica stationary phase in packed capillary electrochromatography I. Enantiomer separation of basic compounds. *Electrophoresis* **2001**, *22*, 535–543. [CrossRef]
53. Fanali, S.; Catarcini, P.; Quaglia, M.G. Use of vancomycin silica stationary phase in packed capillary electrochromatography: III. Enantiomeric separation of basic compounds with the polar organic mobile phase. *Electrophoresis* **2002**, *23*, 477–485. [CrossRef]
54. Karlsson, C.; Wikström, H.; Armstrong, D.W.; Owens, P.K. Enantioselective reversed-phase and non-aqueous capillary electrochromatography using a teicoplanin chiral stationary phase. *J. Chromatogr. A* **2000**, *897*, 349–363. [CrossRef]
55. Wikström, H.; Svensson, L.A.; Torstensson, A.; Owens, P.K. Immobilisation and evaluation of a vancomycin chiral stationary phase for capillary electrochromatography. *J. Chromatogr. A* **2000**, *869*, 395–409. [CrossRef]
56. Si Ahmed, K.; Tazerouti, F.; Badjah-Hadj-Ahmed, A.Y.; Meklati, B.Y. Preparation and chromatographic properties of a multimodal chiral stationary phase based on phenyl-carbamate-propyl-β-CD for HPLC. *J. Sep. Sci.* **2007**, *30*, 2025–2036. [CrossRef] [PubMed]

Sample Availability: Samples of the compounds used in this research are available from the authors.

molecules

MDPI

Article

New Brush-Type Chiral Stationary Phases for Enantioseparation of Pharmaceutical Drugs

Anamarija Knežević [1,*], Jurica Novak [2,3] and Vladimir Vinković [1]

1 Division of Organic Chemistry and Biochemistry, Ruđer Bošković Institute, Bijenička cesta 54, Zagreb 10000, Croatia; vvink@irb.hr
2 Division of Physical Chemistry, Ruđer Bošković Institute, Bijenička cesta 54, Zagreb 10000, Croatia; Jurica.Novak@irb.hr
3 South Ural State University, 20-A, Tchaikovsky Str., Chelyabinsk 454080, Russia
* Correspondence: anamarija.knezevic@irb.hr; Tel.: +385-145-610-29

Academic Editors: Maria Elizabeth Tiritan, Madalena Pinto and Carla Sofia Garcia Fernandes
Received: 25 January 2019; Accepted: 21 February 2019; Published: 25 February 2019

Abstract: The importance of chirality in drug development is unquestionable, with chiral liquid chromatography (LC) being the most adequate technique for its analysis. Among the various types of chiral stationary phases (CSPs) for LC, brush-type CSPs provide the base for interaction analysis of CSPs and enantiomers, which provide valuable results that can be applied to interaction studies of other CSP types. In order to analyze the influence of aromatic interactions in chiral recognition, we designed a set of ten new brush-type CSPs based on (*S*)-*N*-(1-aryl-propyl)-3,5-dinitrobenzamides which differ in the aromatic unit directly linked to the chiral center. Thirty diverse racemates, including several nonsteroidal anti-inflammatory drugs and 3-hydroxybenzodiazepine drugs, were used to evaluate the prepared CSPs. Chromatographic analysis showed that the three new CSPs separate enantiomers of a wide range of compounds and their chromatographic behavior is comparable to the most versatile brush-type CSP—Whelk-O1. The critical role of the nonbonding interactions in positioning of the analyte (naproxen) in the cleft of **CSP-6**, as well as the analysis of interactions that make enantioseparation possible, were elucidated using computational methods. Furthermore, the influence of acetic acid as a mobile phase additive, on this enantiorecognition process was corroborated by calculations.

Keywords: chiral chromatography; chiral recognition; intermolecular interactions; chiral drugs; Whelk-O1 column; mobile phase additives

1. Introduction

Chirality is an essential property in the development of pharmaceutical drugs, as well as in agrochemistry, food science, etc. It was put into the foreground in 1992 when the FDA issued its policy statement concerning the development of stereoisomeric drugs [1]. Even though the development of enantiopure compounds was not mandatory, it became the de facto standard in the pharmaceutical industry, also enabling the new possibilities in drug development, i.e., the chiral switch concept [2]. Accordingly, the need for fast, simple and reliable methods for separation of enantiomers, determination of enantiomeric excess (*ee*) and absolute configuration (AC) has increased. Although chiral liquid chromatography (chiral LC) has a great influence on the determination of absolute configuration [3], its main role is being the most powerful method for the separation of enantiomers [4] and for measuring the enantiopurity of organic compounds. Since the 1990s, when the availability of chiral stationary phases (CSPs) became widespread, chiral LC using CSPs has been the most widely used technique for chiral separations [5]. Besides being simple, versatile and reliable method, chiral LC has also became an ultrafast technique with enantioseparations in the sub-minute time frame [6–12].

Over time, various types of CSPs have been developed, including more than a hundred which have been commercialized. Their properties and mechanism of enantioseparation depend on the nature of the chiral selector [5,13]. Polysaccharide-based CSPs are the most broadly applied for chiral LC separations [14]. However, there are also several Pirkle-type (or brush-type) CSPs, e.g., Whelk-O1 and ULMO, whose versatility has enabled their wide application [13,15]. These CSPs, which consist of small organic molecules covalently bound to the support, are also the most widely investigated CSPs regarding the chiral recognition mechanism [13,16]. The understanding of chiral separation mechanism is essential for estimating elution order, predicting the types of analytes which can be separated on a certain CSP, and improving the design of new selectors. Clearly, the enantioseparation process is easier to elucidate by analyzing interactions of the analyte and a small selector compared to polysaccharide CSPs, whether using experimental or computational methods [5]. Chiral recognition of brush-type CSPs is based on well-documented interactions between the selector and enantiomers of the analyte—hydrogen bonds, dipole-dipole interactions, van der Waals interactions and, in particular, aromatic interactions [16].

The goal of this research was threefold. First, ten new brush-type CSPs were prepared, with a molecular design based on previous studies of chiral recognition in CSP [13,17–19]. The objective was to obtain a CSP with a versatility as similar as possible to the Whelk-O1 column, the brush-type CSP with the broadest application in industry and academia. We opted for a structure with one amide bond, analogous to the Whelk-O1 column, and two aromatic groups—a 3,5-DNB aromatic unit and a substituted aromatic moiety (Figure 1). We tested the prepared CSPs using thirty diverse racemates and determined the CSPs with the best performance. Second, the variability of the aromatic moiety enabled us to elucidate the influence of aromatic substituent on the enantiorecognition process. Keeping that in mind, we investigated the role of nonbonding interactions relevant for chiral recognition on CSPs using computational methods, with special attention to hydrogen bond networks and aromatic interactions. Third, we explored the influence of acetic acid as an additive in the mobile phase using extensive ab initio methods. To the best of our knowledge, this is the first study that examines the possibility of positioning a minor additive within the chiral binding site and its influence on enantiospecific interactions.

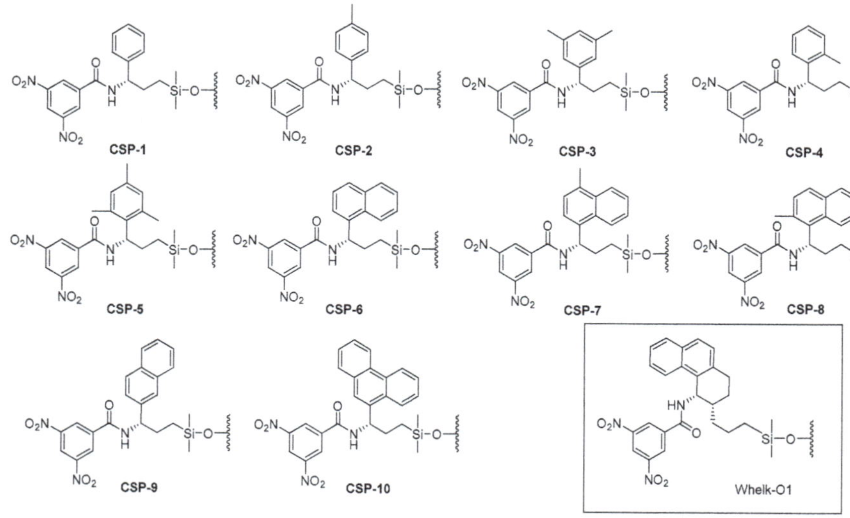

Figure 1. Structures of new brush-type CSPs and Whelk-O1.

2. Results and Discussion

2.1. Preparation of New Brush-Type Chiral Stationary Phases (CSPs)

We previously described the synthesis and determination of an absolute configuration of (*S*)-*N*-(1-aryl-allyl)-3,5-dinitrobenzamides (**DNB-1–DNB-10**) [20]. These compounds are excellent candidates for the preparation of a new Pirkle-type CSPs for HPLC since they possess all properties of a good selector—they are rigid and contain a strong π-acceptor acid (DNB) as well as a π-donor base aromatic group. Furthermore, allyl group in the structure of DNBs enables their binding to silica gel using a simple and straightforward procedure.

For the binding of the chiral DNB selectors onto silica gel we chose a method that does not introduce additional polar moieties into the CSP structure (Scheme 1) [21]. Hydrosilylation of DNBs was conducted using chlorodimethylsilane and Speier's catalyst, followed by the replacement of a chloride with an ethoxy group using trimethylamine and ethanol. Obtained compounds were passed through short silica column and attached to 5 µm silica gel in boiling toluene. Before the end-capping procedure, we analyzed prepared CSPs using elemental analysis and infrared spectroscopy in order to determine the amount of selector attached onto the silica gel. We performed end-capping in the last step to protect free silanol groups with hexamethyldisilazane [22]. Ten new CSPs were thus prepared from the corresponding DNBs (Figure 1) with the optimal 0.2 mmol selector per 1 g of CSP loading [21]. The as-prepared CSPs were finally packed in steel columns using the slurry packing technique.

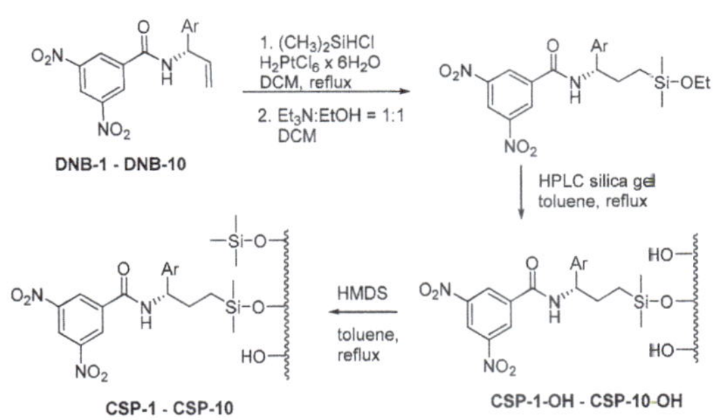

Scheme 1. Preparation of new brush-type chiral stationary phases (CSPs) from corresponding (*S*)-*N*-(1-aryl-allyl)-3,5-dinitrobenzamides (DNBs).

2.2. Evaluation of Prepared CSPs

In order to test the performance of the prepared CSPs we used a diverse set of analytes (Figure 2) which is usually used by our group to evaluate new CSPs [18,19]. The evaluation of prepared CSPs was performed in normal phase chromatography mode using the mixtures of hexane and 2-propanol as the mobile phase at room temperature and with UV detection at 254 nm. Compounds **1–7** and **9** were analyzed in the mobile phase containing 10% of 2-propanol in hexane, while the mobile phase for the analysis of compounds **8** and **10–21** contained 20% of 2-propanol due to their longer retention times. Our goal was to evaluate CSPs with respect to their difference in molecular structure and consequently to neglect the influence of quality of the column packing, column dimensions, etc. Therefore the relevant value for consideration is the separation factor (selectivity, α), which depends the most on the structure of the CSP and has the greatest influence on overall resolution.

Tested set of compounds can be subdivided into four subsets. Compounds **1–4** are small molecules with one or none carbonyl groups which could be the acceptor but not the donor of hydrogen bonds. Because of these characteristics, the satisfactory enantioseparation of compounds **1–4** is usually hard to achieve on brush-type CSPs. The second subset consists of compounds **5–9** which possess bigger aromatic groups and hydroxyl or amino groups capable of hydrogen bond formation. These compounds are usually well separated on brush-type CSPs. Finally, there are two subsets consisting of aromatic amides, **10–21**, with the last subset (**15–21**) also bearing an ester group. These compounds can form strong hydrogen bonds and aromatic interactions which makes them easier to separate using brush-type CSPs.

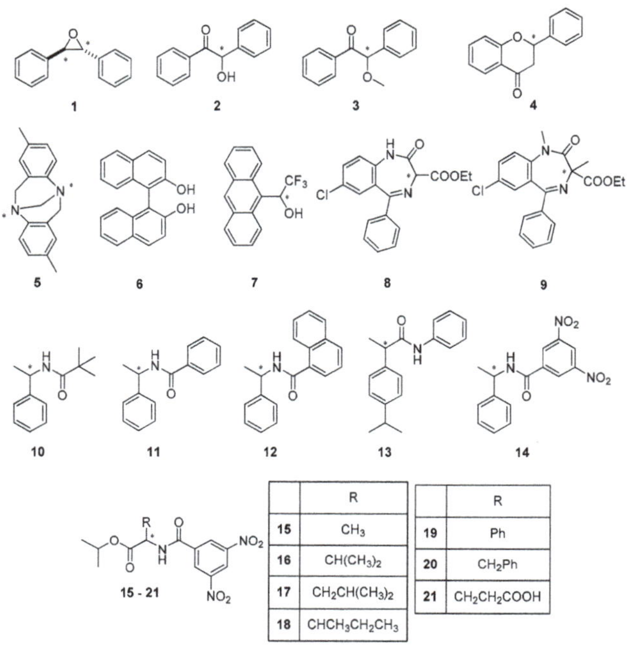

	R
15	CH_3
16	$CH(CH_3)_2$
17	$CH_2CH(CH_3)_2$
18	$CHCH_3CH_2CH_3$
19	Ph
20	CH_2Ph
21	CH_2CH_2COOH

Figure 2. Structures of the racemic analytes used for the evaluation of prepared CSPs.

The results indicate that CSPs which contain substituted phenyl aromatic groups, in most cases do not resolve compounds **1–4** (Figure 3). Enantioseparation of these compounds is better using the naphthyl series of CSPs, although the separation of compounds **3** and **4** is achieved on just a few CSPs. In general, enantioseparation results obtained for compounds **5–9** are better. Overall, the lowest separation factor is observed with CSPs like **CSP-3**, **CSP-4** and **CSP-5**, which have more than one methyl substituent or methyl at the *ortho* position of the phenyl core.

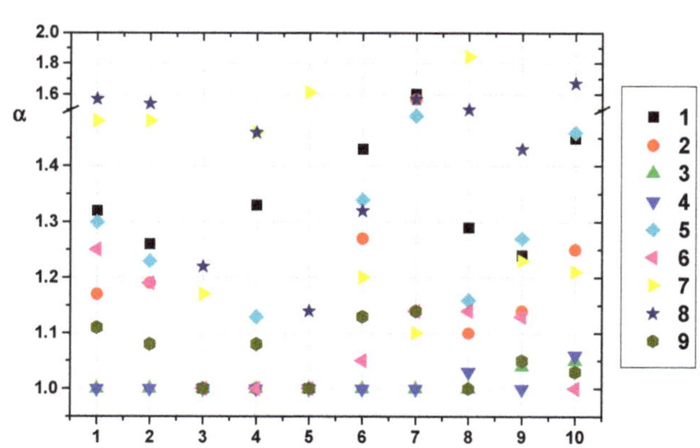

Figure 3. Separation factors of compounds **1–9** on prepared **CSP-1–CSP-10**, mobile phase hexane-2-propanol = 90:10. For compound **8**, the mobile phase ratio was 80:20.

Compounds **10–14** are resolved well using all CSPs, with the only exceptions being **CSP-3** and **CSP-5** (Figure 4a). These CSPs demonstrate the lowest separation factors due to their phenyl aromatic group in the combination with two or three methyl substituents near the chiral carbon atom, which disturb the enantiorecognition process due to steric hindrance. Also, the retention times of these compounds are shorter for the phenyl series of CSPs than the naphthyl series (Table S2). This means that interactions are weaker, as expected for smaller aromatic systems.

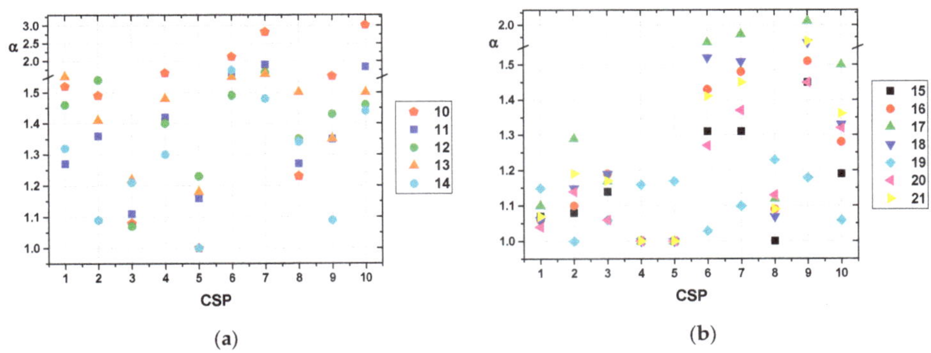

Figure 4. Separation factors of compounds **10–14** (**a**) and **15–21** (**b**) on prepared **CSP-1–CSP-10**, mobile phase hexane-2-propanol = 80:20.

The initial assumption was that 3,5-dinitrobenzoyl derivatives of amino acids with an isopropyl protecting group **15–21** would be resolved well on the prepared CSPs since they possess aromatic, ester and amide groups potentially capable of forming interactions with CSPs which may result in chiral recognition. However, the difference between the phenyl series of CSPs and larger aromatic substituents is even more pronounced in these examples (Figure 4b). Here, it is demonstrated that only CSPs with larger aromatic groups (naphthyl and phenantryl) show very good enantio-separations of these compounds. The lowest results in the naphthyl series were obtained for **CSP-8** which possesses a methyl group at position 2 of the naphthyl ring.

It is interesting to point out that the enantioseparation of racemate **19** on the prepared CSPs is quite unexpected (Figure 5). On the CSPs which show the overall lowest enantioseparation capabilities, its separation factor is the highest in the **15–21** subset. On the contrary, on the CSPs that demonstrate the best performance, this separation factor is the lowest of the subset. This result exemplifies how chiral recognition is difficult to predict and often depends on subtle details.

The above presented results demonstrate that both size and substitution of aromatic group play an important role in the enantioseparation capabilities of prepared CSP. Larger aromatic groups (naphthyl and phenanthryl) overall display higher separation factors for the tested set of racemates. This is expected since intermolecular aromatic interactions, which positively influence the chiral recognition, are stronger for larger aromatic groups. Substitution of more than one methyl group on the phenyl ring decreases the separation factor for all the tested racemates. On the other hand, the influence of the monosubstitution of the aromatic group shows two opposite trends. When the substitution is at the 4-position, there is no noticeable difference compared to the aromatic group without substituents (**CSP-1** compared to **CSP-2** and **CSP-6** to **CSP-7**). Contrarily, substitution at the 2-position greatly influences the enantioseparation capabilities of the prepared CSPs (**CSP-2** compared to **CSP-4** and **CSP-7** to **CSP-8**). The methyl substituent in this case is near the chiral center and can influence the chiral recognition in two ways—it sterically hinders the chiral center which decreases enantiorecognition and it increases the rigidity of the CSPs by reducing the rotational freedom around C*-C^{Ar} bond (Figure 6). The influence of the rigidity of CSP on enantioseparation capability is substantial. **CSP-9** compared to **CSP-6** demonstrates higher flexibility and somewhat lower enantioseparation characteristics (see Figures 3 and 4). On the other hand, **CSP-8** compared to **CSP-6** has higher rigidity and considerably decreased enantioseparation results (which are partially due to steric hindrance of the chiral center). This indicates that a good CSP should be sufficiently rigid to strongly interact with one sterically compatible enantiomer, but also flexible enough to accommodate the analyte and achieve a maximum number of interactions for a wide range of compounds.

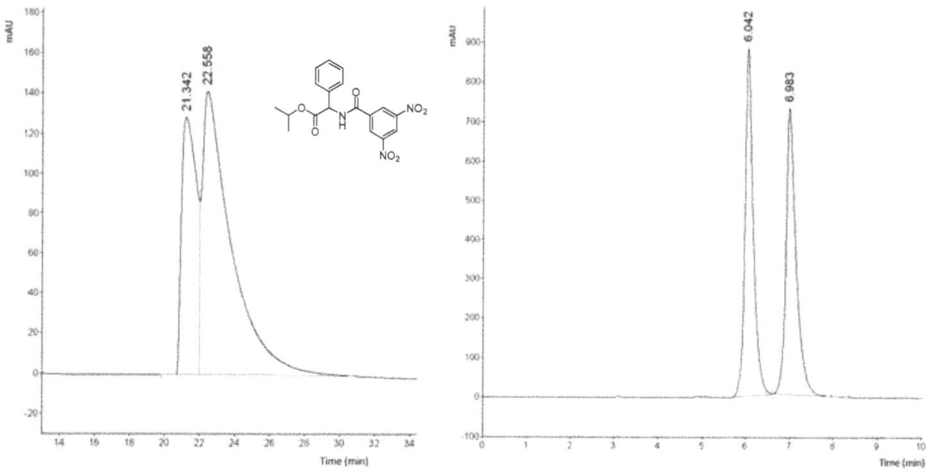

Figure 5. Chromatogram of the separation of racemate **19** on **CSP-10** (**a**) and **CSP-8** (**b**); mobile phase hexane-2-propanol = 80:20, flow 1mL/min, UV detection at 254 nm.

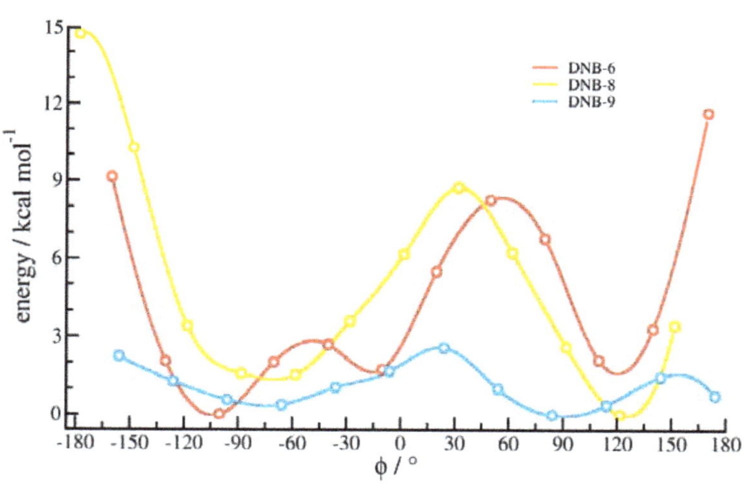

Figure 6. Relaxed potential energy surface scan around C*-C$^{\alpha}$ bond of **DNB-6** (red), **DNB-8** (yellow) and **DNB-9** (blue). Dihedral angle Φ is defined by atoms C$^{\beta}$-C$^{\alpha}$-C*-N. Calculations were performed on M06-2X/aug-cc-pVDZ level of theory.

2.3. Enantioseparation of Pharmaceutical Drugs on Prepared CSPs

Since CSPs bearing naphthyl and phenanthryl aromatic groups were demonstrated as more versatile than phenyl CSPs, **CSP-6–CSP-10** were chosen for testing the enantioseparation of several pharmaceutical drugs including non-steroidal anti-inflammatory drugs (NSAIDs) and 3-hydroxy-benzodiazepine drugs (Figure 7).

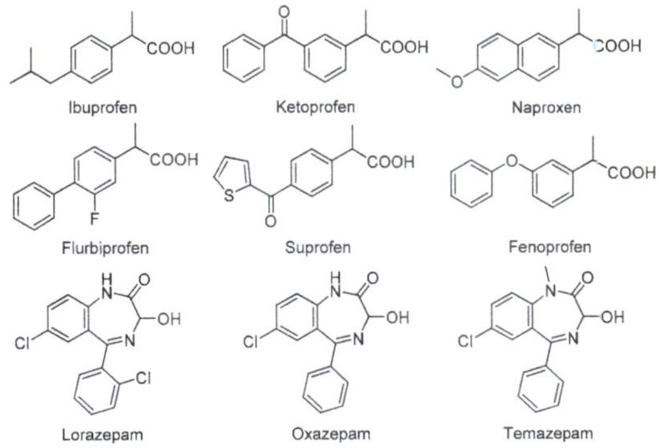

Figure 7. Structures of the tested non-steroidal anti-inflammatory drugs (NSAIDs) and 3-hydroxy-benzodiazepine drugs.

Given that NSAIDs are organic acids, acetic acid (0.1%) was selected as the mobile phase additive. Acidic additives are frequently used in the mobile phases for the analysis of acidic analytes because of their positive influence on the peak shape and retention time [23]. Results of the enantioseparation of NSAIDs on the prepared **CSP-6–CSP-10** (Figure 8a) showed that only naproxen is resolved well on prepared CSPs, while some of the drugs from our test set are only partially resolved on **CSP-7**.

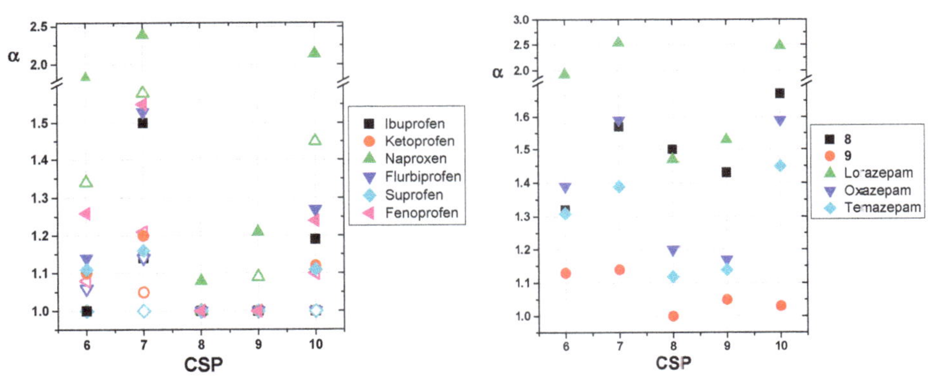

(**a**) (**b**)

Figure 8. (**a**) Separation factor of NSAIDs on prepared **CSP-6–CSP-10** using a mobile phase hexane-2-propanol-CH_3COOH = 90:10:0.1 (hollow symbols) and hexane-2-propanol = 80:20 + 1 g dm^{-3} NH_4OAc (filled symbols); (**b**) Separation factor of compounds **8**, **9** and 3-hydroxy-benzodiazepine drugs on prepared **CSP-6–CSP-10** using a mobile phase hexane-2-propanol = 80:20 + 1 g dm^{-3} NH_4OAc.

It was demonstrated that in some cases neutral salts, such as ammonium acetate or ammonium formate, have a positive influence on the enantioseparation of racemates [23]. For example, NSAIDs showed excellent enantioseparation results on Whelk-O1 columns using a mobile phase with ammonium acetate as an additive [24]. Therefore, we investigated the enantioseparation performance of the newly prepared **CSP-6–CSP-10** on NSAIDs also using ammonium acetate as a mobile phase additive (Figure 8a). Although **CSP-8** and **CSP-9** still only separate the enantiomers of naproxen, the performance of the remaining CSPs was substantially improved. This is especially evident in the case of **CSP-10** (Figure 9) which can separate all of the tested NSAIDs using the abovementioned conditions.

Compounds **8** and **9**, which are structural analogs of 3-hydroxybenzodiazepines with an ester group at the 3-position, were resolved well using the prepared CSPs. Therefore, we decided to test the naphthyl series of CSPs for the enantioseparation of 3-hydroxybenzodiazepine drugs (Figure 8b). Since these compounds possess acidic and basic groups, ammonium acetate was also used as a mobile phase additive. 3-Hydroxybenzodiazepine drugs show very good enantioseparation results on the prepared CSPs, with the best results obtained on **CSP-6**, **CSP-7** and **CSP-10**.

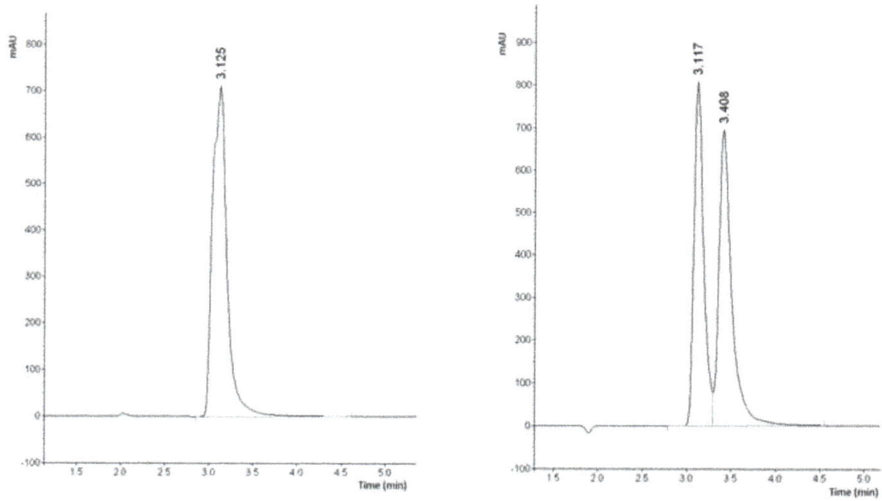

(**a**) (**b**)

Figure 9. Chromatograms of the separation of flurbiprofen on **CSP-10** using mobile phase: hexane-2-propanol-CH_3COOH = 90:10:0.1 (**a**) and hexane-2-propanol = 80:20 + 1 g dm^{-3} NH_4OAc (**b**).

Overall, the best separation factors for the enantioseparation of the tested drugs were obtained on **CSP-6**, **CSP-7** and **CSP-10**. Given the importance of these compounds in the pharmaceutical industry, their enantioseparation was previously analyzed on several commercial brush-type CSP, including Whelk-O1 [24]. In order to compare the results obtained on our CSP with the results obtained for Whelk-O1, we analyzed these pharmaceutical drugs using the same hexane—2-propanol = 80:20 + 1 g dm^{-3} NH_4OAc mobile phase (Table 1).

Table 1. Comparison of the separation factors (α) of tested pharmaceutical drugs obtained on **CSP-6**, **CSP-7**, **CSP-10** and Whelk-O1 [1].

Compound	Whelk-O1	CSP-6	CSP-7	CSP-10
Ibuprofen	1.71	1.00	1.50	1.19
Ketoprofen	1.30	1.10	1.20	1.12
Naproxen	1.48	1.82	2.39	2.14
Flurbiprofen	1.68	1.14	1.53	1.27
Suprofen	1.27	1.11	1.16	1.11
Fenoprofen	1.60	1.26	1.55	1.24
Lorazepam	2.29	1.93	2.54	2.48
Oxazepam	1.97	1.39	1.59	1.59
Temazepam	1.38	1.31	1.39	1.45

[1] For the analyses (*S*,*S*)-Whelk-O1 column (5 μm particle size, 250 × 4.6 mm I.D.) was used.

From the obtained results it is evident that these three new CSPs have promising enatioseparation capabilities which are comparable to those of Whelk-O1. Furthermore, it must be emphasized that the column Whelk-O1, being a commercial one, is technologically optimized regarding the choice of silica gel used as the support, the binding procedure, as well as the packing procedure. It is known that all of these conditions influence the enantioseparation capabilities of CSPs [25,26]. The prepared CSPs haven't been optimized at this point, leaving plenty of room for improvement of these CSPs. These columns have one more advantage over Whelk-O1. In order to prepare Whelk-O1 enantiopure selector, preparative chiral chromatography must be used which has low productivity due to the low solubility of the compound [27,28]. On the other hand, the enantiopure selectors of our CSPs (**DNB-6**, **DNB-7** and **DNB-10**) can be obtained by enzyme resolution using a cheap commercial lipase—*Candida antarctica* lipase B (CAL-B) [20,29].

2.4. Analysis of Interactions Responsible for Chiral Recognition between Naproxen and **CSP-6**

To explore the nature of chiral binding site, a set of molecular dynamics simulations in vacuum of **CSP-6** with naproxen was performed. Since we are interested in distinguishing the details that enable enantiomeric separation, simulations were run separately for (*S*)- and (*R*)-enantiomers of naproxen. According to a ^{1}H-NMR study of chiral recognition of (*S*)- and (*R*)-naproxen in the presence of Whelk-O1 selector analog [30] and to the molecular dynamics simulations of the interface of modified Whelk-O1 selector and naproxen in *n*-hexane [31], both enantiomers of naproxen dock inside the cleft, most probably by an M1 mechanism [32,33]. The main feature of the M1 mechanism is a hydrogen bond between the drug and the amide hydrogen, while aromatic interactions with the dinitrophenyl moiety introduce additional stability to the complex. For simplicity in our computational models we have neglected the non-polar solvent and a linker, the interface to silica gel, substituting it with a hydrogen atom.

The complex of **CSP-6** and (*S*)-naproxen ((*S*)-**A**) has lower energy (by 1.8 kcal mol^{-1}) than the lowest energy complex of **CSP-6** and (*R*)-naproxen ((*R*)-**A**) (Figure 10). This result is in accordance with the experimental results which demonstrate that (*R*)-naproxen is the first eluting enantiomer on **CSP-6** (Figure S1). Ten additional complexes within 2.9 kcal mol^{-1} were examined (Figure S2). In (*S*)-A complex the naphthyl and DNB subunits are almost perpendicular to each other. Naproxen is hydrogen bonded to the amide hydrogen (2.00 Å), with the carbonyl oxygen being a H-bond acceptor and in the vicinity of the β-hydrogen of DNB, being only 2.44 Å apart. The second most important H-bridge is between an oxygen of the DNB nitro group and the hydrogen from the naproxen carboxyl group. The weakest interaction of this type is responsible for anchoring the opposite side of naproxen, the methoxy moiety, to the second nitro group of the DNB subunit. The parallel face-centered stacking arrangement between the two aromatic rings with the distance between the centroids of the naproxen and DNB rings of 4.11 Å, represents a good example of aromatic interaction between strongly electron-deficient (DNB) and neutral aromatic rings (Figure 10b). The next intriguing space oriented interaction is the H-aromatic interaction [34], where three hydrogens of naproxen, including the hydrogen on the chiral carbon atom, are oriented toward the naphthyl ring, forming a recognizable T-shaped motif whose rings' planes form an angle of 75°. Alternative modes of complexation are also found, like the one where naproxen is outside the cleft ((*S*)-**A**-4) or a complex where naproxen is hydrogen bonded to the carbonyl oxygen of **CSP-6** ((*S*)-**A**-5). Those complexes are 2.5 kcal mol^{-1} higher in energy than (*S*)-**A**, and will not be discussed further.

Just by reversing groups at the chiral carbon on naproxen in the (*S*)-**A** complex and running optimization, a new complex is identified, (*R*)-**A**-2, 2.8 kcal mol^{-1} higher in energy (see Supplementary Materials). The O-H···O bond is shorter by 0.10 Å, while O···H-N is 0.03 Å longer than the analogous bonds in (*S*)-**A**, but the orientation of the methyl group toward the naphthyl part of **CSP-6** perturbs the parallel face-centered stacking arrangement almost perfectly possible in the (*S*)-enantiomeric complex. The H-aromatic interaction network is also perturbed, which can be seen from the fact that only hydrogens from the α ring are pointing toward naphthyl moiety.

The (*R*)-**A** complex, more stable than (*R*)-**A**-2 by 0.9 kcal mol^{-1}, is characterized by even stronger hydrogen bonds between the naproxen carboxyl group and DNB (1.86 Å) and the amide hydrogen (1.99 Å). The anchoring interaction between the methoxy hydrogen and the nitro group is missing, which is reflected in the increased distance between the centroids of an aromatic ring of naproxen and the DNB subunit to 5.29 Å. By comparing the (*R*)-**A** and (*R*)-**A**-2 structures it is possible to conclude that the main contribution to the stabilization of the complexes comes from hydrogen bonding with aromatic and H-aromatic interactions playing a minor, but significant role in the enantiorecognition process.

Our findings are in agreement with similar theoretical studies on the chiral recognition of Whelk-O1. For a series of complexes of naproxen with Whelk-O1 and modified CSPs, the M1 mechanism is identified as the most probable docking mechanism [31]. Furthermore, the formation of the hydrogen bond was found to be critical for the successful separation of enantiomers [35].

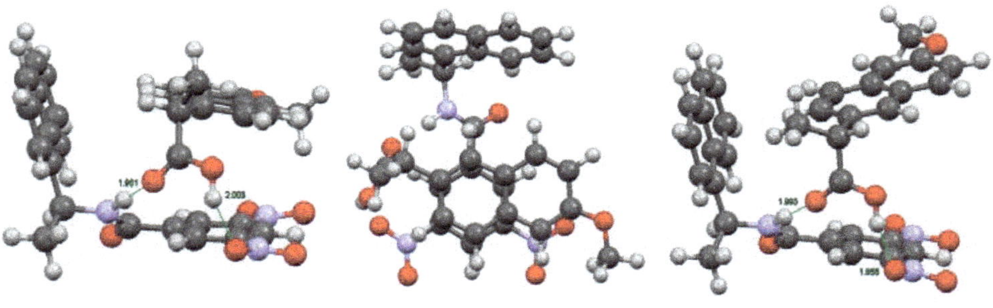

(a) (b) (c)

Figure 10. The lowest energy structures of (*S*)-**A** (**a**), (*S*)-**A**, top view (**b**) and (*R*)-**A** (**c**). Selected bond lengths are indicated in Å. B3LYP/def2-TZVPP//M06-2X/def2-TZVPP (COSMO, ε = 2.78).

2.5. *Influence of Acetic Acid as the Mobile Phase Additive on The Enantioseparation of Naproxen on* **CSP-6**

Although it is empirically known that minor additives (organic acids, bases or neutral salts) may have positive effect on the shape of peaks, retention times and selectivity [23], experimental studies on the mechanism of how additives influence enantioseparation are scarce. Several systematic experimental studies were performed, mostly on polysaccharide CSPs under normal phase conditions [36–38], as well as using polar organic mobile phases [38–40]. Generally, the consensus is that additives in the mobile phase improve the solubility of acidic analytes in nonpolar solvents and suppress non-chiral interactions of analyte with the CSP surface (masking effect), and therefore reduce tailing and can increase selectivity. However, an explanation for the way in which additives could affect the chiral recognition process between the analyte and CSP remains elusive. Some studies [36,39] hypothesize that additives have an ability to displace the analyte or to influence the selector-analyte complex by interfering with the interactions that stabilize the complex, however, these speculative explanations of experimental results have not been substantiated so far.

Our goal was to investigate the possibility of influence of additives on intermolecular interactions between analyte and CSP during enantiorecognition. We opted for acetic acid, a simple acidic additive, and a system we had already investigated—naproxen and **CSP-6**. Acetic acid is known to form dimers in the gas phase and in nonpolar solvents characterized by a strong double hydrogen bond [41,42]. Since naproxen primarily interacts via its carboxylic group with CSPs, we explored the possibility of an interplay of naproxen, acetic acid and **CSP-6**.

The most stabile complex of naproxen, acetic acid and **CSP-6** is (*S*)-**B**, which is 3.9 kcal mol^{-1} lower in energy than (*R*)-**B** (Figure 11). A common motif of both structures is an 8-membered ring formed by a double hydrogen bridge connecting acetic acid and the carboxylic acid group of naproxen. In the (*S*)-**B** complex, the N-H···O bond length is 2.90 Å, while naproxen's carbonyl oxygen serves as a double H-bond acceptor, associating with acetic acid as well. The analogous bridge to (*R*)-naproxen in (*R*)-**B** is longer (3.14 Å) and of a different nature – the H-acceptor is a carboxylic acid oxygen, which is simultaneously an H-donor to acetic acid. Although acetic acid is not connected to naproxen via enantioselective interactions, it modifies H-bond network compared to a complex without the acid ((*S*)-**A** and (*R*)-**A**). Firstly, the interaction between the carboxylate hydrogen and the nitro group is missing. Secondly, the methyl group of acetic acid is a weak H-donor to the nitro subunit. For (*R*)-**B**, the presence of additive makes the contact between naproxen and **CSP-6** along the carboxyl oxygen more favorable.

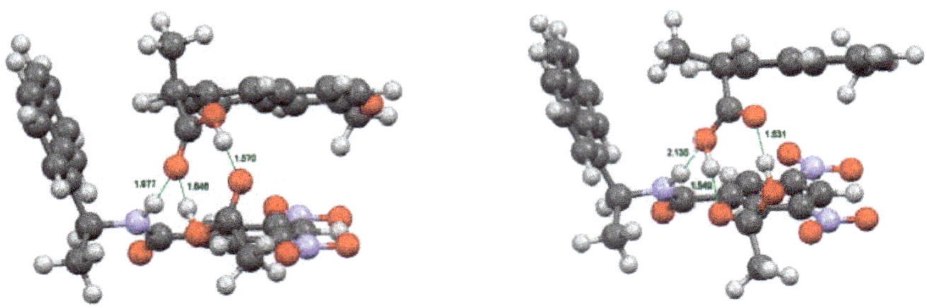

(**a**) (**b**)

Figure 11. The lowest energy structures of (*S*)-**B** (**a**) and (*R*)-**B** (**b**). Selected bond lengths are indicated in Å. B3LYP/def2-TZVPP//M06-2X/def2-TZVPP (COSMO, ε = 2.78).

Competitive interaction patterns are also found (Figure S3, Supplementary Materials). Only one structure is missing an 8-membered double H-bond ring ((*R*)-**B**-2) where a more complex H-bond network can be seen. An amide is connected to acetic acid, which is a H-donor to naproxen, whose carboxylic acid group is an H-donor to the carbonyl oxygen of **CSP-6**. Again, naproxen is positioned above the DNB subunit but outside the cleft.

Computational insights into the influence of acetic acid on the enantioseparation of NSAIDs, like naproxen, reveal that additives can interact with both CSP and carboxylic acid analytes. Even non-enantiospecific interactions with the carboxylate moiety of the drug change the interaction arrangements. In the system we studied, the entire binding scheme is shuffled, which is reflected in the increase of N-H···O distance and weakening of both the hydrogen bond and aromatic interactions. The acetic acid also influences the energetics of complexes, increasing the energy difference between the most stable (*S*)- and (*R*)-complex, enabling better separation. Our theoretical consideration of complexes of naproxen and CSPs, with and without acetic acid, indicate that higher selectivity is possible due to interactions of acetic additive with the analyte and CSP.

3. Materials and Methods

3.1. General Information

All the solvents were puriss. p.a. or HPLC grade and were used directly as supplied by Sigma-Aldrich Chemie GmbH (Munich, Germany), Alfa Aesar (Karlsruhe, Germany) or Acros (part of Thermo Fisher Scientific, Geel, Belgium). For the preparation of chiral stationary phases HPLC silica gel Separon SGX particle size 5 µm and pore size 80 Å from Tessek Ltd. (Prague, Czech Republic) was used. IR spectra were recorded on an MB102 instrument (ABB Bomem, Zurich, Switzerland). Elemental analyses were done on a 2400 CHNS Elemental Analyzer (Perkin-Elmer, Waltham, MA, USA). Chiral HPLC analysis were performed using a Prominence System (Pump LC-20AT, DGU-20A5 Degasser, UV detector SPD-20A, Shimadzu, Kyoto, Japan) or a Knauer system (Pump Knauer 64, 4-Port Knauer Degasser, UV detector Knauer Variable Wavelength Monitor, Interface Knauer, Knauer, Berlin, Germany, and a CD-2095 detector, Jasco, Easton, MD, USA). The packing of prepared CSP into stainless steel columns (150 × 4.6 mm I.D.) was achieved using Knauer Pneumatic HPLC Pump. The (*S*,*S*)-Whelk-O1 column, Regis Technologies, Inc. (Morton Grove, IL, USA), was 5 µm particle size and had dimensions of 250 × 4.6 mm I.D.

Racemates **8–21** used for the evaluation of prepared columns were previously synthesized in the Laboratory for Stereoselective Catalysis and Biocatalysis at the Ruđer Bošković Institute [18,19].

All other compounds were purchased from commercial sources and used without further purification. Unless otherwise noted, the analyses of abovementioned racemates were performed at room temperature using the flow of 1 mL/min and UV detection of the compounds at 254 nm.

3.2. General Procedure for the Preparation of CSP-OHs

The corresponding DNB amide (0.75 mmol) was suspended in dry DCM (10 mL) under an inert atmosphere. A solution of hexachloroplatinic(IV) acid hydrate (20 mg) in isopropanol (0.5 mL) was added to the reaction mixture, followed by chlorodimethylsilane (10 mL). The mixture was refluxed for 5 h, cooled to room temperature and concentrated under reduced pressure. Dark residue was once again dissolved in DCM (5 mL), concentrated under reduced pressure and used without further purification. To the solution of the crude product in dry DCM (8 mL) under inert atmosphere, the 1:1 mixture of triethylamine and absolute ethanol (10 mL) was added dropwise and stirred at room temperature for half an hour. The solvent was evaporated to give a dark residue which was filtered through a short column of silica gel (eluent DCM-MeOH = 100:1). The resulting yellow residue was dissolved in dry toluene (5 mL) and added to a suspension of 5 μm HPLC silica gel (1.50 g) in dry toluene (50 mL). The silica gel was dried prior use for 24 h in Dean-Stark apparatus. The suspension was refluxed for 24 h, then filtered using a G4 sinter and washed with toluene (30 mL) and methanol (2 × 30 mL). Prepared CSP-OHs were dried for 4 h at 60 °C and elemental analysis results and IR spectra were recorded.

CSP-1-OH CHN analysis: C 5.05; H 0.63; N 1.11 (0.23 mmol/1 g); IR (ν/cm^{-1}): 3442 (SiO_2), 1642, 1540, 1343, 1220–1031 (SiO_2), 807 (SiO_2), 465 (SiO_2).

CSP-2-OH CHN analysis: C 4.20; H 0.85; N 0.62 (0.18 mmol/1 g); IR (ν/cm^{-1}): 3450 (SiO_2), 1641, 1544, 1346, 1219–1034 (SiO_2), 807 (SiO_2), 731, 466 (SiO_2).

CSP-3-OH CHN analysis: C 4.91; H 0.93; N 0.75 (0.20 mmol/1 g); IR (ν/cm^{-1}): 3434 (SiO_2), 1640, 1539, 1219–1034 (SiO_2), 808 (SiO_2), 464 (SiO_2).

CSP-4-OH CHN analysis: C 5.39; H 0.82; N 0.51 (0.24 mmol/1 g); IR (ν/cm^{-1}): 3439 (SiO_2), 1640, 1542, 1349, 1227–1030 (SiO_2), 808 (SiO_2), 731, 465 (SiO_2).

CSP-5-OH CHN analysis: C 5.43; H 0.80; N 0.70 (0.22 mmol/1 g); IR (ν/cm^{-1}): 3445 (SiO_2), 1638, 1540, 1214–1029 (SiO_2), 806 (SiO_2), 467 (SiO_2).

CSP-6-OH CHN analysis: C 5.45; H 0.55; N 0.77 (0.21 mmol/1 g); IR (ν/cm^{-1}): 3448 (SiO_2), 1645, 1543, 1347, 1217–1034 (SiO_2), 805 (SiO_2), 729, 465 (SiO_2).

CSP-7-OH CHN analysis: C 6.13; H 1.15; N 1.10 (0.22 mmol/1 g); IR (ν/cm^{-1}): 3452 (SiO_2), 1643, 1545, 1346, 1235–1023 (SiO_2), 805 (SiO_2), 731, 460 (SiO_2).

CSP-8-OH CHN analysis: C 6.73; H 1.17; N 0.41 (0.24 mmol/1 g); IR (ν/cm^{-1}): 3469 (SiO_2), 1638, 1545, 1348, 1230–1034 (SiO_2), 807 (SiO_2), 731, 463 (SiO_2).

CSP-9-OH CHN analysis: C 5.36; H 0.63; N 0.96 (0.20 mmol/1 g); IR (ν/cm^{-1}): 3451 (SiO_2), 1636, 1544, 1350, 1228–1036 (SiO_2), 807 (SiO_2), 730, 468 (SiO_2).

CSP-10-OH CHN analysis: C 6.41; H 1.07; N 1.06 (0.21 mmol/1 g); IR (ν/cm^{-1}): 3439 (SiO_2), 1630, 1542, 1347, 1223–1029 (SiO_2), 806 (SiO_2), 728, 464 (SiO_2).

3.3. End-Capping and Packing Procedure

To a yellow suspension of prepared CSP-OHs in dry toluene (20 mL) in an inert atmosphere, hexamethyldisilazane (2 mL) was added and resulting mixture was refluxed for 20 h. The suspension was cooled, filtered using a G4 sinter and washed with toluene (30 mL) and methanol (2 × 30 mL). After drying of prepared CSP overnight at 60 °C, they were packed into stainless steel columns (150 × 4.6 mm I.D.) using the slurry packing technique (1.5 g of CSP was suspended in 25 mL of solvent hexane-2-propanol = 2:8).

3.4. Computational Methods

Minimum energy structures of **DNB-6**, **DNB-8** and **DNB-9** are taken from our previous publication [20]. Relaxed potential energy surface scan was performed in Gaussian 16 [43] with Minnesota global hybrid functional M06-2X [44] with 54% of the exact exchange and aug-cc-pVDZ functional. The ethenyl moiety of (*S*)-**DNB-6** from our previous work [20] was replaced by hydrogen. Ground state force field parameters of **CSP-6**, naproxen and acetic acid were evaluated using the *antechamber* module of the Amber 16 package and the generalized Amber force field (GAFF) [45]. After geometry optimization of the complex of (*S*)-**DNB-6** and naproxen (**A**) and (*S*)-**DNB-6**, naproxen and acetic acid (**B**), molecular dynamics simulations at 300 K in vacuum were run, with a time step of 1 fs and a time simulation of 10 ns, for both the (*S*)- and (*R*)-enantiomers of naproxen. Geometries were saved every 5 ps. The 20 lowest energy structures from the trajectories were optimized at the B3LYP/def-SVP level of theory. The 10 most stable complexes were then re-optimized using the same functional and larger basis set (def2-TZVPP), while final energies were evaluated on M06-2X/def2-TZVPP level with solvation effects incorporated via COSMO model [46] and dielectric constant that equals to 2.78. Geometry optimizations and single point energy calculations were run in Turbomole [47].

4. Conclusions

In conclusion, we prepared ten new Pirkle-type chiral stationary phases based on (*S*)-N-(1-aryl-propyl)-3,5-dinitrobenzamide selectors. We evaluated the prepared CSPs using thirty diverse racemates, including several nonsteroidal anti-inflammatory and 3-hydroxybenzodiazepine drugs. Our aim was to design CSPs with similar versatility to Whelk-O1 and three of our CSPs (**CSP-6**, **CSP-7** and **CSP-10**) indeed displayed chromatographic behavior comparable to this widely used CSP. The prepared CSPs differ in the aromatic unit directly linked to the chiral center, which enabled us to elucidate the influence of the size and substitution of aromatic moiety on the enantiorecognition process. In order to substantiate experimental results, we investigated the role of nonbonding interactions relevant for chiral recognition on CSPs using computational methods. The model system was enantioseparation of naproxen on **CSP-6** where we elucidated the influence of hydrogen bond network and aromatic interactions on enantiorecognition process. Furthermore, the stability of the complexes is in accordance with the experimentally determined elution order. Finally, we investigated the influence of acetic acid as an additive in the mobile phase in the abovementioned system. To the best of our knowledge, this study is the first one to investigate the possibility of positioning a minor additive within the chiral binding site. We showed that non-enantiospecific interactions with the carboxylic moiety of the analyte can change the interaction arrangements and influence the energetics of the complexes responsible for chiral recognition.

Supplementary Materials: The Supplementary Materials are available online. Table S1: Enantioseparation results of racemates **1–7** and **9** on **CSP-1–CSP-10**, Table S2: Enantioseparation results of racemates **8** and **10–14** on **CSP-1–CSP-10**, Table S3: Enantioseparation results of racemates **15–21** on **CSP-1–CSP-10**, Table S4: Enantioseparation results of NSAIDs on **CSP-6–CSP-10**, Table S5: Enantioseparation results of 3-hydroxy-benzodiazepine drugs on **CSP-6–CSP-10**, Figure S1: Enantioseparation of naproxen on **CSP-6**, Figure S2: Higher energy structures of (*S*)-**A** and (*R*)-**A**, Figure S3: Higher energy structures of (*S*)-**B** and (*R*)-**B**.

Author Contributions: Conceptualization, A.K. and V.V.; Methodology, A.K. and V.V.; Formal analysis, A.K. and J.N.; Investigation, A.K and J.N..; Resources, V.V.; Writing—original draft preparation, A.K and J.N..; Writing—review and editing, V.V.; Visualization, A.K. and J.N.; Supervision, V.V.; funding acquisition, V.V.

Funding: This research was funded by Croatian Science Foundation, grant number IP-2016-06-1142 (LightMol).

Acknowledgments: The authors would like to acknowledge Goran Landek for preliminary results on CSPs, Darko Kontrec for performing the packing of columns, Irena Dokli and Andreja Lesac for technical support and Vladimir A. Potemkin for performing conformational analysis of naproxen. The authors acknowledge generous computer time provided by the Croatian National Grid Infrastructure (CRONGI).

Conflicts of Interest: The authors declare no conflict of interest.

References

1. FDA'S policy statement for the development of new stereoisomeric drugs. *Chirality* **1992**, *4*, 338–340. [CrossRef] [PubMed]
2. Calcaterra, A.; D'Acquarica, I. The market of chiral drugs: Chiral switches versus de novo enantiomerically pure compounds. *J. Pharm. Biomed. Anal.* **2018**, *147*, 323–340. [CrossRef] [PubMed]
3. Roussel, C.; Rio, A.D.; Pierrot-Sanders, J.; Piras, P.; Vanthuyne, N. Chiral liquid chromatography contribution to the determination of the absolute configuration of enantiomers. *J. Chromatogr. A* **2004**, *1037*, 311–328. [CrossRef] [PubMed]
4. Lorenz, H.; Seidel-Morgenstern, A. Processes to Separate Enantiomers. *Angew. Chem. Int. Ed.* **2014**, *53*, 1218–1250. [CrossRef] [PubMed]
5. Scriba, G.K.E. Chiral recognition in separation science—An update. *J. Chromatogr. A* **2016**, *1467*, 56–78. [CrossRef] [PubMed]
6. Wahab, M.F.; Wimalasinghe, R.M.; Wang, Y.; Barhate, C.L.; Patel, D.C.; Armstrong, D.W. Salient Sub-Second Separations. *Anal. Chem.* **2016**, *88*, 8821–8826. [CrossRef] [PubMed]
7. Ismail, O.H.; Pasti, L.; Ciogli, A.; Villani, C.; Kocergin, J.; Anderson, S.; Gasparrini, F.; Cavazzini, A.; Catani, M. Pirkle-type chiral stationary phase on core–shell and fully porous particles: Are superficially porous particles always the better choice toward ultrafast high-performance enantioseparations? *J. Chromatogr. A* **2016**, *1466*, 96–104. [CrossRef] [PubMed]
8. Zawatzky, K.; Barhate, C.L.; Regalado, E.L.; Mann, B.F.; Marshall, N.; Moore, J.C.; Welch, C.J. Overcoming "speed limits" in high throughput chromatographic analysis. *J. Chromatogr. A* **2017**, *1499*, 211–216. [CrossRef] [PubMed]
9. Welch, C.J. Are We Approaching a Speed Limit for the Chromatographic Separation of Enantiomers? *ACS Cent. Sci.* **2017**, *3*, 823–829. [CrossRef] [PubMed]
10. Barhate, C.L.; Joyce, L.A.; Makarov, A.A.; Zawatzky, K.; Bernardoni, F.; Schafer, W.A.; Armstrong, D.W.; Welch, C.J.; Regalado, E.L. Ultrafast chiral separations for high throughput enantiopurity analysis. *Chem. Commun.* **2017**, *53*, 509–512. [CrossRef] [PubMed]
11. Ismail, O.; Felletti, S.; Luca, C.; Pasti, L.; Marchetti, N.; Costa, V.; Gasparrini, F.; Cavazzini, A.; Catani, M. The Way to Ultrafast, High-Throughput Enantioseparations of Bioactive Compounds in Liquid and Supercritical Fluid Chromatography. *Molecules* **2018**, *23*, 2709. [CrossRef] [PubMed]
12. Khundadze, N.; Pantsulaia, S.; Fanali, C.; Farkas, T.; Chankvetadze, B. On our way to sub-second separations of enantiomers in high-performance liquid chromatography. *J. Chromatogr. A* **2018**, *1572*, 37–43. [CrossRef] [PubMed]
13. Lämmerhofer, M. Chiral recognition by enantioselective liquid chromatography: Mechanisms and modern chiral stationary phases. *J. Chromatogr. A* **2010**, *1217*, 814–856. [CrossRef] [PubMed]
14. Ali, I.; Saleem, K.; Hussain, I.; Gaitonde, V.D.; Aboul-Enein, H.Y. Polysaccharides Chiral Stationary Phases in Liquid Chromatography. *Sep. Purif. Rev.* **2009**, *38*, 97–147. [CrossRef]
15. Fernandes, C.; Tiritan, M.E.; Pinto, M. Small Molecules as Chromatographic Tools for HPLC Enantiomeric Resolution: Pirkle-Type Chiral Stationary Phases Evolution. *Chromatographia* **2013**, *76*, 871–897. [CrossRef]
16. Fernandes, C.; Phyo, Y.Z.; Silva, A.S.; Tiritan, M.E.; Kijjoa, A.; Pinto, M.M.M. Chiral Stationary Phases Based on Small Molecules: An Update of the Last 17 Years. *Sep. Purif. Rev.* **2018**, *47*, 89–123. [CrossRef]
17. Forjan, D.M.; Gazić, I.; Vinković, V. Role of the weak interactions in enantiorecognition of racemic dihydropyrimidinones by novel brush-type chiral stationary phases. *Chirality* **2007**, *19*, 446–452. [CrossRef] [PubMed]
18. Ranogajec, A.; Kontrec, D.; Vinkovic, V.; Sunjic, V. Enantiomer Separation and Molecular Recognition with New Chiral Stationary Phases on 4-Chloro-3,5-dinitrobenzoic Acid Amides of α,β-Aminoalcohols and α-Arylethylamines. *J. Liq. Chromatogr. Relat. Technol.* **2003**, *26*, 63–83. [CrossRef]
19. Kontrec, D.; Vinkovic, V.; Sunjic, V. Preparation and evaluation of chiral stationary phases based on N,N-2,4-(or 4,6)-disubstituted 4,5-(or 2,5)-dichloro-1,3-dicyanobenzene. *Chirality* **2000**, *12*, 63–70. [CrossRef]
20. Knežević, A.; Novak, J.; Pescitelli, G.; Vinković, V. Determination of the Absolute Configuration of (*S*)-*N*-(1-Aryl-allyl)-3,5-dinitrobenzamides and Their Elution Order on Brush-Type Chiral Stationary Phases. *Eur. J. Org. Chem.* **2018**, *2018*, 3982–3991. [CrossRef]

21. Ihara, T.; Sugimoto, Y.; Asada, M.; Nakagama, T.; Hobo, T. Influence of the method of preparation of chiral stationary phases on enantiomer separations in high-performance liquid chromatography. *J. Chromatogr. A* **1995**, *694*, 49–56. [CrossRef]
22. Pirkle, W.H.; Readnour, R.S. The influence of end-capping on the enantioselectivity of a chiral phase. *Chromatographia* **1991**, *31*, 129–132. [CrossRef]
23. Snyder, L.R.; Kirkland, J.J.; Glajch, J.L. *Practical HPLC Method Development*; John Wiley & Sons: New York, NY, USA, 2012; ISBN 978-1-118-59151-2.
24. Chiral Handbook for HPLC & SFC Separations. Available online: http://www.registech.com/literature-library/technical-resources/chiral-handbook-for-hplc-sfc-separations (accessed on 26 September 2018).
25. Wahab, M.F.; Patel, D.C.; Wimalasinghe, R.M.; Armstrong, D.W. Fundamental and Practical Insights on the Packing of Modern High-Efficiency Analytical and Capillary Columns. *Anal. Chem.* **2017**, *89*, 8177–8191. [CrossRef] [PubMed]
26. Unger, K.K.; Skudas, R.; Schulte, M.M. Particle packed columns and monolithic columns in high-performance liquid chromatography-comparison and critical appraisal. *J. Chromatogr. A* **2008**, *1184*, 393–415. [CrossRef] [PubMed]
27. Pirkle, W.H.; Welch, C.J.; Lamm, B. Design, synthesis, and evaluation of an improved enantioselective naproxen selector. *J. Org. Chem.* **1992**, *57*, 3854–3860. [CrossRef]
28. Pirkle, W.H.; Welch, C.J. An Improved Chiral Stationary Phase for the Chromatographic Separation of Underivatized Naproxen Enantiomers. *J. Liq. Chromatogr.* **1992**, *15*, 1947–1955. [CrossRef]
29. Knežević, A.; Landek, G.; Dokli, I.; Vinković, V. An efficient enzymatic approach to (S)-1-aryl-allylamines. *Tetrahedron Asymmetry* **2011**, *22*, 936–941. [CrossRef]
30. Pirkle, W.H.; Welch, J. Chromatographic and 1H NMR support for a proposed recognition model. *J. Chromtogr. A* **1994**, *683*, 347–353. [CrossRef]
31. Zhao, C.F.; Diemert, S.; Cann, N.M. Rational optimization of the Whelk-O1 chiral stationary phase using molecular dynamics simulations. *J. Chromatogr. A* **2009**, *1216*, 5968–5978. [CrossRef] [PubMed]
32. Del Rio, A.; Hayes, J.M.; Stein, M.; Piras, P.; Roussel, C. Theoretical reassessment of Whelk-O1 as an enantioselective receptor for 1-(4-halogeno-phenyl)-1-ethylamine derivatives. *Chirality* **2004**, *16*, S1–S11. [CrossRef] [PubMed]
33. Pirkle, W.H.; Welch', C.J. Use of Simultaneous Face to Face and Face to Edge π-π Interactions to Facilitate Chiral Recognition. *Tetrahedron Asymmetry* **1994**, *5*, 777–780. [CrossRef]
34. Martinez, C.R.; Iverson, B.L. Rethinking the term "pi-stacking". *Chem. Sci.* **2012**, *3*, 2191. [CrossRef]
35. Zhao, C.F.; Cann, N.M. Molecular Dynamics Study of Chiral Recognition for the Whelk-O1 Chiral Stationary Phase. *Anal. Chem.* **2008**, *80*, 2426–2438. [CrossRef] [PubMed]
36. Ye, Y.K.; Stringham, R.W. Effect of mobile phase acidic additives on enantioselectivity for phenylalanine analogs. *J. Chromatogr. A* **2001**, *927*, 47–52. [CrossRef]
37. Ye, Y.K.; Stringham, R.W. The effect of acidic and basic additives on the enantioseparation of basic drugs using polysaccharide-based chiral stationary phases. *Chirality* **2006**, *18*, 519–530. [CrossRef] [PubMed]
38. Jibuti, G.; Mskhiladze, A.; Takaishvili, N.; Karchkhadze, M.; Chankvetadze, L.; Farkas, T.; Chankvetadze, B. HPLC separation of dihydropyridine derivatives enantiomers with emphasis on elution order using polysaccharide-based chiral columns: Liquid Chromatography. *J. Sep. Sci.* **2012**, *35*, 2529–2537. [CrossRef] [PubMed]
39. Mosiashvili, L.; Chankvetadze, L.; Farkas, T.; Chankvetadze, B. On the effect of basic and acidic additives on the separation of the enantiomers of some basic drugs with polysaccharide-based chiral selectors and polar organic mobile phases. *J. Chromatogr. A* **2013**, *1317*, 167–174. [CrossRef] [PubMed]
40. Gogaladze, K.; Chankvetadze, L.; Tsintsadze, M.; Farkas, T.; Chankvetadze, B. Effect of Basic and Acidic Additives on the Separation of Some Basic Drug Enantiomers on Polysaccharide-Based Chiral Columns with Acetonitrile as Mobile Phase: SEPARATION OF SOME BASIC DRUG ENANTIOMERS. *Chirality* **2015**, *27*, 228–234. [CrossRef] [PubMed]
41. Fujii, Y.; Yamada, H.; Mizuta, M. Self-association of acetic acid in some organic solvents. *J. Phys. Chem.* **1988**, *92*, 6768–6772. [CrossRef]
42. Chocholoušová, J.; Vacek, J.; Hobza, P. Acetic Acid Dimer in the Gas Phase, Nonpolar Solvent, Microhydrated Environment, and Dilute and Concentrated Acetic Acid: Ab Initio Quantum Chemical and Molecular Dynamics Simulations. *J. Phys. Chem. A* **2003**, *107*, 3086–3092. [CrossRef]

43. Frisch, M.J.; Trucks, G.W.; Schlegel, H.B.; Scuseria, G.E.; Robb, M.A.; Cheeseman, J.R.; Scalmani, G.; Barone, V.; Petersson, G.A.; Nakatsuji, H.; et al. *Gaussian 16*; Revision B.01; Gaussian, Inc.: Wallingford, CT, USA, 2016.
44. Zhao, Y.; Truhlar, D.G. The M06 suite of density functionals for main group thermochemistry, thermochemical kinetics, noncovalent interactions, excited states, and transition elements: Two new functionals and systematic testing of four M06-class functionals and 12 other functionals. *Theor. Chem. Acc.* **2008**, *120*, 215–241.
45. Case, D.A.; Betz, R.M.; Cerutti, D.S.; Cheatham, T.E.; III, T.A.; Darden, R.E.; Duke, T.J.; Giese, H.; Gohlke, A.W.; Goetz, N.; et al. *AMBER*; University of California: San Francisco, CA, USA, 2016.
46. Klamt, A.; Schüürmann, G. COSMO: A new approach to dielectric screening in solvents with explicit expressions for the screening energy and its gradient. *J. Chem. Soc. Perkin Trans. 2* **1993**, *5*, 799–805. [CrossRef]
47. *TURBOMOLE 7.0.1*; University of Karlsruhe and Forschungszentrum: Karlsruhe, Germany, 2015.

Sample Availability: Samples of the compounds **DNB-1–DNB-10** and **CSP-1–CSP-10** are available from the authors.

molecules

MDPI

Article

Chiral Separation of the Phenylglycinol Enantiomers by Stripping Crystallization

Lie-Ding Shiau [1,2]

[1] Department of Chemical and Materials Engineering, Chang Gung University, Taoyuan 33302, Taiwan; shiau@mail.cgu.edu.tw; Tel.: +886-3-2118800 (ext. 5291); Fax: +886-3-2118700

[2] Department of Urology, Chang Gung Memorial Hospital, Linkou, Taoyuan 33302, Taiwan

Academic Editors: Maria Elizabeth Tiritan, Madalena Pinto and Carla Sofia Garcia Fernandes
Received: 8 October 2018; Accepted: 5 November 2018; Published: 7 November 2018

Abstract: Stripping crystallization (SC) is introduced in this work for chiral purification of *R*-phenylglycinol from the enantiomer mixture with an initial concentration ranging from 0.90 to 0.97. As opposed to the solid–liquid transformation in melt crystallization, the three-phase transformation occurs in SC at low pressures during the cooling process. SC combines melt crystallization and vaporization to produce a crystalline product and mixture vapor from a mixture melt due to the three-phase transformation. Thermodynamic calculations were applied to determine the operating pressure for the three-phase transformation during the cooling process in the SC experiments. To consider the possible deviations between the calculated and the actual three-phase transformation conditions, the product purity and the recovery ratio of *R*-phenylglycinol were investigated within a range of operating pressures during the cooling process.

Keywords: crystallization; vaporization; purification; phenylglycinol

1. Introduction

Pure enantiomer is often needed for the desired therapeutic effect due to different pharmacological and pharmacokinetic processes for the enantiomers of drugs. However, separation of the enantiomers has long been a challenging task as the enantiomers have nearly identical physical and chemical properties [1]. Various enantioselective separation techniques, including enantioselective synthesis, chromatographic separation, and preferential crystallization, have been developed for chemical and pharmaceutical industries [2]. Preferential crystallization generally has been used as an attractive means to separate the conglomerate-forming enantiomers from racemate [3–6]. Although chromatographic separation has been investigated extensively [7–10], the synthesis of efficient chiral stationary phases in chromatographic methods is usually deemed a robust technology. Recently, Didaskalou et al. [11] reported the membrane-grafted asymmetric organocatalyst used as an integrated synthesis–enantioseparation platform. Rukhlenko et al. [12] explored the capabilities of the related enantioseparation method by analytically solving the problem of the force-induced diffusion of chiral nanoparticles in a confined region.

Phenylglycinol, also called 2-Amino-2-phenylethanol, is an important example of a chiral compound. Only *R*-phenylglycinol can be used as an important precursor of HIV-1 protease inhibitor [13]. Enantioseparation, using extractant impregnated resins [14] or liquid–liquid extraction [15,16] has been proposed to separate *R*-Phenylglycinol from the racemic mixture. Fundamentally, an enantioselective solvent is chosen and used as extractant for the enantioseparation of phenylglycinol.

Stripping crystallization (SC) is a new separation technology, which combines melt crystallization and vaporization to produce a crystalline product due to the three-phase transformation. SC has been successfully developed to separate the mixtures with close boiling temperatures, including mixed xylenes [17–19], the styrene/ethylbenzene mixture [20], and the *S*-ibuprofen/*R*-ibuprofen mixture [21]. As opposed to extractant impregnated resins or liquid–liquid extraction, no solvent is added in SC. Thus, no removal of solvent is required at the end of SC.

The objective of this research is to study the feasibility of SC in purification of *R*-phenylglycinol from a phenylglycinol mixture. The thermodynamic calculations are adopted to determine the three-phase transformation conditions for the SC experiments. The effects of various operating conditions on the enantiomeric purity and recovery ratio of *R*-phenylglycinol crystalline product are investigated

2. SC Model

As opposed to the solid–liquid transformation in melt crystallization operated at normal pressure during the cooling process [22–27], the three-phase transformation occurs in SC at low pressures during the cooling process. Thus, SC combines melt crystallization and vaporization to produce a crystalline product and mixture vapor from a mixture liquid or melt [17–21]. The SC process is simulated in a series of N stage operations shown in Figure 1, where each stage is operated at a three-phase transformation state. The SC process starts with a mixture liquid or melt. The vapor formed in each stage is removed, while the crystalline product and the remaining liquid or melt in each stage enter the next stage. Thus, only the crystalline product remains at the end of SC when the liquid or melt is nearly eliminated.

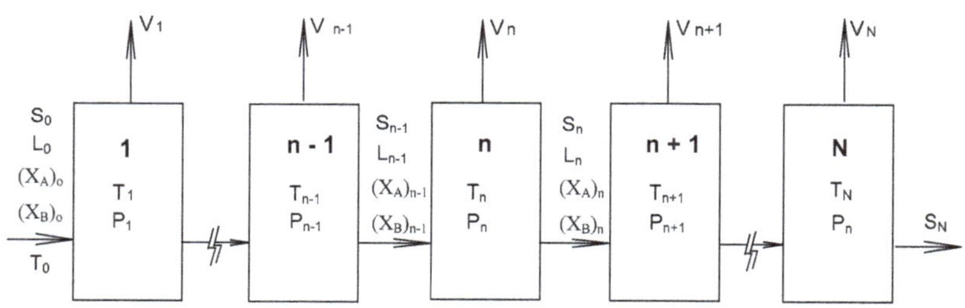

Figure 1. Schematic diagram of the stripping crystallization (SC) operation where each stage is operated at a three-phase transformation state.

When SC is applied to purify *R*-phenylglycinol (B-component) from the mixture of *S*-phenylglycinol (A-component) and *R*-phenylglycinol, the SC process starts with a mixture melt of phenylglycinol. The corresponding three-phase transformation condition in each stage is determined based on the following assumptions: (a) The ideal gas law is assumed for the vapor due to low pressures; (b) The ideal solution for the melt is assumed due to the structure similarity between *S*-phenylglycinol and *R*-phenylglycinol; (c) The Clausius–Clapeyron equation [28,29] is adopted to describe the temperature dependence of the saturated pressure for each component in the melt; (d) The sublimation based on the solid–vapor equilibrium is not considered here as the mixture melt is used in the beginning of the experiments. Some physical properties of *S*-phenylglycinol and *R*-phenylglycinol are listed in Table 1. For simplicity, $\Delta H_V = 2\Delta H_m$ is assumed in the thermodynamic calculations.

As SC is applied to produce *R*-phenylglycinol crystalline product from a mixture melt due to the three-phase transformation, both the solid–liquid equilibrium and the vapor–liquid equilibrium need to be simultaneously satisfied. The solid–liquid equilibrium is described by the Schroder–Van Laar equation [1,28,29], while the vapor–liquid equilibrium is described by Raoult's law [28,29]. Consequently, as similar to a previous work reported by Shiau [21], the three-phase equilibrium equations can be derived in each stage. If T_n is specified in each stage, these equations can be simultaneously solved for P_n, $(X_A)_n$, $(X_B)_n$, $(Y_A)_n$ and $(Y_B)_n$ for n = 1, 2, . . . , N.

Table 1. Some physical properties for phenylglycinol.

Property	Phenylglycinol
molecular structure	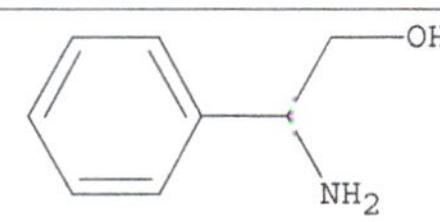
molecular weight	137.2
melting point [a], °C	77
boiling point [a], °C	261
triple-point pressure [b], Pa (N/m^2)	198
heat of melting [c], J/mol	2.57×10^4

[a]: The Merck Index [30]; [b]: Estimated by Clausius–Clapeyron equation [28,29]; [c]: Measured by Differential scanning calorimetry in this work.

Figure 2 displays the thermodynamic calculations of P(T), $X_B(T)$, and $Y_B(T)$ during the cooling process. Thus, the corresponding pressure, P(T), and the corresponding melt composition of *R*-phenylglycinol, $X_B(T)$, decreases during the cooling process for SC. In other words, Figure 2 reveals that, as $X_B(T)$ in a melt decreases, the corresponding temperature and pressure for the three-phase transformation conditions decreases.

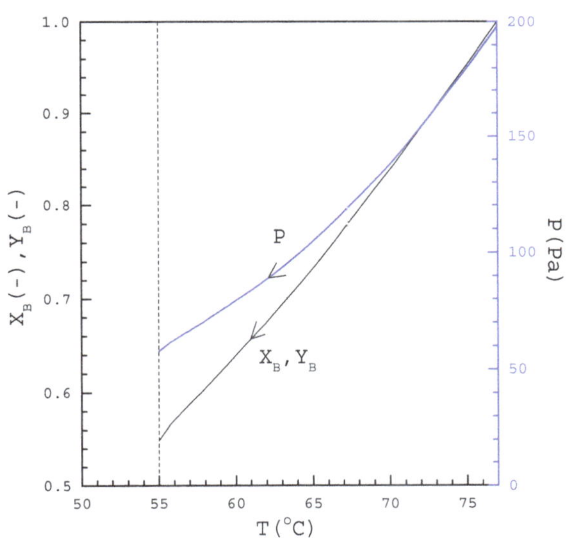

Figure 2. P(T), $X_B(T)$ and $Y_B(T)$ based on the thermodynamic calculations for the three-phase transformation.

As shown in Figure 1, the three-phase transformation occurs in the melt in each stage, leading to the formation of *R*-phenylglycinol crystalline product and mixture vapor. S_n and L_n represent the amount of *R*-phenylglycinol crystalline product and the melt, respectively, remaining in stage n, while V_n represents the amount of the mixture vapor formed and removed in stage n. The entire material balance in stage n can be described by

$$S_{n-1} + L_{n-1} = S_n + L_n + V_n \tag{1}$$

where $S_{n-1} + L_{n-1}$ is the total amount of crystalline product and melt entering stage n. As V_{n-1} represents the amount of vapor formed in stage n − 1 that is subsequently removed, it is not part of the equation for stage n. Thus, the amount of melt decreases and the amount of crystalline product increases during the stage operation.

Although both the melt and the vapor consist of *S*-phenylglycinol and *R*-phenylglycinol, only *R*-phenylglycinol crystalline product is formed in each stage based on the solid–liquid equilibrium described by the Schroder–Van Laar equation [1,28,29]. It is assumed that no impurity trapping occurs in the formation of *R*-phenylglycinol crystalline product based on the thermodynamic calculations. The material balance of *R*-phenylglycinol in stage n can be described by

$$S_{n-1} + L_{n-1}(X_B)_{n-1} = S_n + L_n(X_B)_n + V_n(Y_B)_n \tag{2}$$

It is observed during the experiments that the three-phase transformation occurs in the melt very quickly in each stage, leading to the formation of *R*-phenylglycinol crystalline product and the mixture vapor. Therefore, it is assumed in each stage that the heat released in forming *R*-phenylglycinol crystalline product is quickly removed by vaporizing some portion of the melt. Thus, the energy balance in stage n can be described by

$$(S_n - S_{n-1})\Delta H_{m,B} = V_n \Delta H_{V,B} \tag{3}$$

where $S_n - S_{n-1}$ represents the amount of crystalline product formed in stage n while V_n represents the amount of melt vaporized in stage n. Note that the heat of vaporization is assumed as $\Delta H_{V,B}$ for a mixture melt due to $\Delta H_{V,A} = \Delta H_{V,B}$.

As only the mixture melt L_0 with a known $(X_B)_0$ is injected into the sample container, one obtains $S_0 = 0$. Equations (1) to (3) can be solved simultaneously for three unknown variables- S_n, L_n and V_n. Note that S_N and L_N represents the crystalline product and the melt, respectively, remaining at the end while the total amount vapor formed and removed at the end is given by $\sum_{n=1}^{N} V_n$.

3. Experimental Section

The experimental assembly consisted of a 5-mL sample container in a 1.5-L chamber as shown in Figure 3. The stainless chamber was immersed in a constant temperature bath. A mechanical vacuum pump was used to lower the pressure in the chamber. A temperature probe was positioned in the center of the mixture melt and a pressure gauge was connected to the big chamber. Thus, the operating temperature and pressure could be adjusted mid-experiment. Crystallization and vaporization of the mixture melt during the three-phase transformation could be observed in the chamber via a transparent cover.

R-phenylglycinol (purity >98%) and *S*-phenylglycinol (purity >98%) were purchased from ACROS. In the beginning of the experiment, 1 g mixture melt with a known concentration was injected into the sample container stirred by a magnetic bar at 70 rpm. Then, the temperature was lowered gradually from the melting point (77 °C). The cooling rate generally started at 0.5 °C/min in the beginning and then increased gradually to 1 °C/min in the later stage. As the temperature decreased, pressure was adjusted downward based on Figure 2. Thus, a series of three-phase transformations occurred in the melt, leading to the formation of *R*-phenylglycinol crystalline product and mixture vapor. The experiments were generally ended at around 55 °C and 58 Pa within 25 min when vaporization was no longer observed in the chamber. Figure 4 illustrates the schematic diagram of the batch experiments, in which the melt was simultaneously vaporized and crystallized due to the three-phase transformation. Upon completion, the final product, including the crystals and melt, in the sample container were weighed.

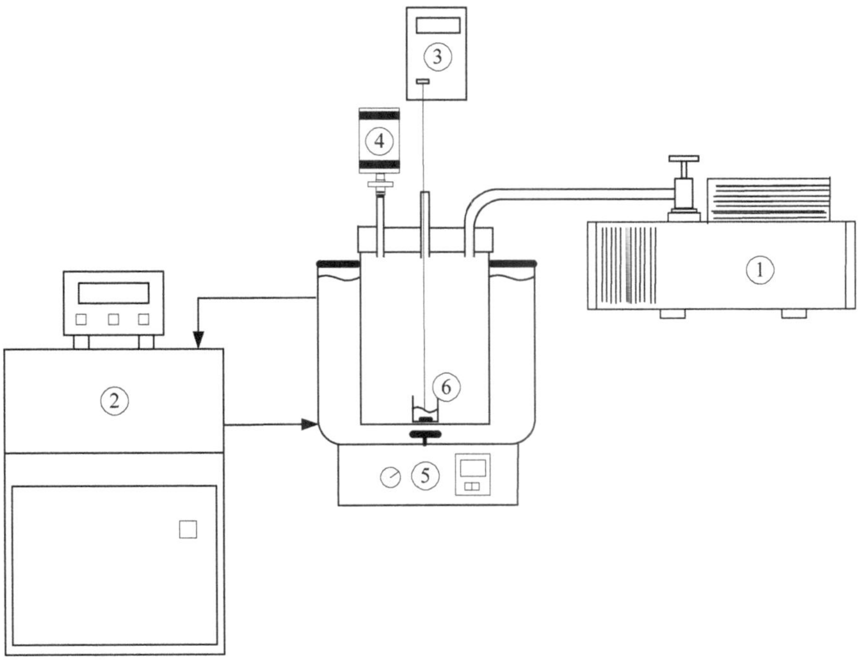

Figure 3. Schematic diagram of the experimental apparatus for the SC operation with the features: (1) Mechanical pump, (2) constant temperature bath, (3) thermocouple, (4) pressure gauge, (5) magnetic stirrer operated at 70 rpm, (6) 5-mL sample container in a 1.5-L chamber.

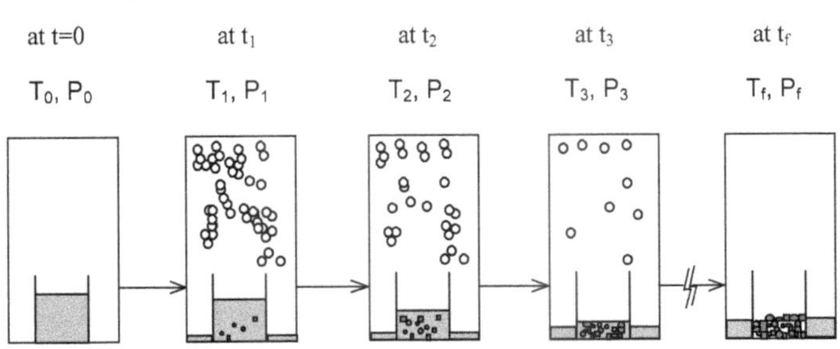

Figure 4. Schematic diagram of a batch SC experiment, where each stage corresponds to a three-phase transformation state at a given time: At t = 0, a mixture melt in the sample container; at $0 < t < t_f$, formation of *R*-phenylglycinol crystalline product and mixture vapor from a mixture melt due to the three-phase transformation; at t_f, only *R*-phenylglycinol crystalline product and the remaining melt left in the sample container. Note that the vapor was condensed and collected outside the sample container in the chamber.

The enantiomeric purity of the final product was analyzed by Polarimeter (Horiba, model: SEPA-300). The polarimetry was measured by dissolving 0.1 g final product in 20 mL 1 M HCl solution. First, a plot of the measured specific optical rotation versus the known enantiomeric purity within the range $X_{B,0}$ = 0.9 to 1.0 was fitted with a linear regression line. Then, by measuring the specific optical rotation of the final sample the enantiomeric purity could be determined. Note that $[\alpha]_D^{20} = -29.9°$ for *R*-phenylglycinol and $[\alpha]_D^{20} = 29.9°$ for *S*-phenylglycinol. It should be noted that, as only crystallization and vaporization occurred during SC, polarimetry could be used to determine the enantiomeric purity of the final product.

From a practical point of view, some solvent might remain in the mixture melt before SC. To elucidate the effects of residual solvent on the final product purity and recovery ratio of *R*-phenylglycinol, 0.1 g ethanol was added into 1 g mixture melt in the beginning of the SC experiments. It was found that the final product purity and recovery ratio for 1 g mixture melt with 0.1 g ethanol were nearly the same as those for 1 g mixture melt without ethanol. Thus, all ethanol was vaporized when SC was operated at low pressures during the cooling process. Solvent inclusion in the formation of *R*-phenylglycinol crystalline product was nearly negligible.

4. Results and Discussion

SC was applied to purify *R*-phenylglycinol for various 1 g feeds: Feed 1 with $(X_B)_0 = 0.90$, feed 2 with $(X_B)_0 = 0.95$, and feed 3 with $(X_B)_0 = 0.97$. Table 2 lists the thermodynamic calculations for 1 g feed 1, where $T_0 = 72.7$ °C is the initial three-phase transformation temperature for the mixture melt. As vaporization was no longer observed in the experiments at around 55 °C, $T_N = 54.6$ °C was chosen for N = 15 with $\Delta T = 1.2$ °C. Thus, T_n was specified in each stage for n = 1, 2, ... , N using $T_{n-1} - T_n = \Delta T$. P_n, $(X_A)_n$, $(X_B)_n$, $(Y_A)_n$, and $(Y_B)_n$ were determined in each stage by solving the thermodynamic equations while S_n, L_n, and V_n were determined in each stage by solving Equations (1) to (3) for $L_0 = 1$ g and $S_0 = 0$. Note that P_n, $(X_B)_n$ and $(Y_B)_n$ in Table 2 were consistent with the results shown in Figure 2. Table 2 also indicates that, as n increased during the cooling process, S_n increased and L_n decreased. As SC was operated from 73 °C and 160 Pa (n = 1) to 55 °C and 58 Pa, (n = 15), only *R*-phenylglycinol crystalline product remain in the last stage ($S_N = 0.606$ g) while the melt was nearly eliminated in the last stage $L_N = 0.098$ g. Similar calculated results were obtained for feed 2 and feed 3.

Table 2. The thermodynamic calculations for 1 g feed with $X_{B,0} = 0.90$ ($\Delta T = 1.2$ °C).

n	T (°C)	P (Pa)	L (g)	S (g)	V (g)	X_B
0	72.7	159.4	1	0	0	0.90
1	71.5	149.5	0.672	0.219	0.109	0.872
2	70.3	140.1	0.496	0.336	0.059	0.845
3	69.1	131.3	0.389	0.407	0.036	0.819
4	67.9	122.9	0.316	0.456	0.024	0.793
5	66.7	115.1	0.265	0.490	0.017	0.768
6	65.5	107.6	0.226	0.516	0.013	0.743
7	64.3	100.7	0.197	0.535	0.010	0.719
8	63.1	94.1	0.175	0.550	0.007	0.696
9	61.9	87.9	0.156	0.562	0.006	0.673
10	60.6	82.1	0.142	0.572	0.005	0.651
11	59.4	76.6	0.129	0.581	0.004	0.629
12	58.2	71.5	0.119	0.588	0.003	0.608
13	57.0	66.6	0.110	0.593	0.003	0.588
14	55.8	62.1	0.102	0.598	0.003	0.568
15	54.6	57.8	0.096	0.603	0.002	0.549

The calculated purity of *R*-phenylglycinol in the final product, including the final crystalline product and the remaining melt, is defined as

$$X_{B,C} = \frac{S_N + L_N(X_B)_N}{S_N + L_N} \tag{4}$$

where S_N, L_N and $(X_B)_N$ are determined in the last stage based on the thermodynamic calculations. The calculated recovery ratio of *R*-phenylglycinol is defined as

$$R_C = \frac{S_N + L_N(X_B)_N}{L_0 X_{B,0}} \tag{5}$$

where L_0 is the initial weight of the mixture melt and $X_{B,0}$ denotes the initial purity of *R*-phenylglycinol in the mixture melt. For example, as shown in Table 2, feed 1 yields S_N = 0.606 g and L_N = 0.098 g with $(X_B)_N$ = 0.549 in the last stage (N = 15), leading to $X_{B,C}$ = 0.937 and R_C = 73% using Equations (4) to (5).

The experimental recovery ratio of *R*-phenylglycinol is defined as

$$R_f = \frac{W_f X_{B,f}}{L_0 X_{B,0}} \tag{6}$$

where W_f refers to the final weight of the product including the crystalline product and the remaining melt obtained at the end of the experiment, and $X_{B,f}$ represents the experimental purity of *R*-phenylglycinol in the final product.

Figure 5 shows $X_{B,f}$ of the final product plotted against $X_{B,0}$ of the initial feed for various feeds. The Solid circles represent the calculated $X_{B,C}$ and the number in parenthesis represents the calculated R_C. Thus, the thermodynamic calculations predict that feed 1 can be purified from $X_{B,0}$ = 0.90 to $X_{B,C}$ = 0.937 with R_C = 73%, feed 2 can be purified from $X_{B,0}$ = 0.95 to $X_{B,C}$ = 0.979 with R_C = 70%, and feed 3 can be purified from $X_{B,0}$ = 0.97 to $X_{B,C}$ = 0.995 with R_C = 69%.

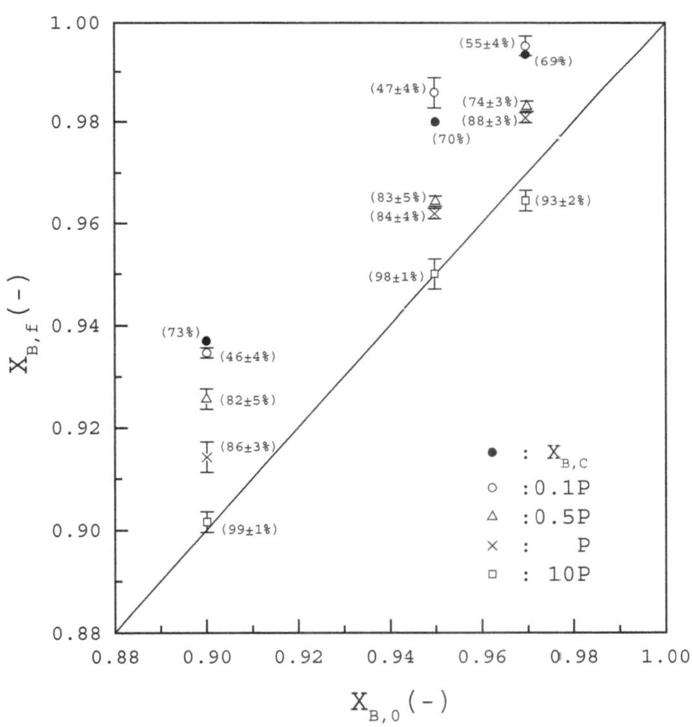

Figure 5. $X_{B,f}$ of the final product plotted against $X_{B,0}$ of the initial feed for various 1 g feeds. The solid line represents $X_{B,f}$ = $X_{B,0}$ indicating no further purification for the initial feed during SC. Solid circle represents the calculated $X_{B,C}$ and the number in parenthesis represents the calculated R_C. Other symbols, including open circle, open triangle, cross sign, and open square, represent the average of the experimental $X_{B,f}$ for four repetitive experiments operated at the specified pressure and the error bar represents the 95% confidence interval for the experimental $X_{B,f}$. The number in parenthesis represents the average of the experimental R_f with the 95% confidence interval for the experimental R_f. Note that no error bar is added for solid circle of the calculated $X_{B,C}$.

Other symbols in Figure 5, including open circle, open triangle, cross sign, and open square, represent the average of the experimental $X_{B,f}$ for four repetitive experiments operated at the specified pressure and the error bar represents the 95% confidence interval for the experimental $X_{B,f}$. The number in parenthesis represents the average of the experimental R_f with the 95% confidence interval for the experimental R_f. For example, cross sign represents the average $X_{B,f}$ when the operating pressures was controlled at P(T) during the cooling process. As a lower $X_{B,f}$ with a higher R_f was observed for each feed compared to the calculated $X_{B,C}$ and R_C, it was speculated that the calculated pressure P(T) in Figure 3 might be higher than the actual three-phase transformation pressure, leading to less impurity (*S*-phenylglycinol) vaporized and more crystalline product formed from the melt during the cooling process.

To consider the possible deviations between the calculated and the actual three-phase transformation pressure, various operating pressures are compared during the cooling process. The open circle in Figure 5 represents the average $X_{B,f}$ when the operating pressures were controlled at $0.1 \times$ P(T) during the cooling process. Similarly, the open triangle represents the average $X_{B,f}$ when the operating pressures were controlled at $0.5 \times$ P(T). The open square represents the average $X_{B,f}$ when the operating pressures were controlled at $10 \times$ P(T).

Figure 5 shows for each feed that $X_{B,f}$ increased with decreasing pressure while R_f decreased with decreasing pressure. For example, when SC was applied for $X_{B,0} = 0.90$, $X_{B,f}$ increased from 0.914 to 0.935 and R_f decreased from 86% to 46% as the operating pressure was decreased from P(T) to $0.1 \times$ P(T). As shown in the figure, when $0.1 \times$ P(T) was adopted for each $X_{B,0}$, $X_{B,f}$ was close to $X_{B,C}$ with R_f (46% to 55%) < R_C (69% to 73%). Consequently, compared to the calculated P(T) in Figure 3, $0.1 \times$ P(T) should be closer to the actual the three-phase transformation pressure. On the other hand, when $10 \times$ P(T) was adopted for each $X_{B,0}$, $X_{B,f}$ was close to $X_{B,0}$ with R_f = 93% to 99%, indicating that the feed was not further purified in the SC experiments.

Discrepancies between the thermodynamic calculations and the experimental results are attributed to (a) the assumption that each stage was operated at the three-phase transformation. However, experimentally, these might not always be achieved; (b) although pure *S*-phenylglycinol crystal should be formed based on the thermodynamic equilibrium, impurity trapping can occur under actual kinetic conditions. The scope of this work was to investigate the feasibility of SC in the purification of *R*-phenylglycinol from a phenylglycinol mixture. In future kinetic studies, the effects of process conditions (e.g., cooling rate) on the crystal growth kinetics and impurity inclusion will be explored based on the impurity trapping correlation proposed by Myerson and Kirwan [31,32].

5. Conclusions

SC was successfully applied for chiral purification of *R*-phenylglycinol from the phenylglycinol enantiomers. A lower pressure during the cooling process generally led to a higher experimental product purity with a lower experimental recovery ratio. When SC was operated under the optimal pressure, which was one-tenth of the pressure based on the thermodynamic calculations, the experimental product purity was close to the calculated product purity while the experimental recovery ratio was slightly lower than the calculated recovery ratio. In other words, when temperature and pressure was lowered from 72.7 °C and 15 Pa to 55 °C and 6 Pa during SC, the purity of *R*-phenylglycinol increased from 0.90 to 0.937, from 0.94 to 0.985, and from 0.97 to 0.995 respectively with the recovery ratio ranging between 46% to 55%.

As no solvent is added into the melt, SC is a clean separation technology. Compared to melt crystallization, neither solid/liquid separation nor crystal washing is required because no mother liquor adheres to the crystal surfaces upon completion. Although a portion of the phenylglycinol enantiomers is lost through the vapor stream of each stage, the vaporized mixture can be recycled for continuous operation or mixed with the feed in the next batch for batch operation. The major difficulty in application of SC lies in the required low pressures during the cooling process. Furthermore, the crystal growth kinetics and impurity trapping during SC need to be elucidated in order to design an apparatus for industrial application.

Funding: This research is funded by Ministry of Science and Technology of Taiwan (MOST106-2221-E-182-053) and Chang Gung Memorial Hospital (CMRPD2G0241).

Acknowledgments: The author would like to thank Ministry of Science and Technology of Taiwan (MOST106-2221-E-182-053) and Chang Gung Memorial Hospital (CMRPD2G0241) for financial support for this research. The author also expresses his gratitude to Yu-Chen Chen and Keng-Fu Liu for their experimental work.

Conflicts of Interest: The author declares no conflict of interest. The funders had no role in the design of the study; in the collection, analyses, or interpretation of data; in the writing of the manuscript, and in the decision to publish the results.

Notation

$\Delta H_{m,i}$	heat of melting for component-i (>0), J/mol
$\Delta H_{V,i}$	heat of vaporization for component-i (>0), J/mol
L_n	mass of the liquid phase out of stage, n, g
N	an integer number (>2), dimensionless
P	pressure, Pa
R_C	calculated recovery ratio, dimensionless
R_f	experimental recovery ratio, dimensionless
S_n	mass of the solid phase out of stage, n, g
T	boiling temperature of component-i, K
$T_{b.i}$	boiling temperature of component-i, K
$T_{m.i}$	melting temperature of component-i, K
$T_{tri.i}$	triple-point temperature of component-i, K
V_n	mass of the vapor phase out of stage, n, g
X_i	mole fraction of component-i in melt, dimensionless
Y_i	mole fraction of component-i in vapor phase, dimensionless

Subscript

0	in the initial feed
n	in stage, n
N	in the last stage

References

1. Jacques, J.; Collet, A.; Wilen, S.H. *Enantiomers, Racemates, and Resolutions*; John Wiley & Sons: New York, NY, USA, 1981.
2. Subramanian, G. (Ed.) *Chiral Separation Techniques*; Wiley-VCH Verlag GmbH: Weinheim, Germany, 2000.
3. Elsner, M.P.; Ziomek, G.; Seidel-Morgenstern, A. Simultaneous preferential crystallization in a coupled batch operation mode. Part II: Experimental study and model refinement. *Chem. Eng. Sci.* **2011**, *66*, 1269–1284. [CrossRef]
4. Qamar, S.; Galan, K.; Elsner, M.P.; Hussain, I.; Seidel-Morgenstern, A. Theoretical investigation of simultaneous continuous preferential crystallization in a coupled mode. *Chem. Eng. Sci.* **2013**, *98*, 25–39. [CrossRef]
5. Kollges, T.; Vetter, T. Model-based analysis of continuous crystallization/reaction processes separating conglomerate forming enantiomers. *Cryst. Growth Des.* **2017**, *17*, 233–247. [CrossRef]
6. Majumder, A.; Nagy, Z.K. A comparative study of coupled preferential crystallizers for the efficient resolution of conglomerate-forming enantiomers. *Pharmaceutics* **2017**, *9*, 55. [CrossRef] [PubMed]
7. Kupai, J.; Rojik, E.; Huszthy, P.; Szekely, G. Role of chirality and macroring in imprinted polymers with enantiodiscriminative power. *ACS Appl. Mater. Interfaces* **2015**, *7*, 9516–9525. [CrossRef] [PubMed]
8. Tanaka, K.; Muraoka, T.; Oyubo, Y.; Takahashi, H.; Ohnishi, A. HPLC enantioseparation on a homochiral MOF-silica composite as a novel chiral stationary phase. *RSC Adv.* **2016**, *6*, 21293–21301. [CrossRef]

9. Han, X.; Huang, J.; Yuan, C.; Liu, Y.; Cui, Y. Chiral 3D covalent organic frameworks for high performance liquid chromatographic enantioseparation. *J. Am. Chem. Soc.* **2018**, *140*, 892–895. [CrossRef] [PubMed]
10. Bruno, R.; Marino, N.; Bartella, L.; Di Donna, L.; De Munno, G.; Pardo, E.; Armentano, D. Highly efficient temperature-dependent chiral separation with a nucleotide-based coordination polymer. *Chem. Commun.* **2018**, *54*, 6356–6359. [CrossRef] [PubMed]
11. Didaskalou, C.; Kupai, J.; Cseri, L.; Barabas, J.; Vass, E.; Holtzl, T.; Szekely, G. Membrane-grafted asymmetric organocatalyst for an integrated synthesis-separation platform. *ACS Catal.* **2018**, *8*, 7430–7438. [CrossRef]
12. Rukhlenko, I.D.; Tepliakov, N.V.; Baimuratov, A.S.; Andronaki, S.A.; Gun'ko, Y.K.; Baranov, A.V.; Fedorov, S.V. Completely chiral optical force for enantioseparation. *Sci. Rep.* **2016**, *6*, 36884. [CrossRef] [PubMed]
13. Nguyen, T.N.T.; Magueur, G.; Ourevitch, M.; Crousse, B.; Begue, J.-P.; Bonnet-Delpon, D. Analogues of Key Precursors of Aspartyl Protease Inhibitors: Synthesis of Trifluoromethyl Amino Epoxides. *J. Org. Chem.* **2005**, *70*, 699–702.
14. Babic, K.; Driessen, G.H.M.; van der Ham, A.G.J.; de Haan, A.B. Chiral separation of amino-alcohols using extractant impregnated resins. *J. Chromatogr. A* **2007**, *1142*, 84–92. [CrossRef] [PubMed]
15. Steensma, M.; Kuipers, N.J.M.; de Haan, A.B.; Kwant, G. Influence of process parameters on extraction equilibria for the chiral separation of amines and amino-alcohols with a chiral crown ether. *J. Chem. Technol. Biotechnol.* **2006**, *81*, 588–597. [CrossRef]
16. Schuur, B.; Steensma, M.; Winkelman, J.G.M.; De Vries, J.G.; De Haan, A.B.; Heeres, H.J. Continuous enantioseparation by liquid-liquid extraction. *Chim. Oggi-Chem. Today* **2009**, *27*, 9–12.
17. Shiau, L.D.; Wen, C.C.; Lin, B.S. Separation and purification of p-xylene from the mixture of m-xylene and p-xylene by distillative freezing. *Ind. Eng. Chem. Res.* **2005**, *44*, 2258–2265. [CrossRef]
18. Shiau, L.D.; Wen, C.C.; Lin, B.S. Application of distillative freezing in the separation of o-xylene and p-xylene. *AIChE J.* **2006**, *52*, 1962–1967. [CrossRef]
19. Shiau, L.D.; Wen, C.C.; Lin, B.S. Separation of p xylene from the multicomponent xylene system by stripping crystallization. *AIChE J.* **2008**, *54*, 337–342. [CrossRef]
20. Shiau, L.D. Purification of Styrene from a Styrene/Ethylbenzene Mixture by Stripping Crystallization. *Ind. Eng. Chem. Res.* **2018**, *57*, 6759–6765. [CrossRef]
21. Shiau, L.D.; Liu, K.F.; Hsu, Y.C. Chiral purification of S-ibuprofen from ibuprofen enantiomers by stripping crystallization. *Chem. Eng. Res. Des.* **2017**, *117*, 301–308. [CrossRef]
22. Ulrich, J. *Melt Crystallization: Fundamentals, Equipment and Applications*; Shaker: Aachen, Germany, 2003.
23. Jiang, X.; Hou, B.; He, G.; Wang, J. Falling film melt crystallization (I): Model development, experimental validation of crystal layer growth and impurity distribution process. *Chem. Eng. Sci.* **2012**, *84*, 120–133. [CrossRef]
24. Beierling, T.; Osiander, J.; Sadowski, G. Melt crystallization of isomeric long-chain aldehydes from hydroformylation. *Sep. Purif. Technol.* **2013**, *118*, 13–24. [CrossRef]
25. Micovic, J.; Beierling, T.; Lutze, P.; Sadowski, G.; Górak, A. Design of hybrid distillation/melt crystallization processes for separation of close boiling mixtures. *Chem. Eng. Process. Process Intensif.* **2013**, *67*, 16–24. [CrossRef]
26. Jiang, X.; Li, M.; He, G.; Wang, J. Research process and model development of crystal layer growth and impurity distribution in layer melt crystallization: A review. *Ind. Eng. Chem. Res.* **2014**, *53*, 13211–13227. [CrossRef]
27. Fukui, K.; Fujikawa, T.; Satone, H.; Yamamoto, T.; Maeda, K. Application of solute distribution theory to melt crystallization of fatty acids. *Chem. Eng. Sci.* **2016**, *143*, 114–121. [CrossRef]
28. Smith, J.M.; Van Ness, H.C.; Abbott, M.M. *Introduction to Chemical Engineering Thermodynamics*; McGraw-Hill Book Co.: Singapore, 2001.
29. Sandler, S.I. *Chemical, Biochemical, and Engineering Thermodynamics*; John Wiley & Sons: Hoboken, NJ, USA, 2006.
30. ONeil, M.J. *The Merck Index: An Encyclopedia of Chemicals, Drugs, and Biologicals*; Merck & Co., Inc.: Whitehouse Station, NJ, USA, 2006.

31. Myerson, A.S.; Kirwan, D.J. Impurity trapping during dendritic crystal growth. 2. Experimental results and correlation. *Ind. Eng. Chem. Fundam.* **1977**, *16*, 420–425. [CrossRef]
32. Myerson, A.S.; Kirwan, D.J. Impurity trapping during dendritic crystal growth. 1. Computer simulation. *Ind. Eng. Chem. Fundam.* **1977**, *16*, 414–420. [CrossRef]

Sample Availability: Samples of the compounds are not available from the authors

molecules

MDPI

Article

Improved Enantioselectivity for Atenolol Employing Pivot Based Molecular Imprinting

Andreea Elena Bodoki [1,†], Bogdan-Cezar Iacob [2,†], Laura Elena Gliga [2], Simona Luminita Oprean [1], David A. Spivak [3], Nicholas A. Gariano [3] and Ede Bodoki [2,*]

1 Department of Inorganic Chemistry, Faculty of Pharmacy, "Iuliu Haţieganu" University of Medicine and Pharmacy, 12 Ion Creangă St., Cluj-Napoca 400010, Romania; abota@umfcluj.ro (A.E.B.); lopream@umfcluj.ro (S.L.O.)

2 Department of Analytical Chemistry, Faculty of Pharmacy, "Iuliu Haţieganu" University of Medicine and Pharmacy, 4 Pasteur St., Cluj-Napoca 400349, Romania; iacob.cezar@umfcluj.ro (B.-C.I.); gliga.laura@umfcluj.ro (L.E.G.)

3 Department of Chemistry, Louisiana State University, Baton Rouge, LA 70803, USA; dspivak@lsu.edu (D.A.S.); nicholas.gariano13@gmail.com (N.A.G.)

* Correspondence: bodokie@umfcluj.ro; Tel.: +40-264-597-256 (ext. 2838)

† These authors contributed equally to this work.

Academic Editors: Maria Elizabeth Tiritan, Madalena Pinto and Carla Sofia Garcia Fernandes

Received: 3 June 2018; Accepted: 26 July 2018; Published: 27 July 2018

Abstract: In the last few decades, molecular imprinting technology went through a spectacular evolution becoming a well-established tool for the synthesis of highly selective biomimetic molecular recognition platforms. Nevertheless, there is still room for advancement in the molecular imprinting of highly polar chiral compounds. The aim of the present work was to investigate the favorable kosmotropic effect of a ternary complex involving a polar chiral template (eutomer of atenolol) and a functional monomer, bridged by a central metal ion through well-defined, spatially directional coordinate bonds. The efficiency of the chiral molecular recognition was systematically assessed on polymers obtained both by non-covalent and metal-mediated molecular imprinting. The influence on the chromatographic retention and enantioselectivity of different experimental variables (functional monomers, cross-linkers, chaotropic agents, metal ions, porogenic systems, etc.) were studied on both slurry packed and monolithic HPLC columns. Deliberate changes in the imprinting and rebinding (chromatographic) processes, along with additional thermodynamic studies shed light on the particularities of the molecular recognition mechanism. The best performing polymer in terms of enantioselectivity ($\alpha = 1.60$) was achieved using 4-vinyl pyridine as functional monomer and secondary ligand for the Co(II)-mediated imprinting of S-atenolol in the presence of EDMA as cross-linker in a porogenic mixture of [BMIM][BF_4]:DMF:DMSO = 10:1:5, *v*/*v*/*v*.

Keywords: metal-mediated molecular imprinting; hydrophilic template; atenolol; chiral separation; β-blockers; molecularly imprinted polymers; molecular recognition

---

1. Introduction

Biological or synthetic receptors selectively recognize their target chemicals based on a combination of weak, short-ranged intermolecular interactions, such as hydrogen bonding, π-π interactions and van der Waals forces; and their selectivity being further refined by additional repulsive steric confinements. Molecularly imprinted polymers (MIPs) able to mimic natural receptors, offer tailored selectivity towards target molecules and better chemical and thermal stability in a simple and cost-effective manner [1]. MIPs have been widely employed for the concentration, separation and analysis of various bioactives, either as stationary phases in chromatography and capillary electrophoresis [2–6], or as recognition elements in chemo- and biosensing [7–9]. Within the extensive body of literature, a considerable part focuses on the use of these polymers for chiral analysis [2,7,10–13] in various pharmaceutical, biomedical or environmental applications. Differentiation between the chiral forms of a molecule (chiral discrimination) is considered the supreme form of molecular recognition. Most often, the needed stereospecific features of MIPs are acquired through the process of non-covalent molecular imprinting using a variety of functional monomers, cross-linkers and porogenic solvents. Even though over the last few decades MIPs have become a well-established analytical tool for the selective recognition and analysis of small molecules, there is still some room for advancement in the molecular imprinting of highly polar compounds and biomacromolecules. Recent attempts at imprinting polar compounds such as polyphenols, e.g., oleuropein [14], polar organic micropollutants, e.g., benzotriazole [15] have been described, especially for sample enrichment applications (SPE adsorbents). The addition of certain additives (hydrophilic functional polymers-oligo- and polyethylene glycol methacrylate [16]; molecular crowding agents—polyethylene glycol; room temperature ionic liquids [17,18]) in the pre-polymerization mixture could improve the overall imprinting efficiency, hydrophilicity, flexibility, morphology and porosity of the resulting polymer. Furthermore, alternative imprinting protocols, such as metal ion-mediated molecular imprinting may further correct some of the observed shortcomings in the molecular imprinting and recognition of polar templates [16].

In the pivot-based or metal ion-mediated molecular imprinting (MMMI) process the metal ions act as a bridge between the functional monomer and the template. The monomers are thus regularly positioned around the template via coordinate bonds restraining the free motion of the species. Consequently, the number of non-specific binding sites decreases, and improved imprinting factors are achieved [19]. Since chiral molecular recognition by MIPs relies on very small energetic differences between the forming transient selector/select and complexes at the polymer's interaction sites, any improvement in the degree of order throughout the polymerization step is beneficial in conveying the template's (i.e., target enantiomer) molecular information to the emerging MIP recognition sites with the highest possible fidelity.

Metal pivot-based imprinted polymers designed for water soluble templates were previously reported by the enhancement of column permeability and affinity towards the polar template [16,18,20,21]. However, the number of metal-mediated imprinted polymers intended for chiral separation is rather scarce [16,17,22]. Various representatives of β-blockers, most often propranolol [23–26], have been employed as model compounds in demonstrating the enantioselectivity of various chiral selectors. Nevertheless, for atenolol (ATNL), as one of the most polar representatives of this class of drugs, usually the poorest enantioselectivity has been reported under optimized binding conditions [26–28].

The aim of the present work was to investigate the favorable kosmotropic effect of a ternary complex involving both the polar chiral template (eutomer of ATNL) and the functional monomer, bridged by the central metal ions through well-defined, spatially directional coordinate bonds. Various aspects of the formation of the ternary metal complex monitored by UV-Vis spectroscopy, as well as particularities regarding the molecular imprinting process (composition of pre-polymerization mixture, initiation of free-radical polymerization, etc.) are also discussed.

2. Results

2.1. Ternary Metal Complexes of ATNL

In order to engage a more rational approach in selecting the appropriate metal ion, functional monomer, molar ratio and porogenic solvent to be tested for subsequent molecular imprinting, UV-Vis spectroscopy provided a simple, fast, cost-effective and relatively straightforward instrumental method which is adjustable to small sample volumes for assessing the formation of the ternary complex. The electronic spectra (350–1100 nm) of binary and ternary complexes of several transition metal ions (Co(II), Cu(II) and Ni(II)) with the ATNL as primary ligand, and various functional monomers (4-vinyl pyridine (4-VPy), 1-vinyl imidazole (VIM), methacrylic acid (MAA), acrylamide (AM), *N,N′*-methylenebis(acrylamide) (BAM), *N,O*-bismethacryloyl ethanolamine (NOBE), vinyl ferrocene

(VFC), 4-vinyl phenylboronic acid (4-VPBA), trans-2-chloromethylvinylboronic acid (CVPBA) and poly(ethylene glycol) methyl ether methacylate (M_n = 300 g mol^{-1}, PEGMA) prepared in a mixture of DMF/DMSO 1:5 (v/v) were recorded.

Bathochromic and hypsochromic shifts or hyperchromic and hypochromic effects occurring in the electronic spectra of ternary metal-template-monomer species with respect to the spectra of binary metal-template analogues were interpreted as evidence of ternary complex formation. Co(II) ions were selected as the pivot for the MIPs using 4-Vpy as a monomer (Figure 1a,b), while Cu(II) ions were preferred for the imprinting process using an acidic monomer, MAA (Figure 1c).

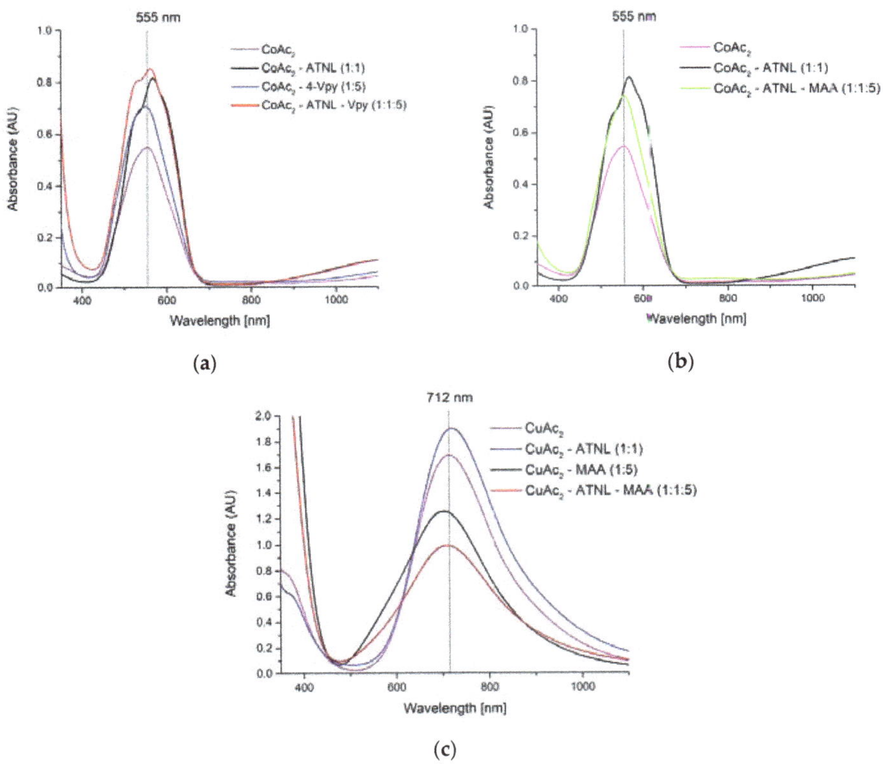

Figure 1. Electronic spectra of Co(II) complexes in DMF/DMSO 1:5 (v/v) with (**a**) binary and ternary Co(II) complexes with ATNL (1:1 molar ratio) and 4-VPy (1:5 molar ratio); (**b**) binary and ternary Co(II) complexes with ATNL (1:1 molar ratio) and MAA (1:5 molar ratio); (**c**) binary and ternary Cu(II) complexes with ATNL (1:1 molar ratio) and MAA (1:5 molar ratio).

2.2. *Preparation of MIPs*

2.2.1. Non-Covalent Molecular Imprinting

Based on the ability of ATNL to interact in a concerted manner with hydrogen-bonding monomers, choices of host functional monomers ranged from the most commonly used donor-acceptor type monomers, such as MAA, AM or VIM to the ones bearing a single hydrogen donor or acceptor motif, such as 4-VPy (Tables 1–3). Choices among the conventionally employed aprotic porogenic solvents for non-covalent imprinting (i.e., toluene, ACN) were often limited by the poor solubility of the highly polar template. Thus, the influence of different aprotic (ACN, DMF, DMSO) and protic (methanol (MeOH), butanol) porogenic solvents on the overall features of the resulting imprinted polymer were also studied. Concomitantly, the influence of different additives (i.e., cross-linkers, ionic liquids) meant to balance the polymeric framework's flexibility and porosity, as well as the selectivity of the imprinted binding sites, were investigated. Both bulk (S4) and monolithic (M6, M16-20) MIPs were obtained using this imprinting approach. Modest to no enantioselectivity (α = 1.0–1.32) was recorded, amongst which the monolithic polymer with 4-VPy as functional monomer (M6, α = 1.30), reticulated with EDMA in the presence of an ionic liquid ([BMIM]BF_4) in DMF/DMSO 1:5 (v/v) as porogenic media was considered the most promising candidate for further investigations by the MMMI approach. Comparable results were obtained employing MAA as a functional monomer in ACN as a porogenic solvent (M18, α = 1.32). The presence of a protic solvent such as MeOH or water, even in trace amounts, added to the solvation of the template in ACN compromised chiral recognition (M20, α = 1.00).

2.2.2. One-Monomer Molecularly Imprinted Polymer (OMNiMIP)

In an effort of reducing to a minimum the number of variables involved in the traditional non-covalent molecular imprinting, the OMNiMIP approach introduced by Sibrian-Vasquez and Spivak [29] was also tested by using a single crosslinking monomer, NOBE, in addition to the chiral template, solvent (DMF) and initiator (Table 1). Under these conditions, the obtained enantioselectivity was somewhat below expectations (S1, α = 1.05). However, when using ACN as porogen, in which case the presence of a small amount of MAA for the solubilization of ATNL was necessary, a certain improvement in enantioselectivity of the slurry packed column (S4, α = 1.17) was observed.

2.2.3. Metal Ion Mediated Molecular Imprinting

The protocols of molecular imprinting were adapted in compliance with the requirements of the metal ion-mediated approach. Three transition metal ions (Co(II), Cu(II) and Ni(II)) were employed as coordination centers (Tables 1–3). Initially, the use of metal ions alongside or together with the single crosslinking monomer (NOBE) did not provide high enough enantioselectivities (S2, α = 1.05; S3, α = 1.01). Furthermore, the use of different molar ratios of 4-VPy as secondary ligand alongside the cross-linker TRIM and porogenic mixture (DMF/DMSO 1:1, v/v) provided low chiral discrimination of the slurry-based columns (S6–10, α = 1.03–1.07). However, changing the crosslinker to EDMA, and adjusting the polymerization media by adding an ionic liquid ([BMIM]BF_4 in DMF/DMSO 1:5, v/v) resulted in a considerable enhancement in chiral selectivity (S11, α = 1.32). Eventually, switching from slurry-based to monolithic columns (with expected gain in column efficiency), alongside the concerted benefits of using MMMI and ionic liquid, led to the fabrication of the best performing MIP monolith (M2, α = 1.60). The optimal composition of the pre-polymerization mixture (Table 2, M2) turned out to be S-ATNL:Co(II):4-VPy (1:1:6 molar ratio), EDMA as cross-linker (C:M ratio = 1:4), [BMIM]BF_4 as ionic liquid in DMF/DMSO 1:5 (v/v) as porogenic system.

2.2.4. Bulk Imprinting vs. MIP Monolith

Initially, bulk imprinting with photochemical initialization at lower temperatures was considered (S1–S11, Table 1). After the removal of excess reagents and template (Figure 2a,b), the polymers were ground and sieved, followed by the slurry packing into HPLC columns. Studies continued on monoliths polymerized in situ in the chromatographic column (M1–24, Tables 2 and 3), but with thermal radical initiation. To further improve the molecular imprinting efficiency and future chromatographic performances of the monoliths, a hydrophilic ionic liquid was also added to the polymerization mixture [18]. Significant differences in the morphology and porosity of the monolith obtained in the presence and absence of the ionic liquid were observed (Figure 2a–d).

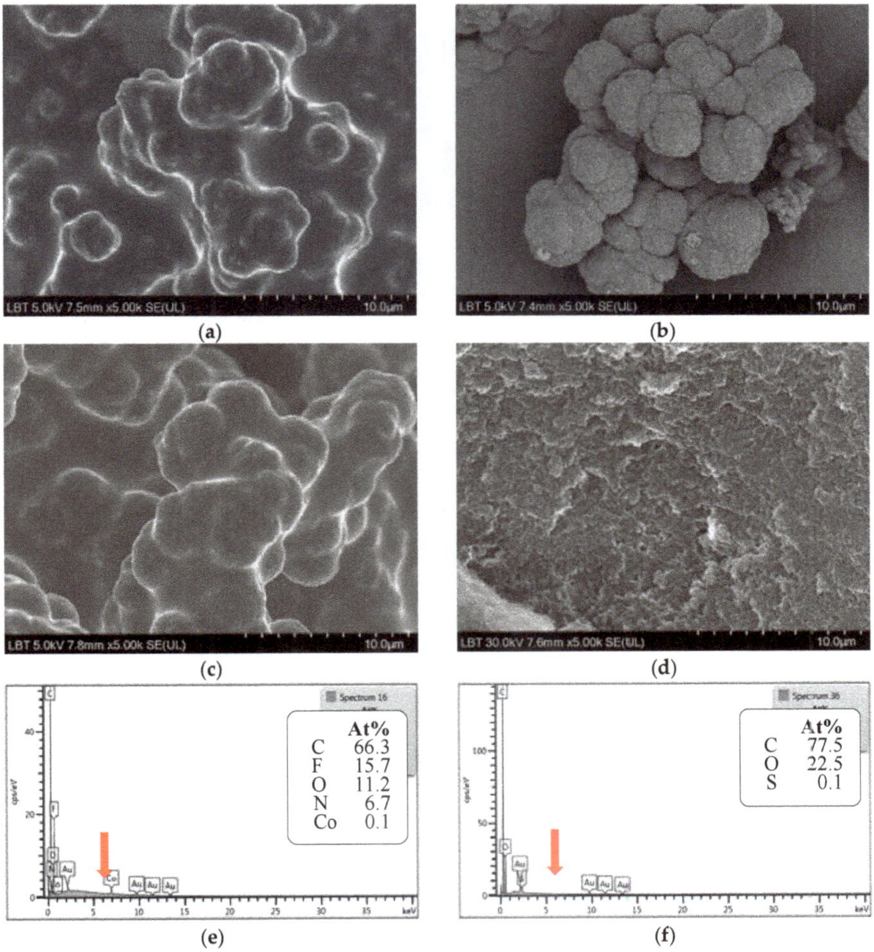

Figure 2. Scanning electronmicrographs of monolithic imprinted polymers of M2 (**a**) before and (**b**) after MeOH:AcOH 9:1(v/v) washing, (**c**) M8 and (**d**) M5 before MeOH:AcOH 9:1(v/v) washing. Elemental mapping spectrum of M2 (**e**) before and (**f**) after template removal indicating the washout of the metal pivot ion (Co(II)).

2.3. Chromatographic Evaluation of the MIPs

The enantioselectivity of MIPs investigated as HPLC stationary phases usually exhibits good interassay reproducibility and adequate efficiency with a high sensitivity of detection. Therefore, the present study used standard HPLC columns (100 × 2.1 mm), either slurry packed with the imprinted polymer samples or as a monolith, to study their distinctive chromatographic behavior and recognition mechanism. The assessment and comparison of imprinting factors ($IF = k'_{MIP}/k'_{NIP}$) was avoided because the observed binding differences are not exclusively due to the existence of specific imprinted cavities, but also due to the significant differences in the morphology (shape, texture, rigidity, porosity, surface area) of the imprinted (MIP) and non-imprinted polymer (NIP). Thus, the efficacy of molecular recognition was assessed based on the highest enantioselectivity achieved for ATNL's enantiomers under different chromatographic conditions (mobile phase composition). The results are synthetically presented in Tables 1–3, along with some of the most representative chromatograms (Figure 3).

Table 1. Chromatographic retention (k′) and enantioselectivity (α) of various MIP-CSPs tested in slurry packed columns (S1–11).

#	T (mmol)	M(s) (mmol)	Me (mmol)	C (mmol)	Molar Ratio T:M:Me:C	A (mmol)	P (mL)	k'_S/k'_R	α
S1	S-ATNL (0.3)	NOBE (7.6)	-	-	1:25:-:-	-	DMF (4)	33.2/31.6	1.05 [a]
S2	S-ATNL (0.3)	NOBE (7.6)	Cu(II) (0.3)	-	1:25:1:-	-	DMF (4)	33.2/31.6	1.05 [a]
S3	S-ATNL (0.3)	NOBE (7.6)	Co(II) (0.3)	-	1:25:1:-	-	DMF (4)	36.8/36.3	1.01 [a]
S4	S-ATNL (0.3)	NOBE/MAA (7.6/1)	-	-	1:25/3:-:-	-	ACN (2)	4.4/3.8	1.17 [b]
S6	S-ATNL (0.3)	4-VPy (0.3)	Co(II) (0.3)	TRIM (7.3)	1:1:1:24	-	DMF/DMSO (2/2)	3.6/3.5	1.03 [a]
S7	S-ATNL (0.3)	4-VPy (1.5)	Co(II) (0.3)	TRIM (6.1)	1:5:1:20	-	DMF/DMSO (2/2)	8.2/7.9	1.04 [a]
S8	S-ATNL (0.3)	4-VPy (3)	Co(II) (0.3)	TRIM (4.6)	1:10:1:15	-	DMF/DMSO (2/2)	8.2/7.9	1.04 [a]
S9	S-ATNL (0.3)	4-VPy/AM (0.3/1.2)	Co(II) (0.3)	TRIM (6.1)	1:1/4:1:20	-	DMF/DMSO (2/2)	0.5/0.5	1.00
S10	S-ATNL (0.3)	4-VPy/BAM (0.3/1.2)	Co(II) (0.3)	TRIM (6.1)	1:1/4:1:20	-	DMF/DMSO (2/2)	0.6/0.5	1.07 [a]
S11	S-ATNL (0.15)	4-VPy (0.9)	Co(II) (0.15)	EDMA (3.6)	1:6:1:24	IL (1.235)	DMF/DMSO (0.12/0.6)	16.2/12.3	1.32 [a]

Mobile phase composition: [a] ACN; [b] ACN:50 mM formate buffer, pH 3 = 86:14, *v*/*v*. In a typical thermal (60 °C) or photo-induced polymerization (24 h), the pre-polymerization mixture also includes 15 mg of AIBN as free radical initiator. T—template, M—monomer, Me—metal ion, C—cross-linker, A—additive, P—porogen, IL—[BMIM]BF_4.

Table 2. Chromatographic retention (k′) and enantioselectivity (α) of various MIP-CSPs tested in monolithic columns (M1–10) with 4-VPy as functional monomer.

#	T (mmol)	M(s) (mmol)	Me (mmol)	C (mmol)	Molar Ratio T:M:Me:C	A (mmol)	P (mL)	k'_S/k'_R	α
M1	S-ATNL (0.2)	4-VPy (1)	Co(II) (0.2)	TRIM (6)	1:5:1:30	IL (0.9)	DMF/DMSO (0.9/0.9)	4.3/4.3	1.00
M2	S-ATNL (0.15)	4-VPy (0.9)	Co(II) (0.15)	EDMA (3.6)	1:6:1:24	IL (1.235)	DMF/DMSO (0.12/0.6)	3.7/2.3	1.60 [a]
M3	S-ATNL (0.15)	4-VPy (0.9)	Ni(II) (0.15)	EDMA (3.6)	1:6:1:24	IL (1.235)	DMF/DMSO (0.12/0.6)	9.4/6.8	1.38 [a]
M4	S-ATNL (0.15)	4-VPy (0.9)	Cu(II) (0.15)	EDMA (3.6)	1:6:1:24	IL (1.235)	DMF/DMSO (0.48/0.6)	6.8/5.3	1.30 [a]
M5	S-ATNL (0.15)	4-VPy (0.9)	Co(II) (0.15)	EDMA (3.6)	1:6:1:24	- *	DMF/DMSO (0.326/1.630)	6.3/5.8	1.08 [a]
M6	S-ATNL (0.15)	4-VPy (0.9)	-	EDMA (3.6)	1:6:-:24	IL (1.235)	DMF/DMSO (0.12/0.6)	3.4/2.6	1.30 [a]
M7	S-ATNL (0.15)	4-VPy (0.9)	Co(II) (0.15)	EDMA (3.6)	1:6:1:24	IL/PEGMA (1.235/0.17)	DMF/DMSO (0.12/0.6)	4.0/4.0	1.00
M8	-	4-VPy (0.9)	Co(II) (0.15)	EDMA (3.6)	-:6:1:24	IL (1.235)	DMF/DMSO (0.12/0.6)	2.7/2.6	1.04 [a]
M9	-	4-VPy (0.9)	-	EDMA (3.6)	-:6:-:24	IL (1.235)	DMF/DMSO (0.12/0.6)	10.4/10.4	1.00
M10	S-ATNL (0.05)	4-VPy (0.15)	Ni(II) (0.05)	TRIM (1.15)	1:3:1:23	-	MeOH (2)	0.3/0.3	1.00

Mobile phase composition: [a] ACN. * Replaced with the corresponding volume of DMF/DMSO 1:5, *v*/*v*. In a typical thermal (60 °C) or photo-induced polymerization (24 h), the pre-polymerization mixture also includes 15 mg of AIBN as free radical initiator. T—template, M—monomer, Me—metal ion, C—cross-linker, A—additive, P—porogen, IL—[BMIM]BF_4.

Table 3. Chromatographic retention (k′) and enantioselectivity (α) of various MIP-CSPs tested in monolithic columns with 4-VPy and various co-monomers (M11–13), or different functional monomers, other than 4-VPy (M14–24).

#	T (mmol)	M(s) (mmol)	Me (mmol)	C (mmol)	Molar Ratio T:M:Me:C	A (mmol)	P (mL)	k'_S/k'_R	α
M11	S-ATNL (0.15)	4-VPy/AM (0.15/0.75)	Co(II) (0.15)	EDMA (3.6)	1:1/5:1:24	IL (1.235)	DMF/DMSO (0.12/0.6)	12.9/10.0	1.29 [a]
M12	S-ATNL (0.15)	4-VPy/4-PBA (0.9/0.15)	Co(II) (0.15)	EDMA (3.6)	1:6/1:1:24	IL (1.235)	DMF/DMSO (0.12/0.6)	4.6/4.4	1.04 [a]
M13	S-ATNL (0.15)	4-VPy/ CVPBA (0.9/0.15)	Co(II) (0.15)	EDMA (3.6)	1:6/1:1:24	IL (1.235)	DMF/DMSO (0.12/0.6)	0.4/0.4	1.00
M14	S-ATNL (0.15)	1-VIM (0.9)	Co(II) (0.15)	EDMA (3.6)	1:6:1:24	IL (1.235)	DMF/DMSO (0.12/0.6)	6.2/4.3	1.43 [a]
M15	S-ATNL (0.15)	MAA/AM (0.15/0.75)	Cu(II) (0.15)	EDMA (3.6)	1:1/5:1:24	IL (1.235)	DMF/DMSO (0.12/0.6)	2.5/2.2	1.14 [b]
M16	S-ATNL (0.15)	MAA/AM (0.15/0.75)	-	EDMA (3.6)	1:1/5:-:24	IL (1.235)	DMF/DMSO (0.12/0.6)	15.3/14.5	1.06 [b]
M17	S-ATNL (0.13)	MAA (0.7)	-	EDMA (0.7)	1:5:-:5	-	ACN (2)	1.8/1.5	1.15 [c]
M18	S-ATNL (0.13)	MAA (0.7)	-	EDMA (0.7)	1:5:-:5	IL (1.235)	ACN (0.72)	6.2/4.7	1.32 [d]
M19	S-ATNL (0.13)	MAA (0.7)	-	PETRA(0.7)	1:5:-:5	IL (1.235)	ACN (0.72)	8.7/7.8	1.11 [e]
M20	S-ATNL (0.2)	VFC (0.1)	-	TRIM (1.25)	2:1:-:12.5	-	ACN (1%H_2O) (5)	0.3/0.3	1.00
M21	S-ATNL (0.2)	1-VIM (0.6)	Ni(II) (0.2)	TRIM (1.25)	1:3:1:6	-	MeOH (5)	0.9/0.9	1.00
M22	S-ATNL (0.04)	1-VIM (0.04)	Ni(II) (0.04)	PETEA (0.25)	1:1:1:6	-	MeOH (1)	0.5/0.5	1.00
M23	S-ATNL (0.02)	1-VIM/BAM (0.02/0.08)	Cu(II) (0.02)	-	1:1/4:1:-	-	MeOH	2.4/2.4	1.00
M24	S-ATNL (0.2)	1-VIM/MAA (0.6/0.2)	Cu(II) (0.2)	PETRA (3.6)	1:3/1:1:18	-	BuOH (5)	3.6/3.5	1.03 [c]

Mobile phase composition: [a] ACN; [b] ACN:50 mM formate buffer, pH 3 = 95:5, *v*/*v*; [c] ACN:50 mM formate buffer, pH 3 = 90:10, *v*/*v*; [d] ACN:50 mM formate buffer, pH 3 = 80:20, *v*/*v*; [e] ACN:50 mM acetate buffer, pH 5 = 80:20, *v*/*v*. In a typical thermal (60 °C) or photo-induced polymerization (24 h), the pre-polymerization mixture also includes 15 mg of AIBN as free radical initiator. T—template, M—monomer, Me—metal ion, C—cross-linker, A—additive, P—porogen, IL—[BMIM]BF_4.

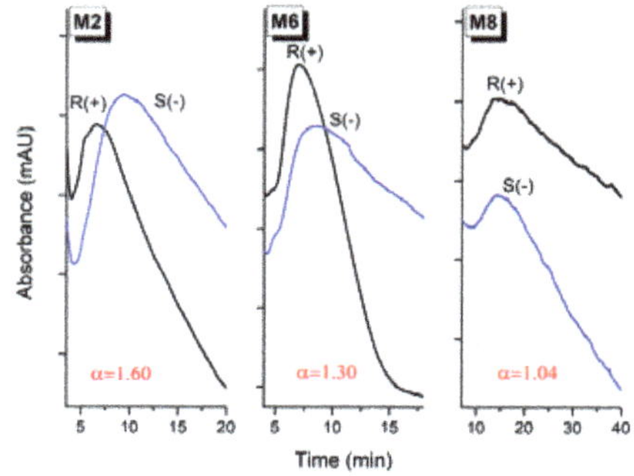

Figure 3. Changes in chromatographic selectivity of metal ion-mediated imprinted (M2), non-covalent imprinted (M6) and non-imprinted (M8) monolithic polymer columns. R(+)—R enantiomer of ATNL, S(−)—S enantiomer of ATNL. For chromatographic conditions see footnote of Table 2.

To gain better insight into the prevalent chromatographic retention mechanism, the influence of the imprinting approach and polymerization mixture constituents, in addition to the mobile phase composition (organic solvent; ratio, nature and pH of the aqueous buffer) on the polymers' retention properties were investigated. The nature of the metal ion used in the pivot-based molecular imprinting approach had an important effect on the monolith enantioselectivity (Figure 4a). In 4-VPy-based polymers selectivity is mainly controlled by hydrogen bonding interactions, which are disrupted even by minute amounts of protic solvent (i.e., MeOH, isopropanol, water) added to the mobile phase. In MAA-based polymers, the partition equilibrium is controlled both by ion-exchange and hydrogen bonding interactions, their contribution being dependent on the ratio and pH of the aqueous buffer. Thermodynamic retention studies (Figure 4b) performed on Co(II)-mediated 4-VPy-based imprinted (M2) and non-imprinted (M8) polymers indicate that the binding of ATNL enantiomers involves an important component of the enthalpic (e.g., hydrogen bonding) term.

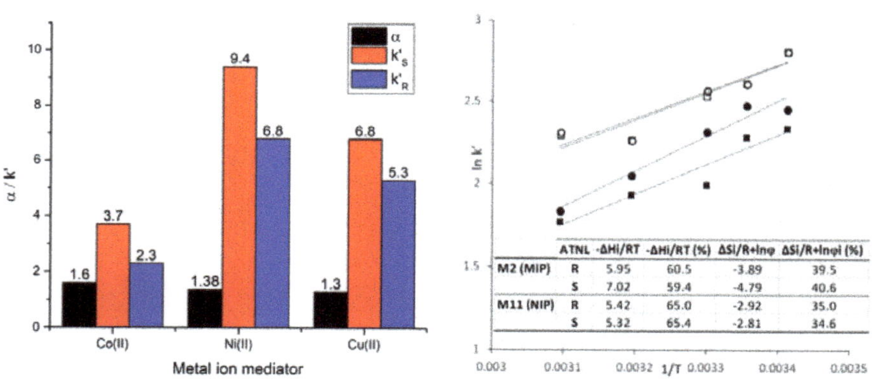

(**a**) (**b**)

Figure 4. (**a**) Enantioselectivity (α) and chromatographic retention (k'_S/k'_R) of metal ion-mediated imprinted monoliths (M2—Co(II), M3—Ni(II), M4—Cu(II)). (**b**) Thermodynamic retention study of ATNL's enantiomers on metal ion-mediated molecularly imprinted (M2, ●—S-ATNL, ■—R-ATNL) and non-imprinted (M8, ○—S-ATNL, □—R-ATNL) polymers.

3. Discussion

3.1. Ternary Metal Complexes of ATNL

ATNL, (R,S)2-(4-[2-Hydroxy-3-(isopropylamino)propoxy]phenyl)acetamide (Figure 5), is a beta-adrenergic antagonist, a cardioselective drug with a prolonged effect. Its S(-) stereoisomer exhibits a significantly higher affinity for the β_1-adrenergic receptors [30]. ATNL is a good chelating agent that may act as a bidentate ligand through the secondary alcohol and amine as electron pair donor moieties and allows the formation of five membered rings that include the central metal ion [31–33]. It is also noteworthy that one of the functional groups typically involved in metal ion coordination, the secondary alcohol moiety, is bound to the ATNL's chiral center. Mononuclear or binuclear binary complexes where the metal to ligand (ATNL) ratio is 1:1, 1:2 or 1:4 have been reported for several first-row transition metal ions (Me = Co(II), Ni(II), Cu(II), Zn(II)) [31,32].

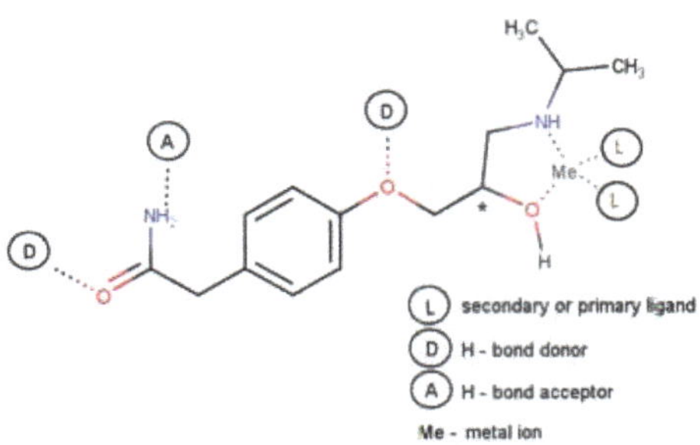

Figure 5. Potential interaction sites (coordinate and/or hydrogen bonding) of ATNL during non-covalent and MMMI.

Mononuclear tetrahedral complexes of $[Me(ATNL)_2]^{2+}$ type with ATNL acting as bidentate, mononuclear octahedral complexes of $[Me(ATNL)_4]^{2+}$ type [31,34] where two ATNL molecules act as bidentate and the ATNL molecules in the axial position act as monodentate, or an O-bridged binuclear complex, $[Cu_2(ATNL)_2Cl_2]$, where ATNL acts as (O, NH) bridging ligand [33], have been reported for first row transition metal ions (Me = Co(II), Ni(II), Cu(II), Zn(II)).

Ternary complexes of ATNL and ligands with N and O atom donor sites from amine and carboxylic moieties have also been reported for first-row transition metal ions [31,35]. The stability of mixed ligand complexes depends on the characteristics of the approaching secondary ligand (e.g., chelating properties, size and spatial configuration of chelate ring etc.), but also on the possible interactions outside the coordination sphere (hydrogen bonding between coordinated ligands, charge neutralization, chelate effect, and electrostatic interaction between non-coordinated charged moieties of ligands). Ternary complexes of metal ion—primary ligand—secondary ligand in the ratio of 1:1:1, 1:1:2 and 1:2:1, where ATNL was either the primary ligand or the secondary ligand, were obtained. Results indicated the preferential formation of ternary complexes in the 1:1:2 ratio over binary complexes for Co(II) as coordination center [31,35].

Thus, one of the hypotheses of the present study relied on the potential engagement of the hydroxyl group linked to ATNL's chiroptic center (most likely in its non-deprotonated form [31,33]) in a specific, spatially well-oriented interaction with an appropriate functional monomer mediated by a central metal ion through coordinate bonds. Such a metal ion-mediated molecular self-assembly of the template-monomer in the polymerization mixture should promote beneficial effects in the chiral molecular imprinting process of this hydrophilic enantiomer.

A favorable outcome using MMMI for the synthesis of enantioselective polymers can only occur if in the pre-polymerization step a stable, well-defined and soluble ternary metal complex exists. If in noncovalent imprinting a nonpolar, aprotic porogen, such as toluene or chloroform, is the ideal choice for promoting template-functional monomer associations [19], then in MMMI polar solvents (i.e., DMF, DMSO, MeOH) are required. Polar media may concomitantly offer the prerequisites of a favorable preorganized state (soluble ternary coordination complex): (i) Solvation of the metal ion and polar template and (ii)deprotonation of ligands (both the chiral template and monomer [36]) essential for metal coordination and (iii)dissolution of the resulting ternary metal complex. The stronger and more-defined interactions within the ternary complex are anticipated to lead to more specific recognition sites upon molecular imprinting. Contingent on the nature of the metal ion, its affinity towards the template and the selected monomer and molar ratio of ligands, different ternary metal complexes may also arise. Therefore, the success or failure of MMMI depends on the state of equilibrium established between these coordination complexes.

In the electronic spectra of transition metal complexes d-d transitions, charge transfer transitions, internal ligand transitions, combination and overtone vibrations of the ligands, and intervalence charge transfer transitions, materialize as bands in the region that spans the near infrared, visible and UV region (4000–30,000 cm^{-1}). Bathochromic and hypsochromic shifts or hyperchromic and hypochromic effects occurring in the electronic spectra of ternary metal-template-monomer species with respect to the spectra of binary metal-template analogues is an indication of a change in the ligand field environment around the coordination center [37–39] for the co-existing species in equilibrium. Based on the data provided by the electronic spectra, Co(II) ions (λ_{max} = 555 nm) were selected as pivot for the MIPs using 4-VPy as a monomer (Figure 1a), while Cu(II) ions were preferred for the imprinting process using an acidic monomer, MAA (Figure 1c).

The electronic spectra of the binary mixtures of ATNL and Co(II), Ni(II) and Cu(II), respectively, recorded in a DMF:DMSO (1:5, v/v) exhibited relatively weak, low-energy bands in the 13,330–19,230 cm^{-1} (750–520 nm) range which could be assigned to the d-d electronic transitions in a distorted octahedral or tetrahedral coordination environment. The electronic spectra also exhibited higher intensity bands at higher energy regions of the spectra that may be assigned to internal ligand transitions or d–π*, L→M or M→L charge transfer bands. The collected data are in agreement with the bands observed in the electronic spectra of previously reported Co(II), Ni(II) or Co(II) complexes of ATNL [31–33].

The d-d electronic transitions translate into two low intensity bands in the electronic spectrum of the binary Co(II)–ATNL mixture, one centered at 17,605 cm^{-1} (568 nm) with a shoulder at 18,868 cm^{-1} (530 nm) and the other at a lower energy region 9090 cm^{-1} (around 1100 nm) (Figure 1a). Upon addition of 4-VPy to the binary mixture, the local coordination environment around the Co(II) ion changes, and this translates into a broader multiple structured band centered at 18,348 cm^{-1} (545 nm) that appears in the spectra of the ternary mixture. The hypsochromic shift of the bands is associated to a hyperchromic effect for the shoulder initially at 18,868 cm^{-1} (530 nm) and to a hypochromic effect for the band initially at 17,605 cm^{-1} (568 nm) (Figure 1a).

Nevertheless, the spectra of binary mixtures of Cu(II)–ATNL and Ni(II)–ATNL changes to a much lesser extent upon the addition of 4-VPy (data not shown). In these cases, it appears that the primary ligand, ATNL, gives rise to a more stable binary complex and that the local environment around Cu(II) and Ni(II), respectively, is significantly less influenced by the presence of the secondary ligand (data not shown).

Acidic monomers such as MAA interact with the Brönsted-basic template ATNL, and the protonation of the amine moiety may alter the beta-blocker's chelating properties. Such a phenomenon was observed in case of the binary Co(II)-ATNL mixture. The local environment around the Co(II) ion changes significantly upon addition of MAA. As shown by the electronic spectra, ATNL is apparently displaced from the coordination sphere of Co(II) (Figure 1b). In contrast, the electronic spectra indicates no alteration of the local environment around Cu(II) when the acidic monomer is added to the binary Cu(II)–ATNL binary mixture (data not shown).

3.2. Preparation of MIPs

3.2.1. Non-Covalent Molecular Imprinting

ATNL has several hydrogen donor and acceptor atoms (Figure 5) which during the non-covalent molecular imprinting could interact in a concerted manner with hydrogen-bonding monomers. Choices of host functional monomers ranged from the most commonly used donor-acceptor type monomers, such as MAA, AM or VIM to the ones bearing a single hydrogen donor or acceptor motif, such as 4-VPy. Furthermore, being a Brönsted-basic template, during the ATNL interaction with MAA a partial or full proton transfer is expected to occur. Based on the nature of the employed porogen, contact hydrogen bonded assemblies may be formed in aprotic solvents such as ACN. Nevertheless, polar protic solvents tend to disrupt such electrostatic interactions having a negative impact on the imprinting factors, but also on the rebinding mechanism of the resulting imprinted material [40]. Different additives (i.e., cross-linkers, ionic liquids [36]) meant to balance the polymeric framework's flexibility and porosity, as well as the selectivity of the imprinted binding sites, were used in the pre-polymerization mixture (Tables 1–3). Modest to no enantioselectivity (α = 1.0–1.32) was recorded, both for the slurry-packed and monolithic columns. Monoliths with different functional monomers (i.e., 4-VPy, MAA) and significantly different porogenic media (DMF/DMSO, ACN) were able to provide similar enantioselectivities (M6, α = 1.30; M18, α = 1.32), as long as the cross-linker and ionic liquid were identical in the polymerization mixture. The ratio of monomer to cross-linker seems to be of less importance in this imprinting approach; however, the presence of a protic solvent such as MeOH or water, even in trace amounts, can compromise the polymer's enantioselectivity for the hydrophilic template. The non-covalent imprinting of ATNL's enantiomer by the OMNiMIP approach, at least under the tested experimental conditions, failed to provide noteworthy results in chiral chromatographic selectivity. One of the possible reasons for the low enantioselectivity recorded in the case of the NOBE-based columns may be the significant swelling effect observed during the sieving of the crushed polymer that may affect the structural integrity of the imprinted cavities.

3.2.2. Metal Ion-Mediated Molecular Imprinting

Particular requirements are to be met for MMMI in terms of components of the pre-polymerization mixtures: polar porogenic solvent (DMF, DMSO, DMF/DMSO) that must provide the solubilization of the metal ion's salt (anhydrous acetates) while keeping the ternary complex in solution, and metal ions that should not compromise the efficiency of the employed free radical initiator (AIBN). Another concern when imprinting polar templates is the chemical and mechanical structure of the polymeric network. A good balance between the polymer's hydrophilicity and its flexibility determined by the nature of the functional monomer and cross-linker must be optimized. Various additives (i.e., cross-linkers, co-monomer, ionic liquids, chaotropic agents) affecting the prototropic forms or the basicity of participating ligands (ATNL enantiomer and functional monomer) may also be decisive in the formation of the ternary metal complex; therefore, any change in the pre-polymerization mixture had to be carefully considered (Tables 1–3).

Metal Ion and Functional Monomer (Secondary Ligand)

Following the effect of the nature of metal ion on the efficiency of pivot-based imprinting using 4-VPy as a secondary ligand; results show a decrease in the enantioselectivity of the resulting monoliths in the order Co(II) $\gg$ Ni(II) $\geq$ Cu(II) (α = 1.60, 1.38 and 1.30, respectively for M2, M3 and M4 respectively). These findings are correlated with the electronic spectra of the metal complexes recorded in the screening step (Section 2.1) that suggest a higher stability of the Co(II) ternary complex (Figure 1a). Since most of the metal ions from the polymeric framework are eliminated during the template removal process (Figure 2e,f), they are no longer involved in the molecular recognition during rebinding. Nevertheless, the chromatographic retention (k') is inversely correlated with the recorded enantioselectivities (Figure 4a); this is most probably due to the higher number of non-specific binding sites emerging when Ni(II) and Cu(II) are employed as mediators. Obviously, changing the nature of the secondary ligand is also critical for the chromatographic performance of the resulting chiral stationary phases (CSP). Keeping Co(II) as a pivot and using 1-VIM as a basic secondary ligand bearing the same electron donor moiety as 4-VPy; selectivity decreases to 1.43 (M14). Evidently, in principle, other metal ion–secondary ligand combinations could equal and possibly surpass the selectivity recorded for the S-ATNL:Co(II):4-VPy (1:1:1)-based monolith. Selecting the "right" metal ion-ligand pair is a matter of a rational choice where the number of potential combinations can be significantly narrowed down by a preliminary spectroscopic screening. Therefore, other potentially promising combinations of metal-secondary ligands (Cu(II)–MAA, Cu(II)–1-VIM, Ni(II)–1-VIM) using various polymerization mixture constituents (cross-linkers, porogenic solvent, with and without ionic liquid) were also tested for the imprinting of S-ATNL, but in all cases suboptimal enantioselectivities were recorded (M10, M15, M21–24, α = 1.00–1.14) as compared to the best performing M2 column (α = 1.60). Unfortunately, even if UV-Vis spectroscopy is able to indicate some promising metal-secondary ligand combinations as starting points, assessing the optimal MMMI conditions is far from being a straightforward process. In addition to the formation of the best stable and soluble ternary metal ion-mediated complex, all the other constituents of the polymerization mixture will collectively play a critical part in the final chromatographic outcome, namely enantioselectivity.

It must be stressed that in the absence of the metal ion (i.e., Co(II)), molecular imprinting is achieved by the conventional non-covalent approach, with a poorer performance in terms of molecular recognition (M6, α = 1.30).

Functional Co-Monomers

Other co-monomers (i.e., AM, BAM, 4-PBA, CVPBA) added alongside the secondary ligand (4-VPy) had a negative effect on enantioselectivity (S9–10 and M11–13, α = 1.00–1.29). The interaction between co-monomers may reduce to a large extent the binding interactions with the template [41] and may also increase the heterogeneity of the resulting binding sites. Adding hydrophilic macromonomers to the polymerization mixture was reported as another convenient strategy to enhance the imprinting factor of polymers in case of water soluble templates [16]. In our case, instead of further boosting the hydrophilicity of the polymeric network, the addition of poly(ethyleneglycol) methyl ether methacrylate (PEGMA) fully compromised enantioselectivity (M7, α = 1.00).

Cross-Linker

Using EDMA instead of TRIM as a cross-linker turned out to be decisive for a more favorable polymer morphology and improved chiral selectivity, regardless of the type of column used (slurry packed–S10, α = 1.07 and S11, α = 1.32; monolith, M1, α = 1.00 and M2, α = 1.60). The beneficial effect of the ionic liquid on the imprinting efficiency was also unequivocally demonstrated by the selectivity value of column M2, α = 1.60 vs. column M5, α = 1.08. Moreover, chromatographic retention is inversely correlated with the recorded selectivity factors in the presence (M2, k'_S = 3.7) and absence M5, k'_S = 6.3) of [BMIM]BF_4 most probably due to the increased stability of the ternary complex.

Ionic Liquid

The presence of the ionic liquid translates into a spectacular change in the morphology of the imprinted material. A highly porous, globular polymeric framework (Figure 2b, M2) is obtained in the presence of the ionic liquid, while a much denser structure is observed in its absence (Figure 2d, M5). Before template removal, a film-like structure of excess reagents covered the outer surface of the polymer (Figure 2a, M2); while upon washing with MeOH:AcOH 9:1 (v/v) a highly indented surface was revealed (Figure 2b, M2).

3.2.3. Bulk Imprinting vs. MIP Monolith

MIPs are frequently prepared by bulk polymerization, in which case the obtained polymer block is ground and sieved before use. Bulk imprinting is more suitable for photochemical initiation at lower temperatures, which in theory should provide more homogenous and better defined imprinted sites. However, obtaining CSPs using this approach is cumbersome, wastes large amounts of material, and is often accompanied by the physical damage of some of the binding cavities. Furthermore, the resulting highly irregular MIP particles with polydisperse granulometry dramatically reduce the efficiency of the HPLC columns packed with such materials [42]. Imprinted monoliths polymerized directly in the chromatographic column offer a much simpler and long known alternative [43], at the cost of being only compatible with thermal radical initiation and only amenable to a few porogens. Nevertheless, such a continuous, but highly porous polymeric bed should in principle provide lower backpressure and greatly reduced mass transfer resistance, and thus higher chromatographic efficiency in comparison with the packed MIP-based columns. Room temperature ionic liquids (RTILs), demonstrating excellent solvation features, represent a more environment-friendly alternative as solvents for the preparation of MIPs [44]. As a result of their particular structure, they tend to promote self-assembly and improve specific template–monomer interactions, thus limiting non-specific binding [44,45]. In addition to offering higher imprinting factors and faster polymerization rates, they seem to reduce polymer swelling and boost the permeability of imprinted polymers, including monoliths [46]. Thus, in the search for the optimal experimental conditions that provide the highest enantioselectivity for ATNL's enantiomers, in our study both imprinting approaches were tested, in the presence or absence of RTIL.

Initially anticipating better molecular imprinting results, bulk imprinting with photochemical initialization at lower temperatures was investigated. Upon polymer grinding and sieving (particle size 25–38 μm) it was expected that all slurry-packed particulate columns demonstrate similar properties in terms of flow, back-pressure and sample load. Given the simplicity and earlier success of the OMNiMIP approach in chiral separation [47], N,O-bismethacryloyl ethanolamine (NOBE) was tested as a single cross-linking monomer for the imprinting of ATNL's enantiomers. At first (S1-3, Table 1), pure DMF was selected as porogenic solvent which enabled the solvation of the template enantiomer (S-ATNL). Unfortunately, neither the non-covalent (only NOBE), nor the metal ion-mediated (NOBE/Cu(II); NOBE/Co(II) imprinting approach gave notable enantioselectivity ($\alpha < 1.04$) in any of the tested mobile phases. Nevertheless, adding a small amount of MAA to the NOBE-based polymerization mixture (S4, Table 1) enabled the solubilization of the template in acetonitrile (porogen), and endowed a certain degree of enantioselectivity ($\alpha < 1.17$) of the resulting MIP using acetonitrile/formate buffer, pH 3.0 (85:15, v/v) as mobile phase. Seeking for improvements in the chiral recognition of the synthesized polymers, other functional monomers able to form ternary metal complexes with ATNL's enantiomer were screened by UV-Vis spectroscopy. One promising monomer candidate, forming soluble ternary metal (Cu(II), Co(II), Ni(II)) complexes with the target enantiomer, is 4-VPy (Figure 1). Therefore, different polymerization mixtures (S6–10, Table 1) combining various molar ratios of the template (T), monomer (M), metal ion (Me) and TRIM as cross-linker (C) were screened by metal ion-mediated bulk imprinting using photochemical initiation in a mixture of DMF:DMSO (1:1, v/v) as porogenic solvent. Unfortunately, none of the slurry-packed columns filled with S6-10 imprinted polymers exhibited noteworthy chiral discrimination ($\alpha = 1.00$–1.04).

Seeking higher chromatographic efficiencies potentially able to distinguish between the enantiomers of ATNL, studies continued on monoliths polymerized in the chromatographic column using the same molecular imprinting approaches (M1–24, Tables 2 and 3), but with thermal radical initiation. The polymerization mixture was further adapted, exploiting the favorable influence of RTILs on the imprinting process and on the future chromatographic performances of the resulting continuous polymeric bed. Therefore, a hydrophilic ionic liquid, namely 1-butyl-3-methylimidazolium tetrafluoroborate ([BMIM][BF_4]), was added to the porogenic solvent system. A higher ratio of IL/porogenic solvent system seem to favor the imprinting efficiency; thus finally a mixture of [BMIM][BF_4]:DMF:DMSO = 10:1:5, $v/v/v$ was selected [16,17]. Moreover, TRIM was replaced by a more polar cross-linker bearing numerous hydrogen bond acceptor motifs, namely ethylene glycol dimethacrylate (EDMA). Eventually, derived from the tested bulk imprinting polymerization mixtures, keeping a molar ratio of S-ATNL:Co(II):4Vp = 1:1:5 and EDMA as cross-linker (molar ratio M:C = 1:4) resulted a MIP monolithic column (M2) offering the best recorded selectivity factor (α = 1.60) in pure acetonitrile as mobile phase. Interestingly, in the absence of [BMIM][BF_4] the resulting MIP monolith, (M5) is fully deprived of enantioselectivity (α = 1.00). Yet again, without the Co(II) ion mediating the interaction between the template and monomer during the radical polymerization, the enantioselectivity of the imprinted monolith (M6) drops to 1.30. Nevertheless, since a certain degree of chiral selectivity is still preserved by the M6 monolith, this suggests that in the current experimental conditions a mixed mechanism of imprinting (both non-chiral and metal ion mediated) is most likely to occur in case of M2 monolith. Keeping the same molar ratio of T:Me:M = 1:1:5 as for M2, the enantioselectivity of the resulting MIP monoliths for Ni(II) and Cu(II) (M3 and M4) were also investigated. Although in these cases a certain degree of selectivity for the template enantiomer has been observed (α = 1.38 for M3 and α = 1.30 for M4), it only matched the selectivity factor registered for the monolith obtained in the absence of the metal pivot (M6). The recorded changes in the electronic spectra of the binary and ternary mixtures of S-ATNL-Me(II)-4-VPy display a tendency in the formation of a stable ternary complex in the order Co(II) > Ni(II) > Cu(II) (data not shown). The latter trend correlates well also with the recorded efficiency of chiral recognition for monoliths M2–4. For the control experiments, the reference, non-imprinted monolith (M8) prepared in the absence of the template, but in the presence of Co(II), did not demonstrate any noticeable enantioselectivity (α = 1.04). Evidently, no enantioselectivity is observed for the non-imprinted monolith in the absence of the metal mediator (M9, α = 1.00).

3.3. Chromatographic Retention Mechanism

The influence of the mobile phase composition on the polymers' retention properties, regardless of the employed functional monomer (MAA, 4-VPy) or imprinting approach, indicated a severe decrease of the column capacity factor for both enantiomers of ATNL upon the addition of protic solvents, such as MeOH, acetic acid, or aqueous buffers (data not shown). This would suggest that hydrogen bonding plays a major role in molecular recognition of ATNL's enantiomers on the tested MIPs.

When using monomers with ionizable functional groups, such as MAA (M17–18), a relatively good efficiency with modest selectivity (α = 1.15–1.32) may be achieved in mixed aqueous(pH $\leq$ 5)-organic mobile phases (i.e., ACN: 50mM formate buffer (pH 3.0) = 85:15, v/v) where the partition equilibrium is most probably controlled by ion-exchange interactions (both specific and non-specific) between the amine of the template and the carboxylic groups of the imprinted polymeric structure. Nevertheless, using pure ACN (aprotic, weak solvent with intermediate polarity) as mobile phase, ATNL is totally retained by the polymer due to the synergistic effect of both the electrostatic and the additionally emerging hydrogen bonding interactions.

In the case of the 4-VPy-based MIPs (S6–11; M1–13, Tables 1–3), the ion-exchange mechanism is absent, thus retention and selectivity are mainly controlled by hydrogen bonding interactions. As already mentioned, using 4-VPy as functional monomer, enantioselectivity (where applicable) is only recorded in pure ACN. Chromatographic retention of ATNL drops dramatically (k'~1.05) and is accompanied by a complete loss of selectivity upon the addition in the mobile phase of minute amounts (0.1%, v/v) of protic solvents able to compete with hydrogen bonding.

Unfortunately, upon addition of the porogenic system to the mobile phase, the "memory effect" of the tested polymers could not be improved in the current experimental setup. Due to its strong molar absorptivity, even 1% (v/v) of DMF:DMSO 1:5 mixture added to ACN hindered the UV signal of the eluting enantiomers (data not shown).

Furthermore, the thermodynamics of the molecular recognition were assessed on the best performing 4-VPy-based monolithic polymer (M2). The retention and sorption selectivities were studied in the temperature range 20–50 °C at a flow rate of 0.2 mL min^{-1} ACN. The influence of temperature on the retention (k′) of enantiomers is shown in Figure 4.

As expected, the experimental van't Hoff plots recorded for the MMMI polymer (M2) for both enantiomers of ATNL are linear ($r^2 > 0.91$), showing decreasing retention with the increase of temperature. The average value of the thermodynamic terms ($-\Delta Hi/RT$ and $\Delta Si/R + \ln\varphi$) [48] for the two enantiomers were also calculated for the studied temperature range (Figure 4 inset). For the imprinted polymer, the percent contribution of the entropic term ($\Delta Si/R + \ln\varphi$) for both enantiomers lays around 40%. The binding of ATNL enantiomers implies more selective energetic interactions with the M2 monolith, rather than the entropically controlled steric interactions with the imprinted memory sites [49,50]. The average percent contribution of the steric term ($\Delta Si/R + \ln\varphi$) for the binding of the template is slightly higher (40.6%) in comparison with the binding of its antipode (R-ATNL, ~39.5%). In case of the non-imprinted polymer (M8), identical van't Hoff plots ($r^2 > 0.86$) were obtained for both enantiomers. For this polymer, the average steric term contribution for the template and its antipode are somewhat smaller, 34.6% and 35%, respectively. These results indicate that energetic interactions (hydrogen bonding) are mainly responsible for recording the chromatographic retention, but the steric complementary (shape and size) of the emerging imprinted cavities also brings their contribution to the selective binding of the ATNL enantiomers in comparison with the reference, the non-imprinted polymer.

4. Materials and Methods

4.1. Reagents

Analytical grade standard S(-)-ATNL 98% was purchased from Toronto Research Chemicals (Toronto, Canada). R(+)-ATNL 99% was provided from Santa Cruz Biotechnology, Inc. (Dallas, TX, USA). 2-(trifluoromethyl)acrylic acid 98% (TFMAA) and vinylferrocene 97% (VFC) were purchased from Tokyo Chemical Industry (Tokyo, Japan). Methacrylic acid (MAA) 99%, pentaerythritol triacrylate (PETRA), pentaerythritol tetraacrylate (PETEA), trimethylolpropane trimethacrylate (TRIM), 2,2′-azobis(2-methylpropionitrile) 98% (AIBN), 4,4′-azobis(4-cyanovaleric acid) 98% (ACVA), 4-vinylpyridine 95% (4-Vpy) and 1-vinylimidazole 99% (1-VIM) were purchased from Aldrich (Steinheim, Germany). Tetrabutylammonium hexafluorophosphate 98% (4BA6FPh) was provided from Fluka (Steinheim, Germany). Ortho-phosphoric acid 85% (*w*/*w*), glacial acetic acid 100%, hydrochloric acid 37% (*w*/*w*), dimethylformamide 99% (DMF) and ammonium hydroxide 25% (*w*/*w*) pro analysi were purchased from Merck (Darmstadt, Germany). Formic acid 95% (*w*/*w*), 1-butyl-3-methylimidazolium tetrafluoroborate 98% ([BMIM][BF_4]), acrylamide 99% (AM), *N*,*N*′-Methylenebis(acrylamide) 99% (BAM) and boric acid were obtained from Sigma-Aldrich (Steinheim, Germany) and sodium hydroxide 99.3% from Lach-Ner (Neratovice, Czech Republic). HPLC grade solvents (acetonitrile (ACN), methanol (MeOH), butanol (BuOH), were obtained from Sigma-Aldrich (Steinheim, Germany) and used without further purification. Anhydrous Co(II) acetate 98%, Cu(II) acetate 98% and Ni(II) acetate 99% were purchased from Alfa Aesar (Kandel, Germany) and dimethyl sulfoxide 99.5% (DMSO) was from Carl Roth (Karlsruhe, Germany).

NOBE was synthesized by a previously published method [29].

All other chemicals were analytical reagent grade and were used as received.

Ultrapure water (18.2 MΩ, Barnstead EASYPure ROdi) was used for the preparation of all samples, buffers and related aqueous solutions. Phosphate buffer at various pHs was prepared by dissolving phosphoric acid in ultrapure water and adjusted accordingly with NaOH (1 M).

Stock solutions of 2 mg mL^{-1} ATNL enantiomer were prepared in 2 mL volumetric flasks using ACN as solvent and were stored in the refrigerator at +4–6 °C.

4.2. Apparatus

Chromatographic experiments were performed with an Agilent 1200 HPLC system (Agilent Technologies, Waldbronn, Germany), equipped with a degassing unit, quaternary pump, autosampler, column oven and a diode-array detector. Signal acquisition and data processing were performed in the Chemstation B03.01 (Agilent Technologies, Waldbronn, Germany) software. The detection was performed at 200 nm and the flow-rate was 0.2 mL min^{-1}. All the mobile phases were filtered through a 0.22 µm membrane filter from Millipore before use. Standard samples of atenolol enantiomers (30 µg mL^{-1}) dissolved in HPLC grade ACN were injected in a volume of 10 µL.

The UV-Vis spectra were recorded in conventional (1 cm optical path) quartz cuvettes by a double-beam SPECORD® 250 PLUS (Analytik Jena, Jena, Germany) spectrophotometer in the range of 350–1100 nm. The electronic spectra of each ligand and the colored metal complexes were measured in the same solvent as the one used for the MIP preparation.

4.3. MIP Preparation

The molecularly imprinted and the non-imprinted polymers were synthesized under the conditions illustrated in Tables 1–3. The pre-polymerization mixtures intended for non-covalent imprinting were prepared by weighing the solid components, followed by the addition of liquid monomers and the adequate solvent(s). In the case of metal ion-mediated imprinting, the anhydrous metal ion salts (acetates) were dissolved in the adequate dry solvent or solvent mixture and ATNL was added to the solution. Subsequently, functional monomers, cross-linkers, ionic liquid and the free radical initiator were admixed. In a typical radical polymerization, 15 mg of AIBN has been employed, whereas in the OMNiMIP approach 24 mg of the same radical initiator has been added. The pre-polymerization mixture was sonicated for 15 min and further degassed by purging a gentle flow of nitrogen for 5 min. Depending on the desired form of the polymer (bulk or monolith), the vials or stainless-steel HPLC columns filled with the pre-polymerization mixtures were sealed and caped. In the case of bulk polymers, photopolymerization was performed both at room temperature (22 ± 2 °C) under a UV lamp for 24 h, in 5 mL sealed glass vials, whereas thermal polymerization was carried out in the BMT Ecocell convection oven at 60 °C for 24 h in 5 mL vials or stainless-steel HPLC columns. Removal of the template was achieved by Soxhlet extraction with MeOH/acetic acid (9:1, *v*/*v*), for 24 h. The polymers were then ground using a laboratory mortar and pestle and then sieved using standard testing sieves (φ = 200 mm, CISA sieves, aperture 38 and 25 µm, respectively), and the fraction between 25 and 38 µm was collected. The particles were slurry packed into stainless-steel columns (length 100 mm, internal diameter 2.1 mm) using an LC-10AT VP pump (Shimadzu, Japan) to full volume for HPLC analysis.

The imprinted monoliths were prepared by in situ thermopolymerization, in stainless-steel HPLC columns (length 100 mm, internal diameter 2.1 mm) sealed with Teflon tape and screwable caps, and kept at 60 °C for 24 h in a BMT Ecocell oven. The removal of the template was achieved by washing with MeOH/acetic acid (9:1, *v*/*v*). Control polymers were synthesized under the same conditions in the absence of the template.

4.4. Microscopic Characterization of MIPs

For scanning electron microscopy (SEM) characterization, the polymers (MIP and NIP) were metalized with gold in a Polaron E–5100 plasma-magnetron sputter coater (Polaron Equipment Ltd., Watford, UK) in the presence of argon (45 s at 2 kV and 20 mA). Ultrastructural images were obtained in a FEI Quanta 3D FEG scanning electron microscope (FEI, Hillsboro, ON, USA)) at 30kV and different magnification powers.

4.5. Chromatographic Evaluation of The Imprinted Polymers

4.5.1. Bulk MIP

The template was removed from the imprinted polymers by Soxhlet extraction with MeOH:acetic acid (90:10, *v*/*v*) for 24 h. After grinding and sieving (25 and 38 μm), the polymer particles were slurry packed using a Shimadzu LC-10AT HPLC pump into steel columns (100 × 2.1 mm) to full volume for chromatographic experiments. As mobile phase various solvent systems were tested isocratically at 20 °C, at a flow rate of 0.2 mL min^{-1}, starting with pure ACN and gradually switching to mixtures of ACN with increasing ratio of aqueous buffers (50 mM formate buffer (pH 3.0); 50 mM acetate buffer (pH 5.0) and 100 mM borate buffer (pH 9.3)).

4.5.2. MIP Monolith

The template and excess reagents were removed from the imprinted polymer monolith by pumping through the column MeOH:acetic acid (90:10, *v*/*v*) at a flow rate of 0.2 mL min^{-1} for 24 h. The columns were equilibrated with the corresponding mobile phase for 12 h at a flow rate of 0.20 mL min^{-1} to remove any remaining template. If not otherwise stated, HPLC analyses were performed isocratically at 20 °C, at a flow rate of 0.2 mL min^{-1} using pure ACN or a mixture of ACN/aqueous buffer in variable proportions as the mobile phase, monitoring the eluted analytes at a wavelength of 200 nm.

For all imprinted polymers the separation factor, α, was measured as a ratio of capacity factors $k'_{S\ enantiomer}/k'_{R\ enantiomer}$, with k′ determined by the following relation: $k' = (t_r - t_0)/t_0$, where t_r is the retention time of the analyte and t_0 is the retention time of the void volume measured using acetone as marker.

5. Conclusions

Improved chiral selectivity of the important β-blocker atenolol was achieved by the addition of a metal pivot which gave an imprinting factor of 1.60 versus the traditional molecular imprinted polymer formulation without the pivot which gave an imprinting factor of 1.32. Atenolol is a hydrophilic drug, and molecular imprinting in polar and/or aqueous phases is difficult for traditional molecular imprinting methods based on non-covalent hydrogen bonding or electrostatic complexes between monomers and template, due to disruption of the complex by the polar/aqueous porogenic solvent. The use of a metal to overcome this complex disruption in polar solvents and to coordinate the template and functional monomer led to approximately 23% improvement in chiral selectivity when using Co(II), and a 16% improvement when using Ni(II). This has an important impact due to the large demand for imprinting templates that are only soluble in highly polar and/or aqueous-based solvents. Furthermore, a 25% enhancement in enantioselectivity was found when using monolithic materials versus ground and sieved bulk imprinted polymers. Investigation of binding parameters showed that better selectivity was not a result of increased binding affinity (i.e., larger k′ values) versus non-metal systems but was due to increase in differential enthalpic contributions of binding between the imprinted polymer and each enantiomer of atenolol. Thus, it can be concluded that the underlying mechanism of improvement of enantioselectivity of the imprinted polymer is due to the metal pivot approach maintaining fidelity of the imprinted site during polymerization. The choice of functional monomer was shown to be important based on the affinity of the functional monomer for the metal-pivot; in particular, 4-vinylpyridine or vinylimidazole did not disrupt important template-metal interactions whereas methacrylic acid displaced at least one of the template-to-metal interactions. This may be general for metal-pivot systems that require close proximity of the template to the metal for creating a selective binding site. In addition, the choice of crosslinker was important for optimum performance, for example, entry M2 shows that an EDMA crosslinked molecularly imprinted material provided 60% enhancement in enantioselectivity versus a nearly identical formulation using TRIM. Other molecularly imprinted materials using the crosslinkers PETRA and PETEA also showed little to no enantioselective performance, supporting the conclusion that the difunctional crosslinker EDMA is required for these systems versus any trifunctional crosslinkers.

Author Contributions: Conceptualization, A.E.B., B.-C.I., D.A.S. and E.B.; Data curation, A.E.B., B.-C.I., S.L.O.O. and E.B.; Investigation, A.E.B., B.-C.I., L.E.G. and N.A.G.; Methodology, A.E.B., B.-C.I., D.A.S. and E.B.; Resources, S.L.O., D.A.S. and E.B.; Writing—original draft, A.E.B., B.-C.I., D.A.S. and E.B.

Acknowledgments: This work was supported under the contract funded by the University of Medicine and Pharmacy "Iuliu Hatieganu" Cluj-Napoca, internal grant No. 4944/11/08.03.2016 and partly by a grant of Ministry of Research and Innovation, CNCS-UEFISCDI, project number PN-III-P1-1.1-TE-2016-0628, within PNCDI III. David Spivak and Nicholas Gariano were supported by the National Science Foundation under grant CHE-1411547. The ultrastructural characterization of the polymers by Lucian Barbu-Tudoran, as well as the help in the spectroscopic studies offered by Tudor Rusan is greatly acknowledged.

Conflicts of Interest: The authors declare no conflict of interest.

References

1. Săndulescu, R.; Cristea, C.; Bodoki, E.; Oprean, R. Chapter 3: Recent Advances in the Analysis of Bioactive Compounds based on Molecular Recognition. In *Frontiers in Bioactive Compounds, Natural Sources, Physicochemical Characterization Applications*; Apetrei, C., Ed.; Bentham Science Publisher: Sharjah, UAE, 2016; Volume 1, pp. 69–126.
2. Iacob, B.-C.; Bodoki, E.; Oprean, R. Recent advances in capillary electrochromatography using molecularly imprinted polymers. *Electrophoresis* **2014**, *35*, 2722–2732. [CrossRef] [PubMed]
3. Cheong, W.; Ali, F.; Choi, J.H.; Lee, J.O.; Sung, K.Y. Recent applications of molecular imprinted polymers for enantio-selective recognition. *Talanta* **2013**, *106*, 45–59. [CrossRef] [PubMed]
4. Cheong, W.J.; Ali, F.; Kim, Y.S.; Lee, J.W. Comprehensive overview of recent preparation and application trends of various open tubular capillary columns in separation science. *J. Chromatogr. A* **2013**, *1308*, 1–24. [CrossRef] [PubMed]
5. Pan, J.; Chen, W.; Ma, Y.; Pan, G. Molecularly imprinted polymers as receptor mimics for selective cell recognition. *Chem. Soc. Rev.* **2018**. [CrossRef] [PubMed]
6. Chen, W.; Ma, Y.; Pan, J.; Meng, Z.; Pan, G.; Sellergren, B. Molecularly Imprinted Polymers with Stimuli-Responsive Affinity: Progress and Perspectives. *Polymers* **2015**, *7*, 1689–1715. [CrossRef]
7. Iacob, B.-C.; Bodoki, E.; Oprean, R. Chiral Electrochemical Sensors Based on Molecularly Imprinted Polymers with Pharmaceutical Applications. In *Handbook of Sustainable Polymers*; Pan Stanford Publishing, CRC Press: Boca Raton, FL, USA, 2016; pp. 587–614.
8. Prasad, B.B.; Tiwari, M.P. Molecularly Imprinted Nanomaterial-Based Highly Sensitive and Selective Medical Devices. In *Biomedical Materials and Diagnostic Devices*; John Wiley & Sons, Inc.: Hoboken, NJ, USA; Scrivener Publishing LLC: Salem, MA, USA, 2012; pp. 339–391.
9. Iacob, B.-C.; Bodoki, E.; Florea, A.; Bodoki, A.E.; Oprean, R. Simultaneous enantiospecific recognition of several β-blocker enantiomers using molecularly imprinted polymer-based electrochemical sensor. *Anal. Chem.* **2015**, *87*, 2755–2763. [CrossRef] [PubMed]
10. Iacob, B.-C.; Bodoki, E.; Farcau, C.; Barbu-Tudoran, L.; Oprean, R. Study of the Molecular Recognition Mechanism of an Ultrathin MIP Film-Based Chiral Electrochemical Sensor. *Electrochim. Acta* **2016**, *217*, 195–202. [CrossRef]
11. Kupai, J.; Rojik, E.; Huszthy, P.; Szekely, G. Role of Chirality and Macroring in Imprinted Polymers with Enantiodiscriminative Power. *ACS Appl. Mater. Interfaces* **2015**, *7*, 9516–9525. [CrossRef] [PubMed]
12. Zhao, Q.-L.; Zhou, J.; Zhang, L.-S.; Huang, Y.-P.; Liu, Z.-S. Coatings of molecularly imprinted polymers based on polyhedral oligomeric silsesquioxane for open tubular capillary electrochromatography. *Talanta* **2016**, *152*, 277–282. [CrossRef] [PubMed]
13. Yang, S.; Wang, Y.; Jiang, Y.; Li, S.; Liu, W. Molecularly Imprinted Polymers for the Identification and Separation of Chiral Drugs and Biomolecules. *Polymers* **2016**, *8*, 216. [CrossRef]
14. Didaskalou, C.; Buyuktiryaki, S.; Kecili, R.; Fonte, C.P.; Szekely, G. Valorisation of agricultural waste with an adsorption/nanofiltration hybrid process: from materials to sustainable process design. *Green Chem.* **2017**, *19*, 3116–3125. [CrossRef]
15. Bakkour, R.; Bolotin, J.; Sellergren, B.; Hofstetter, T.B. Molecularly Imprinted Polymers for Compound-Specific Isotope Analysis of Polar Organic Micropollutants in Aquatic Environments. *Anal. Chem.* **2018**, *90*, 7292–7301. [CrossRef] [PubMed]

16. Li, X.-Y.; Ma, L.; Huang, Y.-P.; Liu, Z.-S.; Aisa, H.A. Preparation of metallic pivot-based imprinted monoliths with a hydrophilic macromonomer. *RSC Adv.* **2015**, *5*, 36753–36761. [CrossRef]
17. Bai, L.H.; Chen, X.X.; Huang, Y.P.; Zhang, Q.W.; Liu, Z.S. Chiral separation of racemic mandelic acids by use of an ionic liquid-mediated imprinted monolith with a metal ion as self-assembly pivot. *Anal. Bioanal. Chem.* **2013**, *405*, 8935–8943. [CrossRef] [PubMed]
18. Zhong, D.-D.; Huang, Y.-P.; Xin, X.-L.; Liu, Z.-S.; Aisa, H.A. Preparation of metallic pivot-based imprinted monolith for polar template. *J. Chromatogr. B* **2013**, *934* (Suppl. C), 109–116. [CrossRef] [PubMed]
19. Iacob, B.-C.; Bodoki, A.E.; Oprean, L.; Bodoki, E. Metal—Ligand interactions in molecular imprinting. In *Ligand*; Chandraleka, S., Biswas, B., Eds.; Intech Open: Rijeka, Croatia, 2018; in press.
20. Zhang, J.; Li, F.; Wang, X.-H.; Xu, D.; Huang, Y.-P.; Liu, Z.-S. Preparation and characterization of dual-template molecularly imprinted monolith with metal ion as pivot. *Eur. Polym. J.* **2016**, *80* (Suppl. C), 134–144. [CrossRef]
21. Zhao, L.; Ban, L.; Zhang, Q.-W.; Huang, Y.-P.; Liu, Z.-S. Preparation and characterization of imprinted monolith with metal ion as pivot. *J. Chromatogr. A* **2011**, *1218*, 9071–9079. [CrossRef] [PubMed]
22. Vidyasankar, S.; Ru, M.; Arnold, F.H. Molecularly imprinted ligand-exchange adsorbents for the chiral separation of underivatized amino acids. *J. Chromatogr. A* **1997**, *775*, 51–63. [CrossRef]
23. Fischer, L.; Mueller, R.; Ekberg, B.; Mosbach, K. Direct enantioseparation of .beta.-adrenergic blockers using a chiral stationary phase prepared by molecular imprinting. *J. Am. Chem. Soc.* **1991**, *113*, 9358–9360. [CrossRef]
24. Tamai, G.; Edani, M.; Imai, H. Chiral separation and determination of propranolol enantiomers in rat or mouse blood and tissue by column switching high performance liquid chromatography with ovomucoid bonded stationary phase. *Biomed. Chromatogr.* **1990**, *4*, 157–160. [CrossRef] [PubMed]
25. Aboul-Enein, H.Y.; Bakr, S.A. Enantiomeric Resolution of Propranolol and Analogs on Two Cellulose (Chiralcel of and OC) AND One Amylose (Chiralpak Ad) Chiral Stationary Phases. *J. Liquid Chromatogr. Relat. Technol.* **1998**, *21*, 1137–1145. [CrossRef]
26. Moldovan, R.-C.; Dascăl, G.-S.; Mirel, V.; Bodoki, E.; Oprean, R. Chiral separation of 16 beta-blockers on immobilized polysaccharide chiral stationary phases. *Farmacia* **2015**, *63*, 909–912.
27. Weng, X.; Bao, Z.; Xing, H.; Zhang, Z.; Yang, Q.; Su, B.; Yang, Y.; Ren, Q. Synthesis and characterization of cellulose 3,5-dimethylphenylcarbamate silica hybrid spheres for enantioseparation of chiral β-blockers. *J. Chromatogr. A* **2013**, *1321*, 38–47. [CrossRef] [PubMed]
28. Younes, A.A.; Ates, H.; Mangelings, D.; Vander Heyden, Y. A separation strategy combining three HPLC modes and polysaccharide-based chiral stationary phases. *J. Pharm. Biomed. Anal.* **2013**, *75*, 74–85. [CrossRef] [PubMed]
29. Sibrian-Vazquez, M.; Spivak, D.A. Enhanced Enantioselectivity of Molecularly Imprinted Polymers Formulated with Novel Cross-Linking Monomers. *Macromolecules* **2003**, *36*, 5105–5113. [CrossRef]
30. Mehvar, R.; Brocks, D.R. Stereospecific pharmacokinetics and pharmacodynamics of beta-adrenergic blockers in humans. *J. Pharm. Pharm. Sci.* **2001**, *4*, 185–200. [PubMed]
31. Zaid, A.A.; Farooqui, M.; Janrao, D.M. Potentiometric study of atenolol as hypertension drug with Co(II), Ni(II), Cu(II) and Zn(II) transition metal ions in aqueous solution. *J. Saudi Chem. Soc.* **2015**, *19*, 92–96. [CrossRef]
32. Gölcü, A.; Yücesoy, C.; Serin, S. Synthesis and characterization of metal complexes of acebutolol, atenolol, and propranolol antihypertension drugs. *Synth. React. Inorg. Metal-Org. Chem.* **2004**, *34*, 1259–1275. [CrossRef]
33. Bontchev, P.R.; Pantcheva, I.N.; Gochev, G.P.; Mehandjiev, D.R.; Ivanov, D.S. Complexes of copper(II) with the β-blocker atenolol. *Transit. Met. Chem.* **2000**, *25*, 196–199. [CrossRef]
34. Cozar, O.; Szabó, L.; Cozar, I.B.; Leopold, N.; David, L.; Căinap, C.; Chiş, V. Spectroscopic and DFT study of atenolol and metoprolol and their copper complexes. *J. Mol. Struct.* **2011**, *993*, 357–366. [CrossRef]
35. Rajbhoj, A.S.; Gaikwad, S.T.; Chondhekar, T.K.; Pirzada, S.A. Stability constants of ternary complexes of drugs and aminoacids. *Int. J. Chem. Sci.* **2005**, *3*, 241–246.

36. Ali, M.; Dutta, P.; Pandey, S. Effect of Ionic Liquid on Prototropic and Solvatochromic Behavior of Fluorescein. *J. Phys. Chem. B* **2010**, *114*, 15042–15051. [CrossRef] [PubMed]
37. Herrmann, W.A. Comprehensive Coordination Chemistry. The Synthesis, Reactions, Properties and Applications of Coordination Compounds (7 Bände). Herausgegeben von G. Wilkinson, R. Gillard and J. A. McCleverty. Pergamon Press: Oxford 1987. Volume 4: Middle Transition Elements. *Angew. Chem.* **1989**, *101*, 809.
38. Lever, A.P.B. *Inorganic Electronic Spectroscopy*, 2nd ed.; Elsevier: Amsterdam, The Netherlands, 1984.
39. Reddy, L.S.; Endo, T.; Reddy, G.S. Chapter 1: Electronic (absorption) spectra of 3D transition metal complexes. In *Advanced Aspects of Spectroscopy*; Farrukh, M.A., Ed.; Intech Open: Rijeka, Croatia, 2012.
40. Ye, L. *Molecular Imprinting: Principles and Applications of Micro- and Nanostructured Polymers*; Pan Stanford Publishing, CRC Press: Boca Raton, FL, USA, 2013.
41. Meng, A.C.; LeJeune, J.; Spivak, D.A. Multi-analyte imprinting capability of OMNiMIPs versus traditional molecularly imprinted polymers. *J. Mol. Recognit.* **2009**, *22*, 121–128. [CrossRef] [PubMed]
42. Lee, W.C.; Cheng, C.H.; Pan, H.H.; Chung, T.H.; Hwang, C.C. Chromatographic characterization of molecularly imprinted polymers. *Anal. Bioanal. Chem.* **2008**, *390*, 1101–1109. [CrossRef] [PubMed]
43. Svec, F.; Frechet, J.M.J. Continuous rods of macroporous polymer as high-performance liquid chromatography separation media. *Anal. Chem.* **1992**, *64*, 820–822. [CrossRef]
44. Wu, X.; Du, J.; Li, M.; Wu, L.; Han, C.; Su, F. Recent advances in green reagents for molecularly imprinted polymers. *RSC Adv.* **2018**, *8*, 311–327. [CrossRef]
45. Greaves, T.L.; Drummond, C.J. Solvent nanostructure, the solvophobic effect and amphiphile self-assembly in ionic liquids. *Chem. Soc. Rev.* **2013**, *42*, 1096–1120. [CrossRef] [PubMed]
46. Wang, C.; Li, X.-J.; Yang, J.; Zhao, Y.-X.; Liu, Z.-S.; Aisa, H.A. Preparation of ionic liquid-mediated imprinted monolith for selective capture and purification of corilagin. *J. Chromatogr. B* **2017**, *1041–1042*, 98–103. [CrossRef] [PubMed]
47. LeJeune, J.; Spivak, D.A. Chiral effects of alkyl-substituted derivatives of *N,O*-bismethacryloyl ethanolamine on the performance of one monomer molecularly imprinted polymers (OMNiMIPs). *Anal. Bioanal. Chem.* **2007**, *389*, 433–440. [CrossRef] [PubMed]
48. Denderz, N.; Lehotay, J.; Čižmárik, J.; Cibulková, Z.; Šimon, P. Thermodynamic study of molecularly imprinted polymer used as the stationary phase in high performance liquid chromatography. *J. Chromatogr. A* **2012**, *1235*, 77–83. [CrossRef] [PubMed]
49. Denderz, N.; Lehotay, J. Application of the van't Hoff dependences in the characterization of molecularly imprinted polymers for some phenolic acids. *J. Chromatogr. A* **2012**, *1268*, 44–52. [CrossRef] [PubMed]
50. Denderz, N.; Lehotay, J. Identification of driving forces for the recognition processes on molecularly imprinted polymers. *Nova Biotechnol. Chim.* **2013**, *12*, 63–69. [CrossRef]

Sample Availability: Not available.

molecules

MDPI

Review

Chiral Stationary Phases for Liquid Chromatography: Recent Developments

Joana Teixeira [1], Maria Elizabeth Tiritan [1,2,3], Madalena M. M. Pinto [1,2] and Carla Fernandes [1,2,*]

1 Laboratório de Química Orgânica e Farmacêutica, Departamento de Ciências Químicas, Faculdade de Farmácia, Universidade do Porto, Rua de Jorge Viterbo Ferreira, 228, 4050-313 Porto, Portugal; jbteixeira@live.com.pt (J.T.); elizabeth.tiritan@iucs.cespu.pt (M.E.T.); madalena@ff.up.pt (M.M.M.P.)

2 Interdisciplinary Centre of Marine and Environmental Research (CIIMAR), Edifício do Terminal de Cruzeiros do Porto de Leixões, Av. General Norton de Matos s/n, 4050-208 Matosinhos, Portugal

3 Cooperativa de Ensino Superior, Politécnico e Universitário (CESPU), Instituto de Investigação e Formação Avançada em Ciências e Tecnologias da Saúde (IINFACTS), Rua Central de Gandra, 1317, 4585-116 Gandra PRD, Portugal

* Correspondence: cfernandes@ff.up.pt; Tel.: +351-22-042-8688

Received: 31 January 2019; Accepted: 26 February 2019; Published: 28 February 2019

Abstract: The planning and development of new chiral stationary phases (CSPs) for liquid chromatography (LC) are considered as continuous and evolutionary issues since the introduction of the first CSP in 1938. The main objectives of the development strategies were to attempt the improvement of the chromatographic enantioresolution performance of the CSPs as well as enlarge their versatility and range of applications. Additionally, the transition to ultra-high-performance LC were underscored. The most recent strategies have comprised the introduction of new chiral selectors, the use of new materials as chromatographic supports or the reduction of its particle size, and the application of different synthetic approaches for preparation of CSPs. This review gathered the most recent developments associated to the different types of CSPs providing an overview of the relevant advances that are arising on LC.

Keywords: liquid chromatography; enantioseparation; chiral stationary phase; chiral selector; chromatographic support

1. Introduction

Now more than ever, analytical and preparative enantiomeric separations play a crucial role in industry and academic research [1]. There are a wide variety of methods to achieve and analyze enantiomerically pure compounds, including liquid chromatography (LC) [2,3], supercritical fluid chromatography [4–6], diastereomeric crystallization [7,8], membranes [9,10], asymmetric catalysis [11], simulated moving bed [12,13], dynamic and enzyme-mediated kinetic resolution [14,15], among others.

LC using chiral stationary phases (CSPs) proved to be an essential tool with a wide range of applications, including preparative separation of enantiomers of diverse analytes [16,17], determination of enantiomeric composition [18,19], monitorization of asymmetric reactions [20,21], analysis of the stereochemistry of natural compounds [22,23], pharmacokinetic [24,25], forensic [26–28], environmental [29–31], and enantioselective studies [32,33], among others.

The development of CSPs for LC combined with the improvement of chromatography instrumentation revolutionized the enantioseparation approaches. LC using CSPs has demonstrated to be extremely useful, accurate, versatile, and it has been a widely used technique in diverse fields and applications, emphasizing, for example, the enantioseparation of underivatized amino acids [34,35], diverse classes of pharmaceuticals [36–40], atropisomers [41], as well as the study of intermolecular interactions between biomolecules and drugs [42], among others.

Over the last decades, several types of CSPs have been developed [43–46] and, among them, more than a hundred are currently commercially available [39]. These comprise Pirkle-type, ligand-exchange-type, molecularly-imprinted, and based on macrocyclic antibiotics, proteins, polysaccharides, cyclodextrins, crown ethers, cyclofructans, synthetic polymers, among others [43–46]. Nevertheless, although many different types of CSPs are described, the development of new CSPs continues to be a field of research with great importance.

2. Chiral Stationary Phases: Recent Developments

Since the first description of a CSP, in 1938, by Henderson and Rule [47], and to follow the constant challenges on different areas as well as the advances in chromatographic instrumentation, the development of new CSPs for LC has been a continuous and evolutionary subject.

In this review, for each type of CSPs, the most recent CSPs were presented describing the strategies used for their development (Figure 1).

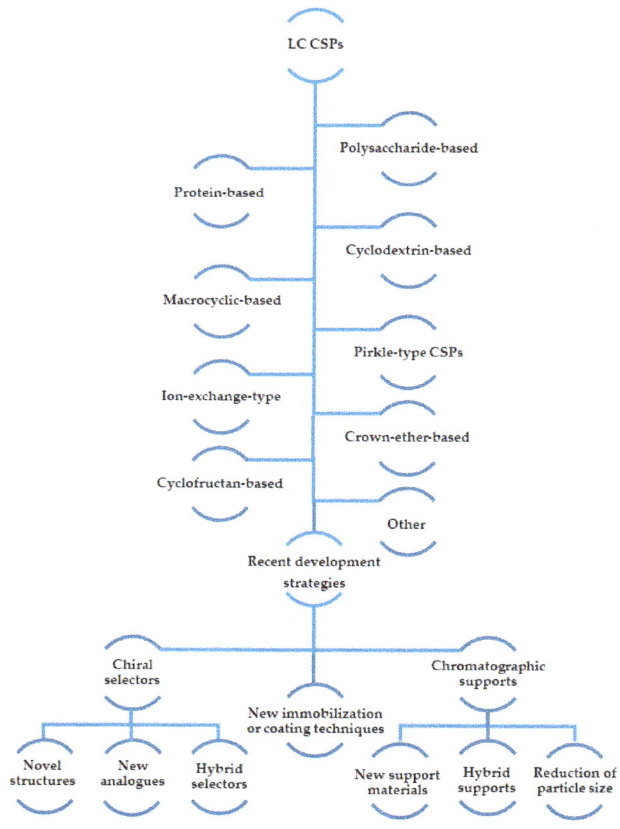

Figure 1. Summary of recent strategies for development of new chiral stationary phases (CSPs) for liquid chromatography (LC).

Herein, only the CSPs that were not reported in previous fundamental reviews will be presented, highlighting the main advantages of the strategy used for their preparation, the best chromatographic results and the objectives to be achieved.

2.1. Polysaccharide-Based CSPs

The first application of a polysaccharide as a chromatographic chiral selector was described by Hessen and Hagel, in 1976 [48]. Since then, different polysaccharides were extensively used as CSPs due to their high enantioselectivity properties after derivatization [49]. Nevertheless, amylose and cellulose are the main polysaccharides used to obtain CSPs [50], followed by chitosan and chitin [51]. The chiral recognition ability of polysaccharide derivatives is dependent on diverse structural features, including sugar units, stereogenic centres of the glucopyranose units, type of the linkage and its position, as well as the adjacent polymer chains [49]. The helical twist of the polymer backbone is also essential for enantioselectivity [43].

Polysaccharide derivatives as efficient chiral selectors can include phenyl, alkyl or benzylcarbamates, esters, benzoate, or aryl or cycloalkyl groups [52]. Benzoate or phenylcarbamate moieties may comprise methyl, methoxy, among other groups, and/or chlorine substituents in the aromatic ring [52], affording different solubility and chiral recognition ability [52]. Moreover, the position of the substituents in the aromatic ring influences the enantioseparation performance of the chiral selector [49].

Polysaccharide derivatives can be coated onto a chromatographic support, as silica or derivatives, by an adsorption process [53,54] allowing a larger surface area [55] and high efficiency [43]. CSPs comprising coated polysaccharide derivatives can operate in normal phase, polar organic and reversed-phase elution mode; however, they have restrictions due to the non-compatibility with "non-standard" solvents, such as dichloromethane, chloroform, toluene, or acetone [43]. The use of those solvents in the mobile phase may cause the dissolution of the adsorbed polymer and, consequently, removal of the selector from the chromatographic column [43]. Immobilized polysaccharides emerged as a reliable alternative allowing the use of a broader selection of solvents as mobile phases [56–60]. Different procedures can be used for covalently bonded the polysaccharide derivatives to the chromatographic support, such as a polymerization reaction and photoinduced and enzymatic polymerization [56]. Nevertheless, despite the solvent versatility, in general, the potential of chiral recognition of immobilized polysaccharide-based CSPs is lower than the coated due to modification of stereospecific conformation that can occur during the immobilization process [56,57].

This type of CSPs is recognized as being the most successful and widely applied for both analytical [61–69] and preparative enantioseparations [17,70–76], being responsible for about 99% of reported chiral separations [50]. Among the developed polysaccharide-based CSPs, the 3,5-dimethylphenyl *tris*-phenylcarbamates of amylose and cellulose proved to have the best enantiorecognition performance [77–80]. In our group, this type of CSPs has proved to be effective for analytical as well as preparative applications [26,81–84].

The chiral recognition mechanisms concerning these CSPs are not yet completely understood. In an attempt to improve the knowledge related with structural features associated with the chiral recognition mechanisms and their chromatographic behavior at a molecular level, several studies concerning to docking, spectroscopy, molecular modelling, and quantum chemical calculations were recently performed and compiled by Scriba et al. [45].

Several reviews have assembled the advances on preparation and evaluation of this type of CSPs over the years [43,45,49–52,57–59,80,85–93]. Nevertheless, this research field is always evolving being the most recent polysaccharide-based CSPs presented on Table S1 (supplementary material).

Recent developments on polysaccharide-based CSPs comprise different approaches, with the general objective being the improvement of the enantioseparation performance. The strategies include the introduction of new polysaccharide derivatives (mainly new chitin and chitosan derivatives but also cellulose derivatives), hybrid selectors, and different chromatographic supports (monoliths, core-shell, microspheres), as well as the application of different methodologies for coating or immobilization procedures. Generally, for the new derivatives, the effect of different substituents on chiral recognition has been discussed [94–97]. Additionally, the effect of mobile phase composition on enantioseparation was also explored [94–96,98,99].

Recently, Han et al. [94] developed two CSPs, using a derivative of cellulose *tris* (3,5-dimethylphenylcarbamate) (CSP1) and the same derivative functionalized with carboxylic acid (CSP2) (Figure 2). They concluded that a large variety of substituents could avoid the chiral recognition properties of the cellulose derivatives, reducing the performance of the CSP. The best chromatographic results were obtained for *trans*-stilbene oxide, with α and Rs values of 1.84 and 9.59, respectively.

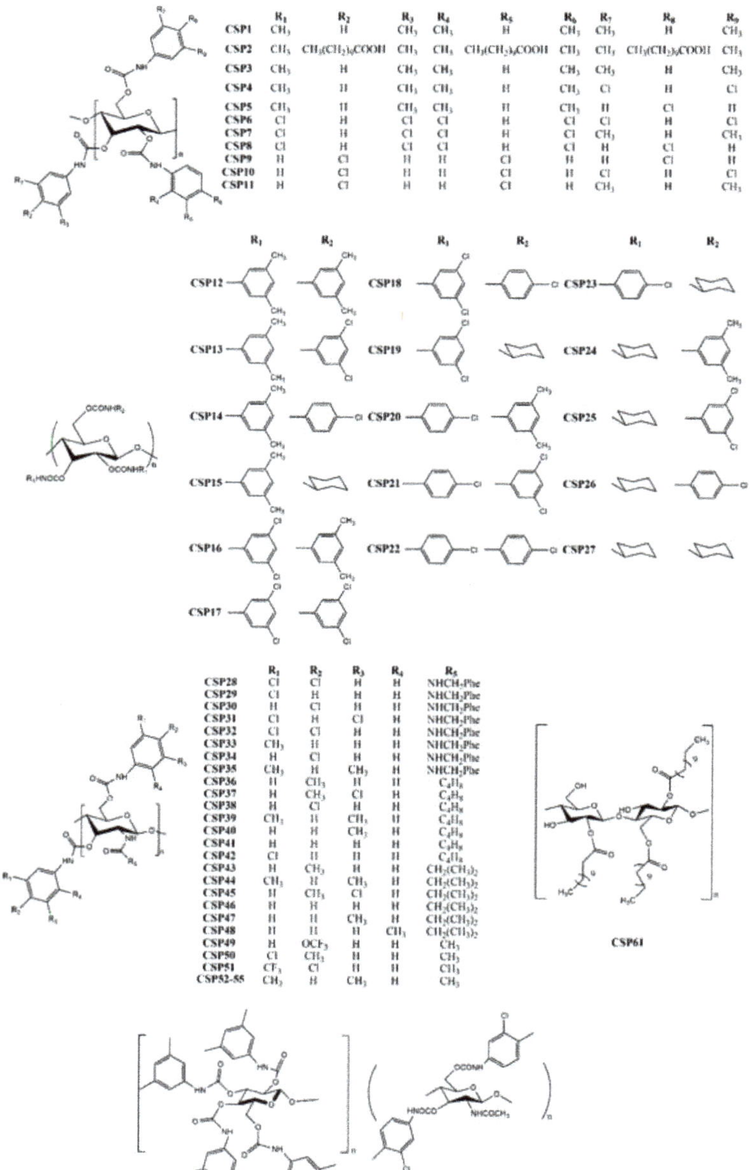

Figure 2. Chemical structures of polysaccharide-based CSP1–56 and CSP61.

Shen et al. [100] synthetized cellulose derivatives with different combination of carbamate substituents and prepared 25 new CSPs (CSP3–27) (Figure 2). The effect of the carbamate substituents at 2,3-positions and 6-position of the glucose moiety were the main focus of the study. It was found that the chiral recognition properties of the CSPs comprising derivatives with two different phenylcarbamates were higher than if CSPs only had one substituent. The resolution was improved by the presence of different carbamate substituents, suggesting that the chiral recognition was dependent on the electronic properties, position and number of substituents of the glucose unit [100]. The highest separation factor obtained by using these recent CSPs was 2.87, for Pirkle alcohol.

Chitin and chitosan-based CSPs have received particular attention in the last few years [51]. Through continuous efforts to develop effective CSPs, other recent reports describing the use of chitin [101,102] and chitosan [95–97] derivatives are rising, with the carbamates as one of the most studied [49]. The growing interest in these polysaccharides comes from the fact that they have low solubility, which allows the use of a wide variety of mobile phases [52]. The influence of substituents on chitin and chitosan derivatives was also investigated. For some analytes, these CSPs possessed an enhanced chiral recognition when compared to cellulose and amylose derivatives, which may be attributed to the variety of solvents that can be used [103].

Tang et al. [97] developed eight CSPs (CSP28–35) comprising chitosan 3,6-*bis*(arylcarbamate)-2 -(*p*-methylbenzylurea) with diverse substituents in the aromatic rings of the carbamates as well as in the amide group (Figure 2). Selectors with electron-donating substituents demonstrated a higher ability of enantioseparation. Previous reports emphasized that an electron-donating substituent at the 4-position of the aromatic ring was beneficial for the chiral separation [44]. Despite the selectors with 4-methyl substituent and 3-chloro-4-methyl portion presented a superior enantioseparation, the highest resolution (Rs = 18.1) and separation factor (α = 6.72) were obtained by the CSP with 3,5-dimethyl substituent [97].

In another study, Zhang et al. [95] prepared seven CSPs (CSP36–42) comprising derivatives of chitosan *bis*(phenylcarbamate)-(*N*-cyclobutylformamide) (Figure 2). The same substituent on different positions resulted on modifications on the suprastructure of the selector leading to different size of cavities, for example, due to different electronic effects. The obtained CSPs proved to have considerable stability on different solvents and a good enantiorecognition, allowing the researchers to obtain a separation factor of 8.64 for voriconazole [95].

Other new chitosan-based CSPs were developed, in this case, comprised of derivatives of chitosan (*bis*(methylphenylcarbamate)-(isobutyrylamide)) (CSP43–48) (Figure 2) [96]. The introduction of some substituents on specific positions of the aromatic ring linked to the carbamate were favorable for enantioseparation, such as methyl substituents. Additionally, the low solubility of chitosan was proved to be an advantage for the solvent tolerance and good enantioresolution performance achieved. As an example of its performance, high enantioselectivity and resolution were obtained for voricomazole, with α and Rs values 4.32 and 11.9, respectively [96].

Zhang et al. [102] synthetized derivatives of chitin using three different phenyl isocyanates (4-trifluoromethoxy, 3-chloro-4-methyl, 4-chloro-3-trifluoromethylphenylcarbamate) to develop three CSPs (CSP49–51) (Figure 2). All CSPs were applied for enantioseparation of tadalafil and its intermediate, demonstrating great enantiorecognition potential, with resolution and separation factor values of 4.72 and of 2.15, respectively [102].

Mei et al. [101] derivatized natural and regenerated chitins with 3,5-dimethyphenyl isocyanate, to prepare CSP52 and CSP53–55, respectively (Figure 2), with the difference between them only related to the raw material. The regenerated chitins were obtained from natural chitins, after a treatment with acetic anhydride, showing a more promising performance. They pointed out that CSP prepared from selectors with lower molecular weight provided an improved resolution [101]. The best chromatographic results were obtained for voricomazole, with Rs = 11.7 and α = 3.06.

Another strategy was the development of hybrid selectors [98]. Hybrid selectors or biselectors comprise two different polysaccharide derivatives coated on a chromatographic support [104]. Zhang et al. prepared the CSP56, which comprised a biselector based on derivatives of amylose and chitin (Figure 2), combining the low solubility of chitin derivatives with the excellent chiral recognition properties of amylose derivatives [98]. The obtained CSP presented an improved resistance against organic solvents with high enantioselectivity, with a Rs value of 8.49 for mephobarbital, and an α value of 4.32 for 2-(5-chloro-2-((4-methoxybenzyl)-amido)phenyl)-4-cyclopropyl-1,1,1-trifluorobut-3-yn-2-ol.

Regarding the use and preparation of chromatographic supports, a new technique of encapsulation in organic polymer monolith was reported, by Fouad et al. [103], as an alternative preparation of the chromatographic support. They functionalized an organic polymer monolith with carbamylated amylose as selector to obtain the CSP57. The synthesis of the amylose derivative was described previously [105]. The encapsulation of amylose was described as an economic methodology and it allowed the conjugation of amylose with reversed-phase elution mode for several analytes [103]. This promising technique allowed good results with a maximum resolution of 2.80 and a separation factor of 3.90 for the testes analytes.

Another different approach for preparation of chromatographic support was reported by Li et al. [99], who functionalized core-shell silver particles with cellulose derivatives through coating and intermolecular polycondensation and developed CSP58. A synergetic effect between silver and cellulose was observed considering the high values of resolution (Rs = 2.61) and enantioseparation (α = 8.42). This CSP demonstrated a particular selectivity toward analytes having the functional group ketone [99].

In another study, Bezhitashvili et al. [106] reported the covalent immobilization of a cellulose derivative (cellulose-(3,5-dichlorophenylcarbamate)) onto the surface of core-shell particles to obtain CSP59. The synthesis of the cellulose derivative was described previously [77]. The click chemistry was the synthetic methodology applied for the immobilization of the cellulose derivative to the chromatographic support [107]. The authors emphasized the short time of analysis achieved with baseline separations [106]. The highest separation factor obtained was 15.3 with a resolution of 11.0 for 2-(4-methylbenzylsulfinyl)-benzamide.

Huang et al. [108] developed a new methodology for coating cellulose *tris*(3,5-dimethylphenyl carbamate) derivative on silica microspheres, without any surface pre-treatment since no aggregation occurred, and prepared CSP60 (range of pore size 10–150 nm). The synthesis of the cellulose derivative was previously described [109]. The silica microspheres with reduced size have functionalized polymeric beads being highly crosslinked [108]. This technique allowed a high-loading of the chiral selector and the obtained CSP provided a good performance, being the best separation factor of 2.41 for 2,2,2-trifluoro-1-(9-anthryl) ethanol.

The CSP61 (Figure 2) was reported by Vieira et al. [110] using a technique of thermal immobilization of cellulose dodecanoate on silica particles without the use of chemical reagents. Despite of the absence of a chemical reagent during the procedure of immobilization, the selector was strongly linked to the chromatographic support allowing an exceptional selectivity. Some advantages of the immobilization technique were highlighted, including its low cost and eco-friendly feature [110]. The separation factor obtained was 3.10.

Besides the recent developments on polysaccharide-based CSPs, it is important to emphasize that this type of CSPs also cover a wide range of recent applications [111–113]. For example, Padró et al. [89] recently reviewed applications of polysaccharide-based CSPs in different fields. Studies comparing the enantioresolution performance of coated and covalently immobilized CSPs based on polysaccharides were also found [114].

Additionally, the influence of mobile phase is a common focus in several studies [81,115].

2.2. *Protein-Based CSPs*

The intrinsic ability of chiral recognition by enzymes, plasma proteins and receptors inspired the application of proteins in enantioseparation techniques [43]. Proteins are complex structures with a large surface area comprising a variety of stereogenic centers and different binding sites, which allow multiple possibilities of intermolecular interactions with small molecules [55]. The first application of a protein as CSP was reported in 1973, describing the separation of tryptophan enantiomers using a bovin serum albumin (BSA)-sepharose CSP [116]. After this first report, many CSPs based on proteins have been developed, with the most used proteins the human serum albumin (HSA), α_1-acid glycoprotein (AGP), crude ovomucoid (OVM), and cellobiohydrolase I (CBH I) [86]. All these proteins as chiral selectors have been well documented for chromatographic enantioseparation for a wide range of chiral compounds and for binding affinity studies [117].

Proteins as CSPs are applied on affinity and pharmacokinetic studies since they can mimic the in vivo systems [118], being this feature very important in drug discovery [43]. The possibility of using aqueous or aqueous-organic as mobile phases was pointed out as other advantage of protein-based CSPs considering its compatibility with mass spectrometric detection. The disadvantages of this type of CSPs are the low capacity and efficiency. Moreover, the possibility of denaturation of protein limit the ranges of pH, ionic strength, temperature, and organic modifier composition of mobile phase [88], which is a result of its reduced chemical and biochemical stabilities.

HSA is the most applied protein-based CSP, and it is used predominantly on studies of drug-protein binding [43]. In separation techniques, it is applied for weakly acidic and neutral compounds [119] as well as zwitterionic molecules [120]. For preparation of this type of CSPs, the protein can be physically adsorbed onto the packing material or it can be covalently bound [121].

A number of reviews on CSPs based on proteins have appeared over the years, focusing on their developments and applications [43,45,50,85,87,88,118,120–122]. Recently, Bocian et al. [117] briefly reported several studies related to protein-based CSPs, including the most common (HSA and AGP) as well as the more unusual, namely, avidin and fatty acid binding proteins. The developed strategies presented on that review were mainly related with the chromatographic support as the introduction of monoliths. The introduction of new selectors was also described [117]. Bertucci and Tedesco [42] recently reviewed the advances concerning the HSA as chiral selector highlighting the application of competitors for a particular binding site of HSA as the greatest advance. Scriba et al. [45] compiled some studies, focusing on the understanding of the binding sites and the main interactions between protein-based CSPs with diverse analytes, by molecular modeling.

The most recent protein-based CSPs was not reported in those reviews; their chromatographic performance are presented on Table S2 (supplementary material). It was found that different proteins as new selectors were not introduced. Nevertheless, diverse techniques of preparation of chromatographic support, namely new techniques of immobilization [42,117,123], entrapment [124], or the application of monoliths [125] have been reported attempting to overcome the stability problems of the proteins.

Proteins are usually immobilized on silica and its pore size can be defined in order to optimize the separation. Matsunaga and Haginaka [126] immobilized AGP on silica particles with different sizes, 5, 3 and 2.1 μm (CSP62–64). The use of this protein as chiral selector was reported, in 1985, by Hellerstein et al. [127]. In another study, Bi et al. [124] entrapped the same protein on a silica support with 100 Å and 300 Å (CSP65–66). Relatively to the first study mentioned, the resolution and efficiency of CSPs with lower particle size were superior. As an example, an excellent resolution was obtained for benzoin (Rs = 14.2) [126]. In the latter study, the described methodology using CSP65–66 proved to be an alternative to high-throughput screening and analysis of biological interactions due to the good affinity results, a maximum of 2.10×10^6 M^{-1} [124].

Matsunaga and Haginaka [128] also studied the effect of particle size, on the efficiency of the column, with cellulase as chiral selector (CSP67). The first application of cellulase as chiral selector was reported by Vandenbosch et al. in 1992 [129]. As in the previous studies, the column with the lowest particle size provided the greatest efficiency and enantioselectivity [128], with a resolution value of 10.7 for propranolol, for example.

Zheng et al. combined a covalent immobilization process with a cross-linking/modification methodology, using HSA as chiral selector, to achieve CSP68. The aim of this new immobilization strategy was to enhance the protein retention [130]. The CSP obtained presented a high binding affinity for warfarin, with a maximum affinity constant of 2.60×10^5 M^{-1}.

A polyclonal antibody CSP (CSP69) was developed by Bi et al. [131] through an alternative methodology, which consisted of the isolation and immobilization of the selector presented on a serum sample (on-line immunoextraction). This methodology was previously described by Matsuda et al. [132]. Once again, it was found that it could be an alternative to the traditional immobilization methodologies since it is not necessary extra steps of protein purification and immobilization [131]. The stability and the robustness of the CSP were also highlighted [131]. This methodology allowed the preparation of a CSP with considerable affinity. For example, a binding affinity value of 90.0×10^6 M^{-1} was obtained for disopyramide.

A new protein-based CSP was reported by Fedorova et al. [133] using a different adsorption methodology, which consisted of BSA adsorbed on eremomycin and grafted on silica (CSP70). An improved resolution, with a good resolution (Rs = 2.14) was provided, in comparison with a CSP comprising only eremomycin.

One of the most recent developments concerning this type of CSPs was the functionalization of monoliths with proteins. Monoliths are based on silica and present the advantage of optimizing the proportion of monomers and cross-linkers. This optimization enables the control of the average size of the throughput channels and the porous [134]. Monoliths can be prepared with different materials and techniques; the advantages comprise a superior flow as well as an enhanced mass transfer resulting on a more efficient separation [121]. Pfaunmiller et al. [125] immobilized HSA on monoliths to obtain CSP71–72. The main objective was to optimize the amount of protein that could be immobilized. As a consequence, the prepared monoliths allowed an improvement on all chromatographic parameters [125].

The recent applications described for this type of CSPs are more diversified than the developments and some of them are related with the optimization of the chromatographic conditions [133–135]. Nevertheless, the number of publications describing the use of protein-based CSPs has been decreasing over the years [117,136]. Binding affinity studies between drugs and proteins and drug-protein interactions were also found [137–139].

2.3. *Cyclodextrin-Based CSPs*

The first application of cyclodextrins as CSPs was described by Armstrong and DeMond in 1984 [140]. Since then, several cyclodextrin-based CSPs have been reported [137,140–142]. Cyclodextrins consist on cyclic oligosaccharides [88]; this type of macrocycles can be divided into three classes, α, β, and γ [43]. The structure of a cyclodextrin consists on a truncated cone [43] with an interior non-polar cavity and free hydroxyl groups located on larger and tiny edges [143]. The hydroxyl groups can be derivatized with diverse polar or apolar substituents [55], which can influence the conformational flexibility of a given cyclodextrin, modifying the size of its cavity and creating additional binding sites [43].

The chiral recognition mechanism is typically based on the formation of an inclusion complex between the analytes and the internal cavity of the cyclodextrin [43]. Additionally, the analytes can establish different types of interactions with the exterior side, including dipole-dipole, hydrogen-bond, ionic, π-π, or London interactions [120]. Cyclodextrins present a considerable number of stereogenic centres, which also contributes to enantiorecognition [55].

Cyclodextrin derivatives can be prepared through physical coating or covalently bonding to a chromatographic support [144]. Covalent bonding of cyclodextrin derivatives is the most applied methodology, since it provides a powerful and resistant linkage to the chromatographic support. The most common linkers are ether, amino, and urea. Recently, a triazole linker was also described [144].

The high stability of this type of CSPs allows the use of an extensive variety of solvents as components of mobile phases, with a wide range of polarities, affording an efficient enantioseparation for different analytes [55]. This type of CSPs can be applied in multimodal elution conditions [142].

Several reviews have been devoted to the developments and applications of cyclodextrin-based CSPs [43,45,85,87,88,140,142,144]. Additionally, Guo et al. [145] reviewed the most recent developments concerning on cyclodextrin functionalized monolithic columns.

Studies related to chiral recognition mechanisms of this type of CSPs using diverse methodologies, such as nuclear magnetic resonance, docking, or molecular modeling, were also addressed to understand the molecular interactions as well as the effect of some chromatographic conditions, such as pH, temperature, or organic modifier, in the enantioseparation [146,147].

The recent cyclodextrin-based CSPs and the evaluation of their chromatographic performance are described on Table S3 (supplementary material). The most recent developments are comprised mainly of the introduction of new derivatives and application of different methodologies of immobilization to the chromatographic support. The preparation of hybrid CSPs to enhance the interactions between the analyte and the stationary phase was also emphasized. It was found that the majority of the new cyclodextrin-based CSPs were prepared based on the most widely used cyclodextrin as CSP, i.e., the β-cyclodextrin [148–151]. Moreover, the immobilization strategy of the chiral selector on the chromatographic support was, mainly, by click chemistry [149–152]. The main advantages of this approach are the mild reaction conditions and the enhanced tolerance of the CSPs to solvents and the range of pH values [153]. The introduction of new methodologies to prepare the chromatographic support was also focused, including the preparation of hybrid supports [154,155], the introduction of monoliths [156] and new chromatographic supports [157], or using a different technique to prepare the support [158].

Zhou et al. [152] reported the linkage of a C6-disubstituted cationic β-cyclodextrin onto an alkynylated β-cyclodextrin bonded to a silica support to afford the CSP73 (Figure 3). The obtained bilayer cationic β-cyclodextrin presented a remarkable enantioselectivity for the tested analytes. As an example of its excellent enantioseparation and resolution, α and Rs values of 2.39 and 4.40, respectively, were obtained for 4-(chlorophenyl) propyl ester [152].

Tang et al. [151] resorted to thiol-ene click chemistry to prepare a sulfoether-bridged cationic per(3,5-dimethyl) phenylcarbamoylated-β-cyclodextrin-based CSP (CSP74) (Figure 3) being able to establish π-π interactions and hydrogen bonding interactions with the tested analytes. Its enantiorecognition ability was demonstrated by a separation factor of 1.70 and resolution of 6.03 for 3-(chlorophenyl) propyl ester.

Zhou et al. prepared a perphenylcarbamate β-cyclodextrin chloride linked by click chemistry to an alkynyl silica support to obtain the CSP75 (Figure 3) [159]. After evaluation of its enantioseparation performance using diverse analytes, they concluded that the introduction of the 3-methoxypropylammonium substituent promoted favorable intermolecular interactions with the analytes. In addition, it was suggested that the mobile phase could cause steric hindrance which prevented the establishment of interactions that were crucial for enantiorecognition [159]. The performance of the CSP was promising with a maximum resolution value of 9.84 and a separation factor of 2.76 for 7-methoxyflavanone and 6-methoxyflavanone, respectively.

A new *N*-benzyl-phenethylamino-β-cyclodextrin was synthetized and bonded to mesoporous nanoparticles of silica obtaining the CSP76 (Figure 3) [149]. The new CSP demonstrated to have a superior flexibility and stability, in comparison with the native β-cyclodextrin-based CSP, being obtained through a more economic process [149]. Relatively to its performance, the higher separation factor and resolution values were 1.30 and 1.97, respectively, for carvedilol.

Four new cyclodextrin-based CSPs (CSP77–80) (Figure 3) were developed by chemical bonding of carboxymethyl-β-cyclodextrin derivatives to silica gel by an amidation reaction on aqueous solution [160]. The carboxymethyl moiety provided additional interactions with the tested analytes, in comparison with the native β-cyclodextrin demonstrating a superior enantioselectivity and resolution [160]. For example, an excellent separation factor value (α = 6.08) was achieved for methyl 2-amino-3-(3-(methylsulfonyl)phenyl)-propanoate hydrochloride. Moreover, for 1-((benzyloxy)carbonyl)-4-hydroxypyrrolidine-2-carboxylic acid, the resolution value was 9.56 [160].

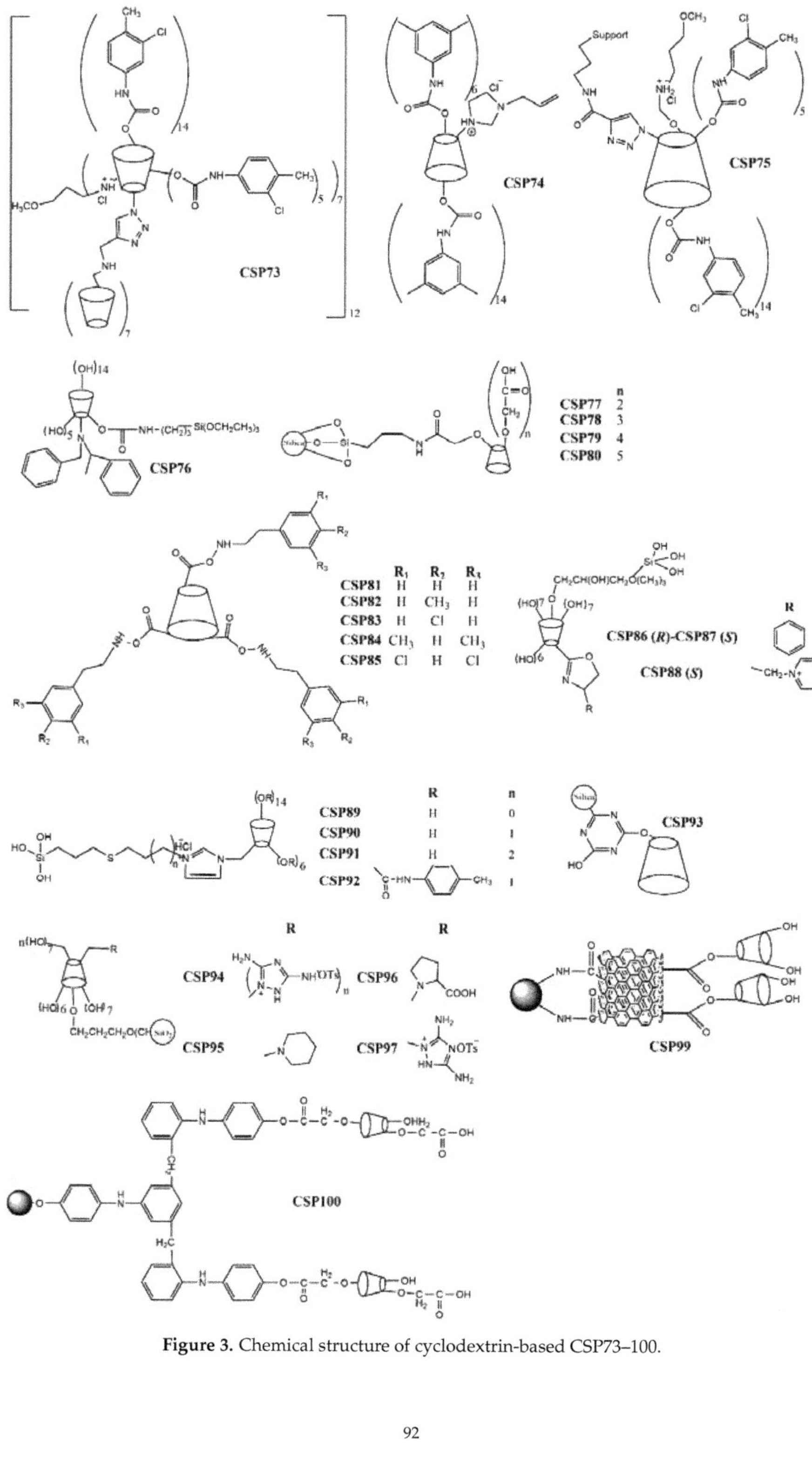

Figure 3. Chemical structure of cyclodextrin-based CSP73–100.

The effect of different substituents on CSPs has also been investigated. Chen et al. [148] synthetized β-cyclodextrin derivatives with a phenylcarbamate moiety with different patterns of substituents, which were subsequently immobilized onto the silica gel through intermolecular polycondensation of the triethoxysilyl groups (CSP81–85) (Figure 3). They reported that the presence of an aromatic ring with electron-withdrawing groups on the β-cyclodextrin improved the chiral recognition for analytes with electron-donor groups since the number of possible π-π interactions was superior [148]. The hydrogen-bond interactions between the carbonyl group or nitrogen of the analytes and the amino group of phenylcarbamate of cyclodextrin-based CSP were also improved [148]. Relatively to the performance of the CSPs, the highest separation factor value achieved was 2.87 for Pirkle alcohol.

Li et al. [150] arrived at similar conclusions after preparing oxazolinyl-functionalized β-cyclodextrins covalently bonded to silica support (CSP86–88) (Figure 3). They described that analytes with electron-donating or hydrogen-bonding groups were easily enantioseparated due to a higher number of π-π and hydrogen-bonding interactions [150]. CSP88 was more suitable for enantioseparation of polar compounds since it promoted electrostatic interactions due to the presence of an ionic group [150]. Additional factors that could influence the enantioseparation performance of the CSPs, such as the spacer length, selector concentration, and rim functionalities, were also investigated [150]. A reduced surface concentration and a superior flexibility of the spacer decreased the enantioselectivity, since it weakened the interactions between the selector and the analyte [150]. Additionally, a superior selector concentration could be beneficial for enantioseparation of some racemates. The performance of the CSP was promising with an excellent resolution for ketoprofen (Rs = 22.0) and a separation factor value of 15.5 for loxoprofen.

The same group developed four thioether bridged cationic cyclodextrin-based CSPs (CSP89–92) (Figure 3) and the influence of the spacer length, selector concentration, and rim functionalities on the performance of the CSP were studied [161]. In this case, it was found that CSPs comprising a spacer with a superior length could compromise their ability of enantiorecognition; however, a higher concentration of the selector was positive [161]. The higher resolution value achieved was 12.7 for 4-nitrophenyl propyl oxide, and the separation factor value was 3.30 for styrene oxide.

Regarding the development of hybrid CSPs, a spherical β-cyclodextrin-silica hybrid CSP (CSP93) (Figure 3) was reported by Wang et al. [154] highlighted by the presence of multiple functional groups, which expanded the spectrum of possible interactions with analytes. The β-cyclodextrin derivative was introduced into the pore channels and pore wall framework, and the linker was attached just into the pore channels [154]. Both the interior and exterior of the CSP participate in the process of the enantiorecognition. For example, separation factor and resolution values of 1.63 and 4.65 were obtained for diclofop and mandelonitrile, respectively.

Regarding the use of new materials as chromatographic supports, a modification of the most common chromatographic support was performed by Zhao et al. [155] to obtain new CSPs. The modified silica gel was named hydride silica, and its surface was covered by silica-hydrogen bond instead of silica-hydroxyl. The hydride silica presents a superior resistance to water, a reduced polarity, and an improved separation rate and stability and it can be used with a wide variety of solvents [155,162]. Zhao et al. [155] prepared polar group derivatives of β-cyclodextrin bonded to hydride silica to obtain CSP94–97 (Figure 3). The higher resolution value achieved was 9.31 for methyl (2*R*,3*R*,4*S*,5*R*)-5-(4-fluorophenyl)-4-nitro-3-phenyl-3-(trifluoromethyl)-pyrrolidine-2-carboxylate; the best separation factor value was 3.65 for methyl (2*R*,3*S*,4*S*,5*R*)-5-(4-fluorophenyl)-4-nitro-3-(*p*-tolyl)-pyrrolidine-2-carboxylate.

Ghanem et al. [156] used a different strategy to prepared new cyclodextrin-based CSPs. They encapsulated the trimethylated-β-cyclodextrin to a polymeric monolithic, to obtain a superior surface area, reduced pore size, and enhanced total pore volume, and developed the CSP98. They also studied the physical characteristics of the CSP to established relationships with the potential of enantiorecognition and concluded that a superior concentration of the selector improved the enantioseparation [156]. The CSP98 demonstrated a suitable mechanical and thermal stability as well as reproducibility [156] with a maximum resolution value of 2.51 for flavanone, and a separation factor of 1.42 for carprofen.

A different chromatographic support was also proposed by Qiang et al. [157], who described a β-cyclodextrin CSP based on graphene oxide (CSP99) (Figure 3), which was covalently linked to amino silica gel by an amide bond. The graphene oxide and cyclodextrin presented a synergetic effect for enantiorecognition being the hydrogen bonding and π-π interactions the main interactions between the CSP and analyte. The CSP99 was also applied for hydrophilic interaction chromatography. Regarding the chromatographic results, a separation factor of 38.8 was achieved for equol, and a resolution value of 2.17 for 1-phenylethanol [157].

A light-assisted preparation of carboxyl methyl β-cyclodextrin-based CSP (CSP100) (Figure 3) was described by Tang et al. [158] who used ultra-violet light to link the chiral selector to silica, which promoted the modification of ionic bonds into covalent bonds. This technique proved to be eco-friendly and efficient [158]. The morphology and chemical composition of CSP100 was characterized. Moreover, it was concluded that its enantiorecognition ability was dependent of hydrogen bonding and dipole-dipole interactions [158]. The maximum resolution value achieved was 8.04 for chlortrimetron.

2.4. *Macrocyclic Antibiotic-Based CSPs*

Macrocyclic antibiotics are the second most versatile group of CSPs, after polysaccharides; their planning was inspired by cyclodextrins. The first report of macrocyclic antibiotics as CSP was in 1994, describing the application of vancomycin, thiostrepton, and rifamycin B as CSPs [163].

Macrocyclic antibiotics are divided into four groups: ansamycins, polypeptides, glycopeptides, and aminoglycosides [164]. Ansamycins comprise an aromatic unit linked to an aliphatic chain, and their classification is based on the aromatic moiety. If the aromatic unit is a naphthalene or naphthoquinone, it is denominated naphthalenic ansamycin, while if it is a benzene or benzoquinone, it is a benzenic ansamycin [164]. The most common ansamycins used as CSPs are rifamycins B and SV; the first one is enantioselective for cationic compounds and the second for neutral and anionic [165]. Polypeptides have few aromatic ring units while aminoglycosides do not have this type of structural feature [165]. Only one polypeptide is used as CSP, thiostrepton, whereas aminoglycoside class comprises more CSPs, such as fradiomycin, kanamycin, and streptomycin [166].

Glycopeptides are the most promising class of macrocyclic antibiotic-based CSPs, including avoparcin, ristocetin A, teicoplanin, vancomycin, and derivatized analogues from vancomycin, among others [165]. The chemical structure of glycopeptides consists on a glycosylated cyclic or polycyclic peptide. The central framework is a heptapeptide, in which five of the seven amino acid residues are common to all glycopeptides [164]. Glycopeptides have some flexibility due to the possibility of rotation of sugar groups [165].

The structure of macrocyclic antibiotics allows a variety of interactions with the analytes, such as hydrophobic, π-π, dipole-dipole, hydrogen-bond, electrostatic, ionic, and Van der Waals interactions [166], being either attractive or repulsive. It is possible for the formation of inclusion complexes to occur [167]. The high number of stereogenic centers in their structures also contributes for its high capacity of chiral recognition [55]. Nevertheless, the chiral recognition mechanism of this type of CSPs it is not currently quite understood [43].

The chromatographic support of this type of CSPs is, predominantly, silica gel [164]. Macrocyclic antibiotic-based CSPs are able to operate in all chromatographic elution modes [88]. Besides that, the macrocyclic antibiotic-based CSPs provide a complementary enantioselective profile [167].

Over the years, the developments carried out to obtain diverse macrocyclic antibiotic-based CSPs as well as their applications have been compiled [43,45,85,87,88,164,165,167,168]. Additionally, some authors focused their studies on the mechanism of chiral recognition [169,170].

The most recent reports related to new macrocyclic antibiotic-based CSPs as well as their chromatographic parameters are presented in Table S4 (supplementary material). The developments did not comprise the introduction of new antibiotics as chiral selectors but rather the use of new chromatographic supports, specifically the use of silica particles with sub-2-μm size. In fact, they are mainly based on the preparation of new teicoplanin and vancomycin-based CSPs by reducing the size of the packaging material [171–173]. The main objective was the improvement on chromatographic performance by reduction on analysis time and enhance of resolution and enantioselectivity.

Min et al. [171] described the preparation of a teicoplanin-based CSP bonded to sub-2 μm superficially porous particles (CSP101). The main focus was to avoid aggregation and to uniformize the size distribution, enhancing the surface area. The high resolution and enantioselectivity obtained in a short time of analysis were highlighted. The maximum resolution value was 5.60 for methionine, and the separation factor was 9.40 for norvaline.

Ismail et al. [172] developed a teicoplanin-based CSP with a sub-2 μm chromatographic support; however, in this case, they used totally porous silica particles (CSP102). They pointed out the flexibility of the CSP to operate on different elution modes. The selectivity, efficiency, and the short analysis time on ultra-high-performance liquid chromatographic (UHPLC) were also emphasized [172]. Its performance was promising achieving a resolution value of 10.7 for alanine, and a separation factor of 3.45 for mandelic acid.

Vancomycin was bonded to sub 2-μm diol hydride-based silica particles by Rocchi et al. [173]. Four new CSPs were developed (CSP103–106) with the same main objective: reduction of analysis time. It was inferred that this technique could be applied to other chiral selectors due to the promising results [173]. The maximum resolution value was 3.36 and the separation factor was 2.69 for haloxyfop.

Despite the reduction of particle size of the support, new materials were introduced as chromatographic support. Recently, Xu et al. [174] described the preparation of a vancomycin-based CSP through the combination of monoliths and polymeric cross-linking (CSP107). The new CSP possessed a good mechanical stability, permeability, and enantioselectivity [174]. The influence of some chromatographic conditions was also investigated. The performance of the CSP was satisfactory, with a resolution value of 1.47 for salbutamol, and a separation factor of 1.23 for carteolol. Hellinghausen et al. [175] prepared the CSP108 through the prior synthesis of vancomycin, by Edman degradation, and further binding it to superficially porous particles through a primary amine group of vancomycin, which resulted from the removal of an *N*-terminus leucine residue. The CSP108 presented promising results with a good resolution and separation factor values for 2-amino-2-phenylbutyric acid (Rs = 2.70 and α = 1.57).

A vancomycin-based CSP was recently prepared (CSP109) by a photochemistry-based method [176]. Additionally, the influence of flow rate, elution mode, buffer, and the mass of analyte were also investigated. The addition of 2-propanol, buffer and an increase on analyte mass improved its enantioresolution performance, since π-π interactions were superior. The chromatographic performance was good, with a maximum resolution of 3.08 and a separation factor of 4.23.

It is also important to highlight that the complementary behavior of the different macrocyclic antibiotic-based CSPs continues to be a subject of great relevance. In fact, several recent studies can be found in literature [166,167,177]. Most of them compared the enantioresolution performance of teicoplanin and teicoplanin aglycone CSPs [166,167] or of vancomycin and teicoplanin CSPs [177].

2.5. Brush-Type or Pirkle-Type CSPs

Brush-type or Pirkle-type CSPs were introduced in 1979, when Pirkle and House described the development and application of a chiral fluoro alcoholic CSP able to enantioseparate diverse classes of analytes [178].

Neutral synthetic chiral low-molecular mass molecules are the base of this type of CSPs [43]. These molecules should promote donor-acceptor interactions as a hydrogen-bond, π-π, or dipole-dipole, in addition to attractive and/or repulsive Van der Waals interactions [45]. As they comprise small molecules as chiral selectors, the mechanism of chiral recognition is, frequently, based on the "three-points" model, which refers that the establishment of at least three interactions between one of the enantiomers to be resolved and the CSP are essential for chiral recognition [179].

The chiral selectors are usually covalently linked to a silica support, which can have monosubstituted or trisubstituted silane groups, through a spacer [55]. Over several years, Pirkle's group has developed successive generations of CSPs [180], based on the principle of reciprocity [181] and on chromatographic [182,183] and spectroscopic [184,185] methods to understand the chiral recognition mechanisms. Among them, Whelk-O1 CSP, created by a rational approach, is the most applied and versatile CSP in both academic and industrial fields [186].

Initially, the preferred elution mode was the normal phase since it provides a favorable environment for the interactions needed to enantioseparate the analytes [88]. Nevertheless, this type of CSPs can also be used in polar organic and reversed-phase elution modes [187–189].

The advantages inherent to this type of CSPs are the compatibility with a wide range of solvents used as mobile phase, the stability to temperature and pressure, the considerable loading capacity and the possibility to be easily scaled up to preparative chromatography [190]. Another key advantage is the possibility of switching the configuration of the chiral selector and to use the inverted configuration column approach [43]. Its kinetic performance is reasonable and the fact that the structure of the chiral selector is relatively "simple" allows an easier knowledge of the chiral recognition mechanisms as well as a consecutive optimization of the selector [191]. Pirkle-type CSPs are characterized by their diversity and versatility since it is possible to use a variety of different small molecules as chiral selectors as well as introduce different substituents that can improve enantioselectivity. In addition, they can be highly specific for certain types of chiral compounds [120].

Pirkle-type CSPs have evolved over the years. Certain types of CSPs have more reported progresses, mainly due to the possibility of use of a wide variety of small molecules as chiral selectors. Several Pirkle-type CSPs can be found in literature comprising chiral selectors closely related to the original Pirkle's group CSPs and others structurally different [43,45,85,87,88,180,186,190,191]. Recently, a literature survey made by our group covering the report on Pirkle-type CSPs developed during the last 17 years was published [190]. We described 226 new CSPs, including a wide diversity of small molecules as chiral selectors, including amine, amino alcohol and amino acids derivatives, peptides, drugs, selectors based on natural products, and xanthone derivatives, among others [190].

The recent developments of this type of CSPs also include the use of new chromatographic supports, such as monolith supports, core-shell particles, or particles with a reduced size (sub 2-μm). The reduction of particle size enables the adaptation of Pirkle-type CSPs to UHPLC, the first ones to be converted, which are associated with the inherent advantages such as the reduction of analysis time and quantity of the solvent, improved efficiency, and enantioresolution [192,193]. Regarding the use of core-shell particles, it was found that the packaging with this type of material or, alternatively, with fully porous particles led to differences on chromatographic performance. The diffusion on core-shell particles is lower than in fully porous particles, which is especially beneficial for large analytes, since it prevents a decrease on efficiency due to an inefficient mass transfer. The distribution of particle size of core-shell particles is nearly unimodal, which increases efficiency on enantioseparation of small analytes [194,195].

Different synthetic methodologies to obtain the chiral selectors as well as for its immobilization on the chromatographic support were also introduced. The synthesis of biselectors was another approach [190]. The most recent Pirkle-type CSPs and the chromatographic parameters obtained after evaluation of their enantioresolution performance are presented on Table S5 (supplementary material).

Qiao et al. [196] developed a CSP based on *N*-ferrocenyl benzoyl-(1*S*,2*R*)-1,2-diphenyl ethanol as a chiral selector (CSP110) (Figure 4). The conjugation of a cyclopentadienyl carbon ring with an aromatic ring demonstrated to improve enantioselectivity. The chiral recognition mechanisms were also explored, revealing that hydrophobic, hydrogen-bond, π-π, and dipole-dipole interactions between the chiral ferrocene CSP and acidic and basic groups of the analytes were crucial. The performance of the CSP was promising achieving a maximum resolution value of 4.13 and a separation factor of 2.43 for 3-nitrophenol.

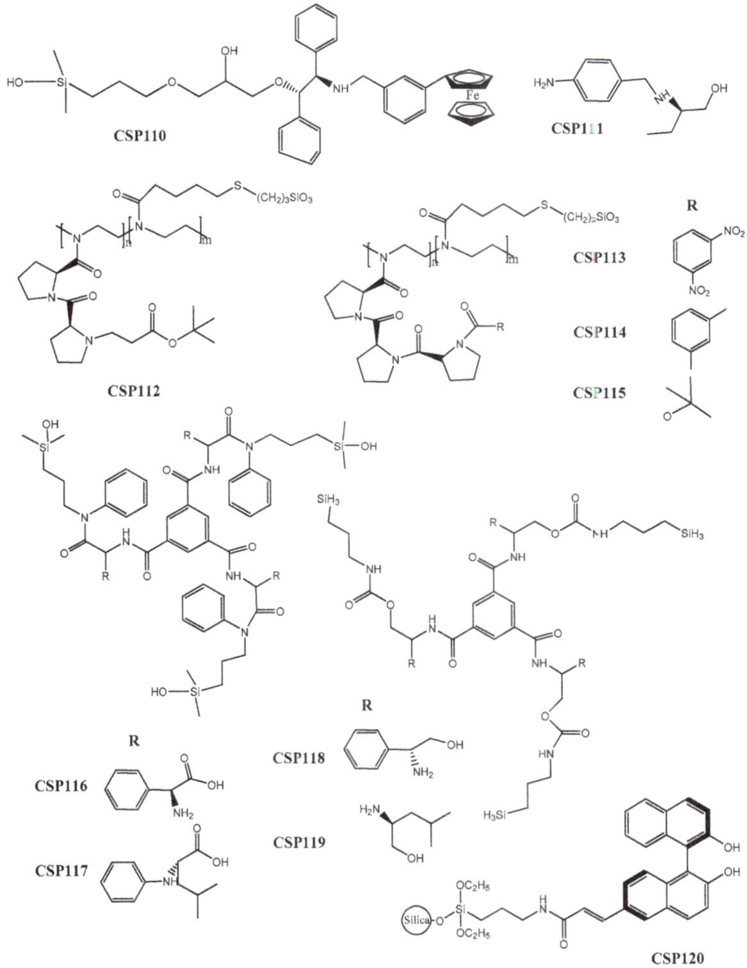

Figure 4. Chemical structures of Pirkle-type CSP110–120.

Çakmak et al. [197] synthetized an aromatic amine derivative of (*R*)-2-amino-1-butanol for the application as chiral selector of a new CSP (CSP111) (Figure 4). In the same study, they used docking, molecular dynamics simulation, and quantum mechanical computation methods to characterize the mechanisms of chiral recognition. The performance of the new CSP was good with a high resolution value of 3.85 for 2-phenylpropionic acid, and a separation factor of 2.75 for mandelic acid.

Four new pseudopeptide-based CSPs were developed (CSP112–115) (Figure 4) inspired by the possibility of enantiorecognition ability of an organocatalyst [198]. It was found that the enantioselectivity of the CSPs was dependent of the degree of derivatization of diproline portion and of the length of polymeric chain. The chromatographic results were promising, achieving, for example, separation factor and resolution values of 9.80 and 2.89, respectively, for 1-phenylethan-1-amine and *N*-(1-(naphthalen-2-yl) ethyl)-3,5-dinitrobenzamide.

Additionally, in another recent work, derivatives of amino acids and amino alcohols as CSPs were prepared by Yu et al. [199], based on C3-symmetric CSPs (CSP116–119) (Figure 4). It was found that a phenyl group linked to amide was crucial for chiral recognition and, despite the chiral selectors did not possess a π-acidic or π-basic group, their performance was promising. For example, a separation factor value of 2.58 was achieved for 2-phenyl-2-pentanol.

Wang et al. [200] synthetized the (*R*)-6-acrylic-binaphtol as chiral selector through addition of the acrylic group to the (*R*)-binaphtol and developed the CSP120 (Figure 4). The mechanisms of chiral recognition, the effect of the temperature and mobile phase composition were also discussed. It was found that the flexibility of the CSP and the π-π stacking event allowed the retention of the analytes without compromise the enantioseparation [200]. Regarding the chromatographic results, a separation factor value of 1.12 was achieved for 3,5-dinitro-*N*-(1-phenylethyl) benzamide.

Along with the continuous developments of this type of CSPs, it is important to emphasize that a broad range of recent applications have also been reported [201–205].

2.6. Ion-Exchange-Type CSPs

Ion-exchange-type CSPs were introduced by Salvadori et al. in 1985, who described the application of cinchona alkaloids as CSPs [206]. Nevertheless, Lindner group developed the majority of this type of CSPs [207]. Ion-exchanger selectors can be subdivided into three groups: anionic, cationic, or zwitterionic [208].

The most common anion-exchangers as chiral selectors are cinchona alkaloids [45] and terguride [43]. Anion-exchanger selectors are appropriate for enantioseparation of acidic compounds; their enantioselectivity are attributed to the five stereogenic centers of the basic nucleus common to quinine and quinidine [209]. Cation-exchanger selectors are useful for enantioseparation of basic analytes, which are structurally based on chiral sulfonic or carboxylic acid compounds as selectors [45]. Zwitterionic selectors were introduced, more recently, by Lindner et al. [210] by merging key cation- and anion-exchange moieties in one single chiral selector [45]. Those CSPs can been applied for the enantioresolution of acid, basic, and amphoteric compounds [209]. Zwitterionic CSPs have overcome the main disadvantage of anion and cation-exchanger CSPs, since these two groups only separate enantiomers with opposite charge [210].

The chiral mechanism of recognition is mainly based on ionic interactions between the charged analytes and the opposite charged groups of the CSPs [208]. Hydrogen bonds as well as π-π interactions are also important for complex formation [45]. The ion-pairing of solvent controls the adsorption and retention of the analytes [43]. Polar-organic and reversed-phase elution modes are the preferential elution modes for this type of CSPs [43]. The retention and enantioselectivity are affected by the pH and the nature and concentration of acid or base added to the mobile phase [43].

The progresses resorting to this type of CSPs have been reviewed over the years [43,45,85,88,207, 209,211–213]. Recently, Ilisz et al. [213] compiled the most recent developments concerning to anionic and zwitterionic-exchange-based CSPs, which are related to the application of different techniques of preparation of chromatographic support to attempt the optimization of chromatographic parameters. The most recent ion-exchanger-type CSPs and their chromatographic parameters are presented on Table S6 (supplementary material). The majority of the recent developments focused on quinine and quinidine derivatives as chiral selectors [214–217], the key structural moiety representative of anion-exchanger selectors.

Todoroki et al. [214] developed a new technique to prepare new ion-exchange CSPs, specifically cinchona alkaloid-based quinine and quinidine-type fluorous-tagged-CSPs (CSP121–125) (Figure 5). The main objective was to improve the enantioseparation properties enabling a sensitive, selective, robust, and reproducible analysis methodology. The versatility of the new CSPs was another advantage, as it was capable to enantioseparate bulky, aromatic compounds, in addition to amino acids, such as threonine with a resolution value of 11.8, and asparaginine, with a separation factor of 4.56.

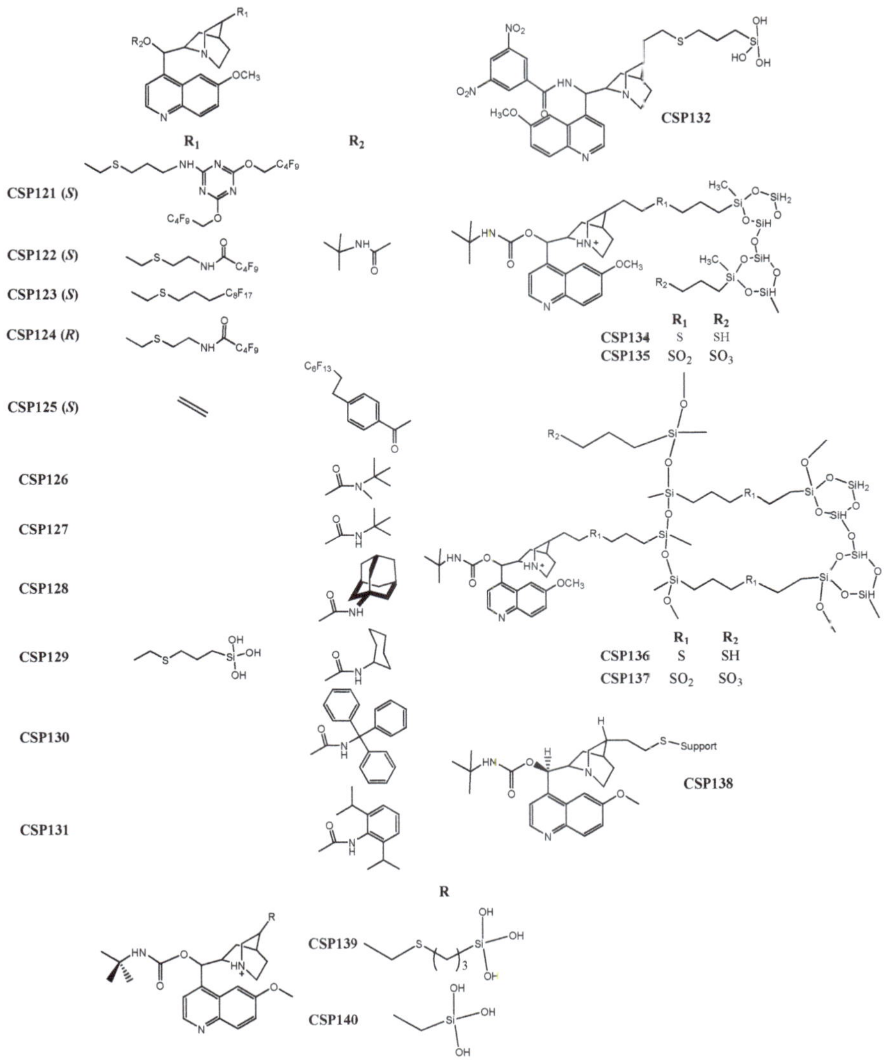

Figure 5. Chemical structures of Ion-exchange-type CSP132–140.

The complementarity profile between anion-exchange-type CSPs was the focus of a recent study reported by Lämmerhofer et al. [218], who prepared several cinchona carbamate selectors with distinct carbamate residues to obtain CSP126–131 (Figure 5). Different structural moieties were introduced to enhance the possibility of complementary; for example, the introduction of bulky groups to create steric hindrance or aromatic rings to provide π-π interactions. The complementary accomplished with the new CSPs allowed the expansion of the enantioselectivity range. The performance of the CSPs was promising, achieving, for example, a separation factor of 17.0 for leucine.

De Martino et al. [216] reported the synthesis of an anion exchange hybrid selector, the 3,5-dinitrobenzoyl-9-amino-9-deoxy-9-epiquinine, to develop CSP132 (Figure 5). The strategy applied was the association of typical moieties of Pirkle-type selectors with key moieties of anionic-exchange-based selectors, enlarging the possibility of multiple interactions with the analytes.

The performance of this hybrid CSP was promising, with separation factor and resolution values of 2.06 and 11.0, respectively, for diazepam *N*-oxide.

A new immobilization technique based on click chemistry, was described by Lämmerhofer et al. [219], who prepared a cross-linked *tert*-butylcarbamoyl quinine-based CSP (CSP133). The technique allowed achieving a CSP with reduced resistance to mass transfer and retention times, as well as an improved stability. During the optimization of the procedure, some features were discussed, such as the amount of polysiloxane, chiral selector, radical initiator, and reaction solvent, as well as reaction time and size of the chromatographic support particles [219]. The performance of the CSP was promising; for example, with separation factor and resolution values of 1.54 and 5.20, respectively, for *N*-[(9*H*-fluoren-9-ylmethoxy) carbonyl]-phenylalanine.

The same group also resorted to click chemistry to prepare other CSPs based on *tert*-butylcarbamoyl quinine (CSP134–137) (Figure 5) [215]. The optimization of the selector's structure allowed the avoidance of non-specific interactions that could reduce chiral recognition [215]. The introduction of a sulfonic group afforded a reduction on the retention times and an improvement, in some cases, of separation factors since its negative charge provided electrostatic interactions, promoting an effect similar to the counterion effect [215]. The performance of the CSP was satisfactory achieving a maximum resolution value of 6.20 for *N*-(9-fluorenylmethoxycarbonyl)-phenylalanine and a separation factor of 1.66 for *N*-acetyl-phenylalanine.

The application of core-shell particles was another strategy. A new CSP based on *tert*-butylcarbamoylquinine selector (CSP138) (Figure 5) was described by the same group, to promote the enantioseparation of several proteinogenic amino acids [217]. Core-shell particles were introduced in order to improve the analysis time, which was a promising methodology for the bioanalytical area, since it could be combined with sensitive fluorescence detection or highly sensitive and selective mass spectrometric detection. The column presented a reasonable performance with good enantioselectivity and resolution. For example, α and Rs values of 1.55 and 4.08, respectively, were achieved for threonine [217].

Armstrong et al. [220] also resorted to core-shell particles to develop two new quinine-based CSPs (CSP139–140) (Figure 5) for ultrafast liquid and supercritical fluid chromatography. The new CSPs allowed fast analysis with high enantioselectivity and efficiency for the tested analytes. The performance of the new CSPs was promising, affording, for example, a maximum resolution of 25.5 and a separation factor of 14.5 for *N*-(3,5-dinitrobenzoyl)-leucine.

It is important to highlight that recent applications of this type of CSPs were diversified, with the focus, mainly, on the application of anion and zwitterionic-type CSPs with amino acids [221–225].

2.7. Crown-Ether-Based CSPs

Crown-ethers as CSPs were firstly described by Sousa et al. [226] for the enantioseparation of primary amine salts. Crown-ethers consist on macrocyclic polyethers, with a cavity of a specific size, being able to form complexes with analytes [227]. CSPs based on crown-ethers can be divided into two major groups: the crown- ethers comprising a 1,1′-binaphthyl group and those containing two tartaric acid groups [228]. The enantiomers of α-amino acids and primary amines may be separated by the first type of crown-ethers CSPs [229]; the second group can be applied for enantioseparation of primary and secondary amino compounds and non-amino compounds [229].

The mechanisms of chiral recognition are typically driven by triple hydrogen bonds established by an ionized ammonium group of the analytes and three oxygen of the CSP, leading to the formation of an inclusion complex [43]. The electron-donor oxygen particles are distributed on inside of the cavity of the crown-ether [88]. Steric hindrance from the substituents of the analyte ions and the functional groups of the crown-ethers can influence the enantioseparation [43]. Additional interactions are essential to complement the formation of the complex, including π-π, hydrogen-bond, and dipole-dipole interactions [230]. Mobile phases should be strong acidic aqueous solutions to achieve the total protonation of the amino group of the analytes [43]. Crown-ether-based CSPs can be obtained by a coating process or by immobilization [231]. To avoid the leaching of the CSP from the column and to allow the analysis of hydrophobic compounds the use of covalently bonded CSPs it can be preferable than the coated [232].

The developments of crown-ether-based CSPs have been revised over the years [43,45,85,87,88,228,229]. Hyun et al. [228] reviewed the most recent developments, related to both classes of crown-ethers as CSPs, highlighting techniques for the preparation of the chromatographic support or the protection of unreacted residues.

Crown-ether-based CSPs have some restrictions related to the target analytes; however, their preparation and the chromatographic conditions can be modifier to improve the chromatographic results. The majority of the recent developments encompassed the use of different strategies for immobilization of the chiral selectors to the chromatographic support (click chemistry) and the introduction of different functional groups on previous described selectors. The most recent crown-ether-based CSPs as well as their chromatographic performance presented on Table S7 (supplementary material). Some of the recent CSPs are based on calix[4]arene derivatives as chiral selectors [233–235]. The new derivatives are mainly prepared using click chemistry [234,235].

Accordingly, the CSP141 (Figure 6) comprising the aza-15-crown-5-capped methylcalix[4] resorcinarene derivative was developed by Ma et al. [233]. Structurally, the CSP possesses two key recognition sites, enhancing the possibility of interactions and, consequently, enantioseparation of analytes. The robustness of the CSP was highlighted since it could operate on different elution modes with a short analysis time (for example, $k_1 = 0.08$ for *m*-nitrophenol).

Yaghoubnejad et al. [234] prepared a calix[4]arene functionalized with two L-alanine units to develop CSP142 (Figure 6), through covalent binding between the allyl groups at the lower rim of the chiral selector and the chromatographic support, using click chemistry. The CSP142 was able to enantioseparate both π-acidic and π-basic analytes. It was suggested that the used technique could easily be adapted to other derivatives to obtain improved CSPs. The maximum resolution value achieved was 1.43 and the separation factor was 2.00 for mandelic acid.

Click chemistry has also been explored, by Li et al. [235], for the preparation of a click-dibenzo-18-crown-6-ether-based CSP (CSP143) (Figure 6). The effect of pH and concentration of salt in the mobile phase on chromatographic parameters was evaluated [235]. It was found that the retention of strong acids decreased with the increment on salt concentration. Regarding the pH values, the retention of both acidic and basic analytes decreased with its reduction. The retention factors were good with a minimum of 0.10 for uracil.

CSPs comprising carboxyl derivatives of crown ethers as chiral selectors were also prepared. Németh et al. [236] synthetized derivatives of acridino-crown ethers containing a carboxyl group to obtain eleven CSPs (CSP144–154) (Figure 6). The CSPs were developed taking into account some structural features that could favor the interactions crucial for chiral recognition mechanisms. Due to the rigidity of the tricyclic ring system, the enantioselectivity was improved [236]. The obtained performance was reasonable with a maximum resolution value of 1.20 for 1-(1-naphthyl)-ethylamine, and a separation factor of 2.05 for 1-(4-nitrophenyl)-ethylamine hydrogen chloride.

The development of hybrid crown-ether-based CSPs was also reported. Deoxycholic-calix[4]arene hybrid-type selectors were synthetized, by Yaghoubnejad et al. [237], aiming to enhance the interactions of the obtained CSPs (CSP155–156) (Figure 6). The calix[4]arene unit was fundamental for the mechanisms of chiral recognition being responsible for the establishment of hydrophobic and π-π interactions, important for inclusion complexes formation. The presence of an acidic or basic modifier in mobile phase was beneficial for enantioresolution of acidic or basic analytes [237]. Relatively to its performance, a maximum resolution value of 3.93 and a separation factor of 4.30 were obtained for mandelic acid.

CSP141

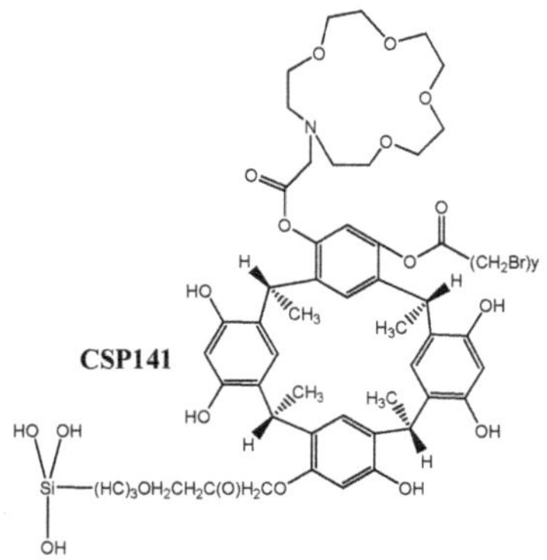

CSP142

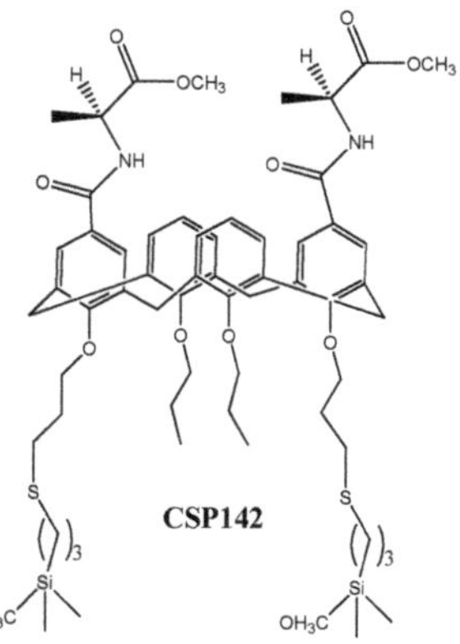

CSP143

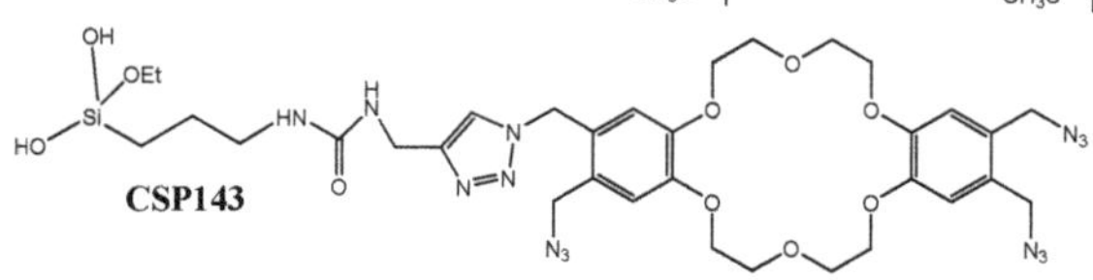

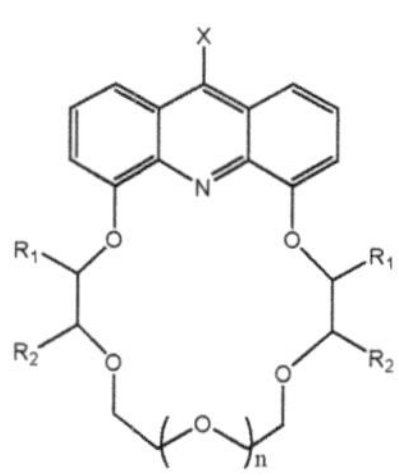

	X	R_1	R_2	n
CSP144 (*R,R*)	COOK	CH_3	H	1
CSP145 (*R,R*)	COOH	CH_3	H	1
CSP146 (*R,R*)	COOH	$(CH_3)_2CHCH_2$	H	1
CSP147 (*R,R*)	$CONH(CH_2)_3S(CH_2)_3$ spherical HPLC silica gel	CH_3	H	1
CSP148 (*R,R*)	$CONH(CH_2)_3$ spherical HPLC silica gel	CH_3	H	1
CSP149 (*R,R*)	$CONH(CH_2)_3$ spherical HPLC silica gel	$(CH_3)_2CHCH_2$	H	1
CSP150 (*S,S*)	H	H	CH_3	1
CSP151 (*S,S*)	COOH	H	CH_3	1
CSP152 (*S,S*)	COOH	H	CH_3	2
CSP153 (*R,R*)	COOH	CH_3	H	2
CSP154 (*R,R*)	$CONH(CH_2)_3$ spherical HPLC silica gel	H	CH_3	1

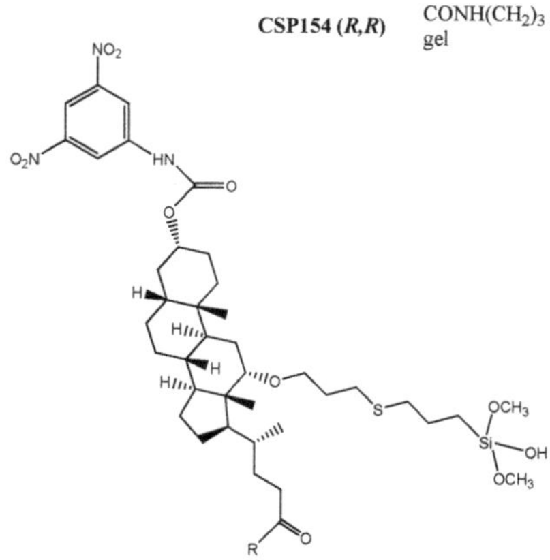

R

CSP155

CSP156

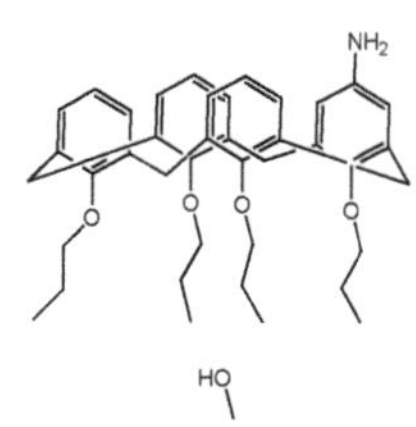

Figure 6. Chemical structures of crown-ether-based CSP141–156.

2.8. Cyclofructan-Based CSPs

Cyclofructans are the most recent type of CSPs being introduced, in 2009, by Armstrong et al. [238]. Moreover, they demonstrated that suitable derivatized of cyclofructans presented a superior enantioselectivity in comparison with native cyclofructans [238].

Cyclofructans are cyclic oligosaccharides formed by units of D-fructofuranose β(2→1) linked together [45]. They are also described as a crown-ether nucleus rounded by fructofuranose units, with its number between 6 and 8 [239]. Each unit has four stereogenic centres [88]. In opposition to cyclodextrins, the interior of the nucleus is hydrophilic [45]. The mechanism of chiral recognition is based on the formation of a complex, which is driven by polar interactions, including dipole-dipole and hydrogen-bond interactions [45]. Therefore, the analytes to enantioseparated should not be hydrophobic and may have hydrogen-acceptor and polarizable groups next to stereogenic center [240]. The acidic hydrogen-bond play an important role on chiral recognition, thus, the presence of a polarizable group that causes steric hindrance to the basic portion of the cyclofructan is favorable [240]. The main advantages of this type of CSPs are their high loadability and versatility, as it is able to enantioseparate basic, acidic, and neutral analytes [239]. Moreover, they can be used in different elution modes [239]. Although cyclofructan-based CSPs are recent, some reviews can be found related to its developments and applications [45,55,85,87,88,90]. The elucidation of their chiral recognition mechanisms has been the focus of some studies to clarify the interactions between the CSP and the analytes [45]. The most recent cyclofructan-based CSPs and its chromatographic behavior are presented in Table S8 (supplementary material). The latest developments include the synthesis of new derivatives of cyclofructan as chiral selectors, the preparation of new chromatographic supports, and the application of different immobilization strategies.

In order to evaluate the effect of electron-donating and electron-withdrawing groups on enantioselectivity, Khan et al. [239] synthetized chlorinated aromatic derivatives of cyclofructan 6 and developed CSP157–166 (Figure 7). The presence of a chlorine proved to be beneficial for enantioselectivity, in opposition to nitro group, especially in the ortho position of the aromatic ring, which negatively affected the chiral recognition [239]. A maximum resolution value of 6.90 for 2-2′-binaphthol, and a separation factor of 2.05 for Tröger's base were obtained.

The influence of the degree of substitution, as well as the size of the substituents, was researched by Padivitage et al. [241] by preparing CSP167–171 (Figure 7), with basic derivatives cyclofructan 6 as a selector. It was concluded that bulky groups caused steric hindrance and that a high degree of substitution (up to six substituents) negatively affected the enantioselectivity. Moreover, it was found that charged cyclofructans did not possess a superior ability of enantiorecognition [241]. Relatively to the performance of the CSPs, as an example, separation factor and resolution values of 1.43 and 3.10 were obtained for warfarin.

An alternative technique for preparation of a CSP was presented by Qiu et al. [242] (CSP172) (Figure 7), using click chemistry to immobilize the chiral selector, cyclofructan 6, to a resin. The resin was chosen as chromatographic support due the advantages inherent to this material, such as high adsorption capacity, high mechanical strength, lower cost, and reduced sensitivity to pH. Although the chromatographic results were only reasonable, with a resolution of 1.40 for *trans*-1-amino-2-indanol, and a separation factor of 1.41 for *N*-*p*-tosyl-1,2-diphenylethylenediamine, the stability and reproducibility of the CSP were emphasized [242].

Similarly to other types of CSPs, the applications of cyclofructan-based CSPs are becoming more prominent [243,244].

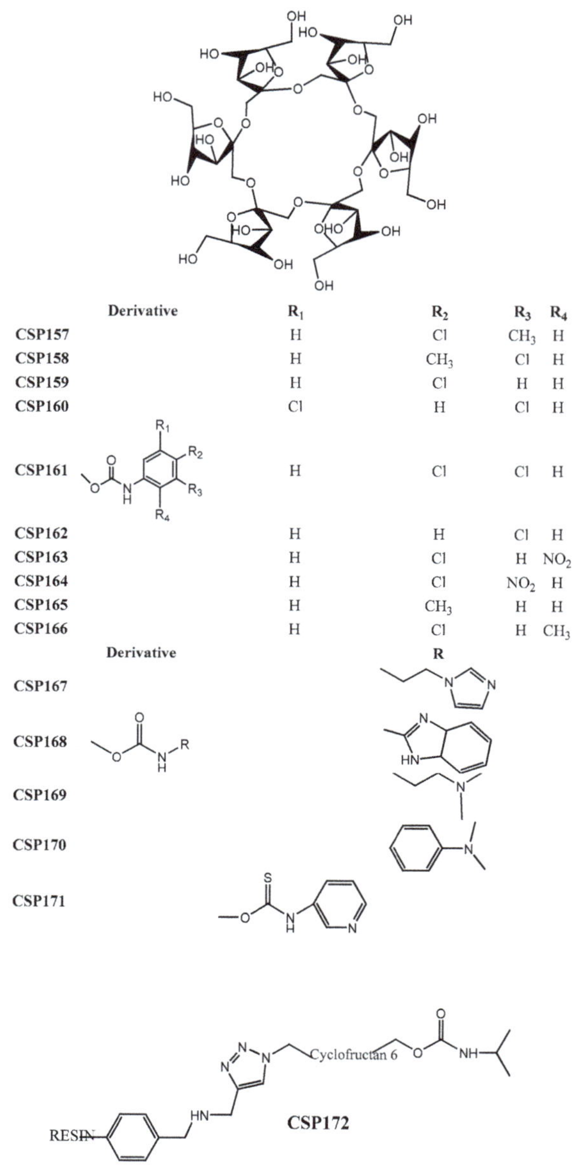

Figure 7. Chemical structures of cyclofructan-based CSP157–172.

2.9. Molecularly-Imprinted CSPs

A different approach to chiral separation has been applied by using molecularly-imprinted CSPs. The synthesis of artificial selectors that are specific for a selected target (template) [245] is the principle of this type of CSPs. Each molecular imprinted CSP can only be applied for a specific type of analytes, as they are frequently applied on preparative enantioseparation and extraction of the desired compounds [246].

Several reviews have been devoted to molecularly-imprinted CSPs, mainly focusing on their developments and the different fields of application [45,88,246–253]. Although the developments concerning this type of CSPs are becoming more usual, their enantioresolution performance is currently not competitive in comparison to the existing CSPs. The most recent developments comprised the adaptation of this type of CSP to monoliths, nanoparticles, and predominantly to polymers. The introduction of different supports such as alginate microspheres [254] or polymer functionalized with quantum dots [255] were also reported as well as the description of different functional monomers and crosslinking agents [256–259].

Recently, Gutierrez-Climente et al. [245] prepared a new CSP by molecularly-imprinted nanoparticles on silica beads to reduce the tailing effect, commonly observed with this type of CSPs, through the reduction of particle size. The influence of some factors, such as the buffer percentage and concentration, pH, temperature, and column length, on chromatographic performance was evaluated. Regarding the chromatographic results, a maximum resolution value of 1.44 and a separation factor of 1.45 were obtained for citalopram.

In another study, Yang et al. [260] prepared a molecularly-imprinted polymer on porous silica gel microspheres to improve the chromatographic performance of a previous developed CSP and to reduce the analysis time. The new CSP demonstrated a higher affinity than the nonimprinted polymer with the silica gel, and selectivity for the target analyte, oseltamivir, with a retention factor of 13.5.

The optimization of the capacity of enantioseparation of a molecularly-imprinted monolith using a molecular crowding agent was recently reported by Wang et al. [261]. The main aim was to enhance the interactions between the CSP and the target analyte (*S*)-amlodipine. The composition of mobile phase, ionic strength, pH, and content of organic modifier were also taken into account when attempting to improve chromatographic performance [261].

2.10. Other CSPs

Despite the CSPs already mentioned, there are other types of CSPs, such as ligand-exchange, based on synthetic polymers, among others. The ligand-exchange CSPs do not present significant recent developments. Regarding polymers, several synthetic polymers can be used as selectors of CSPs [262]. Nevertheless, despite the interest in this type of material, synthetic polymer-based CSPs are not yet commercialized.

Regarding synthetic polymers, their classification can be based on the type of polymerization, as addition or condensation polymers and cross-linked gels, which are prepared resorting to molecular imprinted technique [263]. A synthetic and optically active polymer can be used for preparation of CSPs if it possesses a helical conformation, which contributes to the wide range of applications and effective separations [264]. The chiral recognition mechanism is based on hydrogen-bond, π-π interactions, and steric factors [45].

As for the other types of CSPs, the CSPs comprising synthetic polymers were the focus of several reviews [45,55,262–267]. The most recent developments are related with the introduction of monoliths [267], specifically of nanoparticles and hybrid monoliths. Ding et al. [264] also reported the use of smart polymers.

Recently, Maeda et al. [268] synthetized derivatives of optically active poly(diphenylacetylene) with chiral and achiral substituents. The helical structure of the polymer demonstrated to influence the enantioselectivity. The performance of the CSP was promising, affording excellent enantioselectivity and resolution, with α = 19.3 and Rs = 15.7 for ruthenium (III) acetylacetonate.

In another recent study, optically active π-conjugated polymers formed by alternated units of thieno[3–b]cthiophene and glucose-linked biphenyl were prepared, with its backbone conformation important for the enantioseparation of the obtained CSP [269]. Its performance was satisfactory with a maximum separation factor of 1.56 for cobalt (II) acetylacetonate.

A stable, porous, and crystalline organic polymer was introduced by Zhang et al. [270] highlighting its enhanced stability and resolution. The enantioresolution performance of the obtained CSP was reasonable with a separation factor value of 1.21 for *trans*-metoconazole, and a resolution of 2.56 value for *p*-nitrochlorobenzene.

Additionally, it is important to highlight the introduction of chip-based columns, motivated by the same factors than the transition for UHPLC, i.e., reduction of analysis time and improve the efficiency; the adaptation of LC remains challenging due to technical aspects [271–273]. An advantage focused for this type of columns was the simplicity of the process of its production as referred in the first report by Manz et al. [274]. More recently, they have been applied to extraction methodologies [275,276]. Despite the incorporation of micro and nanoparticles, this remains a challenging issue [273]; the introduction of monolith chip-based columns has already been reported [277].

3. Conclusions

The development of new CSPs for LC is a continuous and challenger issue covering various types of CSPs. This review gathered the most recent developments associated to different types of CSPs providing an overview of the advances that are occurring on this research area.

The most recent strategies, summarized in Figure 1, comprised the introduction of new chiral selectors or new chromatographic supports, and the application of different immobilization or coating methodologies for preparation of the CSPs. Regarding the chiral selectors, novel structures or analogues related to previously reported selectors were described as well as the use of hybrid selectors. The focus in chromatographic supports with lower particle size, the innovation related to the application of new materials such as monoliths and core-shell particles, as well as the use of hybrid supports, were also reported. In addition, several non-conventional approaches for immobilization or coating the chiral selectors to the chromatographic support were included, with particular emphasis to click chemistry as well as new encapsulation techniques or thermal immobilization without the use of chemical reagents, among others. Regardless of the type of CSPs, the main objectives of the development strategies were similar, concerning the improvement of the enantioresolution performance of the CSPs, as well as the increase of versatility and range of applications. Additionally, the transition to UHPLC and the possibility of the new CSPs to be used in all elution modes or using mobile phases compatible with mass spectrometric detection has also been underscored.

Even though several innovative strategies have been applied and several new CSPs were recently developed, they do not yet go beyond the exploratory stage. Nevertheless, the various strategies presented constitute an important trigger aiming to achieve new CSPs, as commercially viable products with high versatility and broad range of analytical and preparative applications or, on the contrary, as exceptionally efficient for the enantioresolution of specific target analytes.

In our opinion, the development of efficient chromatographic tools for LC enantioresolution is a subject that should continue to receive special attention, since it has constructive repercussions in several other research areas, such as biomedical, toxicology and forensic sciences, environment, food and fragrances, industry, among others. Moreover, the goal of developing a universal CSP remains a dream to be reached by those working in this research area.

Supplementary Materials: The following are available online at http://www.mdpi.com/1420-3049/24/5/865/s1, Table S1: Recent developments of polysaccharide-based CSPs, Table S2: Recent developments of protein-based CSPs, Table S3: Recent developments of cyclodextrin-based CSPs, Table S4: Recent developments of macrocyclic-based CSPs, Table S5: Recent developments of donor-acceptor or Pirkle-type CSPs, Table S6: Recent developments of ion-exchange-based CSPs, Table S7: Recent developments of crown-ether-based CSPs, Table S8: Recent developments of cyclofructan-based CSPs.

Author Contributions: J.T. collected the primary data and contributed in writing of the manuscript. C.F., M.E.T., and M.M.M.P. supervised the development of the manuscript, and assisted in data interpretation, manuscript evaluation, and editing.

Funding: This work was supported by the Strategic Funding UID/Multi/04423/2013 through national funds provided by FCT—Foundation for Science and Technology and European Regional Development Fund (ERDF), in the framework of the programme PT2020, the project PTDC/MAR-BIO/4694/2014 (reference POCI-01-0145-FEDER-016790; Project 3599—Promover a Produção Científica e Desenvolvimento Tecnológico e a Constituição de Redes Temáticas (3599-PPCDT)) as well as by Project No. POCI-01-0145-FEDER-028736, co-financed by COMPETE 2020, under the PORTUGAL 2020 Partnership Agreement, through the European Regional Development Fund (ERDF), and CHIRALXANT-CESPU-2018.

Conflicts of Interest: The authors declare no conflict of interest.

References

1. Ward, T.J.; Ward, K.D. Chiral separations: A review of current topics and trends. *Anal. Chem.* **2012**, *84*, 626–635. [CrossRef] [PubMed]
2. Phyo, Y.Z.; Cravo, S.; Palmeira, A.; Tiritan, M.E.; Kijjoa, A.; Pinto, M.M.M.; Fernandes, C. Enantiomeric resolution and docking studies of chiral xanthonic derivatives on chirobiotic columns. *Molecules* **2018**, *23*, 142. [CrossRef] [PubMed]
3. Fernandes, C.; Palmeira, A.; Santos, A.; Tiritan, M.E.; Afonso, C.; Pinto, M.M. Enantioresolution of Chiral Derivatives of Xanthones on (*S,S*)-Whelk-O1 and *L*-Phenylglycine Stationary Phases and Chiral Recognition Mechanism by Docking Approach for (*S,S*)-Whelk-O1. *Chirality* **2013**, *25*, 89–100. [CrossRef] [PubMed]
4. Speybrouck, D.; Lipka, E. Preparative supercritical fluid chromatography: A powerful tool for chiral separations. *J. Chromatogr. A* **2016**, *1467*, 33–55. [CrossRef] [PubMed]
5. Bajpai, L.; Naidu, H.; Asokan, K.; Shaik, K.M.; Kaspady, M.; Arunachalam, P.; Wu, D.R.; Mathur, A.; Sarabu, R. Integrating a post-column makeup pump into preparative supercritical fluid chromatography systems to address stability and recovery issues during purifications. *J. Chromatogr. A* **2017**, *1511*, 101–106. [CrossRef] [PubMed]
6. Zehani, Y.; Lemaire, L.; Millet, R.; Lipka, E. Small scale separation of isoxazole structurally related analogues by chiral supercritical fluid chromatography. *J. Chromatogr. A* **2017**, *1505*, 106–113. [CrossRef] [PubMed]
7. Bishop, R. Aspects of crystallization and chirality. In *Chirality in Supramolecular Assemblies: Causes and Consequences*; John Wiley & Sons, Ltd.: London, UK, 2016; pp. 65–93. [CrossRef]
8. Kovalenko, V.N.; Kozyrkov, Y.Y. A simple method for resolution of endo-/exo-monoesters of trans-norborn-5-ene-2,3-dicarboxylic acids into their enantiomers. *Chirality* **2015**, *27*, 151–155. [CrossRef] [PubMed]
9. Fernandes, C.; Tiritan, M.E.; Pinto, M.M.M. Chiral separation in preparative scale: A brief overview of membranes as tools for enantiomeric separation. *Symmetry* **2017**, *9*, 206. [CrossRef]
10. Afonso, C.A.M.; Crespo, J.G. Recent advances in chiral resolution through membrane-based approaches. *Angew. Chem. Int. Ed.* **2004**, *43*, 5293–5295. [CrossRef] [PubMed]
11. Andrushko, V.; Andrushko, N. *Stereoselective Synthesis of Drugs and Natural Products*; John Wiley & Sons, Inc.: Hoboken, NJ, USA, 2013; Volume 1.
12. Mun, S.; Wang, N.H.L. Improvement of the performances of a tandem simulated moving bed chromatography by controlling the yield level of a key product of the first simulated moving bed unit. *J. Chromatogr. A* **2017**, *1488*, 104–112. [CrossRef] [PubMed]
13. Song, J.Y.; Kim, K.M.; Lee, C.H. High-performance strategy of a simulated moving bed chromatography by simultaneous control of product and feed streams under maximum allowable pressure drop. *J. Chromatogr. A* **2016**, *1471*, 102–117. [CrossRef] [PubMed]
14. Kitamura, M.; Ohkuma, T.; Tokunaga, M.; Noyori, R. Dynamic kinetic resolution in binap-ruthenium(ii) catalyzed hydrogenation of 2-substituted 3-oxo carboxylic esters. *Tetrahedron Asymmetry* **1990**, *1*, 1–4. [CrossRef]
15. Schoemaker, H.E.; Mink, D.; WubboLts, M.G. Dispelling the myths—Biocatalysis in industrial synthesis. *Science* **2003**, *299*, 1694–1697. [CrossRef] [PubMed]
16. Miller, L.; Potter, M. Preparative chromatographic resolution of racemates using HPLC and SFC in a pharmaceutical discovery environment. *J. Chromatogr. B Anal. Technol. Biomed. Life Sci.* **2008**, *875*, 230–236. [CrossRef] [PubMed]
17. Leek, H.; Andersson, S. Preparative scale resolution of enantiomers enables accelerated drug discovery and development. *Molecules* **2017**, *22*, 158. [CrossRef] [PubMed]
18. Kirkpatrick, A.D.; Fain, M.; Yang, J.; Trehy, M. Enantiomeric impurity analysis using circular dichroism spectroscopy with United States Pharmacopeia Liquid Chromatographic Methods. *J. Pharm. Biomed. Anal.* **2018**, *156*, 366–371. [CrossRef] [PubMed]
19. Batista, A.N.L.; dos Santos, F.M.; Batista, J.M.; Cass, Q.B. Enantiomeric mixtures in natural product chemistry: Separation and absolute configuration assignment. *Molecules* **2018**, *23*, 492. [CrossRef] [PubMed]

20. Dawood, A.W.H.; de Souza, R.O.M.A.; Bornscheuer, U.T. Asymmetric synthesis of chiral halogenated amines using amine transaminases. *ChemCatChem* **2018**, *10*, 951–955. [CrossRef]
21. Chen, J.; Xu, X.; He, Z.; Qin, H.; Sun, W.; Fan, B. Nickel/Zinc iodide co-catalytic asymmetric [2 + 2] cycloaddition reactions of azabenzonorbornadienes with terminal alkynes. *Adv. Synth. Catal.* **2018**, *360*, 427–431. [CrossRef]
22. Wu, S.; Harada, S.; Morikawa, T.; Nishidda, A. Enantioselective total synthesis of a natural hydrocarbazolone alkaloid, identification of its stereochemistry, and revision of its spectroscopic data. *Tetrahedron Asymmetry* **2017**, *28*, 1083–1088. [CrossRef]
23. Gholami, H.; Zhang, J.; Anyika, M.; Borhan, B. Absolute stereochemical determination of asymmetric sulfoxides via central to axial induction of chirality. *Org. Lett.* **2017**, *19*, 1722–1725. [CrossRef] [PubMed]
24. Toki, H.; Ichikawa, T.; Mizuno-Yasuhira, A.; Yamaguchi, J. A rapid and sensitive chiral LC-MS/MS method for the determination of ketamine and norketamine in mouse plasma, brain and cerebrospinal fluid applicable to the stereoselective pharmacokinetic study of ketamine. *J. Pharm. Biomed. Anal.* **2018**, *148*, 288–297. [CrossRef] [PubMed]
25. Jiao, H.; Li, Y.; Sun, L.; Zhang, H.; Yu, L.; Yuan, Z.; Xie, L.; Chen, J.; Wang, Y. A chiral LC-MS/MS method for the enantioselective determination of *R*-(+)- and *S*-(−)-pantoprazole in human plasma and its application to a pharmacokinetic study of *S*-(−)-pantoprazole sodium injection. *Biomed. Chromatogr.* **2017**, *31*, e3980. [CrossRef] [PubMed]
26. Silva, B.; Pereira, J.A.; Cravo, S.; Araújo, A.M.; Fernandes, C.; Pinto, M.M.M.; Pinho, P.G.; Remião, F. Multi-milligram resolution and determination of absolute configuration of pentedrone and methylone enantiomers. *J. Chromatogr. B* **2018**, *110*, 158–164. [CrossRef] [PubMed]
27. Ribeiro, C.; Santos, C.; Gonçalves, V.; Ramos, A.; Afonso, C.; Tiritan, M.E. Chiral drug analysis in forensic chemistry: An overview. *Molecules* **2018**, *23*, 262. [CrossRef] [PubMed]
28. Silva, B.; Fernandes, C.; Guedes-de-Pinho, P.; Remião, F. Chiral resolution and enantioselectivity of synthetic cathinones: A brief review. *J. Anal. Toxicol.* **2018**, *42*, 17–24. [CrossRef] [PubMed]
29. Ribeiro, A.R.; Maia, A.S.; Cass, Q.B.; Tiritan, M.E. Enantioseparation of chiral pharmaceuticals in biomedical and environmental analyses by liquid chromatography: An overview. *J. Chromatogr. B Anal. Technol. Biomed. Life Sci.* **2014**, *968*, 8–21. [CrossRef] [PubMed]
30. Ribeiro, C.; Ribeiro, A.R.; Maia, A.S.; Tiritan, M.E. Occurrence of chiral bioactive compounds in the aquatic environment: A review. *Symmetry* **2017**, *9*, 215. [CrossRef]
31. Ribeiro, A.R.; Gonçalves, V.M.; Maia, A.S.; Ribeiro, C.; Castro, P.M.; Tiritan, M.E. Dispersive liquid–liquid microextraction and HPLC to analyse fluoxetine and metoprolol enantiomers in wastewaters. *Environ. Chem. Lett.* **2015**, *13*, 203–210. [CrossRef]
32. Cosimelli, B.; Greco, G.; Laneri, S.; Novellino, E.; Sacchi, A.; Collina, S.; Rossi, D.; Cosconati, S.; Barresi, E.; Taliani, S.; et al. Studies on enantioselectivity of chiral 4-acetylamino-6-alkyloxy-2-alkylthiopyrimidines acting as antagonists of the human adenosine receptor. *MedChemComm* **2018**, *9*, 81–86. [CrossRef] [PubMed]
33. George, R.; Haywood, A.; Good, P.; Hennig, S.; Khan, S.; Norris, R.; Hardy, J. Can saliva and plasma methadone concentrations be used for enantioselective pharmacokinetic and pharmacodynamic studies in patients with advanced cancer? *Clin. Ther.* **2017**, *39*, 1840–1848. [CrossRef] [PubMed]
34. Ilisz, I.; Péter, A.; Lindner, W. State-of-the-art enantioseparations of natural and unnatural amino acids by high-performance liquid chromatography. *TrAC-Trends Anal. Chem.* **2016**, *81*, 11–22. [CrossRef]
35. Tanwar, S.; Bhushan, R. Enantioresolution of Amino Acids: A Decade's Perspective, Prospects and Challenges. *Chromatographia* **2015**, *78*, 1113–1134. [CrossRef]
36. Hancu, G.; Cârcu-Dobrin, M.; Budău, M.; Rusu, A. Analytical methodologies for the stereoselective determination of fluoxetine: An overview. *Biomed. Chromatogr.* **2018**, *32*, e4040. [CrossRef] [PubMed]
37. Ali, I.; Suhail, M.; Asnin, L. Chiral separation of quinolones by liquid chromatography and capillary electrophoresis. *J. Sep. Sci.* **2017**, *40*, 2863–2882. [CrossRef] [PubMed]
38. Batra, S.; Bhushan, R. Resolution of enantiomers of bupropion and its metabolites by liquid chromatography. *Biomed. Chromatogr.* **2016**, *30*, 670–682. [CrossRef] [PubMed]
39. Felix, G.; Berthod, A. Commercial chiral stationary phases for the separations of clinical racemic drugs. *Sep. Purif. Rev.* **2007**, *36*, 285–481. [CrossRef]
40. Shaikh, S.; Muneera, M.S.; Thusleem, O.A. Chiral chromatography and its application to the pharmaceutical industry. *Pharm. Rev.* **2009**, *7*, 1371–1381.

41. Peluso, P.; Mamane, V.; Aubert, E.; Cossu, S. Recent trends and applications in liquid-phase chromatography enantioseparation of atropisomers. *Electrophoresis* **2017**, *38*, 1830–1850. [CrossRef]
42. Bertucci, C.; Tedesco, D. Human serum albumin as chiral selector in enantioselective high-performance liquid chromatography. *Curr. Med. Chem.* **2017**, *24*, 743–757. [CrossRef] [PubMed]
43. Lämmerhofer, M. Chiral recognition by enantioselective liquid chromatography: Mechanisms and modern chiral stationary phases. *J. Chromatogr. A* **2010**, *1217*, 814–856. [CrossRef] [PubMed]
44. Fernandes, C.; Tiritan, M.E.; Cravo, S.; Phyo, Y.Z.; Kijjoa, A.; Silva, A.M.S.; Cass, Q.B.; Pinto, M.M.M. New chiral stationary phases based on xanthone derivatives for liquid chromatography. *Chirality* **2017**, *29*, 430–442. [CrossRef] [PubMed]
45. Scriba, G.K.E. Chiral recognition in separation science—An update. *J. Chromatogr. A* **2016**, *1467*, 56–78. [CrossRef] [PubMed]
46. Kalíková, K.; Riesová, M.; Tesařová, E. Recent chiral selectors for separation in HPLC and CE. *Cent. Eur. J. Chem.* **2011**, *10*, 450–471. [CrossRef]
47. Henderson, G.M.; Rule, H.G. A new method of resolving a racemic compound. *Nature* **1938**, *142*, 917–918. [CrossRef]
48. Hesse, G.; Hagel, R. Über Inclusions-Chromatographie und ein neues Retentionsprinzip für benzolderivative. *Chromatographia* **1976**, *9*, 62–68. [CrossRef]
49. Chen, X.; Yamamoto, C.; Okamoto, Y. Polysaccharide derivatives as useful chiral stationary phases in high-performance liquid chromatography. *Pure Appl. Chem.* **2007**, *79*, 1561–1573. [CrossRef]
50. Ali, I.; Saleem, K.; Hussain, I.; Gaitonde, V.D.; Aboul-Enein, H.Y. Polysaccharides chiral stationary phases in liquid chromatography. *Sep. Purif. Rev.* **2009**, *38*, 97–147. [CrossRef]
51. Ribeiro, J.; Tiritan, M.E.; Pinto, M.M.M.; Fernandes, C. Chiral stationary phases for liquid chromatography based on chitin and chitosan-derived marine polysaccharides. *Symmetry* **2017**, *9*, 190. [CrossRef]
52. Chankvetadze, B. Recent developments on polysaccharide-based chiral stationary phases for liquid-phase separation of enantiomers. *J. Chromatogr. A* **2012**, *1269*, 26–51. [CrossRef] [PubMed]
53. Okamoto, Y.; Kawashima, M.; Hatada, K. Useful chiral packing materials for high-performance liquid chromatographic resolution of enantiomers: Phenylcarbamates of polysaccharides coated on silica gel. *J. Am. Chem. Soc.* **1984**, *106*, 5357–5359. [CrossRef]
54. Park, J.H.; Whang, Y.C.; Jung, Y.J.; Okamoto, Y.; Yamamoto, C.; Carr, P.W.; McNeff, C.V. Separation of racemic compounds on amylose and cellulose dimethylphenylcarbamate-coated zirconia in HPLC. *J. Sep. Sci.* **2003**, *26*, 1331–1336. [CrossRef]
55. Moldoveanu, S.C.; David, V. Stationary phases and columns for chiral chromatography. In *Selection of the HPLC Method in Chemical Analysis*; Elsevier: Amsterdam, The Netherlands, 2017; pp. 363–376.
56. Ali, I.; Aboul-Enein, H.Y. Impact of immobilized polysaccharide chiral stationary phase on enantiomeric separations. *J. Sep. Sci.* **2006**, *29*, 762–769. [CrossRef] [PubMed]
57. Ali, I.; Aboul-Enein, H. Immobilized Polysaccharide CSPs: An Advancement in Enantiomeric Separations. *Curr. Pharm. Anal.* **2007**, *3*, 71–82. [CrossRef]
58. Ikai, T.; Okamoto, Y. Structure control of polysaccharide derivatives for efficient separation of enantiomers by chromatography. *Chem. Rev.* **2009**, *109*, 6077–6101. [CrossRef] [PubMed]
59. Al-Othman, Z.A.; Ali, I.; Asim, M.; Khan, T.A. Recent trends in chiral separations on immobilized polysaccharides CSPs. *Comb. Chem. High Throughput Screen.* **2012**, *15*, 339–346. [CrossRef] [PubMed]
60. Francotte, E.; Huynh, D. Immobilized halogenophenylcarbamate derivatives of cellulose as novel stationary phases for enantioselective drug analysis. *J. Pharm. Biomed. Anal.* **2002**, *27*, 421–429. [CrossRef]
61. Khatiashvili, T.; Kakava, R.; Matarashvili, I.; Tabani, H.; Fanali, C.; Volonterio, A.; Farkas, T.; Chankvetadze, B. Separation of enantiomers of selected chiral sulfoxides with cellulose *tris*(4-chloro-3-methylphenylcarbamate)-based chiral columns in high-performance liquid chromatography with very high separation factor. *J. Chromatogr. A* **2018**, *1545*, 22–24. [CrossRef] [PubMed]
62. Cass, Q.B.; Bassi, A.L.C.; Calafatti, S.A.; Matlin, S.A.; Tiritan, M.E.; de Campos, L.M.M. Carbohydrate carbamate coated onto microporous silica: Application to chiral analysis of commercial pharmaceutical drugs. *Chirality* **1996**, *8*, 143–146. [CrossRef]
63. Matlin, S.A.; Tiritan, M.E.; Cass, Q.B.; Boyd, D.R. Enantiomeric resolution of chiral sulfoxides on polysaccharide phases by HPLC. *Chirality* **1996**, *8*, 147–152. [CrossRef]

64. Cass, Q.B.; Degani, A.L.G.; Tiritan, M.E.; Matlin, S.A.; Curran, D.P.; Balog, A. Enantiomeric resolution by HPLC of axial chiral amides using amylose *tris*[(*S*)-1-phenylethylcarbamate]. *Chirality* **1997**, *9*, 109–112. [CrossRef]
65. Cass, Q.B.; Watanabe, C.S.; Rabi, J.A.; Bottari, P.Q.; Costa, M.R.; Nascimento, R.M.; Cruz, J.E.; Ronald, R.C. Polysaccharide-based chiral phase under polar organic mode of elution in the determination of the enantiomeric purity of emtricitabine an anti-HIV analogue nucleoside. *J. Pharm. Biomed. Anal.* **2003**, *33*, 581–587. [CrossRef]
66. Lourenço, T.C.; Armstrong, D.W.; Cass, Q.B. Enantiomeric Resolution of a Chiral Sulfoxide Series by LC on Synthetic Polymeric Columns with Multimodal Elution. *Chromatographia* **2010**, *71*, 361–372. [CrossRef]
67. Jie, Z.; Qiuzheng, D.; Suzhen, Z.; Fang, S.; Xinyu, L.; Zhenzhong, Z. Enantioseparation of Three Important Intermediates of Tanikolide with Immobilized Cellulose Chiral Stationary Phase. *J. Chromatogr. Sci.* **2015**, *53*, 959–962. [CrossRef]
68. Zacharis, C.K.; Vastardi, E. A Validated LC Method for the Determination of Enantiomeric Purity of Clopidogrel Intermediate Using Amylose-Based Stationary Phase. *Chromatographia* **2015**, *78*, 819–824. [CrossRef]
69. Ramisetti, N.R.; Arnipalli, M.S.; Nimmu, N.V.; Bondigalla, R. UHPLC Determination of Besifloxacin Enantiomers on Immobilized Amylose *tris*(3,5-dichlorophenylcarbamate) Chiral Stationary Phase. *Chromatographia* **2017**, *80*, 1509–1515. [CrossRef]
70. Tiritan, M.E.; Cass, Q.B.; Del Alamo, A.; Matlin, S.A.; Grieb, S.J. Preparative enantioseparation on polysaccharide phase using microporous silica as a support. *Chirality* **1998**, *10*, 573–577. [CrossRef]
71. Lourenco, T.C.; Batista, J.M., Jr.; Furlan, M.; He, Y.; Nafie, L.A.; Santana, C.C.; Cass, Q.B. Albendazole sulfoxide enantiomers: Preparative chiral separation and absolute stereochemistry. *J. Chromatogr. A* **2012**, *1230*, 61–65. [CrossRef] [PubMed]
72. Belaz, K.R.; Coimbra, M.; Barreiro, J.C.; Montanari, C.A.; Cass, Q.B. Multimilligram enantioresolution of sulfoxide proton pump inhibitors by liquid chromatography on polysaccharide-based chiral stationary phase. *J. Pharm. Biomed. Anal.* **2008**, *47*, 81–87. [CrossRef] [PubMed]
73. Cass, Q.B.; Oliveira, R.V. Separation of Multi-Milligram Quantities of Gossypol Enantiomers on Polysaccharide-Based Stationary Phases. *J. Liq. Chromatogr. Relat. Technol.* **2007**, *25*, 819–829. [CrossRef]
74. Heydari, R.; Shamsipur, M. Enantiomeric Separation and Quantitation of Tenofovir Disoproxil Fumarate Using Amylose-Based Chiral Stationary Phases by High-Performance Liquid Chromatography. *Acta Chromatogr.* **2015**, *27*, 583–595. [CrossRef]
75. Lal, M.; Bhushan, R. Analytical and semi-preparative enantioresolution of (*RS*)-ketorolac from pharmaceutical formulation and in human plasma by HPLC. *Biomed. Chromatogr.* **2016**, *30*, 1526–1534. [CrossRef] [PubMed]
76. Sadutto, D.; Ferretti, R.; Zanitti, L.; Casulli, A.; Cirilli, R. Analytical and semipreparative high performance liquid chromatography enantioseparation of bicalutamide and its chiral impurities on an immobilized polysaccharide-based chiral stationary phase. *J. Chromatogr. A* **2016**, *1445*, 166–171. [CrossRef] [PubMed]
77. Okamoto, Y.; Aburatani, R.; Miura, S.; Hatada, K. Chiral stationary phases for HPLC: Cellulose *tris*(3,5-dimethylphenylcarbamate) and tris(3,5-dichlorophenylcarbamate) chemically bonded to silica gel. *J. Liq. Chromatogr.* **1987**, *10*, 1613–1628. [CrossRef]
78. Witte, D.T.; Bruggeman, F.J.; Franke, J.P.; Copinga, S.; Jansen, J.M.; De Zeeuw, R.A. Comparison between cellulose and amylose *tris*(3,5-dimethylphenylcarbamate) chiral stationary phases for enantiomeric separation of 17 amidotetralins. *Chirality* **1993**, *5*, 545–553. [CrossRef]
79. Matlin, S.A.; Tiritan, M.E.; Crawford, A.J.; Cass, Q.B.; Boyd, D.R. HPLC with carbohydrate carbamate chiral phases: Influence of chiral phase structure on enantioselectivity. *Chirality* **1994**, *6*, 135–140. [CrossRef]
80. Franco, P.; Senso, A.; Oliveros, L.; Minguillón, C. Covalently bonded polysaccharide derivatives as chiral stationary phases in high-performance liquid chromatography. *J. Chromatogr. A* **2001**, *906*, 155–170. [CrossRef]
81. Fernandes, C.; Brandão, P.; Santos, A.; Tiritan, M.E.; Afonso, C.; Cass, Q.B.; Pinto, M.M.M. Resolution and determination of enantiomeric purity of new chiral derivatives of xanthones using polysaccharide-based stationary phases. *J. Chromatogr. A* **2012**, *1269*, 143–153. [CrossRef] [PubMed]
82. Sousa, E.P.; Tiritan, M.E.; Oliveira, R.V.; Afonso, C.M.; Cass, Q.B.; Pinto, M.M. Enantiomeric resolution of kielcorin derivatives by HPLC on polysaccharide stationary phases using multimodal elution. *Chirality* **2004**, *16*, 279–285. [CrossRef] [PubMed]

83. Sousa, M.E.; Tiritan, M.E.; Belaz, K.R.; Pedro, M.; Nascimento, M.S.; Cass, Q.B.; Pinto, M.M. Multimilligram enantioresolution of low-solubility xanthonolignoids on polysaccharide chiral stationary phases using a solid-phase injection system. *J. Chromatogr. A* **2006**, *1120*, 75–81. [CrossRef] [PubMed]
84. Silva, B.; Fernandes, C.; Tiritan, M.E.; Pinto, M.M.; Valente, M.J.; Carvalho, M.; de Pinho, P.G.; Remiao, F. Chiral enantioresolution of cathinone derivatives present in "legal highs", and enantioselectivity evaluation on cytotoxicity of 3,4-methylenedioxypyrovalerone (MDPV). *Forensic Toxicol.* **2016**, *34*, 372–385. [CrossRef] [PubMed]
85. Taylor, D.R.; Maher, K. Chiral separations by high-performance liquid chromatography. *J. Chromatogr. Sci.* **1992**, 67–84. [CrossRef]
86. Cavazzini, A.; Pasti, L.; Massi, A.; Marchetti, N.; Dondi, F. Recent applications in chiral high performance chiral chromatography: A review. *Anal. Chim. Acta* **2011**, *706*, 205–222. [CrossRef] [PubMed]
87. Haginaka, J. Pharmaceutical and biomedical applications of enantioseparations using liquid chromatographic techniques. *J. Pharm. Biomed. Anal.* **2002**, *27*, 357–372. [CrossRef]
88. Tang, M.; Zhang, J.; Zhuang, S.; Liu, W. Development of chiral stationary phases for high-performance liquid chromatographic separation. *TrAC-Trends Anal. Chem.* **2012**, *39*, 180–194. [CrossRef]
89. Padró, J.M.; Keunchkarian, S. State-of-the-art and recent developments of immobilized polysaccharide-based chiral stationary phases for enantioseparations by high-performance liquid chromatography. *Microchem. J.* **2018**, *140*, 142–157. [CrossRef]
90. Xie, S.; Yuan, L. Recent development trends for chiral stationary phases based on chitosan derivatives, cyclofructan derivatives and chiral porous materials in high performance liquid chromatography. *J. Sep. Sci.* **2019**, 6–20. [CrossRef] [PubMed]
91. Yashima, E. Polysaccharide-based chiral stationary phases for high-performance liquid chromatographic enantioseparation. *J. Chromatogr. A* **2001**, *906*, 105–125. [CrossRef]
92. Ikai, T.; Yamamoto, C.; Kamigaito, M.; Okamoto, Y. Immobilized-type chiral packing materials for HPLC based on polysaccharide derivatives. *J. Chromatogr. B Anal. Technol. Biomed. Life Sci.* **2008**, *875*, 2–11. [CrossRef] [PubMed]
93. Okamoto, Y.; Ikai, T. Chiral HPLC for efficient resolution of enantiomers. *Chem. Soc. Rev.* **2008**, *37*, 2593–2608. [CrossRef] [PubMed]
94. Han, M.; Jin, X.; Yang, H.; Liu, X.; Liu, Y.; Ji, S. Controlled synthesis, immobilization and chiral recognition of carboxylic acid functionalized cellulose *tris*(3,5-dimethylphenylcarbamate). *Carbohydr. Polym.* **2017**, *172*, 223–229. [CrossRef] [PubMed]
95. Zhang, J.; Wang, X.-C.; Chen, W.; Bai, Z.-W. Synthesis of substituted phenylcarbamates of N-cyclobutylformylated chitosan and their application as chiral selectors in enantioseparation. *Analyst* **2016**, *141*, 4470–4480. [CrossRef] [PubMed]
96. Tang, S.; Bin, Q.; Chen, W.; Bai, Z.-W.; Huang, S.H. Chiral stationary phases based on chitosan bis(methylphenylcarbamate)-(isobutyrylamide) for high-performance liquid chromatography. *J. Chromatogr. A* **2016**, *1440*, 112–122. [CrossRef] [PubMed]
97. Tang, S.; Mei, X.; Chen, W.; Huang, S.H.; Bai, Z.-W. A high-performance chiral selector derived from chitosan (p-methylbenzylurea) for efficient enantiomer separation. *Talanta* **2018**, *185*, 42–52. [CrossRef] [PubMed]
98. Zhang, J.; Wang, Z.Q.; Chen, W.; Bai, Z.-W. Preparation and enantioseparation of biselector chiral stationary phases based on amylose and chitin derivatives. *Anal. Sci.* **2015**, *31*, 1091–1097. [CrossRef] [PubMed]
99. Li, Y.; Zhu, N.; Ma, Y.; Li, Q.; Li, P. Preparation of polysaccharide-based chiral stationary phases on SiO2@Ag core-shell particles by means of coating and intermolecular polycondensation and comparative liquid chromatography enantioseparations. *Anal. Bioanal. Chem.* **2018**, *410*, 441–449. [CrossRef] [PubMed]
100. Shen, J.; Wang, F.; Bi, W.; Liu, B.; Liu, S.; Okamoto, Y. Synthesis of cellulose carbamates bearing regioselective substituents at 2,3- and 6-positions for efficient chromatographic. *J. Chromatogr. A* **2018**, *1572*, 54–61. [CrossRef] [PubMed]
101. Mei, X.M.; Chen, W.; Bai, Z.-W. Enantioseparation characteristics of the chiral stationary phases based on natural and regenerated chitins. *J. Sep. Sci.* **2017**, *40*, 1710–1717. [CrossRef] [PubMed]
102. Zhang, J.; Wang, Z.Q.; Wang, X.-C.; Zhang, J.J.; Bai, Z.-W.; Chen, W. Enantioseparation characteristics of tadalafil and its intermediate on chitin derived chiral stationary phases. *Analyst* **2015**, *140*, 5593–5600. [CrossRef] [PubMed]

103. Fouad, A.; Marzouk, A.A.; Ibrahim, S.M.; El-Adl, S.M.; Ghanem, A. Functionalized polymer monoliths with carbamylated amylose for the enantioselective reversed phase nano-liquid chromatographic separation of a set of racemic pharmaceuticals. *J. Chromatogr. A* **2017**, *1515*, 91–99. [CrossRef] [PubMed]
104. Wang, Z.Q.; Liu, J.D.; Chen, W.; Bai, Z.-W. Enantioseparation characteristics of biselector chiral stationary phases based on derivatives of cellulose and amylose. *J. Chromatogr. A* **2014**, *1346*, 57–68. [CrossRef] [PubMed]
105. Sun, B.; Li, X.; Jin, Z.; Tian, L.; Wang, F.; Liu, G.; Tang, S.; Pan, F. Preparation of Regioselectively Modified Amylose Derivatives and Their Applications in Chiral HPLC. *Chromatographia* **2012**, *75*, 1347–1354. [CrossRef]
106. Bezhitashvili, L.; Bardavelidze, A.; Mskhiladze, A.; Gumustas, M.; Ozkan, S.A.; Volonterio, A.; Farkas, T.; Chankvetadze, B. Application of cellulose 3,5-dichlorophenylcarbamate covalently immobilized on superficially porous silica for the separation of enantiomers in high-performance liquid chromatography. *J. Chromatogr. A* **2018**, *1571*, 132–139. [CrossRef] [PubMed]
107. Yin, C.; Chen, W.; Zhang, J.; Zhang, M.; Zhang, J. A facile and efficient method to fabricate high-resolution immobilized cellulose-based chiral stationary phases via thiol-ene click chemistry. *Sep. Purif. Technol.* **2019**, *210*, 175–181. [CrossRef]
108. Huang, M.; Ma, H.; Niu, M.; Hu, F.; Wang, S.; Li, L.; Lv, C. Preparation of silica microspheres with a broad pore size distribution and their use as the support for a coated cellulose derivative chiral stationary phase. *J. Sep. Sci.* **2018**, *41*, 1232–1239. [CrossRef] [PubMed]
109. Okamoto, Y.; Kawashima, M.; Hatada, K. Chromatographic resolution: XI. Controlled chiral recognition of cellulose triphenylcarbamate derivatives supported on silica gel. *J. Chromatogr.* **1986**, *363*, 173–186. [CrossRef]
110. Vieira, A.T.; Assunc, R.M.N.; Faria, A.M. Stationary phase based on cellulose dodecanoate physically immobilized on silica particles for high-performance liquid chromatography. *J. Chromatogr. A* **2018**, *1572*, 72–81. [CrossRef] [PubMed]
111. Petrie, B.; Camacho Muñoz, M.D.; Martín, J. Stereoselective LC–MS/MS methodologies for environmental analysis of chiral pesticides. *TrAC-Trends Anal. Chem.* **2019**, *110*, 249–258. [CrossRef]
112. D' Orazio, G.; Asensio-Ramos, M.; Fanali, C. Enantiomers separation by capillary electrochromatography using polysaccharide-based stationary phases. *J. Sep. Sci.* **2019**, *42*, 360–384. [CrossRef] [PubMed]
113. Zeng, Q.; Wen, Q.; Xiang, Y.; Zhang, L. Chromatographic enantioseparation of chiral sulfinamide derivatives on polysaccharide-based chiral stationary phases. *J. Chromatogr. A* **2018**, *1571*, 240–244. [CrossRef] [PubMed]
114. Beridze, N.; Tsutskiridze, E.; Takaishvili, N.; Farkas, T.; Chankvetadze, B. Comparative enantiomer-resolving ability of coated and covalently immobilized versions of two polysaccharide-based chiral selectors in high-performance liquid chromatography. *Chromatographia* **2018**, *81*, 611–621. [CrossRef]
115. Bouanini, M.; Belboukhari, N.; Menéndez, J.C.; Sekkoum, K.; Cheriti, A.; Aboul-Enein, H.Y. Chiral separation of novel iminonaringenin derivatives. *Chirality* **2018**, *30*, 484–490. [CrossRef] [PubMed]
116. Stewart, K.K.; Doherty, R.F. Resolution of DL-Tryptophan by affinity chromatography on bovine-serum albumin-agarose columns. *Proc. Natl. Acad. Sci. USA* **1973**, *70*, 2850–2852. [CrossRef] [PubMed]
117. Bocian, S.; Skoczylas, M.; Buszewski, B. Amino acids, peptides, and proteins as chemically bonded stationary phases—A review. *J. Sep. Sci.* **2016**, *39*, 83–92. [CrossRef] [PubMed]
118. Hage, D.S.; Anguizola, J.A.; Bi, C.; Li, R.; Matsuda, R.; Papastavros, E.; Pfaunmiller, E.; Vargas, J.; Zheng, X. Pharmaceutical and biomedical applications of affinity chromatography: Recent trends and developments. *J. Pharm. Biomed. Anal.* **2012**, *69*, 93–105. [CrossRef] [PubMed]
119. Haginaka, J. Protein-based chiral stationary phases for high-performance liquid chromatography enantioseparations. *J. Chromatogr. A* **2001**, *906*, 253–273. [CrossRef]
120. Wallmorth, D.M.; Lee, J.T. Chiral bioanalysis. *Handb. Anal. Sep.* **2003**, *4*, 129–184.
121. Mallik, R.; Hage, D.S. Affinity monolith chromatography. *J. Sep. Sci.* **2006**, *29*, 1686–1704. [CrossRef] [PubMed]
122. Moldoveanu, S.C.; David, V. Stationary phases and columns for immunoaffinity type separations. In *Selection of the HPLC Method in Chemical Analysis*; Elsevier: Amsterdam, The Netherlands, 2017; pp. 387–392.
123. Haginaka, J. Recent progresses in protein-based chiral stationary phases for enantioseparations in liquid chromatography. *J. Chromatogr. B Anal. Technol. Biomed. Life Sci.* **2008**, *875*, 12–19. [CrossRef] [PubMed]
124. Bi, C.; Jackson, A.; Vargas-Badilla, J.; Li, R.; Rada, G.; Anguizola, J.A.; Pfaunmiller, E.; Hage, D.S. Entrapment of α1-acid glycoprotein in high-performance affinity columns for drug-protein binding studies. *J. Chromatogr. B Anal. Technol. Biomed. Life Sci.* **2016**, *1021*, 188–196. [CrossRef] [PubMed]

125. Pfaunmiller, E.; Hartmann, M.; Dupper, C.M.; Soman, S.; Hage, D.S. Optimization of human serum albumin monoliths for chiral separations and high-performance affinity chromatography. *J. Chromatogr. A* **2012**, *1269*, 198–207. [CrossRef] [PubMed]
126. Matsunaga, H.; Haginaka, J. Separation of enantiomers on chiral stationary phase based on chicken α1-acid glycoprotein: Effect of silica particle diameters on column performance. *J. Chromatogr. A* **2014**, *1363*, 96–100. [CrossRef] [PubMed]
127. Hellerstein, M.K.; Sasak, V.; Ordovas, J.; Munro, H.N. Isolation of α1-acid glycoprotein from human plasma using high-performance liquid chromatography. *Anal. Biochem.* **1985**, *146*, 366–371. [CrossRef]
128. Matsunaga, H.; Haginaka, J. Separation of enantiomers on chiral stationary phase based on cellulase: Effect of preparation method and silica particle diameters on chiral recognition ability. *J. Chromatogr. A* **2016**, *1467*, 155–162. [CrossRef] [PubMed]
129. Vandenbosch, C.; Lindner, W.; Massart, D.L. Evaluation of the enantioselectivity of an ovomucoid and a cellulase chiral stationary phase towards a set of β blocking agents. *Anal. Chim. Acta* **1992**, *270*, 1–12. [CrossRef]
130. Zheng, X.; Podariu, M.; Bi, C.; Hage, D.S. Development of enhanced capacity affinity microcolumns by using a hybrid of protein cross-linking/modification and immobilization. *J. Chromatogr. A* **2015**, *1400*, 82–90. [CrossRef] [PubMed]
131. Bi, C.; Matsuda, R.; Zhang, C.; Isingizwe, Z.; Clarke, W.; Hage, D.S. Studies of drug interactions with α1-acid glycoprotein by using on-line immunoextraction and high-performance affinity chromatography. *J. Chromatogr. A* **2017**, *1519*, 64–73. [CrossRef] [PubMed]
132. Matsuda, R.; Donald, J.; Beyersdorf, J.; Hage, D.S. Analysis of drug–protein binding using on-line immunoextraction and high-performance affinity microcolumns: Studies with normal and glycated human serum albumin. *J. Chromatogr. A* **2015**, *1416*, 112–120. [CrossRef] [PubMed]
133. Fedorova, I.A.; Shapovalova, E.N.; Shpigun, O.A.; Staroverov, S.M. Bovine serum albumin adsorbed on eremomycin and grafted on silica as new mixed-binary chiral sorbent for improved enantioseparation of drugs. *J. Food Drug Anal.* **2016**, *24*, 848–854. [CrossRef] [PubMed]
134. Zheng, Y.; Wang, X.; Ji, Y. Monoliths with proteins as chiral selectors for enantiomer separation. *Talanta* **2012**, *91*, 7–17. [CrossRef] [PubMed]
135. Cârje, A.G.; Ion, V.; Muntean, D.; Balint, A.; Imre, S. Enantioseparation of indapamide by high-performance liquid chromatography using ovomucoid glycoprotein as chiral selector. *Farmacia* **2016**, *64*, 181–186.
136. Batra, S.; Bhushan, R. Bioassay, determination and separation of enantiomers of atenolol by direct and indirect approaches using liquid chromatography: A review. *Biomed. Chromatogr.* **2018**, *32*, e4090. [CrossRef] [PubMed]
137. Chrysanthakopoulos, M.; Vallianatou, T.; Giaginis, C.; Tsantili-Kakoulidou, A. Investigation of the retention behaviour of structurally diverse drugs on α1-acid glycoprotein column: Insight on the molecular factors involved and correlation with protein binding data. *Eur. J. Pharm. Sci.* **2014**, *60*, 24–31. [CrossRef] [PubMed]
138. Zi, Z.; Hage, D.S. Analysis of stereoselective drug interactions with serum proteins by high-performance affinity chromatography: A historical perspective. *J. Pharm. Biomed. Anal.* **2017**, *144*, 12–24. [CrossRef]
139. Souza, M.C.O.; Marques, M.P.; Duarte, G.; Lanchote, V.L. Analysis of bupivacaine enantiomers in plasma as total and unbound concentrations using LC-MS/MS: Application in a pharmacokinetic study of a parturient with placental transfer. *J. Pharm. Biomed. Anal.* **2019**, *164*, 268–275. [CrossRef] [PubMed]
140. Armstrong, D.W.; DeMond, W. Cyclodextrin bonded phases for the liquid chromatographic separation of optical, geometrical, and structural isomers. *J. Chromatogr. Sci.* **1984**, *22*, 411–415. [CrossRef]
141. Chmielewska, A.; Konieczna, L.; Baczek, T. A novel two-step liquid-liquid extraction procedure combined with stationary phase immobilized human serum albumin for the chiral separation of cetirizine enantiomers along with M and P parabens. *Molecules* **2016**, *21*, 1654. [CrossRef] [PubMed]
142. Bressolle, F.; Audran, M.; Pham, T.-N.; Vallon, J.-J. Cyclodextrins and enantiomeric separations of drugs by liquid chromatography and capillary electrophoresis: Basic principles and new developments. *J. Chromatogr. B Biomed. Appl.* **1996**, *687*, 303–336. [CrossRef]
143. Berthood, A. Chiral recognition mechanisms in enantiomers separations: A general view. In *Chiral Recognition in Separation Methods*; Springer: Berlin/Heidelberg, Germany, 2010; pp. 1–32.
144. Xiao, Y.; Ng, S.C.; Tan, T.T.Y.; Wang, Y. Recent development of cyclodextrin chiral stationary phases and their applications in chromatography. *J. Chromatogr. A* **2012**, *1269*, 52–68. [CrossRef] [PubMed]

145. Guo, J.; Lin, Y.; Xiao, Y.; Crommen, J.; Jiang, Z. Recent developments on cyclodextrin functionalized monolithic columns for the enantioseparation of chiral drugs. *J. Pharm. Biomed. Anal.* **2016**, *130*, 110–125. [CrossRef] [PubMed]
146. Szabó, Z.-I.; Szőcs, L.; Horváth, P.; Komjáti, B.; Nagy, J.; Jánoska, Á.; Muntean, D.-L.; Noszál, B.; Tóth, G. Liquid chromatography with mass spectrometry enantioseparation of pomalidomide on cyclodextrin-bonded chiral stationary phases and the elucidation of the chiral recognition mechanisms by NMR spectroscopy and molecular modeling(Article). *J. Sep. Sci.* **2016**, *39*, 2941–2949. [CrossRef] [PubMed]
147. Arsad, S.R.; Maarof, H.; Ibrahim, W.A.W.; Aboul-Enein, H.Y. Theoretical and Molecular Docking Study of Ketoconazole on Heptakis(2,3,6-*tri*-O-methyl)-β-cyclodextrin as Chiral Selector. *Chirality* **2016**, *28*, 209–214. [CrossRef] [PubMed]
148. Chen, X.J.; Yang, G.L.; Xu, X.D.; Sheng, J.J.; Shen, J. Preparation and chromatographic evaluation of β-cyclodextrin derivative CSPs bearing substituted phenylcarbamate groups for HPLC. *J. Liq. Chromatogr. Relat. Technol.* **2016**, *39*, 647–657. [CrossRef]
149. Li, L.; Cheng, B.; Zhou, R.; Cao, Z.; Zeng, C.; Li, L. Preparation and evaluation of a novel N-benzyl-phenethylamino-β-cyclodextrin-bonded chiral stationary phase for HPLC. *Talanta* **2017**, *174*, 179–191. [CrossRef] [PubMed]
150. Li, L.; Zhang, M.; Wang, Y.; Zhou, W.; Zhou, Z. Preparation of chiral oxazolinyl-functionalized β-cyclodextrin-bonded stationary phases and their enantioseparation performance in high-performance liquid chromatography. *J. Sep. Sci.* **2016**, *39*, 4136–4146. [CrossRef] [PubMed]
151. Tang, X.; Li, X.; Sun, Y.; Xiao, Y.; Wang, Y. Thiol-ene click derived structurally well-defined per(3,5-dimethyl)phenylcarbamoylated cationic cyclodextrin separation material for achiral and chiral chromatography. *J. Sep. Sci.* **2018**, *41*, 2710–2718. [CrossRef] [PubMed]
152. Zhou, J.; Yang, B.; Tang, J.; Tang, W. A cationic cyclodextrin clicked bilayer chiral stationary phase for versatile chiral separation in HPLC. *New J. Chem.* **2018**, *42*, 3526–3533. [CrossRef]
153. Chu, C.; Liu, R. Application of click chemistry on preparation of separation materials for liquid chromatography. *Chem. Soc. Rev.* **2011**, *40*, 2177–2188. [CrossRef] [PubMed]
154. Wang, L.; Dong, S.; Han, F.; Zhao, Y.; Zhang, X.; Zhang, X.; Qiu, H.; Zhao, L. Spherical β-cyclodextrin-silica hybrid materials for multifunctional chiral stationary phases. *J. Chromatogr. A* **2015**, *1383*, 70–78. [CrossRef] [PubMed]
155. Zhao, B.; Li, L.; Wang, Y.; Zhou, Z. Preparation of polar group derivative β-cyclodextrin bonded hydride silica chiral stationary phases and their chromatography separation performances. *Chin. Chem. Lett.* **2018**. [CrossRef]
156. Ghanem, A.; Adly, F.G.; Sokerik, Y.; Antwi, N.Y.; Shenashen, M.A.; El-Safty, S.A. Trimethyl-β-cyclodextrin-encapsulated monolithic capillary columns: Preparation, characterization and chiral nano-LC application. *Talanta* **2017**, *169*, 239–248. [CrossRef] [PubMed]
157. Qiang, L.I.; Yuan-yuan, L.I.; Nan, Z.H.U.; Zhu-xian, G.A.O.; Tian-jun, L.I.; Tong, Z.; Yu-long, M.A. Preparation of cyclodextrin type stationary phase based on graphene oxide and its application in enantioseparation and hydrophilic interaction chromatography. *Chin. J. Anal. Chem.* **2018**, *46*, 1455–1463. [CrossRef]
158. Tang, Q.; Yu, B.; Gao, L.; Cong, H.; Zhang, S. Light-assisted preparation of cyclodextrin-based chiral stationary phase and its separation performance in liquid chromatography. *New J. Chem.* **2018**, *42*, 1115–1120. [CrossRef]
159. Zhou, J.; Yang, B.; Tang, J.; Tang, W. Cationic cyclodextrin clicked chiral stationary phase for versatile enantioseparations in high-performance liquid chromatography. *J. Chromatogr. A* **2016**, *1467*, 169–177. [CrossRef] [PubMed]
160. Zhou, M.; Long, Y.; Zhi, Y.; Xu, X. Preparation and chromatographic evaluation of a chiral stationary phase based on carboxymethyl-β-cyclodextrin for high-performance liquid chromatography. *Chin. Chem. Lett.* **2018**, *29*, 1399–1403. [CrossRef]
161. Li, X.; Jin, X.; Yao, X.; Ma, X.; Wang, Y. Thioether bridged cationic cyclodextrin stationary phases: Effect of spacer length, selector concentration and rim functionalities on the enantioseparation. *J. Chromatogr. A* **2016**, *1467*, 279–287. [CrossRef] [PubMed]
162. Pesek, J.J.; Matyska, M.T. Silica Hydride-Based Packing Materials: HPLC Stationary Phases for a Global Approach to Complex Sample Analysis. *Curr. Chrom.* **2018**, *5*, 33–42. [CrossRef]

163. Armstrong, D.W.; Tang, Y.; Chen, S.; Zhou, Y.; Bagwlll, C.; Chen, J. Macrocyclic antibiotics as a new class of chiral selectors for liquid chromatography. *Anal. Chem.* **1994**, 1473–1484. [CrossRef]
164. Ilisz, I.; Pataj, Z.; Aranyi, A.; Péter, A. Macrocyclic antibiotic selectors in direct HPLC enantioseparations. *Sep. Purif. Rev.* **2012**, *41*, 207–249. [CrossRef]
165. Ward, T.J.; Farris, A.B. Chiral separations using the macrocyclic antibiotics: A review. *J. Chromatogr. A* **2001**, *906*, 73–89. [CrossRef]
166. Ilisz, I.; Berkecz, R.; Péter, A. Retention mechanism of high-performance liquid chromatographic enantioseparation on macrocyclic glycopeptide-based chiral stationary phases. *J. Chromatogr. A* **2009**, *1216*, 1845–1860. [CrossRef] [PubMed]
167. Aburachid Cardoso, P.; Costa César, I. Chiral method development strategies for HPLC using macrocyclic glycopeptide-based stationary phases. *Chromatographia* **2018**, *81*, 841–850. [CrossRef]
168. Beesley, T.E.; Lee, J.T. Method development strategy and applications update for CHIROBIOTIC chiral stationary phases. *J. Liq. Chromatogr. Relat. Technol.* **2009**, *32*, 1733–1767. [CrossRef]
169. Fernandes, C.; Tiritan, M.E.; Cass, Q.; Kairys, V.; Fernandes, M.X.; Pinto, M.M. Enantioseparation and chiral recognition mechanism of new chiral derivatives of xanthones on macrocyclic antibiotic stationary phases. *J. Chromatogr. A* **2012**, *1241*, 60–68. [CrossRef] [PubMed]
170. Hrobonová, K.; Lehotay, J.; Cizmárik, J. HPLC enantioseparation of phenylcarbamic acid derivatives by using macrocyclic chiral stationary phases. *Nov. Biotechnol. Chim.* **2016**, *15*, 12–22. [CrossRef]
171. Min, Y.; Sui, Z.; Liang, Z.; Zhang, L.; Zhang, Y. Teicoplanin bonded sub-2 micrometer superficially porous particles for enantioseparation of native amino acids. *J. Pharm. Biomed. Anal.* **2015**, *114*, 247–253. [CrossRef] [PubMed]
172. Ismail, O.H.; Ciogli, A.; Villani, C.; De Martino, M.; Pierini, M.; Cavazzini, A.; Bell, D.S.; Gasparrini, F. Ultra-fast high-efficiency enantioseparations by means of a teicoplanin-based chiral stationary phase made on sub-2 μm totally porous silica particles of narrow size distribution. *J. Chromatogr. A* **2016**, *1427*, 55–68. [CrossRef] [PubMed]
173. Rocchi, S.; Rocco, A.; Pesek, J.J.; Matyska, M.T.; Capitani, D.; Fanali, S. Enantiomers separation by nano-liquid chromatography: Use of a novel sub-2 micrometer vancomycin silica hydride stationary phase. *J. Chromatogr. A* **2015**, *1381*, 149–159. [CrossRef] [PubMed]
174. Xu, D.; Shao, H.; Luo, R.; Wang, Q.; Sánchez-López, E.; Fanali, S.; Marina, M.L.; Jiang, Z. A facile and efficient single-step approach for the fabrication of vancomycin functionalized polymer-based monolith as chiral stationary phase for nano-liquid chromatography. *J. Chromatogr. A* **2018**, *1557*, 43–50. [CrossRef] [PubMed]
175. Hellinghausen, G.; Lopez, D.A.; Lee, J.T.; Wang, Y.; Weatherly, C.A.; Portillo, A.E.; Berthod, A.; Armstrong, D.W. Evaluation of the Edman degradation product of vancomycin bonded to core-shell particles as a new HPLC chiral stationary phase. *Chirality* **2018**, *30*, 1067–1078. [CrossRef] [PubMed]
176. Yu, B.; Zhang, S.; Li, G.; Cong, H. Light-assisted preparation of vancomycin chiral stationary phase based on diazotized silica and its enantioseparation evaluation by high-performance liquid chromatography. *Talanta* **2018**, *182*, 171–177. [CrossRef] [PubMed]
177. Dolzan, M.D.; Shu, Y.; Smuts, J.P.; Petersen, H.; Ellegaard, P.; Micke, G.A.; Armstrong, D.W.; Breitbach, Z.S. Enantiomeric separation of citalopram analogues by HPLC using macrocyclic glycopeptide and cyclodextrin based chiral stationary phase. *J. Liq. Chromatogr. Relat. Technol.* **2016**, *39*, 154–160. [CrossRef]
178. Pirkle, W.H.; House, D.W. Chiral high-performance liquid chromatographic stationary phases. 1. Separation of the enantiomers of sulfoxides, amines, amino acids, alcohols, hydroxy acids, lactones and mercaptans. *J. Org. Chem.* **1979**, *44*, 1957–1960. [CrossRef]
179. Davankov, V.A. The nature of chiral recognition: Is it a three-point interaction? *Chirality* **1997**, *9*, 99–102. [CrossRef]
180. Welch, C.J. Evolution of chiral stationary phase design in the Pirkle laboratories. *J. Chromatogr. A* **1994**, *666*, 3–26. [CrossRef]
181. Pirkle, W.H.; Däppen, R. Reciprocity in Chiral Recognition. Comparison of several chiral stationary phases. *J. Chromatogr. A* **1987**, *404*, 107–115. [CrossRef]
182. Pirkle, W.H.; Hyun, M.H.; Tsipouras, A.; Hamper, B.C.; Banks, B. A rational approach to the design of highly effective chiral stationary phases for the liquid chromatographic separation of enantiomers. *J. Pharm. Biomed. Anal.* **1984**, *2*, 173–181. [CrossRef]

183. Pirkle, W.H.; Hyun, M.H. Preparation and use of hydantion-based chiral stationary phases. *J. Chromatogr. A* **1985**, *322*, 309–320. [CrossRef]
184. Pirkle, W.H.; Murray, P.G.; Wilson, S.R. X-ray crystallographic evidence in support of a proposed chiral recognition mechanism. *J. Org. Chem.* **1996**, *61*, 4775–4777. [CrossRef] [PubMed]
185. Pirkle, W.H.; Selness, S.R. Chiral recognition studies: Intra- and intermolecular 1H{1H}-nuclear overhauser effects as effective tools in the study of bimolecular complexes. *J. Org. Chem.* **1995**, *60*, 3252–3256. [CrossRef]
186. Fernandes, C.; Tiritan, M.E.; Pinto, M.M.M. Small molecules as chromatographic tools for HPLC enantiomeric resolution: Pirkle-type chiral stationary phases evolution. *Chromatographia* **2013**, *76*, 871–897. [CrossRef]
187. Dungelova, J.; Lehotay, J.; Cizmarik, J.; Armstrong, D.W. Study of the mechanism of enantioseparation. IV. Study of enantioseparation of some derivatives of phenylcarbamic acid using p-complex stationary phase in HPLC. *J. Liq. Chromatogr. Relat. Technol.* **2003**, *26*, 2331–2350. [CrossRef]
188. Asnin, L.D.; Guiochon, G. The adsorption of naproxen enantiomers on the chiral stationary phase (*R*, *R*)-Whelk-O1 under reversed-phase conditions: The effect of mobile phase composition. *J. Chromatogr. A* **2010**, *1217*, 2871–2878. [CrossRef] [PubMed]
189. Welch, C.J.; Szczerba, T.; Perrin, S.R. Some recent high-performance liquid chromatography separations of the enantiomers of pharmaceuticals and other compounds using the Whelk-O 1 chiral stationary phase. *J. Chromatogr. A* **1997**, *758*, 93–98. [CrossRef]
190. Fernandes, C.; Phyo, Y.Z.; Silva, A.S.; Tiritan, M.E.; Kijjoa, A.; Pinto, M.M.M. Chiral stationary phases based on small molecules: An update of the last 17 years. *Sep. Purif. Rev.* **2018**, *47*, 89–123. [CrossRef]
191. Gasparrini, F.; Misiti, D.; Villani, C. High-performance liquid chromatography chiral stationary phases based on low-molecular mass selectors. *J. Chromatogr. A* **2001**, *906*, 35–50. [CrossRef]
192. Villani, C.; D'Acquarica, I.; Gasparrini, F.; Ritchi, H.; Kotoni, D.; Pierini, M.; Ciogli, A.; Kocergin, J. The evolution of chiral stationary phases from HPLC to UHPLC. *LC-GC Eur.* **2014**, *27*, 128–137.
193. Waterlot, C.; Ghinet, A.; Lipka, E. Core-shell Particles: A Way to Greening Liquid Chromatography in Environmental Applications. *Curr. Chromatogr.* **2018**, *5*. [CrossRef]
194. Spudeit, D.A.; Dolzan, M.D.; Breitbach, Z.S.; Barber, W.E.; Micke, G.A.; Armstrong, D.W. Superficially porous particles vs fully porous particles for bonded high-performance liquid chromatographic chiral stationary phases: Isopropyl cyclofructan 6. *J. Chromatogr. A* **2014**, *1363*, 89–95. [CrossRef] [PubMed]
195. Catani, M.; Ismail, O.; Gasparrini, F.; Antonelli, M.; Pasti, L.; Marchetti, N.; Fellettia, S.; Cavazzini, A. Recent advancements and future directions of superficially porous chiral stationary phases for ultrafast high-performance enantioseparations. *Analyst* **2017**, *142*, 555–566. [CrossRef] [PubMed]
196. Qiao, L.; Zhou, X.; Li, X.; Du, W.; Yu, A.; Zhang, S.; Wu, Y. Synthesis and performance of chiral ferrocene modified silica gel for mixed-mode chromatography. *Talanta* **2017**, *163*, 94–101. [CrossRef] [PubMed]
197. Çakmak, R.; Ercan, S.; Sünkür, M.; Yilmaz, H.; Topal, G. Design, preparation and application of a Pirkle-type chiral stationary phase for enantioseparation of some racemic organic acids and molecular dynamic studies. *Org. Commun.* **2017**, *10*, 216–227. [CrossRef]
198. Shen, H.; Du, G.; Liu, K.; Ye, L.; Xie, S.; Jiang, L. Synthesis and evaluation of pseudopeptide chiral stationary phases for enantioselective resolution. *J. Chromatogr. A* **2017**, *1521*, 53–62. [CrossRef] [PubMed]
199. Yu, J.; Armstrong, D.W.; Ryoo, J.J. Synthesis of new C3 symmetric amino acid- and aminoalcohol-containing chiral stationary phases and application to HPLC enantioseparations. *Chirality* **2018**, *30*, 74–84. [CrossRef] [PubMed]
200. Wang, Y.; Liu, D.; Zhang, Y.; Tang, Y.; Zhao, J.; Shen, B. Synthesis of a Novel Chiral Stationary Phase by (R)-1,1′-Binaphthol and the Study on Mechanism of Chiral Recognition. *Symmetry* **2018**, *10*, 704. [CrossRef]
201. Yang, J.; Carmella, S.G.; Hecht, S.S. Analysis of N′-nitrosonornicotine enantiomers in human urine by chiral stationary phase liquid chromatography–nanoelectrospray ionization–high resolution tandem mass spectrometry. *J. Chromatogr. B* **2017**, *1044*, 127–131. [CrossRef] [PubMed]
202. Yip, S.H.; Wu, D.-R.; Li, P.; Sun, D.; Watterson, S.H.; Zhao, R.; Tino, J.; Mathur, A. Separation of Bruton's tyrosine kinase inhibitor atropisomers by supercritical fluid chromatography. *J. Chromatogr. A* **2019**, *1586*, 106–115. [CrossRef] [PubMed]
203. Felletti, S.; De Luca, C.; Ismail, O.H.; Pasti, L.; Costa, V.; Gasparrini, F.; Cavazzini, A.; Catani, M. On the effect of chiral selector loading and mobile phase composition on adsorption properties of latest generation fully- and superficially-porous Whelk-O1 particles for high-efficient ultrafast enantioseparations. *J. Chromatogr. A* **2018**, *1579*, 41–48. [CrossRef] [PubMed]

204. Tumashov, A.A.; Gruzdev, D.A.; Vigorov, A.Y.; Musiyak, V.V.; Chulakov, E.N.; Levit, G.L.; Krasnov, V.P.; Charushin, V.N. Analysis of racemic conjugates of purine with heterocyclic amines by chiral high-performance liquid chromatography. *Russ. Chem. Bull.* **2018**, *67*, 1704–1709. [CrossRef]
205. Knežević, A.; Novak, J.; Pescitelli, G.; Vinković, V. Determination of the Absolute Configuration of (*S*)-N-(1-Aryl-allyl)-3,5-dinitrobenzamides and Their Elution Order on Brush-Type Chiral Stationary Phases. *Eur. J. Org. Chem.* **2018**, *2018*, 3982–3991. [CrossRef]
206. Rosini, C.; Bertucci, C.; Pini, D.; Altemura, P.; Salvadori, P. Cinchona alkaloids for preparing new, easily accessible chiral stationary phases. I. 11-(10,11-dihydro-6′-methoxy-cinchonan-9-OL)-tiopropylsilanized silica. *Tetrahedron Lett.* **1985**, *26*, 3361–3364. [CrossRef]
207. Lämmerhofer, M. Liquid chromatographic enantiomer separation with special focus on zwitterionic chiral ion-exchangers. *Anal. Bioanal. Chem.* **2014**, *406*, 6095–6103. [CrossRef] [PubMed]
208. Vander Heyden, Y.; Mangelings, D.; Matthijs, N.; Perrin, C. Chiral separations. In *Handbook of Pharmaceuticals Analysis by HPLC*; Elsevier: Amsterdam, The Netherlands, 2005; Volume 6, pp. 447–498.
209. Natalini, B.; Sardella, R. Chromatographic separations and analysis: Chiral ion and ligand exchange stationary phases. *Compr. Chir.* **2012**, *8*, 115–152.
210. Hoffmann, C.V.; Pell, R.; Lämmerhofer, M.; Lindner, W. Synergistic effects on enantioselectivity of zwitterionic chiral stationary phases for separations of chiral acids, bases and amino acids by HPLC. *Anal. Chem.* **2008**, *80*, 8780–8789. [CrossRef] [PubMed]
211. Nesterenko, P.N.; Haddad, P.R. Zwitterionic ion-exchangers in liquid chromatography. *Anal. Sci.* **2000**, *16*, 565–574. [CrossRef]
212. Gübitz, G.; Schmid, M.G. Chiral separation by chromatographic and electromigration techniques. A review. *Biopharm. Drug Dispos.* **2001**, *22*, 291–336. [CrossRef] [PubMed]
213. Ilisz, I.; Bajtai, A.; Lindner, W.; Péter, A. Liquid chromatographic enantiomer separations applying chiral ion-exchangers based on *Cinchona* alkaloids. *J. Pharm. Biomed. Anal.* **2018**, *159*, 127–152. [CrossRef] [PubMed]
214. Todoroki, K.; Ishii, Y.; Ide, T.; Min, J.Z.; Inoue, K.; Huang, X.; Zhang, W.; Hamashima, Y.; Toyo'oka, T. Advanced dress-up chiral columns: New removable chiral stationary phases for enantioseparation of chiral carboxylic acids. *Anal. Chim. Acta* **2015**, *882*, 101–111. [CrossRef] [PubMed]
215. Woiwode, U.; Sievers-Engler, A.; Zimmermann, A.; Lindner, W.; Sánchez-Muñoz, O.L.; Lämmerhofer, M. Surface-anchored counterions on weak chiral anion-exchangers accelerate separations and improve their compability for mass-spectrometry-hyphenation. *J. Chromatogr. A* **2017**, *1503*, 21–31. [CrossRef] [PubMed]
216. De Martino, M.; Bencivenni, G.; Mazzanti, A.; Menta, S.; Ismail, O.H.; Sabia, R.; Ciogli, A. 3,5-dinitrobenzoyl-9-amino-9-deoxy-9-epiquinine as Pirkle-anion exchange hybrid-type chiral selector in high-performance liquid chromatography. *Chromatographia* **2017**, *80*, 751–762. [CrossRef]
217. Woiwode, U.; Neubauer, S.; Lindner, W.; Buckenmaier, S.; Lämmerhofer, M. Enantioselective multiple heartcut two-dimensional ultra-high-performance liquid chromatography method with a coreshell chiral stationary phase in the second dimension for analysis of all proteinogenic amino acids in a single run. *J. Chromatogr. A* **2018**, *1562*, 69–77. [CrossRef] [PubMed]
218. Woiwode, U.; Ferri, M.; Maier, N.M.; Lindner, W.; Lämmerhofer, M. Complementary enantioselectivity profiles of chiral cinchonan carbamate selectors with distinct carbamate residues and their implementation in enantioselective two-dimensional high-performance liquid chromatography of amino acids. *J. Chromatogr. A* **2018**, *1558*, 29–36. [CrossRef] [PubMed]
219. Schmitt, K.; Lämmerhofer, M. Optimization of the surface modification process of cross-linked polythiol-coated chiral stationary phases synthesized by a two-step thiol-ene click reaction. *J. Sep. Sci.* **2018**, *41*, 1338–1345. [CrossRef] [PubMed]
220. Patel, D.C.; Breitbach, Z.S.; Yu, J.; Nguyen, K.A.; Armstrong, D.W. Quinine bonded to superficially porous particles for high-efficiency and ultrafast liquid and supercritical fluid chromatography. *Anal. Chim. Acta* **2017**, *963*, 164–174. [CrossRef] [PubMed]
221. Orosz, T.; Forró, E.; Fülöp, F.; Lindner, W.; Ilisz, I.; Péter, A. Effects on N-methylation and amidination of cyclic β-amino acids on enantioselectivity and retention characteristics using Cinchona alkaloid- and sulfonic acid-based chiral zwitterionic stationary phases. *J. Chromatogr. A* **2018**, *1535*, 72–79. [CrossRef] [PubMed]

222. Bajtai, A.; Fekete, B.; Palkó, M.; Fülöp, F.; Lindner, W.; Kohout, M.; Ilisz, I.; Péter, A. Comparative study on the liquid chromatographic enantioseparation of cyclic β-amino acids and the related cyclic β-aminohydroxamic acids on *Cinchona* alkaloid-based zwitterionic chiral stationary phases. *J. Sep. Sci.* **2018**, *41*, 1216–1223. [CrossRef] [PubMed]
223. Calderón, C.; Santi, C.; Lämmerhofer, M. Chiral separation of disease biomarkers with 2-hydroxycarboxylic acid structure. *J. Sep. Sci.* **2018**, *41*, 1224–1231. [CrossRef] [PubMed]
224. Nakano, Y.; Taniguchi, M.; Fukusaki, E. High-sensitive liquid chromatography-tandem mass spectrometry-based chiral metabolic profiling focusing on amino acids and related metabolites. *J. Biosci. Bioeng.* **2018**. [CrossRef] [PubMed]
225. Iannia, F.; Pucciarinia, L.; Carottia, A.; Gioielloa, A.; Galarinib, R.; Natalinic, S.; Sardellaa, R.; Lindnerd, W.; Natalini, B. Improved chromatographic diastereoresolution of cyclopropyl dafachronic acid derivatives using chiral anion exchangers. *J. Chromatogr. A* **2018**, *1557*, 20–27. [CrossRef] [PubMed]
226. Sousa, L.R.; Sogah, G.D.Y.; Hoffmann, D.H.; Cram, D.J. Host-Guest Complexation. 12. Total optical resolution of amine and amino ester salts by chromatography. *J. Am. Chem. Soc.* **1978**, *100*, 4569–4576. [CrossRef]
227. León-González, M.E.; Rosales-Conrado, N.; Pérez-Arribas, L.V.; Guillén-Casla, V. Two-dimensional liquid chromatography for direct chiral separations: A review. *Biomed. Chromatogr.* **2014**, *28*, 59–83. [CrossRef] [PubMed]
228. Hyun, M.H. Liquid chromatographic enantioseparations on crown ether-based chiral stationary phases. *J. Chromatogr. A* **2016**, *1467*, 19–32. [CrossRef] [PubMed]
229. Hyun, M.H. Development of HPLC chiral stationary phases based on (+)-(18-Crown-6)-2,3,11,12 -tetracarboxylic acid and their applications. *Chirality* **2015**, *27*, 576–588. [CrossRef] [PubMed]
230. Bakó, P.; Keglevich, G.; Rapi, Z.; Töke, L. The enantiomeric differentiation ability of chiral crown ethers based on carbohydrates. *Curr. Org. Chem.* **2012**, *16*, 297–304. [CrossRef]
231. Mohammadzadeh Kakhki, R. Application of crown ethers as stationary phase in the chromatographic methods. *J. Incl. Phenom. Macrocycl. Chem.* **2013**, *75*, 11–22. [CrossRef]
232. Paik, M.J.; Kang, J.S.; Huang, B.S.; Carey, J.R.; Lee, W. Development and application of chiral crown ethers as selectors for chiral separation in high-performance liquid chromatography and nuclear magnetic resonance spectroscopy. *J. Chromatogr. A* **2013**, *1274*, 1–5. [CrossRef] [PubMed]
233. Ma, M.; Wei, Q.; Meng, M.; Yin, J.; Shan, Y.; Du, L.; Zhu, X. Preparation and application of aza-15-crown-5-capped methylcalix [4] resorcinarene-bonded silica particles for use as chiral stationary phase in HPLC. *Chromatographia* **2017**, *80*, 1007–1014. [CrossRef]
234. Yaghoubnejad, S.; Ahmadi, S.H. Preparation and evaluation of a chiral HPLC stationary phase based on cone calix[4]arene functionalized at the upper rim with L-alanine units. *Biomed. Chromatogr.* **2018**, *32*, e4122. [CrossRef] [PubMed]
235. Li, Y.; Sheng, Z.; Zhu, C.; Yin, W.; Chu, C. Silica based click-dibenzo-18-crown-6-ether high-performance liquid chromatography stationary phase and its application in separation of fullerenes. *Talanta* **2018**, *178*, 195–201. [CrossRef] [PubMed]
236. Németh, T.; Dargó, G.; Petró, J.L.; Petrik, Z.; Lévai, S.; Krámos, B.; Béni, Z.; Nagy, J.; Balogh, G.T.; Huszthy, P.; et al. Synthesis and pK_a determination of new enantiopure dimethyl-substituted acridino-crown ethers containing a carboxyl group: Useful candidates for enantiomeric recognition studies. *Chirality* **2017**, *29*, 522–535. [CrossRef] [PubMed]
237. Yaghoubnejad, S.; Ahmadi, S.H. Preparation and evaluation of a deoxycholic-calix[4]arene hybrid-type receptor as a chiral stationary phase for HPLC. *J. Sep. Sci.* **2018**, *41*, 1903–1912. [CrossRef] [PubMed]
238. Sun, P.; Wang, C.; Breitbach, Z.S.; Zhang, Y.; Armstrong, D.W. Development of new HPLC chiral stationary phases based on native and derivatized cyclofructans. *Anal. Chem.* **2009**, *81*, 10215–10226. [CrossRef] [PubMed]
239. Khan, M.M.; Breitbach, Z.S.; Berthood, A.; Armstrong, D.W. Chlorinated aromatic derivatives of cyclofructan 6 as HPLC chiral stationary phases. *J. Liq. Chromatogr. Relat. Technol.* **2016**, *39*, 497–503. [CrossRef]
240. Janecková, L.; Kalíková, K.; Vozka, J.; Armstrong, D.W.; Bosáková, Z.; Tesarová, E. Characterization of cyclofructan-based chiral stationary phase by linear free energy relationship. *J. Sep. Sci.* **2011**, *34*, 2639–2644. [CrossRef] [PubMed]

241. Padivitage, N.L.; Smuts, J.P.; Breitbach, Z.S.; Armstrong, D.W.; Berthood, A. Preparation and evaluation of HPLC chiral stationary phases based on cationic/basic derivatives of cyclofructan 6. *J. Liq. Chromatogr. Relat. Technol.* **2015**, *38*, 550–560. [CrossRef] [PubMed]
242. Qiu, H.; Kiyono-Shimobe, M.; Armstrong, D.W. Native/derivatized cyclofructan 6 bound to resins via "click" chemistry as stationary phases for achiral/chiral separations. *J. Liq. Chromatogr. Relat. Technol.* **2014**, *37*, 2302–2326. [CrossRef]
243. Moskalová, M.; Kozlov, O.; Gondová, T.; Budovská, M.; Armstrong, D.W. HPLC enantioseparation of novel spirobrassinin analogs on the cyclofructan chiral stationary phases. *Chromatographia* **2017**, *80*, 53–62. [CrossRef]
244. Frink, L.A.; Berthood, A.; Xu, Q.L.; Gao, H.; Kurti, L.; Armstrong, D.W. Separation of 2-naphtol atropisomers on cyclofructan-based chiral stationary phases. *J. Liq. Chromatogr. Relat. Technol.* **2016**, *39*, 710–717. [CrossRef]
245. Gutierrez-Climente, R.; Gomez-Caballero, A.; Guerreiro, A.; Garcia-Mutio, D.; Unceta, N.; Goicolea, M.A.; Barrio, R. Molecularly imprinted nanoparticles grafted to porous silica as chiral selectors in liquid chromatography. *J. Chromatogr. A* **2017**, *1508*, 53–64. [CrossRef] [PubMed]
246. Sellergren, B. Imprinted chiral stationary phases in high-performance liquid chromatography. *J. Chromatogr. A* **2001**, *906*, 227–252. [CrossRef]
247. Wei, Z.H.; Mu, L.N.; Huang, Y.P.; Liu, Z.S. Imprinted monoliths: Recent significant progress in analysis field. *TrAC-Trends Anal. Chem.* **2017**, *86*, 84–92. [CrossRef]
248. Rutkowska, M.; Plotka-Wasylka, J.; Morrison, C.; Wieczorek, P.P.; Namiesnik, J.; Marc, M. Application of molecularly imprinted polymers in analytical chiral separations and analysis. *TrAC-Trends Anal. Chem.* **2018**, *102*, 91–102. [CrossRef]
249. Yang, S.; Wang, Y.; Jiang, Y.; Li, S.; Liu, W. Molecularly imprinted polymers for the identification and separation of chiral drugs and biomolecules. *Polymers* **2016**, *8*, 216. [CrossRef]
250. Bitas, D.; Samanidou, V. Molecularly imprinted polymers as extracting media for the chromatographic determination of antibiotics in milk. *Molecules* **2018**, *23*, 316. [CrossRef] [PubMed]
251. Sierra, I.; Marina, M.L.; Pérez-Quintanilla, D.; Morante-Zarcero, S.; Silva, M. Approaches for enantioselective resolution of pharmaceuticals by miniaturised separation techniques with new chiral phases based on nanoparticles and monoliths. *Electrophoresis* **2016**, *37*, 2538–2553. [CrossRef] [PubMed]
252. Liu, Z.S.; Zheng, C.; Yan, C.; Gao, R.Y. Molecularly imprinted polymers as a tool for separation in CEC. *Electrophoresis* **2007**, *28*, 127–136. [CrossRef] [PubMed]
253. Prabhu, P.; Dhawan, V.; Fernandes, C.; Suares, D. Recent progress on nanofabrication of molecularly imprinted polymers. In *Emerging Applications of Nanoparticles and Architecture Nanostructures*; Elsevier: Amsterdam, The Netherlands, 2018; pp. 385–409.
254. Moniera, M.; Shafikb, A.L.; Abdel-Latifa, D.A. Surface molecularly imprinted amino-functionalized alginate microspheres for enantio-selective extraction of L-ascorbic acid. *Carbohydr. Polym.* **2018**, *195*, 652–661. [CrossRef] [PubMed]
255. Khataee, A.; Hassanzadeh, J.; Elmira Kohan, E. Specific quantification of atropine using molecularly imprinted polymer on graphene quantum dots. *Spectrochim. Acta A* **2018**, *205*, 614–621. [CrossRef] [PubMed]
256. Alatawi, R.A.S.; Monier, M.; Elsayed, N.H. Chiral separation of (±)-methamphetamine racemate using molecularly imprinted sulfonic acid functionalized resin. *J. Colloid Interface Sci.* **2018**, *541*, 654–663. [CrossRef] [PubMed]
257. Pisarev, O.A.; Polyakova, I.V. Molecularly imprinted polymers based on methacrylic acid and ethyleneglycol dimethacrylate for L-lysine recognition. *React. Funct. Polym.* **2018**, *130*, 98–110. [CrossRef]
258. Moniera, M.; El-Mekabaty, A.; Abdel-Latifa, D.A. Synthesis and evaluation of enantio-selective L-histidine imprinted salicylic acid functionalized resin. *React. Funct. Polym.* **2018**, *128*, 104–113. [CrossRef]
259. Monier, M.; Youssef, I.; Abdel-Latifa, D.A. Synthesis of imprinted styrene-maleic acid functionalized resin for enantioselective extraction of R-amphetamine. *Chem. Eng. J.* **2019**, *356*, 693–701. [CrossRef]
260. Yang, Y.-J.; Liu, X.-W.; Kong, X.-J.; Qin, Z.; Jiao, Z.-H.; Li, S.-H.; Li, J.-Y. Preparation and evaluation of oseltamivir molecularly imprinted polymer silica gel as liquid chromatography stationary phase. *Molecules* **2018**, *23*, 1881. [CrossRef] [PubMed]
261. Wang, X.H.; Dong, Q.; Ying, L.L.; Chi, S.S.; Lan, Y.H.; Huang, Y.P.; Liu, Z.S. Enhancement of selective separation on molecularly imprinted monolith by molecular crowding agent. *Anal. Bioanal. Chem.* **2017**, *409*, 201–211. [CrossRef] [PubMed]

262. Shen, J.; Okamoto, Y. Efficient separation of enantiomers using stereoregular chiral polymers. *Chem. Rev.* **2016**, *116*, 1094–1138. [CrossRef] [PubMed]
263. Nakano, T. Optically active synthetic polymers as chiral stationary phases in HPLC. *J. Chromatogr. A* **2001**, *906*, 205–225. [CrossRef]
264. Ding, P.; Chang, B.; Qing, G.; Sun, T. New approach for chiral separation: From polysaccharide-based materials to chirality-responsive polymers. *Sci. China Chem.* **2014**, *57*, 1492–1506. [CrossRef]
265. Maeda, K.; Yashima, E. Helical polyacetylenes induced via noncovalent chiral interactions and their applications as chiral materials. *Top. Curr. Chem.* **2017**, *375*, 72. [CrossRef] [PubMed]
266. Mallakpour, S.; Khadem, E. Hybrid optically active polymer/metal oxide composites: Recent advances and challenges. In *Hybrid Polymer Composite Materials*; Woodhead Publishing: Sawston, UK, 2017; pp. 379–406. [CrossRef]
267. Ding, X.; Yang, J.; Dong, Y. Advancements in the preparation of high-performance liquid chromatographic organic polymer monoliths for the separation of small-molecule drugs. *J. Pharm. Anal.* **2018**, *8*, 75–85. [CrossRef] [PubMed]
268. Maeda, K.; Maruta, M.; Sakai, Y.; Ikai, T.; Kanoh, S. Synthesis of optically active poly(diphenylacetylene)s using polymer reactions and an evaluation of their chiral recognition abilities as chiral stationary phases for HPLC. *Molecules* **2016**, *21*, 1487. [CrossRef] [PubMed]
269. Ikai, T.; Awata, S.; Kudo, T.; Ishidate, R.; Maeda, K.; Kanoh, S. Chiral stationary phases consisting of π-conjugated polymers bearing glucose-linked biphenyl units: Reversible switching of resolution abilities based on a coil-to-helix transition. *Polym. Chem.* **2017**, *29*, 4190–4198. [CrossRef]
270. Zhang, K.; Cai, S.L.; Yan, Y.L.; He, Z.H.; Lin, H.M.; Huang, X.L.; Zheng, S.R.; Fan, J.; Zhan, W.G. Construction of a hydrazone-linked chiral covalent organic framework-silica composite as the stationary phase for high-performance liquid chromatography. *J. Chromatogr. A* **2017**, *1519*, 100–109. [CrossRef] [PubMed]
271. De Mello, A. FOCUS. *Lab Chip* **2002**, *3*, 48N–54N. [CrossRef] [PubMed]
272. Grinias, J.P.; Kennedy, R.T. Advances in and prospects of microchip liquid chromatography. *TrAC-Trends Anal. Chem.* **2016**, *81*, 110–117. [CrossRef] [PubMed]
273. Kecskemeti, A.; Gaspar, A. Particle-based liquid chromatographic separations in microfluidic devices—A review. *Anal. Chim. Acta* **2018**, *1021*, 1–19. [CrossRef] [PubMed]
274. Manz, A.; Graber, N.; Widmer, H.M. Miniaturized total chemical analysis systems: A novel concept for chemical sensing. *Sens. Actuator B Chem.* **1990**, *1*, 244–248. [CrossRef]
275. Baharfar, M.; Yamini, Y.; Seidi, S.; Karami, M. Quantitative analysis of clonidine and ephedrine by a microfluidic system: On-chip electromembrane extraction followed by high performance liquid chromatography. *J. Chromatogr. B* **2017**, *1068–1069*, 313–321. [CrossRef] [PubMed]
276. Payán, M.R.; Murillo, E.S.; Coello, J.; López, M.Á.B. A comprehensive study of a new versatile microchip device based liquid phase microextraction for stopped-flow and double-flow conditions. *J. Chromatogr. A* **2018**, *1556*, 29–36. [CrossRef] [PubMed]
277. Zhai, H.; Huang, L.; Chen, Z.; Su, Z.; Yuan, K.; Liang, G.; Pan, Y. Chip-based molecularly imprinted monolithic capillary array columns coated GO/SiO_2 for selective extraction and sensitive determination of rhodamine B in chili powder. *Food Chem.* **2017**, *214*, 664–669. [CrossRef] [PubMed]

molecules

MDPI

Article

Improved Resolution of 4-Chloromandelic Acid and the Effect of Chlorine Interactions Using (*R*)-(+)-Benzyl-1-Phenylethylamine as a Resolving Agent

Yangfeng Peng [1,*], Cai Feng [1], Sohrab Rohani [2] and Quan He [3]

1 School of Chemical Engineering, East China University of Science and Technology, Shanghai 200237, China; cai_erf@126.com

2 Department of Chemical and Biochemical Engineering, Western University, London, ON N6A 5B9, Canada; srohani@uwo.ca

3 Department of Engineering, Faculty of Agriculture, Dalhousie University, Truro, NS B2N 5E3, Canada; quan.he@dal.ca

* Correspondence: yfpeng@ecust.edu.cn; Tel: +86-21-64252345

Received: 1 December 2018; Accepted: 17 December 2018; Published: 18 December 2018

Abstract: In order to avoid the disadvantage of commonly used resolving agent 1-phenylethylamine (hereafter: PEA), which is soluble in water, (*R*)-(+)-benzyl-1-phenylethylamine ((*R*)-(+)-BPA) was used to resolve 4-chloromandelic acid (4-ClMA) in this study. The optimal resolution conditions were determined: absolute ethanol as a solvent, the molar ratio of 4-ClMA to (*R*)-(+)-BPA as 1:1, the filtration temperature as 15 °C, and the amount of solvent as 1.6 mL/1 mmol 4-ClMA. Thermophysical properties, such as melting point, heat of fusion, and solubility, exhibited significant differences between the less and more soluble salts. The single crystals for the pair of diastereomeric salts were cultivated and their crystal structures were examined thoroughly. In addition to commonly observed interactions like hydrogen bonding and CH/π interactions. The chlorine ... chlorine interaction was observed in the less soluble salt presenting as Cl ... Cl between adjacent hydrogen network columns, while the Cl/π interaction was observed in the more soluble salt. It was found that halogen interactions played an important role in chiral recognition of 4-ClMA by (*R*)-(+)-BPA.

Keywords: chlorine interactions; 4-chloromandelic acid; benzyl-1-phenylethylamine; chiral discrimination; resolution

1. Introduction

In response to the new guidelines on drugs with chiral centers launched by the U.S. Food and Drug Administration (FDA) in 1992 [1], significant efforts have been made to develop new methods and/or improve separation efficiency for obtaining enantiopure pharmaceutical intermediates. These methods include catalytic asymmetric synthesis, chromatographic resolution, extraction resolution, membrane resolution, diastereomeric salt resolution, and enzymatic resolution [2–8] (pp. 9–50). Among these established technologies, optical resolution of racemic substrates through diastereomeric salt formation is still one of the most practical and economical approaches for an industrial scale production, and it has been widely used in the pharmaceutical industry. This route holds a number of advantages, such as the simplicity of crystallization, low cost of operation, recyclability of resolving agents, and availability of both enantiomers. Although extensive experience has been garnered with more than 10,000 successful reported resolutions, the underlying chiral recognition mechanism in the resolution is not well understood. There is no concrete principle to guide the selection and design of resolving agents. For a given racemate, screening the most suitable resolving agent involves tedious trial and error procedures, which is time-consuming and laboratory-intensive [7,8] (pp. 9–50).

Investigation was conducted on the crystal structures of a pair of diastereomeric salts formed between the racemic substrate. The resolving agent is a promising way to study the difference in stability between two salts and gain insights into the chiral discrimination mechanism. The rationale has hinged on an expectation of extracting the common and characteristic crystal structure characteristics of diastereomeric salts in successful resolutions, thus guiding resolving agent selection for a given racemate. Most research efforts have focused on experimentally obtaining crystal structures of diastereomeric salts, followed by qualitative analysis of the difference in molecule packing and intermolecular interactions in the crystal lattices. A number of characteristic features were of great interest, including the structural similarity [9], relative molecular length [10–12], hydrogen bonding network [13,14], the CH/π interaction and packing mode of aromatic groups [15,16], and the role of incorporated solvent molecules in diastereomeric salts [17,18]. Recently, two case studies reported the role of halogen bonding interactions in diastereomeric resolution [19,20].

Enantiopure mandelic acid and its derivatives are important chiral building blocks. For example, pure enantiomers of 2-chloromandelic acid (2-ClMA) is used for the production of clopidogrel, a widely administered anticoagulant [21,22]. Enantiopure 4-chloromandelic acid (4-ClMA) is a precursor in the preparation of 2-aryloxy-2-arylalkanoic acids for diabetes and lipid disorders drugs [23]. Phenylethylamine (PEA) is well-recognized as an efficient resolving agent for racemic acidic substrates [24,25]. In ongoing research on the optical resolution of mandelic acid derivatives [26–29], we had interesting observations: Using PEA as the most commonly used base resolving agent, 4-ClMA was resolved efficiently [26] and 3-ClMA was resolved with a moderate efficiency. However, we failed to resolve 2-ClMA. This indicates that the different position of chlorine in substituted benzene ring led to a significant difference in chiral discrimination. Although PEA is widely used, its solubility in water is relatively high (4% at 20 °C) in the context of optical resolution [30]. This results in a low recovery of PEA in downstream processing. Therefore, an additional extraction step is required [31]. To overcome this challenge, PEA can be modified through the introduction of a benzyl group to its amino group, creating a new resolving agent, *N*-benzyl-1-phenylethylamine (BPA). BPA is proven to be an excellent resolving agent with much lower solubility in water compared to PEA [28]. Since the solubility of BPA in water at 20 °C is only 0.014% [32], BPA can be readily separated from water once the resolution is completed. A high recovery rate of BPA improves the overall economic variability of optical resolution. This encouraged us to further explore a resolution of 4-ClMA using optically active (*R*)-(+)-BPA via diastereomeric salts formation.

To understand the mechanism of chiral discrimination in this new attempt (resolution of 4-ClMA by optical active BPA), the single crystals of both less and more soluble diastereomeric salt, (*R*)-(−)-4-ClMA·(*R*)-(+)-BPA and (*S*)-(+)-4-ClMA·(*R*)-(+)-BPA, were cultivated. From crystal structural examination, it was found that hydrogen bonds, CH/π interaction, van der Waals interaction, supramolecular packing mode, and unique interactions related to the chlorine atom were present in both less soluble and more soluble salts. However, these interactions presented differently. Such interactions of chlorine were not found in the diastereomeric salts from either the resolution of 4-ClMA with PEA [26] or the resolution of 2-chloremandelic acid with BPA [28].

The chemical structures of 4-ClMA and the resolving agent (*R*)-(+)-BPA are presented in Figure 1.

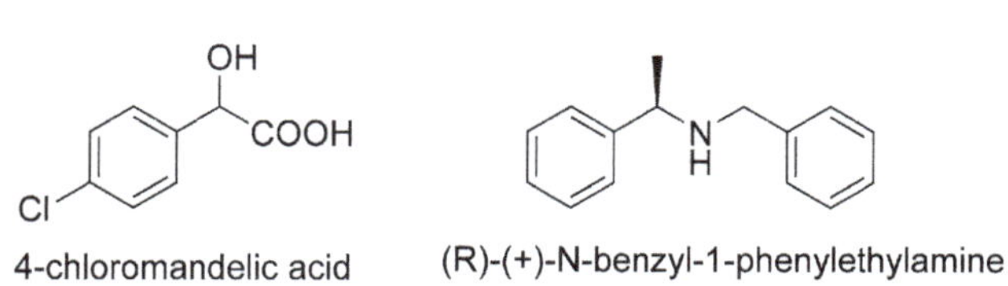

Figure 1. Chemical structure of 4-chloromandelic acid (4-ClMA) and (*R*)-(+)-*N*-benzyl-1-phenylethylamine (BPA).

2. Results and Discussion

2.1. Solvent Screening and Resolution Condition Determination

As shown in Table 1, racemic 4-ClMA can be resolved by (*R*)-(+)-BPA in most solvents described in 3.7. It is difficult to evaluate the efficiency of a resolution process, as the diastereomeric excess comprises with the yield. The product of diastereomeric excess and yield, termed as optical efficiency (*E*), is sometimes adopted to determine the suitable resolution conditions. The resolution efficiency using 50% ethanol, 2-proponal, acetonitrile, and ethyl acetate is in the range of 87.5–93.2%. This is considered somewhat high. However, considering the low %*d.e.* of diastereomeric salts (52.1% to 55.2%), multiple recrystallizations must be performed to obtain enantiomers with 99 %*e.e.* Therefore, ethanol with a relatively high %*d.e.* is chosen as a solvent for the following investigation. Additional benefit of using ethanol is its low price compared to other solvents.

Table 1. Solvent screening on the resolution of (*R,S*)-4-ClMA by (*R*)-(+)-BPA.

4-ClMA/mol	BPA/mol	Solvent [a]	%*d.e.*	Yield [b]/%	*E* [c]/%
0.005	0.005	Methanol	94.5	55.9	52.8
0.005	0.005	Ethanol	94.8	83.1	78.8
0.005	0.005	95% ethanol	96.3	76.3	73.5
0.005	0.005	50% ethanol	55.2	158.5	87.5
0.005	0.005	2-propanol	56.3	156.9	88.4
0.005	0.005	Acetonitrile	52.1	172.0	89.6
0.005	0.005	Ethyl acetate	52.8	176.4	93.2
0.005	0.005	chloroform		No salts	

[a] The volume of all solvents used here was 8 mL. [b] The yield was equal to and resulted at the temperature of filter of 20 °C. [c] Resolution efficiency *E* is the product of yield and diastereomeric purity

Once a suitable solvent was identified, other factors affecting the resolution process were investigated, including the mole ratio of 4-ClMA to (*R*)-(+)-BPA, the amount of ethanol solvent, and the filtration temperature. In order to reduce experiment runs, a three-factor and three-level orthogonal experiment design, 3^3, was used. Table S1 summarized the experimental results. Based on the orthogonal design analysis, the optimal resolution conditions were determined to be the molar ratio of 4-ClMA to (*R*)-(+)-BPA of 1:1, the amount of absolute ethanol of 1.6 mL/1 mmol 4-ClMA, and the filtration temperature of 15 °C. The optimal conditions derived from orthogonal design were verified by experiments. The resolution efficiency *E* reached 84.3% under such conditions, which is significantly higher than the resolution efficiency *E* of 71.4% in the resolution of (*R,S*)-4-ClMA by (*R*)-(+)-PEA, as reported in the literature [26].

2.2. Thermodynamic Properties of Diastereomeric Salts

To explore the chiral recognition ability of (*R*)-(+)-BPA, the corresponding less and more soluble salts, (*R*)-(−)-4-ClMA·(*R*)-(+)-BPA and (*R*)-(−)-4-ClMA·(*S*)-(−)-BPA, were synthesized as described in Sections 3.3 and 3.4. It is generally common that an efficient separation can be expected when the difference in melting point of a pair of diastereomeric salt is larger than 20 °C. The trends of solubility difference between diastereomeric salts parallels include differences in melting points and heat of fusion [33,34]. The thermal properties of the resulting salts were determined and listed in Table 2. The melting point and solubility differences were significant. The melting point of the less soluble salt exceeded that of the corresponding more soluble salt by 34.3 °C. The difference in heat of fusion between the less and more soluble salt was 4.83 kJ·mol^{-1}. Table 3 also showed that the solubility ratio of the more soluble salt (*R*)-(−)-4-ClMA·(*S*)-(−)-BPA to less soluble salt (*R*)-(−)-4-ClMA·(*R*)-(+)-BPA in absolute ethanol at 20 °C is 3.3:1.

Table 2. Thermal properties of diastereomeric salts of (*R*,*S*)-4-ClMA and (*R*)-(+)-BPA.

	Solubility [a]/g	Melting Point/°C	Heat of fusion/kJ·mol^{-1}
Less soluble salt	1.47	166.3	57.41
More soluble salt	4.81	132.0	52.58

[a] Weight (g) of the solute dissolved in 100 g absolute ethanol solvent at 20 °C.

The comparison of the aforementioned thermal properties strongly suggests that the less soluble salt is significantly more stable than the more soluble salt, and the high-resolution efficiency achieved in the Section 2.1 resulted from such difference in stability.

A binary melting point phase diagram of diastereomeric salts was also established. In general, diastereomeric salt mixtures are classified into three types: A eutectic conglomerate, a 1:1 addition compound, and a solid solution [33–36]. The type of diastereomeric salt mixture can be identified on the basis of a binary melting point phase diagram. The formation of eutectic conglomerate is the primary condition for resolution. Figure 2 shows the binary melting point phase diagram for a mixture of (*R*)-(−)-4-ClMA·(*R*)-(+)-BPA and (*R*)-(−)-4-ClMA·(*S*)-(−)-BPA. This mixture was constructed by a detailed DSC analysis of mixtures with different compositions of X_R, which was the fraction of more soluble salt (*R*)-(−)-4-ClMA·(*R*)-(+)-BPA. The system was a eutectic mixture, and the eutectic composition, X_e, was 0.25 from experimental results, which is in agreement with the X_e of 0.25, calculated theoretically from thermodynamic Schroder-van-Laar equation [37]. The fact that the system is a eutectic mixture indicates that the resolution efficiency primarily depends on the difference in stability between the less and more soluble salts.

To further the understanding of the stability difference between the pair of diastereomeric salts of BPA and 4-ClMA, the single crystals of the corresponding diastereomeric salts were grown and examined by X-ray diffraction.

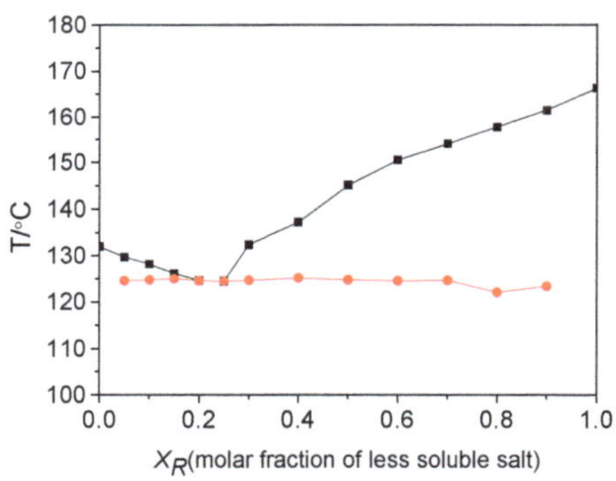

Figure 2. Binary melting point phase diagram of diastereomeric salts. The circle and square represent the temperatures at the beginning and the end of fusion, respectively.

2.3. *Crystal Structure of Diastereomeric Salts*

High-quality single crystals of the less soluble salt (*R*)-(−)-4-ClMA·(*R*)-(+)-BPA were readily produced in the solvent of 2-propanol. The single crystal of the more soluble salt (*S*)-(+)-4-ClMA·(*R*)-(+)-BPA was obtained in a co-solvent solution of acetonitrile and methanol. Crystals of (*R*)-(−)-4-ClMA·(*R*)-(+)-BPA were colorless rods crystallized in a monoclinic C2 space group (Appendix A, ccdc 1030316). Each unit cell contained four (*R*)-(−)-4-ClMA anions and (*R*)-(+)-BPA cations. Crystals of (*S*)-(+)-4-ClMA·(*R*)-(+)-BPA were colorless plates crystallized in an orthorhombic $P2_12_12_1$ space group (Appendix A, ccdc 1885316). Each unit cell contained four (*S*)-(+)-4-ClMA anions and (*R*)-(+)-BPA cations, as shown in Figure 3. Figure S1 graphically displayed the atomic numbering schemes. Detailed crystal data of (*R*)-(−)-4-ClMA·(*R*)-(+)-BPA and (*S*)-(+)-4-ClMA·(*R*)-(+)-BPA were summarized in Table S2.

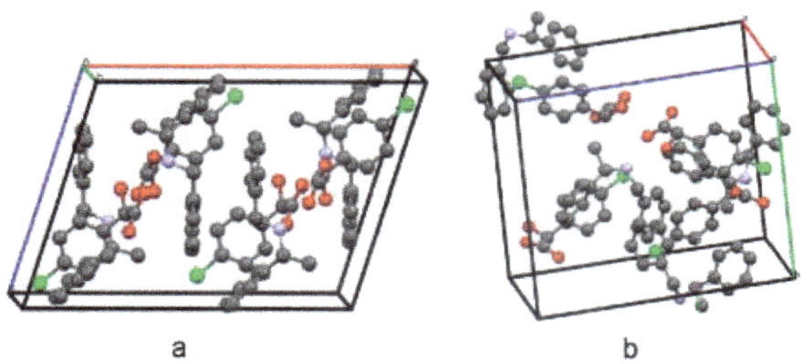

Figure 3. Molecules packing in the unit cells of (*R*)-(−)-4-ClMA·(*R*)-(+)-BPA (**a**) and (*S*)-(+)-4-ClMA·(*R*)-(+)-BPA (**b**).

The arrangement and stacking modes of the organic molecules in the crystal structure mainly depends on a number of weak intra and/or intermolecular interactions [38–41]. In the context of an optical resolution through diastereomeric salt formation, a basic compound and an acidic compound can form various kinds of crystals depending on their molecular structures and functional groups, respectively. In the process of crystallization, it is well-recognized that these molecules arrange as compactly as possible through interactions like hydrogen bonding, CH/π, and van der Waals force to realize the minimization of energy [9–18]. The hydrogen bond was conventionally considered to be the most important factor in accounting for chiral discrimination because it is stronger than other interactions in determining the crystal structure. CH/π interaction, which involves a weak hydrogen bond, has been considered a secondary interaction in comparison with hydrogen bonds. However, accumulated evidence indicates that CH/π interaction in diastereomeric salt crystal structures may contribute to chiral discrimination to a considerable extent. Recently, other intermolecular interactions, like halogen bonds, were found to contribute to chiral discrimination [19,20]. The difference in the stabilities of diastereomeric salts come from such super-molecular interactions. Therefore, they were examined in detail as follows.

2.3.1. Hydrogen-Bonding Network

Tables 3 and 4 show the hydrogen-bonding geometry of (*R*)-(−)-4-ClMA·(*R*)-(+)-BPA and (*S*)-(+)-4-ClMA·(*R*)-(+)-BPA, respectively.

Table 3. Hydrogen-bonding geometry of (*R*)-(−)-4-ClMA·(*R*)-(+)-BPA [Å and °].

D-H ... A	d(D-H)	d(H ... A)	d(D ... A)	<(DHA)
O3A-H3A ... O1A [a]	0.87(2)	1.84(2)	2.6878(15)	164.9(17)
O3A-H3A ... O2A [a]	0.87(2)	2.52 (2)	3.1629(14)	130.6(16)
N1B-H1BA ... O2A [a]	0.85(2)	1.90(2)	2.7457(16)	176.1(16)
N1B-H1BB ... O1A	0.98(2)	1.78(2)	2.7337(15)	163.7(18)
N1B-H1BB ... O3A	0.98(2)	2.42(2)	3.0019(14)	117.4(15)

Symmetry codes: (a) −x + 3/2, y − 1/2, −z + 1.

Table 4. Hydrogen-bonding geometry of (*S*)-(+)-4-ClMA·(*R*)-(+)-BPA [Å and °].

D-H ... A	d(D-H)	d(H ... A)	d(D ... A)	<(DHA)
O(3)-H(3)...O(1) [a]	0.90(3)	1.81(3)	2.707(3)	170(3)
N(1)-H(1A)...O(1) [b]	1.04(3)	1.84(3)	2.867(3)	169(2)
N(1)-H(1B)...O(2) [c]	0.93(3)	1.80(3)	2.690(3)	158(3)

Symmetry codes: (a) x − 1/2, −y + 1/2, −z + 1; (b) x,y + 1,z; (c) x − 1/2, −y + 3/2, −z + 1.

In the single crystal of the less soluble salt (*R*)-(−)-4-ClMA·(*R*)-(+)-BPA, two intramolecular hydrogen bonds were found: N1B-H1BB ... O1A and N1B-H1BB ... O3A. There also were three intermolecular hydrogen bonds: N1B-H1BA ... O2A1, O3A-H3A ... O2A1, and O3A-H3A ... O1A1. The two intramolecular hydrogen bonds were fundamental for the formation of less soluble salts. Three intermolecular hydrogen bonds, one from the ammonium hydrogen of (*R*)-(+)-BPA and neighboring carboxylic oxygen of (*R*)-(−)-4-ClMA and the other two from the carboxylic hydrogens of (*R*)-(−)-4-ClMA and the adjacent carboxylic oxygens, formed the hydrogen network along the C-axis. As shown in Figure S2, such intermolecular hydrogen bonds formed the one-dimensional (1D) double-helix chains in the H-bonding network. These results are similar to those reported in the resolution of 2-ClMA with BPA [28], and different from those formed between primary amines and carboxylic acids [13,26,27].

There are three types of hydrogen bonds present in crystal structure of the more soluble salt of (*S*)-(+)-4-ClMA·(*R*)-(+)-BPA. Two were ammonium hydrogens of (*R*)-(+)-BPA with carboxylate oxygens from adjacent (*S*)-(+)-4-ClMA molecules: N1-H1A ... O1 and N1-H1B ... O2. One was O3-H3 ... O1, which was from the hydroxygen hydrogen of (*S*)-(+)-4-ClMA and carboxylate oxygen from the neighboring (*S*)-(+)-4-ClMA. The H-bonding network also presented a pattern of 1D double-helix chains. There was no significant difference observed in the H-bonding networks between the less and more soluble salts, implying that chiral recognition did not primarily originate from hydrogen bond motifs.

2.3.2. CH/π and π/π Interaction

CH/π interaction is considered a weak hydrogen bond and plays an important role in areas such as crystal packing and molecular recognition. The contribution of the CH/π interaction to chiral discrimination was recently recognized [15,16,27,42,43]. In this study, we found CH/π interactions not only within the hydrogen bonding columns, but in adjacent hydrophobic layers in the crystal structure of the less soluble salt (*R*)-(−)-4-ClMA·(*R*)-(+)-BPA. Specifically, as shown in Figure S3, CH/π interactions were present between the benzene ring of C3-C4 ... C8 of (*R*)-(−)-4-ClMA and the CH groups C2B-H2B and C5B-H5B at neighboring benzene rings of (*R*)-(+)-BPA within the hydrogen-bonding network columns. The CH/π distances were 2.624 Å and 2.740 Å, respectively. In Figure S4, the CH group at benzyl group of (*R*)-(+)-BPA in one hydrophobic layer interacted with the benzene ring of benzyl group of (*R*)-(+)-BPA from an adjacent layer. The CH/π distance was 2.847 Å. Additionally, π/π interactions were observed. Interactions formed between the two benzene rings of (*R*)-(+)-BPA from the two hydrophobic layers with a distance of 3.365 Å.

However, in the crystal structure of the more soluble salt (*S*)-(+)-4-ClMA·(*R*)-(+)-BPA, no such CH/π interactions were discovered, indicating that the CH/π interactions played a dominant role in realizing the close packing among hydrophobic layers. This conclusion is consistent with the observations reported in other studies [15,16,27,28].

2.3.3. Chlorine ... Chlorine/π Interactions

Excluding hydrogen bonds and CH/π interactions, there are other weak interactions in crystal structure of salts, such as halogen bonds. Halogen-bonding is a type of weak interaction, and its strength is comparable to that of hydrogen bonding. The halogen bond has an important role in molecular assembly in many areas, such as crystal engineering, material science, molecular recognition, medical design, and organic reaction [40,41]. Although halogen-bonding has attracted much attention in supramolecular chemistry, there is limited research regarding the role of halogen bonds in the chiral discrimination of optical resolutions. In 1999, Farina resolved 1,2-dibromohexfluoropropane with sparteine in the solvent of chloroform, and halogen bonds were first observed to contribute to the success of this resolution [20]. Kobayashi et al. purposely designed racemic substrates and resolving agents with halogen atoms and found that halogen-bonding interactions could effectively stabilize diastereomeric salt crystals [19]. In addition, there are Cl-Cl or Cl/π interactions that cannot be considered halogen bonds [40,41,44]. However, they play similar roles in the assembly of molecules. In this study, chlorine interactions similar to halogen bonds were observed. They presented differently in less soluble and more soluble salts. This chlorine-involved interaction in the less soluble salt was the Cl ... Cl type with a distance of 3.193 Å, as shown in Figure S5. This distance was smaller than the addition of van der Waal radii of two chlorine atoms (2 × 1.75 Å). Both angles of the C-Cl1 ... Cl2 and the Cl1 ... Cl2-C were 165.27°, forming a type I Cl ... Cl interaction, but only type II are defined as halogen bonds [40]. This Cl ... Cl interaction connected the hydrogen-bonding network columns A and C, or B and D, which helped realize a compact packing of these columns. This was due to the chlorine of the A or B columns, located exactly opposite the neighboring chlorine of the C or D columns. Therefore, a type I Cl ... Cl interaction was found to contribute to chiral recognition.

In the more soluble salt (*S*)-(+)-4-ClMA·(*R*)-(+)-BPA, there was another weak interaction, Cl/π, with a distance of 3.328 Å, as shown in Figure S6. This interaction was formed between the chlorine of (*S*)-(+)-4-ClMA and the benzene group of (*R*)-(+)-BPA from the adjacent column, and connected the hydrogen-bonding network columns A and B, or C and D. C-X ... π halogen bonds have been reported in the literature, in which halogen was connected to aliphatic hydrocarbon [45,46], forming C-X/π angle of 180 °C. However, in the more soluble salt, Cl was connected to the benzene ring, as shown in Figure S6. The angle of C21-Cl1 ... C13 was 159.52°, while the angle of C-Cl ... π was smaller than 159.52°, both of which were far from 180°.

Generally, the length (3.193 Å) of the Cl ... Cl interaction in the less soluble salt was shorter than that (3.328 Å) in the more soluble salt, and significantly shorter than those of halogen bonds (3.3 Å to 3.46 Å) reported in the literature [19,20]. This favored a compact packing of the less soluble salt in addition to the contribution from CH/π or π/π interactions, discussed in Section 2.3.2. The combined effects of halogen interactions, CH/π and/or π/π interactions significantly increased the stability of the less soluble salts, and eventually facilitated an efficient chiral discrimination in the resolution of 4-ClMA with (*R*)-(+)-BPA.

2.3.4. Stacking Mode

In the hydrophobic region, the packing of aromatic groups of less soluble salt (*R*)-(−)-4-ClMA·(*R*)-(+)-BPA and more soluble salt (*S*)-(+)-4-ClMA·(*R*)-(+)-BPA was different, as shown in Figures S7 and S8, respectively.

In the less soluble salt, the benzene rings of (*R*)-(−)-4-ClMA were arranged in T-shapes relative to the neighboring benzene ring of (*R*)-(+)-BPA with an interplanar angle of 93.63°. This type of T-shaped packing of aromatic groups has been well-recognized as energetically favorable to crystal stability [39]. It resulted in a closer packing of columns and benefited the CH/π hydrogen-bonding interaction between columns in the less soluble salts. The benzene rings from (*R*)-(−)-4-ClMA were parallel to one another with a distance of 9.699 Å. This distance is not short enough to account for π/π interactions. Neither were the benzene rings of (*R*)-(+)-BPA. Between hydrophobic layers, a "key and lock" pattern was observed, as shown in Figure S7b. Generally, a planar boundary surface between hydrophobic layers is favorable, since it is able to tightly stack these layers and stabilize the crystal structure [9,11,13]. Unfortunately, in this study, the molecular length of BPA was significantly longer than that of 4-ClMA. Thus, a corrugated boundary surface was formed. The molecules were arranged in a "key and lock" mode, minimizing steric hindrance. This observation is similar to the packing mode in the resolution of 2-ClMA and (*R*)-(+)-BPA [28].

In the more soluble salts, no T-shaped arrangements of aromatic groups were found. The benzene rings of (*R*)-(−)-4-ClMA inclined to the neighboring benzene ring of (*R*)-(+)-BPA. The benzene rings from (*R*)-(−)-4-ClMA were parallel to one another with a distance of 9.179 Å, as shown in Figure S8. A "key and lock" mode was also formed, and the boundary surface was corrugated. Clearly, the "key and lock" packing in the more soluble salt was less compact compared to that in the less soluble salt. A relatively larger space and more voids were seen.

3. Materials and Methods

3.1. Materials

(*RS*)-4-ClMA with a purity of 99% was purchased from Alfa Aesar Johnson Matthey Company (WardHill, MA, USA). ®-(+)-BPA and (*S*)-(−)-BPA, with an optical purity of 99%, were purchased from Aldrich-Sigma Canada (Oakville, ON, Canada). All solvents were analytically pure, as purchased from Aldrich-Sigma Canada or Tiechem (Shanghai, China). The methanol/acetonitrile used for HPLC was HPLC grade and was purchased from Tiechem. Hydroxypropyl-β-cyclopram was purchased from Shandong Binzhou Zhiyuan Biotechnology Company Ltd. (Binzhou, China).

3.2. Analytical Methods

The melting points and heat of fusion of diastereomeric salts were determined by Mettler Toledo DSC 822 (Grei-fensee, Switzeland). The samples (3–6 mg) were prepared in a covered aluminum crucible with a pierced lid to allow the escape of volatiles. The sensors and samples were under nitrogen purge during the experiments. A heating rate of 5 °C/min was employed. The optical purity for the less soluble salt of (*R*)-(−)-4-ClMA·(*R*)-(+)-BPA and more soluble salt of (*R*)-(−)-4-ClMA·(*S*)-(+)-BPA or (*S*)-(−)-4-ClMA·(*R*)-(+)-BPA were determined t 25 °C by a reverse-phase HPLC equipped with a cyanopropyl column (UF-CN: 220 × 4.6 mm × 5 μm, Zhongpu Sci. Dalian, China), with hydroxypropyl-β-cyclodextrin as a chiral mobile phase additive. The HPLC system (Hegong Instrument, Shanghai, China) consisted of a Vertex STI 5000 pump, a STI UV detector (wavelength range of 190–700 nm), a 10 μL sample loop (Rheodyne), and a 7725i sampler. The mobile phase was a mixture of an aqueous buffer (pH of 2.8, 8 mmol/L sodium dihydrogen phosphate and 5 mmol/L hydroxypropyl-β-cyclodextrin) and methanol with a volumetric ratio of 95:5. The flow rate was 1.0 mL/min and the detector wavelength was set at 220 nm.

Crystal structures of the diastereomeric salts were determined by Single Crystal X-ray Diffraction. X-ray crystallography for the less soluble salt (*R*)-(−)-4-ClMA·(*R*)-(+)-BPA was carried out at the Western University in London, Ontario, Canada. Data collection: APEX2 (Bruker, Karlsruhe, Germany, 2009). Cell refinement: SAINT (Bruker, 2009); Data reduction (SAINT); Programs used to solve structure: SHELXT (Sheldrick, Göttingen, Germany, 2008); Programs used to refine structure: SHELXL2014 (Sheldrick, 2008); Molecular graphics: PLATON (Spek, Utrecht, The Netherlands, 2009). X-ray crystallography for the more soluble salt (*S*)-(−)-4-ClMA·(*R*)-(+)-BPA was carried out at Fudan University in Shanghai, China. Data collection: Bruker SMART; Cell refinement: Bruker SMART; Data reduction: Bruker SHELXTL; Programs used to solve structure: Bruker SHELXTL; Programs used to refine structure: SHELXL-2014/7 (Sheldrick, 2014); Molecular graphics: Bruker SHELXTL.

3.3. Preparation of the Less Soluble Salt (R)-(−)-4-ClMA·(R)-(+)-BPA

The less soluble salt was synthesized by enantiopure (*R*)-(−)-4-ClMA and (*R*)-(+)-BPA. (*R*)-(+)-BPA (1.2 mL, 5.9 mmol) was added dropwise to a solution of (*R*)-(−)-4-ClMA (1.1 g, 5.9 mmol) in 20 mL 2-propanol, forming white crystals. The slurry was heated to reflux and was kept at the temperature for 15 min. Subsequently, the mixture was cooled to room temperature. Then, 5 mL hexane was added to the mixture to suspend the solid, and the crystals were collected by filtration and washed with 2-propanol (1.5 mL × 2) twice to give 2.0 g enantiopure (*R*)-(−)-4-ClMA·(*R*)-(+)-BPA with a yield of 85%. The melting point of the salt was 166 °C, specific rotation $[\alpha]_D^{22} = -32.0°$ (c = 1, methanol). ^{1}HNMR (400 MHz, DMSO) δ: 1.37–1.39 (d, 3H, CH_3), 2.50 (s, 2H, CH_2), 3.91–3.96 (m, 1H, CH), 4.85 (s, 1H, CH), 7.27–7.44 (m, 14H, $C_6H_4 + C_6H_5 + C_6H_5$). Elemental analysis: Calculated for $C_{23}H_{24}ClNO_3$ (FW397.88) C: 69.43, H: 6.04, N: 3.52; Found C: 69.26, H: 6.24; N: 3.42.

3.4. Preparation of the More Soluble Salt (R)-(−)-4-ClMA·(S)-(−)-BPA

The more soluble salt (*R*)-(−)-4-ClMA·(*S*)-(−)-BPA was synthesized by optical purity (*R*)-(−)-4-ClMA and (*S*)-(−)-BPA. (*R*)-(+)-BPA (1.1 g, 5.0 mmol) was added dropwise to a solution of

(*R*)-(−)-4-ClMA (0.935 g, 5.0 mmol) in 6 mL 2-propanol, and the white crystals appeared. The slurry was heated to 65 °C, the solid was dissolved completely. The solution was kept at 65 °C for 0.5 h, then cooled slowly to 20 °C. The crystals were collected by filtration and washed with 2-propanol (0.75 mL × 2) twice to give 1.72 g enantiopure (*R*)-(−)-4-ClMA·(*S*)-(−)-BPA with a yield of 86.3%. Its melting point was 138.1 °C; specific rotation $[\alpha]_D^{22} = -51.1°$ (c = 1, methanol), ^{1}HNMR (400 MHz, DMSO) δ: 1.36–1.38 (d, 3H, CH_3), 2.50 (s, 2H, CH_2), 3.92–3.94 (m, 1H, CH), 4.84 (s, 1H, CH), 7.21–7.43 (m, 14H, C_6H_4 + C_6H_5 + C_6H_5). Elemental analysis: Calculated for $C_{23}H_{24}ClNO_3$ (FW397.88) C: 69.43, H: 6.04, N: 3.52; Found C: 69.39, H: 6.25; N: 3.42. The more soluble salt (*S*)-(+)-4-ClMA·(*R*)-(+)-BPA, the enantiomer of (*R*)-(−)-4-ClMA·(*S*)-(−)-BPA, was prepared in a similar way, its melting point was 131.2 °C, and specific rotation $[\alpha]_D^{22} = +51.4°$ (c = 1, methanol).

3.5. Solubility Determination

The solubilities of less and more soluble salts were measured by a synthetic method [47]. The diastereomeric salts were weighed and added in the barrel-type test tube with a thermometer, then the tube was set in a jacket glass flask with a temperature control. A certain amount of absolute ethanol solvent was added dropwise into the tube, and the mixture was stirred at a constant temperature of 20 °C for at least 15 min. Ethanol was continuously added until the diastereomeric salts were completely dissolved. The quantity of added absolute ethanol were weighed accurately to determine the solubilities of two diastereomeric salts. The measurement was performed in duplicate and the average value was reported in this paper. The accuracy of balance was ±1 mg.

3.6. Determination of Binary Phase Diagram of Diastereomeric Salts

A weighed amount of pure less soluble salt (*R*)-(−)-4-ClMA·(*R*)-(+)-BPA and more soluble salt (*R*)-(−)-4-ClMA·(*S*)-(−)-BPA were mixed in a mortar and dissolved in a small amount of absolute methanol. The solid appeared in the mortar and the slurry was crushed until the methanol evaporated completely to give a uniform mixture with a different diastereomeric composition. The resolution of the electronic balance was 0.01 mg. The binary melting point phase diagram was established by using Mettler Toledo DSC 822e differential scanning calorimeter (Greifense, Switzerland) by measuring the temperature at the beginning and the end of fusion of diastereomeric salt mixtures.

3.7. Resolution Procedure

A typical resolution process is described as follows: 4-ClMA (0.935 g, 0.005 mol) and 8 mL absolute ethanol were mixed in a 20 mL bottle under magnetic stirring at room temperature to form a complete dissolution of 4-ClMA. (*R*)-(+)-BPA (1.05 g, 0.005 mol) was added dropwise into the solution and white crystals, namely diastereomeric salts, soon appeared. The mixture was heated by a water bath to 72 °C and stayed at the temperature for 15 min, and the solid was completely dissolved. The solution was then cooled slowly to 20 °C and stayed for 15 min, the precipitated crystalline salts were collected by filtration and washed with cold absolute ethanol (1 mL × 2) twice. The obtained diastereomeric salt was 0.82 g with a yield of 81.8% (on the basis of (*R*)-(−)-4-ClMA·(*R*)-(+)-BPA in the mixture). Its melting point was 164.6 °C, specific rotation $[\alpha]_D^{22} = -27.7°$ (c = 1, methanol), the optical purity was 94.8 %*d.e.* based on the following formula.

$$\text{Diastereomeric excess}: \%d.e. = \frac{[\alpha]_D^{22} - 51.1°}{-32.0° - 51.1°} \tag{1}$$

Here, $[\alpha]_D^{22}$ represents the specific rotation of the precipitated salts from a resolution process, the specific rotation of (*R*)-(−)-4-ClMA·(*R*)-(+)-BPA (less soluble salt) was −32.0°, and that of (*R*)-(−)-4-ClMA·(*S*)-(−)-BPA (more soluble salt) was −51.1°. The optical purity calculated by the formula above was in a good agreement with the optical purity determined on a chiral column using

HPLC. Therefore, diastereomeric purity of the salts determined by the specific rotation was used in this study due to the convenience in its measurement compared to the measurement using HPLC.

4. Conclusions

In the resolution of 4-Chloromandelic acid with (*R*)-(+)-benzyl-1-phenylethylamine, the resulting less soluble and more soluble diastereomeric salts exhibited significant differences in their thermodynamic properties, including melting points of 166.3 °C and 132.0 °C, enthalpy of fusion of 57.41 KJ/mol and 52.58 KJ/mol, and solubility of 1.47 g/100 g ethanol and 4.82 g/100 g ethanol, respectively. These differences originated from the distinct supramolecular interactions in the crystal lattice of the pair of diastereomeric salts. In addition to well-recognized hydrogen-bonding, CH/π interactions and aromatic groups packing, halogen involved interactions, such as Cl ... Cl and Cl/π. These interactions were observed, and they demonstrated significant contributions to chiral discrimination.

Supplementary Materials: The following are available online, Table S1: The Orthogonal Experiment Result for the Resolution of (*R*,*S*)-4-ClMA by (*R*)-(+)-BPA; Table S2: Crystal Structure Data of (*R*)-(−)-4-ClMA·(*R*)-(+)-BPA and (*S*)-(+)-4-ClMA·(*R*)-(+)-BPA; Figure S1: Atomic-numbering Schemes of (*R*)-(−)-4-ClMA·(*R*)-(+)-BPA (a) and (*S*)-(+)-4-ClMA·(*R*)-(+)-BPA (b); Figure S2: The H-bonding network in the less soluble salt (a and b) and more soluble salt (c and d). The red parts represent carboxylate anions of 4-ClMA and the blue parts represent ammonium cations of BPA; Figure S3: The CH/π interactions within hydrogen column of less soluble salt; Figure S4: The CH/π interactions between adjacent hydrophobic layers of less soluble salt. Viewed from b-axis.; Figure S5: The Cl ... Cl interactions between adjacent hydrogen bonding net work columns and the view of adjacent four columns from b-axis in the less soluble salt (a) viewed from a-axis; (b) viewed from b-axis; Figure S6: The Cl/πhalogen bonds interactions between columns and the view of adjacent four columns from a-axis in the more soluble salt (a) viewed from b-axis; (b) viewed from a-axis.; Figure S7: Packing mode of the less soluble salt in hydrophobic region, (a) The distance of benzene rings, (b) the packing mode of hydrophobic layers (viewed from b-axis); Figure S8: Packing mode of the more soluble salt in hydrophobic region, a. The distance between benzene rings, b. the packing mode of hydrophobic layers (viewed from b-axis).

Author Contributions: Conceptualization, Q.H. and Y.P.; Investigation, Y.P. and C.F.; Resources, S.R.; Resource acquisition, Q.H.; Writing-Original Draft Preparation, Y.P.; Writing-Review & Editing, Q.H. and S.R.

Funding: This research was funded by the NSERC Discovery grant number [RGPIN 04211].

Conflicts of Interest: The authors declare no conflict of interest.

Appendix A

The crystal data of (*R*)-(−)-4-ClMA·(*R*)-(+)-BPA and (*S*)-(+)-4-ClMA·(*R*)-(+)-BPA is available for reference at CCDC 1030316 and 1885316.

References

1. *FDA's Policy Statement for the Development of New Stereoisomeric Drugs*; Wiley-Liss, Inc.: Hoboken, NJ, USA, 21 September 2004; pp. 338–340. [CrossRef]
2. Pellissier, H. Recent advances in enantioselective vanadium-catalyzed transformations. *Corordination Chem. Rev.* **2015**, *284*, 93–110. [CrossRef]
3. Tang, K.W.; Yi, J.M.; Liu, Y.B. Enantioselective separation of *R*,*S*-phenylsuccinic acid by biphasic recognition chiral extraction. *Chem. Eng. Sci.* **2009**, *64*, 4081–4088. [CrossRef]
4. Xie, R.; Chu, L.Y.; Deng, J.G. Membranes and membrane process for chiral resolution. *Chem. Soc. Rev.* **2008**, *37*, 1243–1263. [CrossRef] [PubMed]
5. Bocian, S.; Skoczylas, M.; Buszewski, B. Amino acids, peptides, and proteins as chemically bonded stationary phases—A review. *J. Sep. Sci.* **2016**, *39*, 83–92. [CrossRef] [PubMed]
6. Chen, X.; Wu, Q.; Zhu, Du. Enzymatic synthesis of chiral 2-hydroxy carboxylic acids. *Process Biochem.* **2015**, *50*, 759–770. [CrossRef]
7. Faigl, F.; Fogassy, E.; Nógrádi, M.; Pálovics, E.; Schindler, J. Strategies in optical resolution: A practical guide. *Tetrahedron Asymmetry* **2008**, *19*, 519–536. [CrossRef]
8. Kozma, D. Resolution by Formation and Fractional Crystallization of Diastereomeric Salts. In *CRC Handbook of Optical Resolutions via Diastereomeric Salt Formation*, 1st ed.; CRC Press: Boca Raton, FL, USA, 2001; pp. 9–50.

9. Faigl, F.; Schindler, J.; Fogassy, E. Advantages of structural similarities of reactants in optical resolution. *ChemInform* **2007**, *38*. [CrossRef]
10. Szeleczky, Z.; Semsey, S.; Bagi, P.; Fódi, B.; Faigl, F.; Pálovics, E.; Fogassy, E. An aspect of selecting resolving agents: The role of differences in molecule length in diastereomeric salt resolutions. *Sep. Sci. Technol.* **2016**, *51*, 727–732. [CrossRef]
11. Kodama, K.; Kobayashi, Y.; Saigo, K. Role of the relative molecular length of the component in ternary inclusion crystals in the chiral recognition and assembly of superamolecular helical architectures. *Cryst. Growth Des.* **2007**, *7*, 935–939. [CrossRef]
12. Saigo, K.; Sakai, K. Toward efficient optical resolution by diastereomeric salt formation. *J. Syn. Org. Chem. Jpn.* **2011**, *69*, 499–505. [CrossRef]
13. Kinbara, K. Design of resolving agents based on crystal engineering. *Synlett* **2005**, *5*, 732–743. [CrossRef]
14. Kobayashi, Y.; Kokubo, Y.; Aisaka, T.; Saigo, K. Hydrogen-bonding sheets in crystals for chirality recognition: Synthesis and application of (2*S*,3*S*)-2,3-dihydroxy- and (2*S*,3*S*)-2,3-dibenzyloxy-1,4-bis(hydroxyamino)butanes. *Tetrahedron Asymmetry* **2008**, *19*, 2536–2541. [CrossRef]
15. Suezawa, H.; Ishihara, S.; Umezawa, Y.; Tsunoyama, S.; Nishio, M. The aromatic CH/pi hydrogen bond as an important factor in determining the relative stability of diastereomeric salts relevant to enantiomeric resolution—A crystallographic database study. *Eur. J. Org. Chem.* **2004**, *23*, 4816–4822. [CrossRef]
16. Kobayashi, Y.; Saigo, K. The role of CH/π interaction in the stabilization of less-soluble diastereomeric salt crystals. *Chem. Rec.* **2007**, *7*, 47–56. [CrossRef]
17. Sakai, K.; Sakurai, R.; Hirayama, N. Molecular mechanisms of dielectrically controlled resolution (DCR). *Top. Curr. Chem.* **2007**, 233–271. [CrossRef]
18. Sakai, K.; Sakurai, R.; Hirayama, N. Chiral discrimination controlled by the solvent dielectric constant. *Tetrahedron Asymmetry* **2004**, *15*, 1073–1076. [CrossRef]
19. Kobayashi, Y.; Maeda, J.; Ando, T.; Saigo, K. Halogen-bonding interaction stabilizing cluster-type diastereomeric salt crystals. *Cryst. Growth Des.* **2010**, *10*, 685–690. [CrossRef]
20. Farina, A.; Meille, S.V.; Messina, M.T.; Metrango, P.; Resnati, G.; Vecchio, G. Resolution of racemic 1,2-dibromohexafluoropropane through halogen-bonded supramolecular helices. *Angrew. Chem. Int. Ed.* **1999**, *38*, 2433–2434. [CrossRef]
21. Yin, L.; Shan, W.; Jia, X.; Li, X.; Chan, A.S.C. Ru-catalyzed enantioselective preparation of methyl (*R*)-*O*-chloromandelate and its application in the synthesis of (*S*)-clopidogrel. *J. Organomet. Chem.* **2009**, *694*, 2092–2095. [CrossRef]
22. Bousquet, A.; Musolino, A. Hydroxyacetic Ester Derivatives, Preparation Method and Use as Synthesis Intermediates. U.S. Patent 6,573,381, 3 June 2003.
23. Adams, A.D.; Jones, A.B.; Berger, J.P. Preparation of 2-Aryloxy-2-Arylalkanoic Acids for Diabetes and Lipid Disorders. WO Patent 2,002,064,094, 2 May 2002.
24. Bereczki, L.; Marthi, K.; Pokol, G. Structure-resolvability relationship of several diastereomeric salt pairs of 1-phenylethylamine—A cambridge structural database study. *Period. Polytech. Chem. Eng.* **2009**, *53*, 71. [CrossRef]
25. Bandala, Y.; Juaristi, E. Recent advances in the application of alpha-phenylethylamine (alpha-PEA) in the preparation of enantiopure compounds. *Aldrichimica Acta* **2010**, *43*, 65–78.
26. He, Q.; Peng, Y.F.; Rohani, S. Diastereomeric resolution of p-chloromandelic acid with (*R*)-phenylethylamine. *Chirality* **2010**, *22*, 16–23. [CrossRef] [PubMed]
27. He, Q.; Gomma, H.; Rohani, S.; Zhu, J.; Jennings, M. Chiral discrimination in diastereometric salts of chlorine substituted mandelic acid and phenylethyamine. *Chirality* **2010**, *22*, 707–716. [CrossRef] [PubMed]
28. Peng, Y.F.; He, Q.; Rohani, S.; Jenkins, H. Resolution of 2-chloromandelic acid with (*R*)-(+)-N-benzyl-1-phenylethylamine: Chiral discrimination mechanism. *Chirality* **2012**, *24*, 349–355. [CrossRef] [PubMed]
29. He, Q.; Rohani, S.; Zhu, J.; Gomaa, H. Crystallization of racemic compound and conglomerate of (*RS*)-2-chloromandelic acid. *Cryst. Growth Des.* **2010**, *10*, 5136–5145. [CrossRef]
30. Peng, Y.F.; Cai, S.H. Research on Extraction for (*S*)-(−)-α-Methylbenzylamine. *J. ECUST* **1999**, *25*, 329–332. [CrossRef]

31. Wen, L.F.; Zhang, N.; Li, H.C.; Huang, Q.; Wu, X.; Hao, X.; Wu, M.; Ban, C.; Zhao, J. Ternary liquid-liquid equilibrium for mixtures of water + (±)α-phenylethylamine + n-hexane at T = 298.2, 318.2, and 333.2 K. *J. Chem. Eng. Data* **2017**, *62*, 1819–1824. [CrossRef]
32. Suzuki, M.; Sakai, K. Optical active *N*-benzyl-1-phenylethylamine. *J. Synth. Org. Chem.* **2000**, *58*, 147–148. [CrossRef]
33. Yoshioka, R.; Hiramatsu, H.; Okamura, K.; Tsujioka, I.; Yamada, S. Crystal structure-solubility relationships in optical resolution by diastereomeric salt formation of dl-phenylglycine with (1*S*)-(+)-camphor-10-sulfonic acid. *J. Chem. Soc. Perkin Trans.* **2000**, *10*, 2121–2128. [CrossRef]
34. Leusen, F.J.J.; Noordik, J.H.; Karfunkel, H.R. Racemate resolution via crystallization of diastereomeric salts: Thermodynamic considerations and molecular mechanics calculations. *Tetrahedron* **1993**, *49*, 5377–5396. [CrossRef]
35. Bolchi, C.; Pallavicini, M.; Fumagalli, L. Highly efficient resolutions of 1, 4-benzodioxane-2-carboxylic acid with para substituted 1-phenylethylamines. *Tetrahedron Asymmetry* **2005**, *16*, 1639–1643. [CrossRef]
36. Silvestro, G.D.; Palmisano, G.; Pellegata, R. Phase diagram of (*R*)-and (*S*)-4-hydroxy-2-pyrrolldone mixtures: A new calculation of a conglomerate forming system. *J. Pharm. Sci.* **1993**, *82*, 758–760. [CrossRef] [PubMed]
37. Lorenz, H.; Seidel-Morgenstern, S. Binary and ternary phase diagrams of two enantiomers in solvent systems. *Thermochim. Acta* **2002**, *382*, 129–142. [CrossRef]
38. Dunitz, J.D.; Gavezzotti, A. How molecules stick together in organic crystals: The weak intermolecular interaction. *Chem. Soc. Rev.* **2009**, *38*, 2622–2633. [CrossRef] [PubMed]
39. Nishio, M. The CH/π hydrogen bond in chemistry. Conformation, supramolecules, optical resolution and interactions involving carbohydrates. *Phys. Chem. Chem. Phys.* **2011**, *13*, 13873–13900. [CrossRef] [PubMed]
40. Gilday, L.C.; Robinson, S.W.; Barends, T.A.; Langton, M.J.; Mullaney, B.R.; Beer, P.D. Halogen bonding in supramolecular chemistry. *Chem. Rev.* **2015**, *115*, 7118–7195. [CrossRef] [PubMed]
41. Desiraju, G.R.; Ho, P.S.; Kloo, L.; Legon, A.C.; Marquardt, R.; Metrangolo, P.; Polotzer, P.; Resnati, G.; Rissanen, K. Definition of the halogen bond. *Pure Appl. Chem.* **2013**, *85*, 1711–1713. [CrossRef]
42. Ribeiro, N.; Kobayashi, Y.; Maeda, J.; Saigo, K. Enantiopure cyclic *O*-substitrted phenylphosphonothioic acid-synthesis and chirality recognition ability. *Chirality* **2011**, *23*, 438–448. [CrossRef] [PubMed]
43. Ichikawa, A.; Ono, H.; Mikata, Y. Characteristic conformations and molecular packings in crystal structures of diastereomeric esters prepared from (*S*)-2-methoxy-2-(1-naphthy) propanoic acid. *Tetrahedron Asymmetry* **2008**, *19*, 2693–2698. [CrossRef]
44. Donatella, A.; Adrian, S.P.; Carlos, R.D.; Jose, M.L. Halogen … Halogen interaction in the self-assembly of one-dimensional 2,2′-bipyrimidine-based $Cu^{II}Re^{IV}$ systems. *CrystEngComm* **2018**, *20*, 4575–4581. [CrossRef]
45. Nagels, N.; Hauchecorne, D.; Herrebout, W.A. Ecploring the C-X … π halogen bonding motif: An infared and raman study of the complexes of CF3X(X = Cl, Br, and I) with the aromatic model compounds benzene and toluene. *Molecules* **2013**, *18*, 6829–6851. [CrossRef] [PubMed]
46. Hauchecorne, D.; Nagels, N.; van der Veken, B.J.; Herrebout, W.A. C-X … π halogen and C-H … π hydron bonding: Interactions of CF3X(X=Cl, Br, I or H) with ethane and propene. *Phys. Chem. Chem. Phys.* **2012**, *14*, 681–690. [CrossRef] [PubMed]
47. Lee, M.J.; Chang, Y.K.; Zin, H.M.; Chen, C.H. Solid liquid equilibria for 4-methoxyphenol with catechol, ethylenediamine, or piperazine. *J. Chem. Eng. Data* **1997**, *42*, 349–352. [CrossRef]

Sample Availability: Samples of the compounds (*R*)-(−)-4-ClMA·(*R*)-(+)-BPA and (*S*)-(+)-4-ClMA·(*R*)-(+)-BPA are not available from the authors.

molecules

MDPI

Article

Enantioseparation, Stereochemical Assignment and Chiral Recognition Mechanism of Sulfoxide-Containing Drugs

Fei Xiong, Bei-Bei Yang, Jie Zhang and Li Li *

Beijing Key Laboratory of Active Substances Discovery and Druggability Evaluation, Institute of Materia Medica, Chinese Academy of Medical Sciences & Peking Union Medical College, Beijing 100050, China; xxf@imm.ac.cn (F.X.); yangbetty@imm.ac.cn (B.-B.Y.); zhd@imm.ac.cn (J.Z.)

* Correspondence: annaleelin@imm.ac.cn; Tel.: +86-10-6316-5247

Received: 15 September 2018; Accepted: 15 October 2018; Published: 18 October 2018

Abstract: The distinct pharmacodynamic and pharmacokinetic properties of enantiopure sulfoxide drugs have stimulated us to systematically investigate their chiral separation, stereochemical assignment, and chiral recognition mechanism. Herein, four clinically widely-used sulfoxide drugs were chosen and optically resolved on various chiral stationary phases (CSPs). Theoretical simulations including electronic circular dichroism (ECD) calculation and molecular docking were adopted to assign the stereochemistry and reveal the underlying chiral recognition mechanism. Our results showed that the sequence of calculated mean binding energies between each pair of enantiomers and CSP matched exactly with experimentally observed enantiomeric elution order (EEO). It was also found that the length of hydrogen bond might contribute dominantly the interaction between two enantiomers and CSP. We hope our study could provide a fresh perspective to explore the stereochemistry and chiral recognition mechanism of chiral drugs.

Keywords: chiral sulfoxide drugs; enantiomer elution order; electronic circular dichroism; quantum chemical calculation; molecular docking

1. Introduction

Chirality has a profound impact on the chemical processes of biology. Sulfur is an important bio-element, whose certain compounds can exhibit chirality such as sulfoxides, or sulfoximines. Although chiral sulfur compounds have received much less attention than chiral carbon compounds, chiral sulfoxides possessing a pyramidal structure comprise a unique class of drugs containing a chiral heteroatom [1]. The high enantioselectivity of biological events has prompted researchers to develop enantiopure sulfoxide drugs due to their distinct pharmacodynamic and pharmacokinetic properties [2,3]. In addition to its application in medicinal chemistry, sulfoxide is also significant functional group for its use as a ligand and chiral auxiliary in asymmetric synthesis [4,5].

Compounds **1–4** (Figure 1) are clinically widely-used sulfoxide-containing drugs, and represent two typical structures of sulfoxides, i.e., aromatic alkyl sulfoxides (**1**, **2**, **4**) and alkyl sulfoxide (**3**). Both **1** and **2** earned enough attention because of marvelous market success, and **3** is a wake-promoting drug used in the war [6–8]. Recently, the non-steroidal anti-inflammatory drug **4** and its derivatives aroused the interest of researchers for its potent anticancer effect [9,10].

In the last two decades, hundreds of papers reported the chiral separation methods of sulfoxides including capillary electrophoresis [11,12], chemical resolution [13,14], HPLC [15–18], and supercritical fluid chromatography [19,20]. Amongst all, we focused on chiral HPLC, which might be a favorable and effective way for both analytical and preparative separations of chiral compounds [21]. There are several HPLC methods using different types of CSPs for **1**, **2**, and 3 [7,16,22–30]. Relatively few studies have been reported on the synthesis and chiral enantioseparation of **4** [31–36]. Although the piecemeal chiral separation of these compounds was previously reported, a systematic and comprehensive study is beneficial to seek for general clues to the separation of chiral sulfoxide drugs.

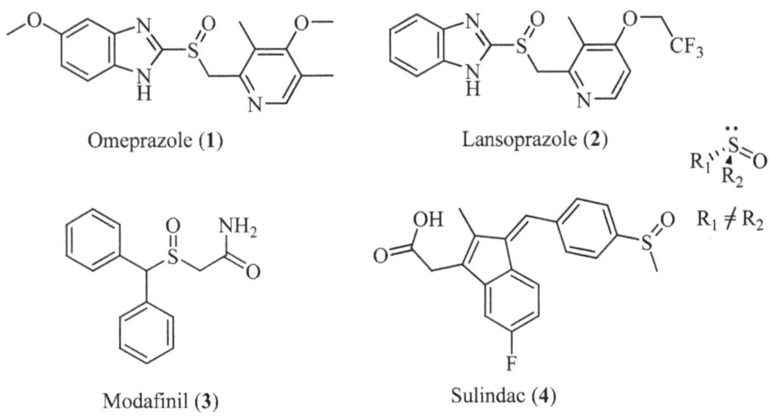

Figure 1. Chemical structures of chiral sulfoxides.

Absolute configuration (AC) assignment is crucial for the stereochemical characterization of chiral compounds. For these four drugs, this was previously fulfilled by single-crystal X-ray diffraction, chemical synthesis, or comparison of circular dichroism spectra with their strongly correlatives [7,24,29,34,37–39]. However, it might be interesting and useful to establish directly their stereochemistry without reference samples and spectra. Presently, ECD spectra together with quantum chemical calculations using time-dependent density functional theory (TDDFT) have provided a feasible and reliable way to facilitate the AC assignments of chiral drugs and natural products [15,18,40–44]. Hence, TDDFT calculations will be adopted to simulate ECD spectra of sulfoxides in this study.

Recently, rapid development of binding energy computation software has revolutionized the study and elucidation of chiral separation mechanism, and binding energy computation could be applied to reproduce the elution orders of enantiomers [44,45]. In this work, we selected these four sulfoxide drugs to take a systematic experimental and theoretical investigation, for the aim of disclosing some potential regulation during the process of chiral separation and characterization of sulfoxides. Specially, chiral resolution of **1–4** will be tested on commercially available and cost-effective Chiralpak AD-H, AS-H, Chiralcel OD-H, and OJ-H chiral columns, followed by AC assignment and chiral separation mechanism analysis.

2. Results and Discussion

2.1. *Chiral Resolution of* **1–4**

2.1.1. Factors Affecting Optical Resolution

The influence of the flow rate, mobile phase composition, column temperature, and acidic modifier on resolution was taken into consideration. With the increasing proportion of n-hexane (n-Hex) in the mobile phase, Rs increased in a non-linear manner (Figure S1A). It was manifested that Rs changed slightly with increasing temperature (Figure S1B) and steadily decreased when the flow rate increased from 0.5 to 1.0 mL/min (Figure S1C). Additionally, the column temperature can also affect the protonation of analytes and the CSP, which can then alter the chiral interactions between the solute and the stationary phase.

The great concern about the tailing phenomenon of acid compound **4** can be overcome using an acidic additive like formic acid (FA). Preliminary study showed that resolution of racemic **4** could only be attained on the Chiralpak AD-H column herein, and FA was then added to make a flagrant contrast. Fortunately, FA could achieve better separation of **4** through decreasing tailing, sharpening the peaks, and improving the selectivity (Table S1 and Figure S2). FA in the mobile phase might protonate both **4** and residual -NH- groups on CSP arising from the weak protonation capability of FA. Hence, non-enantioselective association between **4** and the CSP was partly inhibited. Subsequently, it improved the mass transfer, which greatly enhanced the chiral recognition ability of the stationary phase.

2.1.2. The Optimal Resolution and Enantiomeric Elution Order

The separation conditions for compounds **1–4** on four chiral columns are listed in Table 1, and the optimal UV and ECD chromatograms are shown in Figure 2 and Figures S3–S6. The more satisfactory results of optimal resolution of racemic **1** on the Chiralpak AD-H column, **2** on the Chiralcel OD-H column, **3** on the Chiralpak AS-H column, and **4** on the Chiralpak AD-H column were 6.63, 3.48, 7.46, and 8.39, respectively. Most of the separations were obtained by using *n*-Hex/EtOH as the mobile phase. The different elution abilities between EtOH and IPA may be due to the polarity and capability to form hydrogen bonds with CSPs.

Table 1. Comparison of sulfoxide drugs on chiral columns.

CSP	Compound	Mobile Phase (*v*/*v*)	t_{R1} (min)	t_{R2} (min)	k_1	k_2	α	Rs	Sign [e]
AD-H	1	*n*-Hex:EtOH (60:40) [a,c]	12.97	20.26	3.32	5.80	1.75	6.63	(*S*)-(−)
	2	*n*-Hex:IPA (80:20) [a,d]	18.89	21.04	5.30	6.01	1.13	2.02	(*S*)-(−)
	3	*n*-Hex:EtOH (80:20) [a,c]	19.88	26.00	2.98	4.20	1.45	5.85	(*R*)-(−)
	4	*n*-Hex:FA:EtOH (80:0.1:20) [a,c]	14.96	22.87	1.99	3.57	1.79	8.39	(*S*)-(−)
AS-H	1	*n*-Hex:EtOH (60:40) [b,c]	8.11	12.45	1.70	3.15	1.85	5.54	(*R*)-(+)
	3	*n*-Hex:EtOH (60:40) [b,d]	13.64	21.83	1.73	3.37	1.95	7.46	(*S*)-(+)
OD-H	1	*n*-Hex:EtOH (90:10) [a,d]	18.16	22.40	5.05	6.47	1.28	2.54	(*S*)-(−)
	2	*n*-Hex:IPA (80:20) [a,c]	18.48	27.43	5.16	8.14	1.58	3.48	(*R*)-(+)
OJ-H	2	*n*-Hex:EtOH (95:5) [a,d]	33.95	36.68	10.31	11.22	1.09	1.67	(*R*)-(+)
	3	*n*-Hex:EtOH (55:45) [b,d]	16.06	21.01	2.21	3.20	1.46	5.43	(*S*)-(+)

Retention factor $k = (t_1 - t_0)/t_0$, Resolution factor $Rs = 2(t_2 - t_1)/(w_1 + w_2)$. t_1, t_2 is retention time of enantiomer. t_0 is dead time. w_1, w_2 is peak width of enantiomer. Selectivity factor $\alpha = k_2/k_1 = (t_2 - t_0)/(t_1 - t_0)]$. Detection wavelengths of **1–4** are 275, 275, 240, and 285 nm, respectively. [a] Flow rate: 1 mL/min. [b] Flow rate: 0.8 mL/min. [c] Column temperature: 30 °C. [d] Column temperature: 25 °C. [e] Sign of the absolute configuration of the first eluted enantiomer.

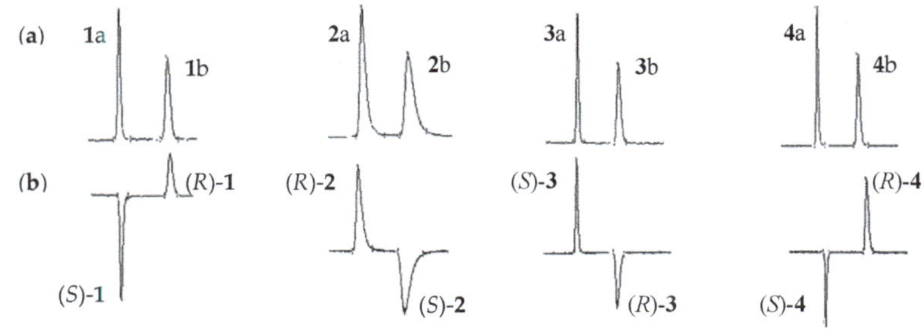

Figure 2. The UV (**a**) and ECD chromatograms (**b**) of racemic **1–4** on various columns. **1** on the ChiralPak AD-H column, **2** on the Chiralcel OD-H column, **3** on the ChiralPak AS-H column, and **4** on the ChiralPak AD-H column. Detection wavelengths of **1–4** are 275, 275, 240, and 285 nm, respectively.

It is an important issue to determine the EEO in chiral HPLC separations [46]. As demonstrated in Table 1 and Figure S7, the EEO on polysaccharide-based CSP was reversed (the details shown in Section 2.2) when the type of chiral columns or mobile phase composition were changed, which was consistent with the literature [24,47]. For compound **1**, the EEO changed when EtOH was replaced with IPA as the polar modifier on Chiralpak AD-H column (Figure S7). Similar situation was reported earlier for several chiral sulfoxides, showing that the change in the structure of chiral selector and composition of mobile phase might cause an opposite affinity pattern of enantiomers [16,23,24,39].

2.2. *AC Assignments of* **1–4**

2.2.1. Experimental UV and ECD Spectra

It is significant to directly assign the AC of each peak in the HPLC chromatograms. Without available ECD spectra of enantiopure standard samples, it is essential to establish their AC using a reliable method. In this work, we fell back on ECD spectroscopy together with quantum chemical calculations, which has been widely applied in the AC assignments of chiral organic molecules.

Experimental ECD and UV spectra of all enantiomers of **1–4** in MeOH were obtained (Figure 3). The experimental UV spectra of aromatic alkyl sulfoxides (**1**, **2**) seem to be similar. The UV spectrum of **1** manifests two absorption bands at 205 nm and 300 nm, and that of **2** is slightly blue shifted with an absorption peak at 285 nm, but a shoulder peak at 222 nm. Only one shoulder peak at 225 nm appeared in the UV spectrum of alkyl sulfoxide **3**. For aromatic methyl sulfoxide **4**, there are two absorption bands at 290 nm and 330 nm in its UV spectrum.

It is well known that a pair of enantiomers would share the same UV spectra and have mirrored ECD spectra. Though aromatic alkyl sulfoxides (**1**, **2**, **4**) show the broad peak in the ECD spectra, all these ECD spectra possess an obvious couplet-like feature, which is common in sulfoxide-containing compounds [48]. In the ECD spectra of enantiomeric **1**, peak **1a** presented two negative Cotton effects (CEs) at 300 nm and 268 nm, followed by an intense positive CE at 232 nm, which could be regarded as two branches of a negative, non-degenerate couplet-like feature. Peak **2a** shows a strong positive CE at 270 nm and a negative CE at 222 nm in the ECD spectrum. The ECD spectrum of peak **3a** shows a positive couplet-like feature consisting of two bands, with the first positive CE being at 238 nm and the second negative peak at 216 nm. Nevertheless, peak **4a** gives a wide negative CE signal over a broad range of 235–375 nm, followed by an obvious positive CE peak at 217 nm.

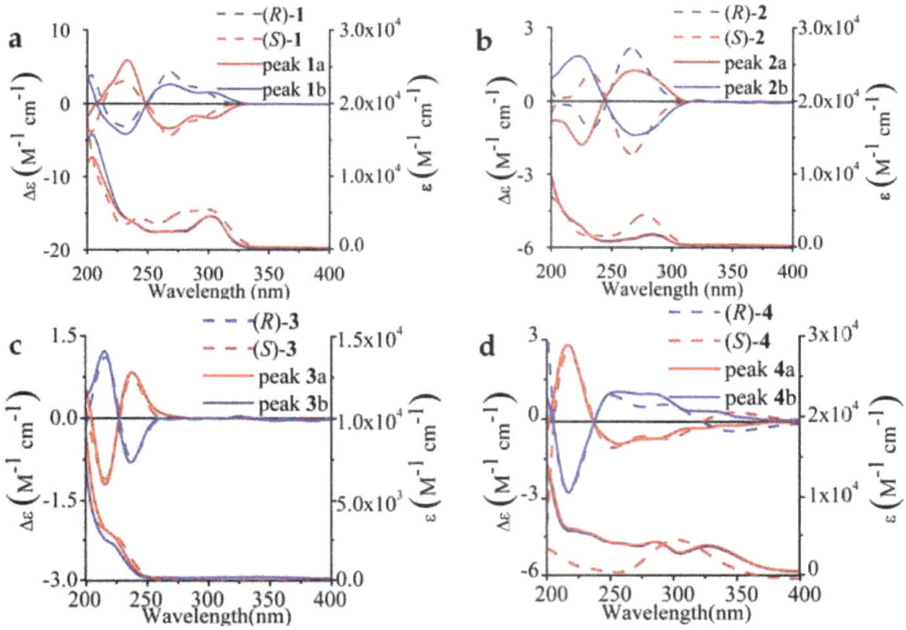

Figure 3. Comparison of experimental and calculated ECD (top) and UV (bottom) spectra of the stereoisomers (**a**–**d**). Solid: experimental in methanol, Dashed: theoretical values, and UV-corrected, bandwidth of **1**–**4** is 0.20 eV, 0.30 eV, 0.30 eV, and 0.40 eV, respectively. The enantiopure isomers of **1**–**4** were prepared under the optimal separation condition of our study, on ChiralPak AD-H, Chiralcel OD-H, ChiralPak AS-H and AD-H columns, respectively. Calculated spectra are Boltzmann averages from calculated spectra of each single conformer.

2.2.2. TDDFT Calculation of UV/ECD Spectra

To simulate UV and ECD spectra of **1**–**4**, it is vital to choose suitable combinations of hybrid functional and basis sets [49]. These calculation parameters might greatly affect TDDFT calculated spectra, thus leading to ambiguous AC judgment. Hence, four different hybrid functional/basis set combinations were adopted to verify the consistency of AC assignments. Geometry optimizations and frequency calculations were run at the B3LYP/6-31G(d) and B3LYP/6-31+G(d,p) levels. ECD spectra were predicted employing various combinations of B3LYP and Cam-B3LYP hybrid functionals and 6-31G(d) and 6-311+G(d,p) basis sets. Fortunately, the calculation did give an unambiguous answer to the AC assignment after UV correction if necessary (Figure S8).

Typically, both UV corrections and intensity scaling are applied when the calculated spectra is compared with an experimentally collected one. The best performing calculation method of sulfoxides **1**–**3** was the B3LYP/6-31G(d)//B3LYP/6-31G(d) basis set level, and that of **4** was B3LYP/6-311+G(d,p)//CAM- B3LYP/6-311+G(d,p).

The optimal match with the experimental spectra is shown in Figure 3. The first eluted enantiomer of **1** on AD-H, **3** on AS-H and **4** on AD-H were thus assigned as *S*, and **2** on OD-H was assigned as *R*. The agreement between the calculated and experimental ECD spectra of compounds **1**–**3** is almost perfect. The maximum absorption peak of the calculated UV spectra of **4** is at 370 nm, and the experimental UV spectra of **4** shows two absorption bands at 290 nm and 330 nm. In the ECD spectra of enantiomer of **4**, two separate CEs over the range of 240–350 nm are merged into one broad CE.

Moreover, the AC of each peak of **1–4** on different chiral columns was also assigned (). The AC judgments of **1**, **2**, and **4** obtained by the TDDFT calculation are consistent with the previously reported results [24,34].

2.2.3. Electron Transitions

Chemical structures of both alkyl sulfoxides **1** and **2** include a substituted pyridyl ring, a benzimidazole ring, and a methyl sulfoxide group. In their ECD spectra, the CEs at 270 nm are associated with two electronic transitions from the sulfoxide chiral center to the benzimidazole ring and the pyridine ring itself. The CE at 230 nm is assignable to $\pi \rightarrow \pi^*$ transitions from π-type S=O orbital to the benzene ring and the charge transfer transition in the pyridine ring itself. Compared with **1**, the ECD spectrum of **2** was slightly blue-shifted, which was due to the electron-withdrawing effect of CF_3 group.

The first CE of alkyl sulfoxide **3** at 238 nm resulted from benzene ^{1}La transition, and the second band at 216 nm might be attributed to the sulfinyl group n-π^* transition. The oxygen lone pair and π^*-type S=O orbitals are heavily mixed with σ and σ^*-type S-C orbitals, respectively [50]. The large conjugated aromatic ring leads to a UV spectrum of aromatic methyl sulfoxide **4**, which is distinct from the obtained spectra of the other sulfoxide drugs. The wide absorption band may be assigned to the phenyl ^{1}La and ^1L_b transitions. The experimental ECD spectra of **4** appeared a broad peak in the range of 235–375 nm. The second CE of **4** at 220 nm may correspond to the sulfoxide-centered (O=S<) n$\rightarrow\pi^*$ transition.

2.3. *Chiral Separation Mechanism and Molecular Docking*

The chiral recognition mechanism of polymer-based CSPs is much more complex, because their chiral recognition usually depends on their higher-order structure and the steric fit of the analyte inside the chiral cavity [51,52]. The sulfinyl (>S=O) group, the carbonyl (C=O) group, -OH and -NH groups, the phenyl moiety, benzimidazole, and the pyridyl group in the structure of chiral sulfoxide drugs may form a hydrogen bond, a dipole–dipole bond, π-π interactions and hydrophobic interaction with C=O, -NH- group and the aromatic ring of CSPs. The π-electron density, steric hindrance, various conformations and spatial structures of the polysaccharide glucose unit on CSPs will affect the interaction between the CSPs and chiral molecules [53]. The different types and intensity of the interactions between the sulfoxide compounds and CSPs bring about different chiral separation phenomena. However, recent studies and real understanding of the recognition mechanisms of polysaccharide-based CSPs are far behind their practical applications.

2.3.1. Common Structural Features

During the molecular docking process of **1–4**, 100 conformers of each enantiomer were generated and divided into different clusters (Figure S9). The common features of the most populated conformational cluster of **1–4** (Figure 4) were created to describe the structural characteristics and better explain the chiral separation mechanisms. There are three types of groups in **1** and **2** which may interact with the CSP, namely, hydrogen bond acceptors (HBA) O=S<, hydrogen bond donors (HBD) -NH-, and hydrophobic moieties (HY) -phenyl/-CF_3. The features of **3** include HBA O=S<, HBD NH_2 and aromatic ring (AR) phenyl. The features of **4** are HBA (O=S<), HY (-CH_3), AR (phenyl) and negative ionizable (NI, -COOH). The apparent difference between the *R* and *S* enantiomer of **1–4** is the distance of pharmacophores shown in Figure 4. The differences of distance may influence the stability of the complexes during separation, resulting in different elution of enantiomers.

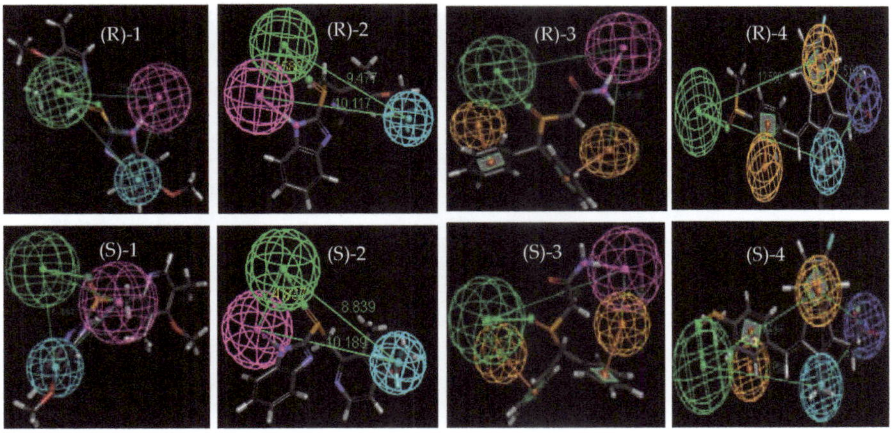

Figure 4. Pharmacophores of **1–4** conformers over the most populated cluster. HBA: green, HBD: purple, HY: light blue, AR: orange, NI: mazarine.

2.3.2. Molecular Interactions between Analytes and CSPs

The interactions between the enantiomers of **1–4** and CSP of the Chiralpak AD-H column are shown in Figure S10. The ligand is depicted as sticks, surrounded by a molecular surface, which is colored according to the interaction with the CSP. Meanwhile, hydrogen bonds are shown as a string of small green spheres. Both enantiomers of **1–4** can form one hydrogen bond with the CSP of the Chiralpak AD-H column.

For the enantiomers of **1**, the methoxy group forms a hydrogen bond with an -NH- group of CSP. As for the enantiomers of **2**, one hydrogen bond exists between the O=S< and the amino group of CSP. The carbonyl of **3** interacts with amino moiety of the CSP through a hydrogen bond. There is a hydrogen bond existing between -COOH of **4** and -NH- of the CSP. For **1**, **2**, and **4**, the length of the hydrogen bond of the (*R*)-stereoisomer is shorter than that of the (*S*)-stereoisomer and this is in accordance with the elution sequence on the Chiralpak AD-H column. On the contrary, the length of the hydrogen bond of (*R*)-**3** (2.195 Å) is longer than that of (*S*)-**3** (2.09 Å), and the elution time of (*S*)-**3** is longer on the Chiralpak AD-H column. These results suggest that a shorter hydrogen bond may ensure the stability of the enantiomer in the stationary phase.

2.3.3. Mean Binding Energy

As we know, the enantiomer with lower binding energy could bind more closely to the CSPs and will subsequently be eluted after the other enantiomer. Mean binding energy between the two enantiomers of **1–4** and the CSP of the Chiralpak AD-H column are listed in Table 2. The results are concurrent with the elution orders observed in the experiment.

It is obvious that the enantiomer with a shorter hydrogen bond is more stable than the other enantiomer during the process of elution. Consequently, hydrogen bonding plays a highly important role in the chromatographic separation of sulfoxides on Chiralpak AD-H column. Our result is consistent with high-resolution magic angle spinning nuclear magnetic resonance spectroscopy experiments that the molecular bases for the chiral recognition were the proton interactions between **1** and amylose-based CSP [54]. This also suggests that docking simulations could be adopted to reproduce the elution orders of the enantiomers by the mean binding energy calculations and explain the chiral separation mechanisms visually through the image. Since the mechanisms involved in chiral recognition are complex, a perfect molecular docking method is still a challenging task. We believe our results will aid the development of molecular docking in the application of chiral sulfoxide drugs separation on polysaccharide-based columns.

Table 2. The calculated average binding energies for compounds **1–4** on the ChiralPak AD-H column.

Entry	Mean Binding Energy (kcal/mol)	The Length of Hydrogen Bond (Å)	Elution Time (min)	Rs
(*R*)-**1**	−5.87	1.932	20.26	6.63
(*S*)-**1**	−5.43	2.003	12.97	
(*R*)-**2**	−6.71	2.054	21.04	2.02
(*S*)-**2**	−6.01	2.232	18.89	
(*R*)-**3**	−7.19	2.195	19.88	5.85
(*S*)-**3**	−7.82	2.090	26.00	
(*R*)-**4**	−7.92	1.762	22.87	8.39
(*S*)-**4**	−6.52	2.016	14.96	

3. Materials and Methods

3.1. Materials and Reagents

Sulfoxides **1–4** were purchased from National Institutes for Food and Drug Control of China (Beijing, China). All solvents including n-hexane (n-Hex), iso-propyl alcohol (IPA), methanol (MeOH), and ethanol (EtOH) were of HPLC grade. Formic acid (FA) is used as an acidic additive.

3.2. Chromatographic Conditions

Compounds **1–4** were separated on a Jasco HPLC system consisting of a PU-2089 pump, an AS-2055 sampler, a CO-2060 column thermostat, a MD-2010 detector and a CD-2095 detector. All the chiral columns including Chiralpak AD-H, AS-H, Chiralcel OD-H, and OJ-H (Daicel Chemical Industries, Tokyo, Japan) were 250 × 4.6 mm i.d. with a 5 μm particle size. The detection wavelength of **1–4** was set at 275, 275, 240, and 285 nm, respectively. The chromatographic parameters for the enantioselectivity evaluation of the CSPs including retention factor (k), separation factor (α), and resolution factor (Rs) for the enantiomers were calculated.

The samples dissolved in EtOH: n-Hex (80:20, *v*/*v*) were formulated as the solution of 2 mg/mL for analysis and 5 mg/mL for preparation. The enantiopure isomers of **1–4** were prepared under the optimal separation condition of our study, on Chiralpak AD-H, Chiralcel OD-H, Chiralpak AS-H and AD-H columns, respectively. Also, the high enantiomeric purities (enantiomeric excesses >99%) of the isolated sulfoxide enantiomers were verified by the enantioselective HPLC system.

3.3. ECD Experiments

ECD analysis of the sulfoxide enantiomers was carried out in MeOH on a Jasco J-815 spectrometer. A quartz cuvette with a 1 mm path length was used. The detection wavelength was at 200–400 nm. The spectra were baseline-corrected against MeOH.

3.4. Computational Methods

3.4.1. TDDFT Computations

All calculations have been performed on S configuration of **1–4**. Preliminary conformational analysis was carried out with the use of the MMFF94 molecular mechanics force field via the MOE software package (Chemical Computing Group, Montreal, QC, Canada) [55]. Geometry optimization and frequency calculation of the MMFF94 conformers were then performed at the B3LYP/6-31G(d) and B3LYP/6-311+G(d,p) levels by using Gaussian 09 (Gaussian, Wallingford, CT, USA) [56]. ECD spectra were predicted employing various combinations of B3LYP or Cam-B3LYP hybrid functionals and 6-31G(d), 6-311+G(d,p) basis sets. All calculations were conducted with the PCM solvation model for MeOH. Calculated UV and ECD spectra of each conformer were simulated at the bandwidth of 0.20–0.40 eV, and the overall spectra were obtained according to the Boltzmann weighting of all conformers. Theoretical ECD and UV spectra were blue-shifted to facilitate the comparison with experimental data if necessary.

3.4.2. Molecular Docking

The 3D-polymer structure of Chiralpak AD-H amylose derivative CSP (AD-12mer.pdb) was downloaded from http://pubs.acs.org [57]. The structure of chiral selector is composed of two AD-12mer.pdb molecules to form "tube-mode" [44]. The structure of AD-H was minimized by means of R2 Dreiding force field using Discovery Studio 2017 software. The structures of the enantiomers of **1–4** were minimized in CHARMM force field. AutoDock 4.2.6 (Scripps Research Institute, La Jolla, CA, USA) [58] was adopted to simulate molecular docking. During the docking process, the grid box was set to 30 × 30 × 30 (Å) with 0.375 Å spacing for **1**, **3**, and **4**, 40 × 40 × 40 (Å) with 0.375 Å spacing for **2**.

In consideration of the solvent effect of the mobile phase, the dielectric constant of **1–4** was respectively set as 10.668, 5.248, 6.124, and 6.124 based on the weighted average of the mixed mobile phase on the AD-H chiral column. Lamarckian genetic algorithm was used to generate 100 conformations.

4. Conclusions

Herein, a systematic study including chromatographic resolution, stereochemical assignment and chiral recognition mechanism of four chiral sulfoxide drugs (**1–4**) is conducted. All four compounds have been completely separated on a Chiralpak AD-H column (Figure S11). Among the tested conditions, n-Hex/EtOH as mobile phase is shown to be favorable for the separation of sulfoxides. EEO inversions were observed when the types of chiral columns or mobile phase composition changed. In this work, comparison of the ECD spectra with the TDDFT calculated data provided a robust way to assign the AC of their stereoisomers. Moreover, the observed EEOs were found to be in accord with mean binding energy calculations. The docking simulation could also explain visually the underlying chiral separation mechanisms. This work has the potential to build up an experimental and theoretical methodology to facilitate the stereochemistry and chiral recognition mechanism of chiral drugs.

Supplementary Materials: The following materials are available online. (1) Figure S1. Plots showing resolution factors of the enantiomers of **1–3** as a function of the n-Hex content in the mobile phase (A), temperature (B) and flow rate (C). (2) Table S1. Effect of the acidic additive on the resolution of **4** on the AD-H column. (3) Figure S2. Comparison of the UV (upper) and ECD (lower) chromatograms of **4** on ChiralPak AD-H column with acidic additive. (4) Figure S3. The UV (upper) and ECD (lower) chromatograms of **1** on ChiralPak AD-H column under the optimal condition. (5) Figure S4. The UV (upper) and ECD (lower) chromatograms of **2** on Chiralcel OD-H column under the optimal condition. (6) Figure S5. The UV (upper) and ECD (lower) chromatograms of **3** on ChiralPak AS-H column under the optimal condition. (7) Figure S6. The UV (upper) and ECD (lower) chromatograms of **4** on ChiralPak AD-H column under the optimal condition. (8) Figure S7. Comparison of the UV (upper) and ECD (lower) chromatograms of **1** on ChiralPak AD-H column with EtOH and IPA. (9) Figure S8. Conformational distribution of enantiomers **1–4** during the docking process. (10) Figure S9. Comparison of TDDFT-calculated ECD and UV spectra. (11) Figure S10. Interactions between two enantiomers of **1–4** and the CSP of the AD-H column. (12) Figure S11. Graphic illustrating the resolution of chiral sulfoxides on the chiral columns.

Author Contributions: Conceptualization, L.L.; Methodology, B.-B.Y.; Investigation, F.X., J.Z., and B.-B.Y.; Resources, L.L.; Funding acquisition, L.L.; Writing-Original Draft Preparation, F.X.; Writing-Review & Editing, L.L.

Funding: This study was financially supported by the CAMS Innovation Fund for Medical Sciences (CIFMS, No. 2016-I2M-3-009).

Conflicts of Interest: The authors have declared no conflict of interest.

References

1. Bentley, R. Role of sulfur chirality in the chemical processes of biology. *Chem. Soc. Rev.* **2005**, *34*, 609–624. [CrossRef] [PubMed]
2. Matsui, T.; Dekishima, Y.; Ueda, M. Biotechnological production of chiral organic sulfoxides: Current state and perspectives. *Appl. Microbiol. Biotechnol.* **2014**, *98*, 7699–7706. [CrossRef] [PubMed]

3. Goundry, W.R.F.; Adams, B.; Benson, H.; Demeritt, J.; McKown, S.; Mulholland, K.; Robertson, A.; Siedlecki, P.; Tomlin, P.; Vare, K. Development and Scale-up of a Biocatalytic Process to Form a Chiral Sulfoxide. *Org. Process Res. Dev.* **2017**, *21*, 107–113. [CrossRef]
4. Otocka, S.; Kwiatkowska, M.; Madalińska, L.; Kiełbasiński, P. Chiral organosulfur ligands/catalysts with a stereogenic sulfur atom: Applications in asymmetric synthesis. *Chem. Rev.* **2017**, *117*, 4147–4181. [CrossRef] [PubMed]
5. Trost, B.M.; Rao, M. Development of chiral sulfoxide ligands for asymmetric catalysis. *Angew. Chem. Int. Ed.* **2015**, *54*, 5026–5043. [CrossRef] [PubMed]
6. Andersson, T.; Weidolf, L. Stereoselective disposition of proton pump inhibitors. *Clin. Drug Investig.* **2008**, *28*, 263–279. [CrossRef] [PubMed]
7. Katsuki, H.; Yagi, H.; Arimori, K.; Nakamura, C.; Nakano, M.; Katafuchi, S.; Fujioka, Y.; Fujiyama, S. Determination of R (+)-and S (−)-lansoprazole using chiral stationary-phase liquid chromatography and their enantioselective pharmacokinetics in humans. *Pharm. Res.* **1996**, *13*, 611–615. [CrossRef] [PubMed]
8. Donovan, J.L.; Malcolm, R.J.; Markowitz, J.S.; DeVane, C.L. Chiral analysis of d-and l-modafinil in human serum: Application to human pharmacokinetic studies. *Ther. Drug Monit.* **2003**, *25*, 197–202. [CrossRef] [PubMed]
9. Bravi, L.; Rudini, N.; Cuttano, R.; Giampietro, C.; Maddaluno, L.; Ferrarini, L.; Adams, R.H.; Corada, M.; Boulday, G.; Tournier-Lasserve, E.; et al. Sulindac metabolites decrease cerebrovascular malformations in CCM3-knockout mice. *Proc. Natl. Acad. Sci. USA* **2015**, *112*, 8421–8426. [CrossRef] [PubMed]
10. Hwu, J.R.; Tsay, S.C.; Chuang, K.S.; Kapoor, M.; Lin, J.Y.; Yeh, C.S.; Su, W.C.; Wu, P.C.; Tsai, T.L.; Wang, P.W.; et al. Syntheses of Platinum-Sulindac Complexes and Their Nanoparticles as Targeted Anticancer Drugs. *Chemistry (Easton)* **2016**, *22*, 1926–1930. [CrossRef] [PubMed]
11. Scriba, G.K. Cyclodextrins in capillary electrophoresis enantioseparations–recent developments and applications. *J. Sep. Sci.* **2008**, *31*, 1991–2011. [CrossRef] [PubMed]
12. Jac, P.; Scriba, G.K. Recent advances in electrodriven enantioseparations. *J. Sep. Sci.* **2013**, *36*, 52–74. [CrossRef] [PubMed]
13. Hein, J.E.; Cao, B.H.; van der Meijden, M.W.; Leeman, M.; Kellogg, R.M. Resolution of Omeprazole Using Coupled Preferential Crystallization: Efficient Separation of a Nonracemizable Conglomerate Salt under Near-Equilibrium Conditions. *Org. Process Res. Dev.* **2013**, *17*, 946–950. [CrossRef]
14. Bortolini, O.; Fantin, G.; Fogagnolo, M.; Medici, A.; Pedrini, P. Optical resolution of sulfoxides by inclusion in host dehydrocholic acid. *Chem. Commun.* **2000**, 365–366. [CrossRef]
15. Ferretti, R.; Carradori, S.; Guglielmi, P.; Pierini, M.; Casulli, A.; Cirilli, R. Enantiomers of triclabendazole sulfoxide: Analytical and semipreparative HPLC separation, absolute configuration assignment, and transformation into sodium salt. *J. Pharm. Biomed. Anal.* **2017**, *140*, 38–44. [CrossRef] [PubMed]
16. Gegenava, M.; Chankvetadze, L.; Farkas, T.; Chankvetadze, B. Enantioseparation of selected chiral sulfoxides in high-performance liquid chromatography with polysaccharide-based chiral selectors in polar organic mobile phases with emphasis on enantiomer elution order. *J. Sep. Sci.* **2014**, *37*, 1083–1088. [CrossRef] [PubMed]
17. Sardella, R.; Ianni, F.; Di Michele, A.; Di Capua, A.; Carotti, A.; Anzini, M.; Natalini, B. Enantioresolution and stereochemical characterization of two chiral sulfoxides endowed with COX-2 inhibitory activity. *Chirality* **2017**, *29*, 536–540. [CrossRef] [PubMed]
18. Cavazzini, A.; Pasti, L.; Massi, A.; Marchetti, N.; Dondi, F. Recent applications in chiral high performance liquid chromatography: A review. *Anal. Chim. Acta* **2011**, *706*, 205–222. [CrossRef] [PubMed]
19. del Nozal, M.J.; Toribio, L.; Bernal, J.L.; Alonso, C.; Jiménez, J.J. Chiral separation of omeprazole and several related benzimidazoles using supercritical fluid chromatography. *J. Sep. Sci.* **2004**, *27*, 1023–1029. [CrossRef] [PubMed]
20. West, C.; Konjaria, M.L.; Shashviashvili, N.; Lemasson, E.; Bonnet, P.; Kakava, R.; Volonterio, A.; Chankvetadze, B. Enantioseparation of novel chiral sulfoxides on chlorinated polysaccharide stationary phases in supercritical fluid chromatography. *J. Chromatogr. A* **2017**, *1499*, 174–182. [CrossRef] [PubMed]
21. Shen, J.; Okamoto, Y. Efficient separation of enantiomers using stereoregular chiral polymers. *Chem. Rev.* **2016**, *116*, 1094–1138. [CrossRef] [PubMed]

22. Dixit, S.; Dubey, R.; Bhushan, R. Normal and polar-organic-phase high-performance liquid chromatographic enantioresolution of omeprazole, rabeprazole, lansoprazole and pantoprazole using monochloro-methylated cellulose-based chiral stationary phase and determination of dexrabeprazole. *Biomed. Chromatogr.* **2014**, *28*, 112–119. [CrossRef] [PubMed]
23. Balmér, K.; Persson, B.-A.; Lagerström, P.-O. Stereoselective effects in the separation of enantiomers of omeprazole and other substituted benzimidazoles on different chiral stationary phases. *J. Chromatogr. A* **1994**, *660*, 269–273. [CrossRef]
24. Cirilli, R.; Ferretti, R.; Gallinella, B.; De Santis, E.; Zanitti, L.; La Torre, F. High-performance liquid chromatography enantioseparation of proton pump inhibitors using the immobilized amylose-based Chiralpak IA chiral stationary phase in normal-phase, polar organic and reversed-phase conditions. *J. Chromatogr. A* **2008**, *1177*, 105–113. [CrossRef] [PubMed]
25. Ferretti, R.; Zanitti, L.; Casulli, A.; Cirilli, R. Green high-performance liquid chromatography enantioseparation of lansoprazole using a cellulose-based chiral stationary phase under ethanol/water mode. *J. Sep. Sci.* **2016**, *39*, 1418–1424. [CrossRef] [PubMed]
26. Cass, Q.B.; Degani, A.L.G.; Cassiano, N.M. Effects on enantioselectivity by the use of polysaccharide-based columns by multimodal elution. *J. Liq. Chromatogr. Relat. Technol.* **2003**, *26*, 2083–2101. [CrossRef]
27. Cass, Q.B.; Kohn, C.K.; Calafatti, S.A.; Aboul-Enein, H.Y. An enantioselective assay for (±)-modafinil. *J. Pharm. Biomed. Anal.* **2001**, *26*, 123–130. [CrossRef]
28. Harvanová, M.; Gondová, T. New enantioselective LC method development and validation for the assay of modafinil. *J. Pharm. Biomed. Anal.* **2017**, *138*, 267–271. [CrossRef] [PubMed]
29. Nageswara Rao, R.; Shinde, D.D.; Kumar Talluri, M.V. Enantioselective HPLC resolution of synthetic intermediates of armodafinil and related substances. *J. Sep. Sci.* **2008**, *31*, 981–989. [CrossRef] [PubMed]
30. Hauck, W.; Adam, P.; Bobier, C.; Landmesser, N. Use of large−scale chromatography in the preparation of armodafinil. *Chirality* **2008**, *20*, 896–899. [CrossRef] [PubMed]
31. Naso, F.; Cardellicchio, C.; Affortunato, F.; Capozzi, M.A.M. Asymmetric synthesis of Sulindac esters by enantioselective sulfoxidation in the presence of chiral titanium complexes. *Tetrahedron-Asymmetr.* **2006**, *17*, 3226–3229. [CrossRef]
32. Liao, S.; Čorić, I.; Wang, Q.; List, B. Activation of H_2O_2 by chiral confined Brønsted acids: A highly enantioselective catalytic sulfoxidation. *J. Am. Chem. Soc.* **2012**, *134*, 10765–10768. [CrossRef] [PubMed]
33. Zhong, Q.; Han, X.; He, L.; Beesley, T.E.; Trahanovsky, W.S.; Armstrong, D.W. Chromatographic evaluation of poly (trans-1,2-cyclohexanediyl-bis acrylamide) as a chiral stationary phase for HPLC. *J. Chromatogr. A* **2005**, *1066*, 55–70. [CrossRef] [PubMed]
34. Slovakova, A.; Von Maltzan, X.F.; Patel, B.; Drake, A.; Hutt, A. Chromatographic resolution, chiroptical characterization and urinary excretion of the enantiomers of sulindac. *Chromatographia* **1998**, *48*, 369–376. [CrossRef]
35. Barnhart, W.W.; Gahm, K.H.; Hua, Z.; Goetzinger, W. Supercritical fluid chromatography comparison of the poly (trans-1,2-cyclohexanediyl-bis acrylamide) (P-CAP) column with several derivatized polysaccharide-based stationary phases. *J. Chromatogr. B* **2008**, *875*, 217–229. [CrossRef] [PubMed]
36. West, C.; Guenegou, G.; Zhang, Y.; Morin-Allory, L. Insights into chiral recognition mechanisms in supercritical fluid chromatography. II. Factors contributing to enantiomer separation on tris-(3,5-dimethylphenylcarbamate) of amylose and cellulose stationary phases. *J. Chromatogr. A* **2011**, *1218*, 2033–2057. [CrossRef] [PubMed]
37. Von Unge, S.; Langer, V.; Sjölin, L. Stereochemical assignment of the enantiomers of omeprazole from X-ray analysis of a fenchyloxymethyl derivative of (+)-(R)-omeprazole. *Tetrahedron-Asymmetr.* **1997**, *8*, 1967–1970. [CrossRef]
38. Prisinzano, T.; Podobinski, J.; Tidgewell, K.; Luo, M.; Swenson, D. Synthesis and determination of the absolute configuration of the enantiomers of modafinil. *Tetrahedron-Asymmetr.* **2004**, *15*, 1053–1058. [CrossRef]
39. Andersson, S.; Nelander, H.; Öhlén, K. Preparative chiral chromatography and chiroptical characterization of enantiomers of omeprazole and related benzimidazoles. *Chirality* **2007**, *19*, 706–715. [CrossRef] [PubMed]
40. Berova, N.; Di Bari, L.; Pescitelli, G. Application of electronic circular dichroism in configurational and conformational analysis of organic compounds. *Chem. Soc. Rev.* **2007**, *36*, 914–931. [CrossRef] [PubMed]
41. Li, L.; Wang, L.; Si, Y.K. Electronic circular dichroism behavior of chiral Phthiobuzone. *Acta Pharm. Sin. B* **2014**, *4*, 167–171. [CrossRef] [PubMed]

42. Li, L.; Si, Y.K. Study on the absolute configuration of levetiracetam via density functional theory calculations of electronic circular dichroism and optical rotatory dispersion. *J. Pharm. Biomed. Anal.* **2011**, *56*, 465–470. [CrossRef] [PubMed]
43. Li, L.; Si, Y.K. Study on the absolute configurations of 3-alkylphthalides using TDDFT calculations of chiroptical properties. *Chirality* **2012**, *24*, 987–993. [CrossRef] [PubMed]
44. Rossi, D.; Nasti, R.; Collina, S.; Mazzeo, G.; Ghidinelli, S.; Longhi, G.; Memo, M.; Abbate, S. The role of chirality in a set of key intermediates of pharmaceutical interest, 3-aryl-substituted-γ-butyrolactones, evidenced by chiral HPLC separation and by chiroptical spectroscopies. *J. Pharm. Biomed. Anal.* **2017**, *144*, 41–51. [CrossRef] [PubMed]
45. Zhang, A.; Lai, W.; Sun, J.; Hu, G.; Liu, W. Probing the chiral separation mechanism and the absolute configuration of malathion, malaoxon and isomalathion enantiomers by chiral high performance liquid chromatography coupled with chiral detector-binding energy computations. *J. Chromatogr. A* **2013**, *1281*, 26–31. [CrossRef] [PubMed]
46. Persson, B.-A.; Andersson, S. Unusual effects of separation conditions on chiral separations. *J. Chromatogr. A* **2001**, *906*, 195–203. [CrossRef]
47. Wang, T.; Chen, Y.W.; Vailaya, A. Enantiomeric separation of some pharmaceutical intermediates and reversal of elution orders by high-performance liquid chromatography using cellulose and amylose tris (3,5-dimethylphenyl- carbamate) derivatives as stationary phases. *J. Chromatogr. A* **2000**, *902*, 345–355. [CrossRef]
48. Donnoli, M.I.; Giorgio, E.; Superchi, S.; Rosini, C. Circular dichroism spectra and absolute configuration of some aryl methyl sulfoxides. *Org. Biomol. Chem.* **2003**, *1*, 3444–3449. [CrossRef] [PubMed]
49. Pescitelli, G.; Bruhn, T. Good computational practice in the assignment of absolute configurations by TDDFT calculations of ECD spectra. *Chirality* **2016**, *28*, 466–474. [CrossRef] [PubMed]
50. Pescitelli, G.; Di Pietro, S.; Cardellicchio, C.; Capozzi, M.A.M.; Di Bari, L. Systematic investigation of CD spectra of aryl benzyl sulfoxides interpreted by means of TDDFT calculations. *J. Org. Chem.* **2010**, *75*, 1143–1154. [CrossRef] [PubMed]
51. Lammerhofer, M. Chiral recognition by enantioselective liquid chromatography: Mechanisms and modern chiral stationary phases. *J. Chromatogr. A* **2010**, *1217*, 814–856. [CrossRef] [PubMed]
52. Chankvetadze, B. Recent developments on polysaccharide-based chiral stationary phases for liquid-phase separation of enantiomers. *J. Chromatogr. A* **2012**, *1269*, 26–51. [CrossRef] [PubMed]
53. Pirkle, W.H.; Pochapsky, T.C. Considerations of chiral recognition relevant to the liquid chromatographic separation of enantiomers. *Chem. Rev.* **1989**, *89*, 347–362. [CrossRef]
54. Barreiro, J.C.; de Campos Lourenço, T.; Silva, L.M.A.; Venâncio, T.; Cass, Q.B. High resolution magic angle spinning NMR as a tool for unveiling the molecular enantiorecognition of omeprazole by amylose-based chiral phase. *Analyst* **2014**, *139*, 1350–1354. [CrossRef] [PubMed]
55. MOE2009.10, Chemical Computing Group Inc. Available online: www.chemcomp.com (accessed on 15 October 2018).
56. Frisch, M.J.; Trucks, G.W.; Schlegel, H.B.; Scuseria, G.E.; Robb, M.A.; Cheeseman, J.R.; Scalmani, G.; Barone, V.; Mennucci, B.; Petersson, G.A.; et al. *Gaussian 09*; Revision C.01; Gaussian, Inc.: Wallingford, CT, USA, 2010.
57. Ye, Y.K.; Bai, S.; Vyas, S.; Wirth, M.J. NMR and computational studies of chiral discrimination by amylose tris (3,5-dimethylphenylcarbamate). *J. Phys. Chem. B* **2007**, *111*, 1189–1198. [CrossRef] [PubMed]
58. AutoDock. Available online: http://autodock.scripps.edu (accessed on 15 October 2018).

Sample Availability: Samples of the enantiopure isomers of compounds **1–4** are available from the authors.

molecules

MDPI

Review

Enantioselective Drug Recognition by Drug Transporters

Yuichi Uwai

School of Pharmacy, Aichi Gakuin University, Nagoya 464-8650, Japan; yuuwai@dpc.agu.ac.jp; Tel.: +81-52-757-6785

Academic Editor: Maria Elizabeth Tiritan
Received: 25 October 2018; Accepted: 22 November 2018; Published: 22 November 2018

Abstract: Drug transporters mediate the absorption, tissue distribution, and excretion of drugs. The cDNAs of P-glycoprotein, multidrug resistance proteins (MRPs/ABCC), breast cancer resistance protein (BCRP/ABCG2), peptide transporters (PEPTs/SLC15), proton-coupled folate transporters (PCFT/SLC46A1), organic anion transporting polypeptides (OATPs/SLCO), organic anion transporters (OATs/SLC22), organic cation transporters (OCTs/SLC22), and multidrug and toxin extrusions (MATEs/SLC47) have been isolated, and their functions have been elucidated. Enantioselectivity has been demonstrated in the pharmacokinetics and efficacy of drugs, and is important for elucidating the relationship with recognition of drugs by drug transporters from a chiral aspect. Enantioselectivity in the transport of drugs by drug transporters and the inhibitory effects of drugs on drug transporters has been summarized in this review.

Keywords: drug transporter; enantioselectivity; transport; inhibition; pharmacokinetics

---

1. Introduction

Many drugs are chiral, and each enantiomer may exhibit specific therapeutic efficacy. For example, several nonsteroidal antiinflammatory drugs (NSAIDs) have an asymmetric carbon in their chemical structures, and the (*S*)-enantiomers exhibit stronger inhibitory potencies against cyclooxygenases [1]. The (*S*)-enantiomer of naproxen is commercially available. We examined the enantioselective effects of flurbiprofen on the disposition of lithium in rats, and demonstrated that the (*S*)-enantiomer, but not the (*R*)-enantiomer, decreased the renal clearance of lithium with the impairment of renal function [2]. The anticoagulant drug warfarin is also chiral. Warfarin is orally administered as the racemate and exhibits enantioselectivity not only in its pharmacological effects, but also in its pharmacokinetics. (*S*)-Warfarin has greater anticoagulant potency than the (*R*)-enantiomer [3] (*S*)-Warfarin is metabolized by cytochrome P450 (CYP) 2C9, and the metabolism of the (*R*)-enantiomer is mediated by CYP1A2 and CYP3A4 [4]. The development of drugs with enantioselective pharmacokinetic properties, such as esomeprazole, has recently been accomplished. This proton pump inhibitor is the (*S*)-enantiomer of omeprazole, which is the racemate. Both enantiomers exert identical effects on H^+, K^+-ATPase [5], but exhibit different pharmacokinetic characteristics. The clearance of esomeprazole is lower than that of the (*R*)-enantiomer [6]. Furthermore, the metabolism of the (*R*)-enantiomer is mainly mediated by CYP2C19, which exhibits a genetic polymorphism, and the contribution of CYP2C19 to the metabolism of esomeprazole is negligible [6].

In 1988, CYP2D6 was the first cytochrome P450 enzyme whose cDNA was identified [7]. The cDNAs of organic ion transporters have been identified since the 1990s. These findings promoted research on the function and expression of drug transporters. Drug transporters are expressed in the plasma membranes of a number of cells including enterocytes, hepatocytes, brain microvascular endothelial cells, and renal epithelial cells, and facilitate the transport of drugs across the plasma membrane.

Drug transporters are now known to play important roles in the absorption, tissue distribution, and excretion of drugs. Previous studies reported enantioselective drug transport by drug transporters. Furthermore, the enantioselective inhibitory effects of drugs on drug transporters has been demonstrated.

Representative drug transporters functioning in the intestinal absorption, biliary excretion, and renal tubular secretion of drugs have been summarized in this review. The interactions between chiral drugs and drug transporters have also been discussed. Figure 1 shows the chemical structures of the chiral drugs described herein.

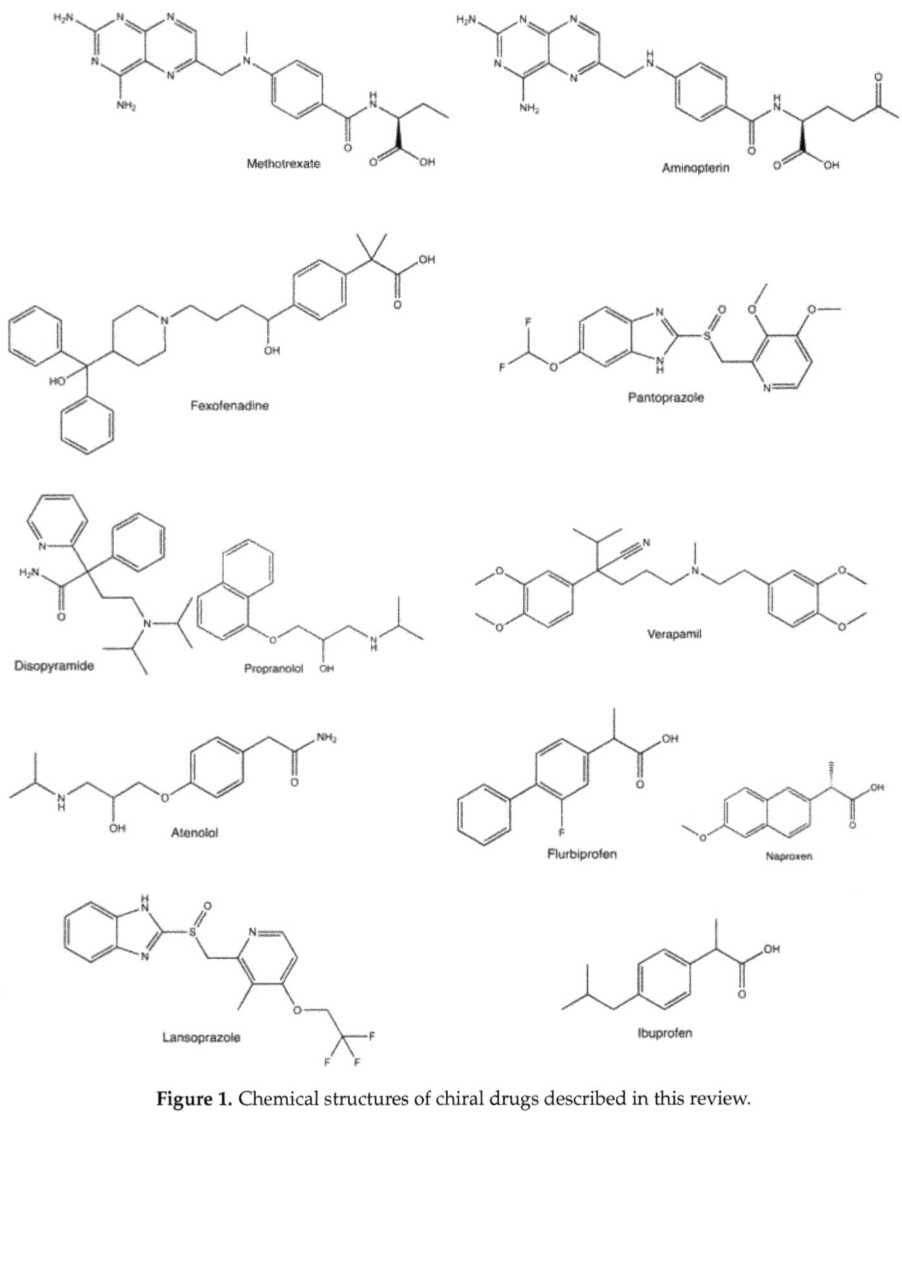

Figure 1. Chemical structures of chiral drugs described in this review.

2. Role of Drug Transporters in the Intestinal Absorption, Biliary Excretion, and Renal Tubular Secretion of Drugs

2.1. Intestinal Drug Absorption by Drug Transporters

Drugs are absorbed by the small intestine after their oral administration. Figure 2 shows drug transporters that are involved in the intestinal absorption of drugs. Peptide transporter PEPT1 (SLC15A1) is one of the most extensively studied drug transporters in the small intestine. It is expressed in the brush-border membrane of intestinal epithelial cells and is responsible for the uptake of dipeptides and tripeptides from the lumen using an inward H^+-electrochemical gradient [8]. PEPT1 recognizes peptide-like β-lactam antibiotics that are orally administered [8]. PEPT1 is used as a target molecule for improving the absorption of poorly absorbed drugs. Valganciclovir, the valine ester prodrug of ganciclovir, was developed to enhance the low bioavailability of ganciclovir, and the mechanism responsible for improved absorption was identified as drug recognition by PEPT1 [9].

The proton-coupled folate transporter PCFT (SLC46A1) is localized to the brush-border membrane of enterocytes, and mediates the absorption of folate [10]. PCFT recognizes the antifolates, methotrexate and aminopterin, as its substrates [11,12]. Aminopterin has not yet been approved as a medicine, and methotrexate is used in the treatment of neoplasia and rheumatoid arthritis. PCFT is considered to play a role in the intestinal absorption of orally administered methotrexate [13].

The organic anion transporting polypeptide OATP2B1 (SLCO2B1), which is expressed in the brush-border membrane of intestinal epithelial cells, is crucially involved in the uptake of fexofenadine, celiprolol, and montelukast from the lumen [14,15].

P-glycoprotein (P-gp; ABCB1) and the breast cancer resistance protein BCRP (ABCG2) are ATP-binding cassette (ABC) transporter proteins that are expressed in the brush-border membrane of enterocytes. These transporters are efflux pumps. P-gp transports various types of drugs, including anticancer agents, antihypertensive agents, antiarrhythmics, antidepressants, antimicrobial agents, anti-human immunodeficiency virus agents, anticonvulsants, antiemetics, immunosuppressants, neuroleptics, and opioids, and there is a broad range of overlapping substrate specificities for CYP3A4 and P-gp [16]. BCRP has also been shown to transport anticancer agents [17]. The tyrosine kinase inhibitors, imatinib, gefitinib, and nilotinib, have also been identified as substrates of BCRP [18–20]. P-gp and BCRP prevent drug absorption in the intestine.

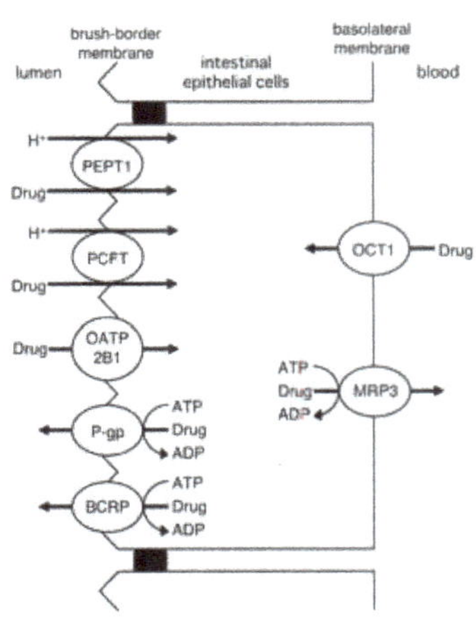

Figure 2. Drug transporters in the small intestine.

In the basolateral membrane, the organic cation transporter OCT1 (SLC22A1) and the multidrug resistance protein MRP3 (ABCC3) mediate drug transport [21]. In OCT1 knockout mice, the intestinal excretion of the typical substrate, tetraethylammonium, was reduced [22]. The serosal efflux of the glucuronide conjugates of 7-ethyl-10-hydroxycamptothecin (SN-38: the active metabolite of irinotecan) and acetaminophen in the jejunal everted sacs was decreased in Mrp3 knockout mice [23].

2.2. Hepatic Transport of Drugs by Drug Transporters

Drug transporters in the plasma membrane of hepatocytes are shown in Figure 3. The OATP family members OATP1B1 (SLCO1B1) and OATP1B3 (SLCO1B3) mediate the hepatic uptake of various drugs, such as HMG-CoA reductase inhibitors, angiotensin II receptor antagonists, nateglinide, asunaprevir, pemafibrate, and bosentan, from the circulation [14,24]. OCT1 functions in the sinusoidal uptake of organic cations, including metformin, tropisetron, sumatriptan and fenoterol, from the circulation [25–28]. P-gp, MRP2 (ABCC2), BCRP, and multidrug and toxin extrusion MATE1 (SLC47A1), which are expressed in the canalicular membrane, mediate the efflux of drugs into bile [24]. MRP2, a member of the ABC family of transporters, excretes monoglucuronosyl bilirubin and monoglucuronosyl bilirubin into bile, and genetic mutations in MRP2 cause Dubin-Johnson syndrome, an autosomal recessive disease characterized by conjugated hyperbilirubinemia [29]. MRP3 is localized to the sinusoidal membrane of hepatocytes [30]. Kitamura et al. reported decreased plasma levels and increased clearance for the biliary excretion of methotrexate in MRP3 knockout mice, suggesting that MRP3 transports methotrexate from hepatocytes into plasma [31].

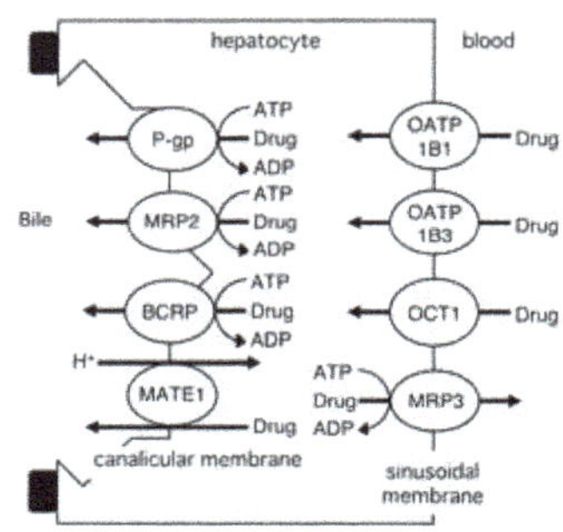

Figure 3. Drug transporters in the liver.

2.3. Renal Tubular Secretion of Drugs

The renal tubular secretion of drugs is mediated by drug transporters, as shown in Figure 4. Organic anion and cation transport systems are present in renal proximal epithelial cells. The organic anion transporter system is involved in the tubular secretion of anionic drugs, including anticancer agents, β-lactam antibiotics, antivirals, and diuretics [24,32]. The organic anion transporters OAT1 (SLC22A6) and OAT3 (SLC22A8) are responsible for the basolateral uptake of anionic drugs from the circulation via an exchange with intracellular α-ketoglutarate, an intermediate in the Krebs cycle. Information on the drug transporters responsible for mediating the efflux of drugs via the brush-border membrane of proximal epithelial cells is more limited than that on basolateral transporters. Studies using rat renal brush-border membrane vesicles indicated that a potential-sensitive organic anion transporter and an anion/organic anion exchange transporter function [33–35]. However, their genes have not yet been identified. Previous studies reported reductions in the urinary excretion of adefovir, tenofovir, hydrochlorothiazide, furosemide, ceftizoxime, and cefazolin in MRP4 knockout mice [36–38]. Based on these findings, MRP4 has been suggested to play an important role in drug transport in the brush-border membrane. MRP2 and OAT4 are known to be expressed in the brush-border membrane of proximal epithelial cells [39,40]. Although their interactions with drugs have been investigated in in vitro experiments, the roles of MRP2 and OAT4 in the tubular transport of drugs currently remain unclear.

The organic cation transport system consists of the uptake type of OCT2 (SLC22A2) and the efflux modes of MATE1 and MATE2-K (SLC47A2) [24,32,41]. Drug transport by OCT2 is electrogenic, and is driven by an internal negative membrane potential. The efflux of drugs by MATE1 and MATE2-K is dominated by a H^+/organic cation antiport, which involves electroneutral transport [41]. Cationic drugs, including cimetidine, metformin, procainamide, memantine, and amantadine, are secreted into urine by the organic cation transport system [32].

P-gp functions in the renal tubular secretion of the cardiac glycoside, digoxin [42]. Although a wide variety of drugs are transported by P-gp [16], most are not recovered into urine. The substrate drugs of P-gp may be reabsorbed by the distal tubules via simple diffusion after tubular secretion.

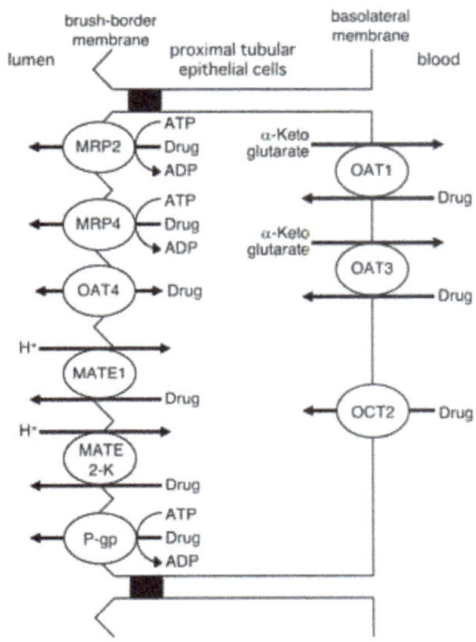

Figure 4. Drug transporters mediating renal tubular secretion drugs.

3. Enantioselective Drug Transport by Drug Transporters

3.1. Enantioselective Transport of Antifolates by PCFT

Previous studies reported enantioselectivity in drug transport by drug transporters, with PCFT as a representative. The antifolates methotrexate and aminopterin have an asymmetric carbon in their structures (Figure 1), and their enantioselective transport was examined. Narawa et al. constructed stably transfected human embryonic kidney cells 293 (HEK293 cells) expressing PCFT, and conducted a methotrexate uptake experiment using these cells [43]. Menter et al. examined aminopterin transport using Chinese hamster ovary (CHO) cells expressing PCFT [12]. The transport of both antifolates showed enantioselectivity, and kinetic parameters were summarized in Table 1. The (*S*)-enantiomers were found to have greater affinity to PCFT than each (*R*)-enantiomer. Menter et al. performed a pharmacokinetic study on aminopterin in dogs and patients with psoriasis, and the findings obtained revealed a correlation with enantioselective absorption and in vitro findings [12].

Table 1. Kinetic parameters of the PCFT-mediated transport of methotrexate and aminopterin.

Drug	K_m or K_t (*R*)-enantiomer	K_m or K_t (*S*)-enantiomer	K_m or K_t (*R*)/(*S*)	V_{max} (*R*)-enantiomer	V_{max} (*S*)-enantiomer	V_{max} (*R*)/(*S*)	Ref.
Methotrexate	211 μM	4.98 μM	42.4	909 pmol/mg/min	891 pmol/mg/min	1.02	[43]
Aminopterin	15.0 μM	0.69 μM	21.7	42.9 pmol/mg/2 min	68.8 pmol/mg/2 min	0.624	[12]

3.2. Enantioselectivity in the Pharmacokinetics of Fexofenadine and Its Transport by OATP2B1

Fexofenadine, a histamine H_1-receptor antagonist, has non-sedative properties that have been attributed to the restriction of its brain penetration by P-gp [44]. Enantioselectivity has been reported in its pharmacokinetics. Miura et al. demonstrated that the maximum plasma concentration and area under the plasma concentration-time curve of (*R*)-fexofenadine were higher than those of the (*S*)-enantiomer after the single oral administration of racemic fexofenadine to healthy volunteers [45]. Sakugawa et al. examined effect of verapamil, an inhibitor of P-gp, on the disposition of each enantiomer of fexofenadine in healthy volunteers, and suggested that the other mechanisms in addition to P-gp contribute to the stereoselective pharmacokinetics of fexofenadine [46]. To my knowledge, there is no reports representative of enantioselective transport of fexofenadine by P-gp from in vitro experiments. Most of the dosage of fexofenadine administered is excreted into urine in its unmetabolized form, and various drug transporters have been shown to contribute to its pharmacokinetics. The OATP family members, OATP1A2, OATP1B1, OATP1B3, and OATP2B1, were proposed to be responsible for the intestinal uptake of fexofenadine or its distribution into the liver [47–52]. MRP2 and the bile salt export pump BSEP (ABCB11) mediate the efflux of fexofenadine from hepatocytes into bile, while MRP3 transports it into plasma [52,53]. OAT3 plays a role in the renal tubular uptake of fexofenadine, and MATE1 contributes to its efflux into urine [54,55]. At least one of the drug transporters described above may be responsible for the enantioselective pharmacokinetics of fexofenadine. The enantioselective transport of fexofenadine has only been reported to occur by OATP2B1. Akamine et al. found greater uptake amounts of (*R*)-fexofenadine in *Xenopus* oocytes injected with OATP2B1 cRNA than that of the (*S*)-enantiomer [56]. They also demonstrated that apple juice decreased the absorption of fexofenadine orally administered to healthy volunteers, and that the juice inhibited its transport by OATP2B1 [56]. Accordingly, OATP2B1 appears to mediate the enantioselective absorption of fexofenadine by the small intestine. Akamine et al. did not describe the kinetic parameters of the transport of enantiomers by OATP2B1. The renal clearance of (*S*)-fexofenadine was higher than that of the (*R*)-enantiomer [45,56], whereas Kusuhara et al. reported no enantioselective transport of fexofenadine by OAT3 and MATE1 [57]. In addition, they showed the similar transport of both enantiomers by OATP1B3 [57]. Unidentified drug transporter(s) may be responsible for the enantioselective disposition of fexofenadine.

3.3. Enantioselective Secretion of Pantoprazole into Milk by BCRP

Enantioselective drug transport was demonstrated with the combination of pantoprazole and BCRP. BCRP affects the absorption, distribution, and excretion of drugs, and BCRP actively secretes xenobiotics, including drugs and carcinogens, and riboflavin into milk [58–60]. Wang et al. showed the greater accumulation of (−)-pantoprazole in the milk of lactating rats infused with racemic pantoprazole than that of (+)-pantoprazole [61], and the higher affinity of the (−)-enantiomer with BCRP [62].

4. Enantioselective Inhibitory Effects of Drugs on Drug Transporters

4.1. Enantioselectivity in Inhibitory Effects of Drugs on OCT1, and Binding Affinities

Enantioselectivity was shown in the inhibitory effect of disopyramide and propranolol on OCT1. The (*R*)-disopyramide inhibited the uptake of tetraethylammonium by HeLa cells expressing OCT1 more strongly than the (*S*)-enantiomer [63], and also for propanol the inhibitory effect of (*S*)-enantiomer is stronger than for the (*R*)-enantiomer [64]. The IC_{50} values of each enantiomer were described in Table 2.

Table 2. IC_{50} values of each enantiomer of disopyramide and propranolol for OCT1.

Drug	IC_{50} Value (μM) (*R*)-enantiomer	IC_{50} Value (μM) (*S*)-enantiomer	(*R*)/(*S*)	Ref.
Disopyramide	15.4	29.9	0.515	[63]
Propranolol	41.7	15.1	2.76	[64]

Moaddel et al. studied the binding of drugs to OCT1 with a liquid chromatography stationary phase containing immobilized membranes obtained from a cell line that expresses OCT1, and estimated their binding affinities using frontal displacement chromatography with [^{3}H]1-methyl 4-phenyl pyridinium as the marker ligand. The significant enantioselectivity on the binding to OCT1 was recognized in verapamil, atenolol, and propranolol [64,65]. (*R*)-Verapamil, (*S*)-atenolol, and (*S*)-propranolol showed the higher affinities than each enantiomer. In Table 3, their K_d values are summarized.

Table 3. K_d values of each enantiomer of verapamil, atenolol, and propranolol for OCT1.

Drug	K_d Value (μM) (*R*)-enantiomer	K_d Value (μM) (*S*)-enantiomer	(*R*)/(*S*)	Ref.
Verapamil	0.05	3.46	0.0145	[65]
Atenolol	0.98	0.46	2.13	[64]
Propranolol	2.85	0.95	3.00	[64]

4.2. Enantioselective Inhibitory Effects of NSAIDs and Lansoprazole on OAT1 and OAT3

The inhibition of renal organic anion transporters leads to the delayed elimination of their substrates from the circulation. NSAIDs interfere with the renal excretion of methotrexate, and this combination is a representative among drug interactions via OAT1 and OAT3. The interaction is fatal when high-dose methotrexate therapy is given to a patient [66,67]. Previous studies demonstrated the inhibitory effects of NSAIDs, including cyclooxygenase-2 inhibitors, on methotrexate uptake by OAT1 and OAT3 [68–72]. Some NSAIDs are chiral (Figure 1). We conducted a drug transport experiment using the *Xenopus* oocyte expression system in order to examine enantioselectivity in the inhibition of OAT1 and OAT3 by NSAIDs [73]. The findings obtained showed the stronger inhibitory effects of the (*S*)-enantiomers of flurbiprofen, ibuprofen, and naproxen on the transport of *p*-aminohippurate and methotrexate by OAT1 than that of each (*R*)-enantiomer. The inhibitory mechanisms of flurbiprofen were investigated, and both enantiomers were found to competitively inhibit OAT1. Enantioselective differences were not observed in the inhibition of OAT3.

Proton pump inhibitors also interact with methotrexate [74,75]; OAT1 and OAT3 were found to be inhibited by omeprazole, lansoprazole, pantoprazole, rabeprazole, and esomeprazole [76,77]. Chirality exists in the structures of proton pump inhibitors (Figure 1), and enantioselectivity was examined in the inhibition of OAT1 and OAT3 by lansoprazole [78]. The inhibitory effects of (*S*)-lansoprazole on the transport of estrone sulfate, methotrexate, and pemetrexed by OAT3 were stronger than those of the (*R*)-enantiomer. Furthermore, enantioselectivity was not recognized against OAT1. Enantioselectivity has been demonstrated in the pharmacokinetics of lansoprazole. The slower elimination of the (*R*)-enantiomer from plasma has been reported in healthy subjects [79]. The faster metabolism of (*S*)-lansoprazole by human liver microsomes was also shown [80]. This information is useful when considering drug interactions between lansoprazole and the substrate drugs of OAT3.

4.3. Enantioselective Inhibitory Effects of NSAIDs on MRP2 and MRP4

The enantioselective inhibitory effects of NSAIDs on MRP2 and MRP4 have been examined. Kawase et al. performed uptake experiments on methotrexate using MRP2- and MRP4-expressing inside-out vesicles, and showed the stronger inhibition of MRP2 by the (*S*)-enantiomers of flurbiprofen, ibuprofen, and naproxen than by each (*R*)-enantiomer [81]. In the case of MRP4, the (*R*)-enantiomers exerted strong inhibitory effects.

Table 4 summarizes the IC_{50} values of each enantiomer of flurbiprofen, ibuprofen, naproxen, and lansoprazole described above. The most prominent enantioselective difference was noted in the inhibition of MRP2 by naproxen.

Table 4. IC_{50} values of each enantiomer of NSAIDs and lansoprazole for OAT1, OAT3, MRP2, and MRP4.

Drug	Transporter	Substrate	IC_{50} Value (μM) (*R*)-Enantiomer	IC_{50} Value (μM) (*S*)-Enantiomer	(*R*)/(*S*)	Ref.
Flurbiprofen	OAT1	*p*-aminohippurate	2.35	0.615	3.82	[73]
	OAT3	estrone sulfate	2.13	1.80	1.18	[73]
	MRP2	methotrexate	133	58.4	2.28	[81]
	MRP4	methotrexate	10.6	37.2	0.285	[81]
Ibuprofen	OAT1	*p*-aminohippurate	6.14	2.84	2.16	[73]
	OAT3	estrone sulfate	2.04	1.20	1.70	[73]
	MRP2	methotrexate	303	139	2.18	[81]
	MRP4	methotrexate	129	267	0.483	[81]
Naproxen	OAT1	*p*-aminohippurate	5.26	1.93	2.73	[73]
	OAT3	estrone sulfate	8.09	6.79	1.19	[73]
	MRP2	methotrexate	510	7.11	71.7	[81]
	MRP4	methotrexate	8.06	49.8	0.162	[81]
Lansoprazole	OAT1	*p*-aminohippurate	43.8	33.6	1.30	[78]
	OAT3	estrone sulfate	1.75	0.61	2.87	[78]

5. Conclusions

Enantioselectivity has been demonstrated in the transport of methotrexate and aminopterin by PCFT, pantoprazole by BCRP, and fexofenadine by OATP2B1. Enantioselective differences were reported in the inhibitory effects of flurbiprofen, ibuprofen, and naproxen on OAT1, MRP2, and MRP4, and of lansoprazole on OAT3. The number of studies on enantioselective drug recognition by drug transporters is markedly smaller than those on drug metabolism enzymes. In research on enantioselectivity, data on drug metabolism accumulate, and Niwa et al. performed meta-analysis based on the reported values regarding the Michaelis-Menten constant, maximal velocity, intrinsic clearance, and inhibition constants [82]. They considered that there is a limited number of reports regarding stereoselective inhibition and induction in vitro [82]. Because drug transporters are also involved in drug interactions, it is desired to pay attention to in enantioselectivity in inhibition as well as in substrate recognition of drug transporters.

Computational methods such as quantitative structure-activity relationship (QSAR) and pharmacophore approaches have become more widely applied to assess interactions between drugs and drug transporters and predictions for substrates and inhibitors for several transporters were described [83]. Because ignoring stereoselectivity reduces the accuracy of the QSAR and modelling analysis, stereoselectivity should become a key aspect of the modelling of interactions between drugs and drug transporters [84].

In the future, research on drug transporters from the chiral aspect of drugs will provide important insights into pharmacokinetics, pharmacodynamics, and drug toxicity.

Author Contributions: Y.U. conceived the study, reviewed the literature, and wrote the manuscript.

Funding: This work was supported by a grant from the Japan Society for the Promotion of Science through a Grant-in-Aid for Scientific Research (KAKENHI, 16K08419 to Y.U.).

Conflicts of Interest: The author states that there are no conflicts of interest.

References

1. Carabaza, A.; Cabré, F.; Rotllan, E.; Gómez, M.; Gutiérrez, M.; García, M.L.; Mauleón, D. Stereoselective inhibition of inducible cyclooxygenase by chiral nonsteroidal antiinflammatory drugs. *J. Clin. Pharmacol.* **1996**, *36*, 505–512. [CrossRef] [PubMed]
2. Uwai, Y.; Matsumoto, M.; Kawasaki, T.; Nabekura, T. Enantioselective effect of flurbiprofen on lithium disposition in rats. *Pharmacology* **2017**, *99*, 236–239. [CrossRef] [PubMed]
3. Park, B.K. Warfarin: Metabolism and mode of action. *Biochem. Pharmacol.* **1988**, *37*, 19–27. [CrossRef]
4. Kaminsky, L.S.; Zhang, Z.Y. Human P450 metabolism of warfarin. *Pharmacol. Ther.* **1997**, *73*, 67–74. [CrossRef]
5. Kendall, M.J. Review article: Esomeprazole—The first proton pump inhibitor to be developed as an isomer. *Aliment. Pharmacol. Ther.* **2003**, *17*, 1–4. [CrossRef] [PubMed]

6. Andersson, T.; Hassan-Alin, M.; Hasselgren, G.; Röhss, K.; Weidolf, L. Pharmacokinetic studies with esomeprazole, the (*S*)-isomer of omeprazole. *Clin. Pharmacokinet.* **2001**, *40*, 411–426. [CrossRef] [PubMed]
7. Gonzalez, F.J.; Vilbois, F.; Hardwick, J.P.; McBride, O.W.; Nebert, D.W.; Gelboin, H.V.; Meyer, U.A. Human debrisoquine 4-hydroxylase (P450IID1): cDNA and deduced amino acid sequence and assignment of the CYP2D locus to chromosome 22. *Genomics* **1988**, *2*, 174–179. [CrossRef]
8. Terada, T.; Inui, K. Peptide transporters: Structure, function, regulation and application for drug delivery. *Curr. Drug Metab.* **2004**, *5*, 85–94. [CrossRef] [PubMed]
9. Sugawara, M.; Huang, W.; Fei, Y.J.; Leibach, F.H.; Ganapathy, V.; Ganapathy, M.E. Transport of valganciclovir, a ganciclovir prodrug, via peptide transporters PEPT1 and PEPT2. *J. Pharm. Sci.* **2000**, *89*, 781–789. [CrossRef]
10. Qiu, A.; Jansen, M.; Sakaris, A.; Min, S.H.; Chattopadhyay, S.; Tsai, E.; Sandoval, C.; Zhao, R.; Akabas, M.H.; Goldman, I.D. Identification of an intestinal folate transporter and the molecular basis for hereditary folate malabsorption. *Cell* **2006**, *127*, 917–928. [CrossRef] [PubMed]
11. Inoue, K.; Nakai, Y.; Ueda, S.; Kamigaso, S.; Ohta, K.Y.; Hatakeyama, M.; Hayashi, Y.; Otagiri, M.; Yuasa, H. Functional characterization of PCFT/HCP1 as the molecular entity of the carrier-mediated intestinal folate transport system in the rat model. *Am. J. Physiol. Gastrointest. Liver Physiol.* **2008**, *294*, G660–G668. [CrossRef] [PubMed]
12. Menter, A.; Thrash, B.; Cherian, C.; Matherly, L.H.; Wang, L.; Gangjee, A.; Morgan, J.R.; Maeda, D.Y.; Schuler, A.D.; Kahn, S.J.; et al. Intestinal transport of aminopterin enantiomers in dogs and humans with psoriasis is stereoselective: Evidence for a mechanism involving the proton-coupled folate transporter. *J. Pharmacol. Exp. Ther.* **2012**, *342*, 696–708. [CrossRef] [PubMed]
13. Inoue, K.; Yuasa, H. Molecular basis for pharmacokinetics and pharmacodynamics of methotrexate in rheumatoid arthritis therapy. *Drug Metab. Pharmacokinet.* **2014**, *29*, 12–19. [CrossRef] [PubMed]
14. Shitara, Y.; Maeda, K.; Ikejiri, K.; Yoshida, K.; Horie, T.; Sugiyama, Y. Clinical significance of organic anion transporting polypeptides (OATPs) in drug disposition: Their roles in hepatic clearance and intestinal absorption. *Biopharm. Drug Dispos.* **2013**, *34*, 45–78. [CrossRef] [PubMed]
15. Tamai, I.; Nakanishi, T. OATP transporter-mediated drug absorption and interaction. *Curr. Opin. Pharmacol.* **2013**, *13*, 859–863. [CrossRef] [PubMed]
16. Zhou, S.F. Structure, function and regulation of P-glycoprotein and its clinical relevance in drug disposition. *Xenobiotica* **2008**, *38*, 802–832. [CrossRef] [PubMed]
17. Mao, Q.; Unadkat, J.D. Role of the breast cancer resistance protein (BCRP/ABCG2) in drug transport —An update. *AAPS J.* **2015**, *17*, 65–82. [CrossRef] [PubMed]
18. Burger, H.; van Tol, H.; Boersma, A.W.; Brok, M.; Wiemer, E.A.; Stoter, G.; Nooter, K. Imatinib mesylate and nilotinib (AMN107) exhibit high-affinity interaction with ABCG2 on primitive hematopoietic stem cells. *Blood* **2004**, *104*, 2940–2942. [CrossRef] [PubMed]
19. Elkind, N.B.; Szentpétery, Z.; Apáti, A.; Ozvegy-Laczka, C.; Várady, G.; Ujhelly, O.; Szabó, K.; Homolya, L.; Váradi, A.; Buday, L.; et al. Multidrug transporter ABCG2 prevents tumor cell death induced by the epidermal growth factor receptor inhibitor Iressa (ZD1839, Gefitinib). *Cancer Res.* **2005**, *65*, 1770–1777. [CrossRef] [PubMed]
20. Brendel, C.; Scharenberg, C.; Dohse, M.; Robey, R.W.; Bates, S.E.; Shukla, S.; Ambudkar, S.V.; Wang, Y.; Wennemuth, G.; Burchert, A.; et al. Imatinib mesylate and nilotinib (AMN107) exhibit high-affinity interaction with ABCG2 on primitive hematopoietic stem cells. *Leukemia* **2007**, *21*, 1267–1275. [CrossRef] [PubMed]
21. Giacomini, K.M.; Huang, S.M.; Tweedie, D.J.; Benet, L.Z.; Brouwer, K.L.; Chu, X.; Dahlin, A.; Evers, R.; Fischer, V.; Hillgren, K.M.; et al. Membrane transporters in drug development. *Nat. Rev. Drug Discov.* **2010**, *9*, 215–236. [PubMed]
22. Jonker, J.W.; Wagenaar, E.; Mol, C.A.; Buitelaar, M.; Koepsell, H.; Smit, J.W.; Schinkel, A.H. Reduced hepatic uptake and intestinal excretion of organic cations in mice with a targeted disruption of the organic cation transporter 1 (*Oct1* [*Slc22a1*]) gene. *Mol. Cell. Biol.* **2001**, *21*, 5471–5477. [CrossRef] [PubMed]
23. Kitamura, Y.; Kusuhara, H.; Sugiyama, Y. Functional characterization of multidrug resistance-associated protein 3 (Mrp3/*Abcc3*) in the basolateral efflux of glucuronide conjugates in the mouse small intestine. *J. Pharmacol. Exp. Ther.* **2010**, *332*, 659–666. [CrossRef] [PubMed]
24. Kusuhara, H.; Sugiyama, Y. In vitro-in vivo extrapolation of transporter-mediated clearance in the liver and kidney. *Drug Metab. Pharmacokinet.* **2009**, *24*, 37–52. [CrossRef] [PubMed]

25. Wang, D.S.; Jonker, J.W.; Kato, Y.; Kusuhara, H.; Schinkel, A.H.; Sugiyama, Y. Involvement of organic cation transporter 1 in hepatic and intestinal distribution of metformin. *J. Pharmacol. Exp. Ther.* **2002**, *302*, 510–515. [CrossRef] [PubMed]
26. Tzvetkov, M.V.; Saadatmand, A.R.; Bokelmann, K.; Meineke, I.; Kaiser, R.; Brockmöller, J. Effects of OCT1 polymorphisms on the cellular uptake, plasma concentrations and efficacy of the 5-HT_3 antagonists tropisetron and ondansetron. *Pharmacogenomics J.* **2012**, *12*, 22–29. [CrossRef] [PubMed]
27. Matthaei, J.; Kuron, D.; Faltraco, F.; Knoch, T.; Dos Santos Pereira, J.N.; Abu Abed, M.; Prukop, T.; Brockmöller, J.; Tzvetkov, M.V. OCT1 mediates hepatic uptake of sumatriptan and loss-of-function OCT1 polymorphisms affect sumatriptan pharmacokinetics. *Clin. Pharmacol. Ther.* **2016**, *99*, 633–641. [CrossRef] [PubMed]
28. Tzvetkov, M.V.; Matthaei, J.; Pojar, S.; Faltraco, F.; Vogler, S.; Prukop, T.; Seitz, T.; Brockmöller, J. Increased systemic exposure and stronger cardiovascular and metabolic adverse reactions to fenoterol in individuals with heritable OCT1 deficiency. *Clin. Pharmacol. Ther.* **2018**, *103*, 868–878. [CrossRef] [PubMed]
29. Toh, S.; Wada, M.; Uchiumi, T.; Inokuchi, A.; Makino, Y.; Horie, Y.; Adachi, Y.; Sakisaka, S.; Kuwano, M. Genomic structure of the canalicular multispecific organic anion-transporter gene (MRP2/cMOAT) and mutations in the ATP-binding-cassette region in Dubin-Johnson syndrome. *Am. J. Hum. Genet.* **1999**, *64*, 739–746. [CrossRef] [PubMed]
30. König, J.; Rost, D.; Cui, Y.; Keppler, D. Characterization of the human multidrug resistance protein isoform MRP3 localized to the basolateral hepatocyte membrane. *Hepatology* **1999**, *29*, 1156–1163. [CrossRef] [PubMed]
31. Kitamura, Y.; Hirouchi, M.; Kusuhara, H.; Schuetz, J.D.; Sugiyama, Y. Increasing systemic exposure of methotrexate by active efflux mediated by multidrug resistance-associated protein 3 (mrp3/abcc3). *J. Pharmacol. Exp. Ther.* **2008**, *327*, 465–473. [CrossRef] [PubMed]
32. Ivanyuk, A.; Livio, F.; Biollaz, J.; Buclin, T. Renal drug transporters and drug interactions. *Clin. Pharmacokinet.* **2017**, *56*, 825–892. [CrossRef] [PubMed]
33. Martinez, F.; Manganel, M.; Montrose-Rafizadeh, C.; Werner, D.; Roch-Ramel, F. Transport of urate and *p*-aminohippurate in rabbit renal brush-border membranes. *Am. J. Physiol.* **1990**, *58*, F1145–F1153. [CrossRef] [PubMed]
34. Ohoka, K.; Takano, M.; Okano, T.; Maeda, S.; Inui, K.; Hori, R. *p*-Aminohippurate transport in rat renal brush-border membranes: A potential-sensitive transport system and an anion exchanger. *Biol. Pharm. Bull.* **1993**, *16*, 395–401. [CrossRef] [PubMed]
35. Inui, K.I.; Masuda, S.; Saito, H. Cellular and molecular aspects of drug transport in the kidney. *Kidney Int.* **2000**, *58*, 944–958. [CrossRef] [PubMed]
36. Imaoka, T.; Kusuhara, H.; Adachi, M.; Schuetz, J.D.; Takeuchi, K.; Sugiyama, Y. Functional involvement of multidrug resistance-associated protein 4 (MRP4/ABCC4) in the renal elimination of the antiviral drugs adefovir and tenofovir. *Mol. Pharmacol.* **2007**, *71*, 619–627. [CrossRef] [PubMed]
37. Hasegawa, M.; Kusuhara, H.; Adachi, M.; Schuetz, J.D.; Takeuchi, K.; Sugiyama, Y. Multidrug resistance-associated protein 4 is involved in the urinary excretion of hydrochlorothiazide and furosemide. *J. Am. Soc. Nephrol.* **2007**, *18*, 37–45. [CrossRef] [PubMed]
38. Ci, L.; Kusuhara, H.; Adachi, M.; Schuetz, J.D.; Takeuchi, K.; Sugiyama, Y. Involvement of MRP4 (ABCC4) in the luminal efflux of ceftizoxime and cefazolin in the kidney. *Mol. Pharmacol.* **2007**, *71*, 1591–1597. [CrossRef] [PubMed]
39. Smeets, P.H.; van Aubel, R.A.; Wouterse, A.C.; van den Heuvel, J.J.; Russel, F.G. Contribution of multidrug resistance protein 2 (MRP2/ABCC2) to the renal excretion of *p*-aminohippurate (PAH) and identification of MRP4 (ABCC4) as a novel PAH transporter. *J. Am. Soc. Nephrol.* **2004**, *15*, 2828–2835. [CrossRef] [PubMed]
40. Ekaratanawong, S.; Anzai, N.; Jutabha, P.; Miyazaki, H.; Noshiro, R.; Takeda, M.; Kanai, Y.; Sophasan, S.; Endou, H. Human organic anion transporter 4 is a renal apical organic anion/dicarboxylate exchanger in the proximal tubules. *J. Pharmacol. Sci.* **2004**, *94*, 297–304. [CrossRef] [PubMed]
41. Motohashi, H.; Inui, K. Organic cation transporter OCTs (SLC22) and MATEs (SLC47) in the human kidney. *AAPS J.* **2013**, *15*, 581–588. [CrossRef] [PubMed]
42. Hori, R.; Okamura, N.; Aiba, T.; Tanigawara, Y. Role of P-glycoprotein in renal tubular secretion of digoxin in the isolated perfused rat kidney. *J. Pharmacol. Exp. Ther.* **1993**, *266*, 1620–1625. [PubMed]

43. Narawa, T.; Itoh, T. Stereoselective transport of amethopterin enantiomers by the proton-coupled folate transporter. *Drug Metab. Pharmacokinet.* **2010**, *25*, 283–289. [CrossRef] [PubMed]
44. Tahara, H.; Kusuhara, H.; Fuse, E.; Sugiyama, Y. P-glycoprotein plays a major role in the efflux of fexofenadine in the small intestine and blood-brain barrier, but only a limited role in its biliary excretion. *Drug Metab. Dispos.* **2005**, *33*, 963–968. [CrossRef] [PubMed]
45. Miura, M.; Uno, T.; Tateishi, T.; Suzuki, T. Pharmacokinetics of fexofenadine enantiomers in healthy subjects. *Chirality* **2007**, *19*, 223–227. [CrossRef] [PubMed]
46. Sakugawa, T.; Miura, M.; Hokama, N.; Suzuki, T.; Tateishi, T.; Uno, T. Enantioselective disposition of fexofenadine with the P-glycoprotein inhibitor verapamil. *Br. J. Clin. Pharmacol.* **2009**, *67*, 535–540. [CrossRef] [PubMed]
47. Cvetkovic, M.; Leake, B.; Fromm, M.F.; Wilkinson, G.R.; Kim, R.B. OATP and P-glycoprotein transporters mediate the cellular uptake and excretion of fexofenadine. *Drug Metab. Dispos.* **1999**, *27*, 866–871. [PubMed]
48. Dresser, G.K.; Bailey, D.G.; Leake, B.F.; Schwarz, U.I.; Dawson, P.A.; Freeman, D.J.; Kim, R.B. Fruit juices inhibit organic anion transporting polypeptide-mediated drug uptake to decrease the oral availability of fexofenadine. *Clin. Pharmacol. Ther.* **2002**, *71*, 11–20. [CrossRef] [PubMed]
49. Nozawa, T.; Imai, K.; Nezu, J.; Tsuji, A.; Tamai, I. Functional characterization of pH-sensitive organic anion transporting polypeptide OATP-B in humans. *J. Pharmacol. Exp. Ther* **2004**, *308*, 438–445. [CrossRef] [PubMed]
50. Shimizu, M.; Fuse, K.; Okudaira, K.; Nishigaki, R.; Maeda, K.; Kusuhara, H.; Sugiyama, Y. Contribution of OATP (organic anion-transporting polypeptide) family transporters to the hepatic uptake of fexofenadine in humans. *Drug Metab. Dispos.* **2005**, *33*, 1477–1481. [CrossRef] [PubMed]
51. Bailey, D.G.; Dresser, G.K.; Leake, B.F.; Kim, R.B. Naringin is a major and selective clinical inhibitor of organic anion-transporting polypeptide 1A2 (OATP1A2) in grapefruit juice. *Clin. Pharmacol. Ther.* **2007**, *81*, 495–502. [CrossRef] [PubMed]
52. Matsushima, S.; Maeda, K.; Ishiguro, N.; Igarashi, T.; Sugiyama, Y. Investigation of the inhibitory effects of various drugs on the hepatic uptake of fexofenadine in humans. *Drug Metab. Dispos.* **2008**, *36*, 663–669. [CrossRef] [PubMed]
53. Matsushima, S.; Maeda, K.; Hayashi, H.; Debori, Y.; Schinkel, A.H.; Schuetz, J.D.; Kusuhara, H.; Sugiyama, Y. Involvement of multiple efflux transporters in hepatic disposition of fexofenadine. *Mol. Pharmacol.* **2008**, *73*, 1474–1483. [CrossRef] [PubMed]
54. Tahara, H.; Kusuhara, H.; Maeda, K.; Koepsell, H.; Fuse, E.; Sugiyama, Y. Inhibition of oat3-mediated renal uptake as a mechanism for drug-drug interaction between fexofenadine and probenecid. *Drug Metab. Dispos.* **2006**, *34*, 743–747. [CrossRef] [PubMed]
55. Matsushima, S.; Maeda, K.; Inoue, K.; Ohta, K.Y.; Yuasa, H.; Kondo, T.; Nakayama, H.; Horita, S.; Kusuhara, H.; Sugiyama, Y. The inhibition of human multidrug and toxin extrusion 1 is involved in the drug-drug interaction caused by cimetidine. *Drug Metab. Dispos.* **2009**, *37*, 555–559. [CrossRef] [PubMed]
56. Akamine, Y.; Miura, M.; Komori, H.; Saito, S.; Kusuhara, H.; Tamai, I.; Ieiri, I.; Uno, T.; Yasui-Furukori, N. Effects of one-time apple juice ingestion on the pharmacokinetics of fexofenadine enantiomers. *Eur. J. Clin. Pharmacol.* **2014**, *70*, 1087–1095. [CrossRef] [PubMed]
57. Kusuhara, H.; Miura, M.; Yasui-Furukori, N.; Yoshida, K.; Akamine, Y.; Yokochi, M.; Fukizawa, S.; Ikejiri, K.; Kanamitsu, K.; Uno, T.; et al. Effect of coadministration of single and multiple doses of rifampicin on the pharmacokinetics of fexofenadine enantiomers in healthy subjects. *Drug Metab. Dispos.* **2013**, *41*, 206–213. [CrossRef] [PubMed]
58. Jonker, J.W.; Merino, G.; Musters, S.; van Herwaarden, A.E.; Bolscher, E.; Wagenaar, E.; Mesman, E.; Dale, T.C.; Schinkel, A.H. The breast cancer resistance protein BCRP (ABCG2) concentrates drugs and carcinogenic xenotoxins into milk. *Nat. Med.* **2005**, *11*, 127–129. [CrossRef] [PubMed]
59. Van Herwaarden, A.E.; Wagenaar, E.; Karnekamp, B.; Merino, G.; Jonker, J W.; Schinkel, A.H. Breast cancer resistance protein (Bcrp1/Abcg2) reduces systemic exposure of the dietary carcinogens aflatoxin B1, IQ and Trp-P-1 but also mediates their secretion into breast milk. *Carcinogenesis* **2006**, *27*, 123–130. [CrossRef] [PubMed]
60. van Herwaarden, A.E.; Wagenaar, E.; Merino, G.; Jonker, J.W.; Rosing, H.; Beijnen, J.H.; Schinkel, A.H. Multidrug transporter ABCG2/breast cancer resistance protein secretes riboflavin (vitamin B_2) into milk. *Mol. Cell. Biol.* **2007**, *27*, 1247–1253. [CrossRef] [PubMed]

61. Wang, L.; McNamara, P.J. Stereoselective interaction of pantoprazole with ABCG2. I. Drug accumulation in rat milk. *Drug Metab. Dispos.* **2012**, *40*, 1018–1023. [CrossRef] [PubMed]
62. Wang, L.; Leggas, M.; Empey, P.E.; McNamara, P.J. Stereoselective interaction of pantoprazole with ABCG2. II. In vitro flux analysis. *Drug Metab. Dispos.* **2012**, *40*, 1024–1031. [CrossRef] [PubMed]
63. Zhang, L.; Schaner, M.E.; Giacomini, K.M. Functional characterization of an organic cation transporter (hOCT1) in a transiently transfected human cell line (HeLa). *J. Pharmacol. Exp. Ther.* **1998**, *286*, 354–361. [PubMed]
64. Moaddel, R.; Patel, S.; Jozwiak, K.; Yamaguchi, R.; Ho, P.C.; Wainer, I.W. Enantioselective binding to the human organic cation transporter-1 (hOCT1) determined using an immobilized hOCT1 liquid chromatographic stationary phase. *Chirality* **2005**, *17*, 501–506. [CrossRef] [PubMed]
65. Moaddel, R.; Yamaguchi, R.; Ho, P.C.; Patel, S.; Hsu, C.P.; Subrahmanyam, V.; Wainer, I.W. Development and characterization of an immobilized human organic cation transporter based liquid chromatographic stationary phase. *J. Chromatogr. B Analyt. Technol. Biomed. Life Sci.* **2005**, *818*, 263–268. [CrossRef] [PubMed]
66. Thyss, A.; Milano, G.; Kubar, J.; Namer, M.; Schneider, M. Clinical and pharmacokinetic evidence of a life-threatening interaction between methotrexate and ketoprofen. *Lancet* **1986**, *327*, 256–258. [CrossRef]
67. Maiche, A.G. Acute renal failure due to concomitant action of methotrexate and indomethacin. *Lancet* **1986**, *327*, 1390. [CrossRef]
68. Uwai, Y.; Saito, H.; Inui, K. Interaction between methotrexate and nonsteroidal anti-inflammatory drugs in organic anion transporter. *Eur. J. Pharmacol.* **2000**, *409*, 31–36. [CrossRef]
69. Takeda, M.; Khamdang, S.; Narikawa, S.; Kimura, H.; Hosoyamada, M.; Cha, S.H.; Sekine, T.; Endou, H. Characterization of methotrexate transport and its drug interactions with human organic anion transporters. *J. Pharmacol. Exp. Ther.* **2002**, *302*, 666–671. [CrossRef] [PubMed]
70. Uwai, Y.; Taniguchi, R.; Motohashi, H.; Saito, H.; Okuda, M.; Inui, K. Methotrexate-loxoprofen interaction: Involvement of human organic anion transporters hOAT1 and hOAT3. *Drug Metab. Pharmacokinet.* **2004**, *19*, 369–374. [CrossRef] [PubMed]
71. Nozaki, Y.; Kusuhara, H.; Kondo, T.; Iwaki, M.; Shiroyanagi, Y.; Nakayama, H.; Horita, S.; Nakazawa, H.; Okano, T.; Sugiyama, Y. Species difference in the inhibitory effect of nonsteroidal anti-inflammatory drugs on the uptake of methotrexate by human kidney slices. *J. Pharmacol. Exp. Ther.* **2007**, *322*, 1162–1170. [CrossRef] [PubMed]
72. Uwai, Y.; Honjo, H.; Iwamoto, K. Inhibitory effect of selective cyclooxygenase-2 inhibitor lumiracoxib on human organic anion transporters hOAT1 and hOAT3. *Drug Metab. Pharmacokinet.* **2010**, *25*, 450–455. [CrossRef] [PubMed]
73. Honjo, H.; Uwai, Y.; Aoki, Y.; Iwamoto, K. Stereoselective inhibitory effect of flurbiprofen, ibuprofen and naproxen on human organic anion transporters hOAT1 and hOAT3. *Biopharm. Drug Dispos.* **2011**, *32*, 518–524. [CrossRef] [PubMed]
74. Beorlegui, B.; Aldaz, A.; Ortega, A.; Aquerreta, I.; Sierrasesúmega, L.; Giráldez, J. Potential interaction between methotrexate and omeprazole. *Ann. Pharmacother.* **2000**, *34*, 1024–1027. [CrossRef] [PubMed]
75. Suzuki, K.; Doki, K.; Homma, M.; Tamaki, H.; Hori, S.; Ohtani, H.; Sawada, Y.; Kohda, Y. Co-administration of proton pump inhibitors delays elimination of plasma methotrexate in high-dose methotrexate therapy. *Br. J. Clin. Pharmacol.* **2009**, *67*, 44–49. [CrossRef] [PubMed]
76. Chioukh, R.; Noel-Hudson, M.S.; Ribes, S.; Fournier, N.; Becquemont, L.; Verstuyft, C. Proton pump inhibitors inhibit methotrexate transport by renal basolateral organic anion transporter hOAT3. *Drug Metab. Dispos.* **2014**, *42*, 2041–2048. [CrossRef] [PubMed]
77. Ikemura, K.; Hamada, Y.; Kaya, C.; Enokiya, T.; Muraki, Y.; Nakahara, H.; Fujimoto, H.; Kobayashi, T.; Iwamoto, T.; Okuda, M. Lansoprazole exacerbates pemetrexed-mediated hematologic toxicity by competitive inhibition of renal basolateral human organic anion transporter 3. *Drug Metab. Dispos.* **2016**, *44*, 1543–1549. [CrossRef] [PubMed]
78. Hamada, Y.; Ikemura, K.; Iwamoto, T.; Okuda, M. Stereoselective inhibition of renal basolateral human organic anion transporter 3 by lansoprazole enantiomers. *Pharmacology* **2018**, *101*, 176–183. [CrossRef] [PubMed]

79. Miura, M.; Tada, H.; Yasui-Furukori, N.; Uno, T.; Sugawara, K.; Tateishi, T.; Suzuki, T. Pharmacokinetic differences between the enantiomers of lansoprazole and its metabolite, 5-hydroxylansoprazole, in relation to CYP2C19 genotypes. *Eur. J. Clin. Pharmacol.* **2004**, *60*, 623–628. [CrossRef] [PubMed]
80. Kim, K.A.; Kim, M.J.; Park, J.Y.; Shon, J.H.; Yoon, Y.R.; Lee, S.S.; Liu, K.H.; Chun, J.H.; Hyun, M.H.; Shin, J.G. Stereoselective metabolism of lansoprazole by human liver cytochrome P450 enzymes. *Drug Metab. Dispos.* **2003**, *31*, 1227–1234. [CrossRef] [PubMed]
81. Kawase, A.; Yamamoto, T.; Egashira, S.; Iwaki, M. Stereoselective inhibition of methotrexate excretion by glucuronides of nonsteroidal anti-inflammatory drugs via multidrug resistance proteins 2 and 4. *J. Pharmacol. Exp. Ther.* **2016**, *356*, 366–374. [CrossRef] [PubMed]
82. Niwa, T.; Murayama, N.; Yamazaki, H. Stereoselectivity of human cytochrome p450 in metabolic and inhibitory activities. *Curr. Drug Metab.* **2011**, *12*, 549–569. [CrossRef] [PubMed]
83. Ekins, S.; Ecker, G.F.; Chiba, P.; Swaan, P.W. Future directions for drug transporter modelling. *Xenobiotica* **2007**, *37*, 1152–1170. [CrossRef] [PubMed]
84. Bhatia, P.; Kolinski, M.; Moaddel, R.; Jozwiak, K.; Wainer, I.W. Determination and modelling of stereoselective interactions of ligands with drug transporters: A key dimension in the understanding of drug disposition. *Xenobiotica* **2008**, *38*, 656–675. [CrossRef] [PubMed]

molecules

MDPI

Review

Enantiomeric Recognition and Separation by Chiral Nanoparticles

Ankur Gogoi [1,†], Nirmal Mazumder [2,†], Surajit Konwer [3], Harsh Ranawat [2], Nai-Tzu Chen [4] and Guan-Yu Zhuo [4,5,*]

1 Department of Physics, Jagannath Barooah College, Jorhat, Assam 785001, India; ankurgogoi@gmail.com
2 Department of Biophysics, School of Life Sciences, Manipal Academy of Higher Education, Manipal, Karnataka 576104, India; nirmaluva@gmail.com (N.M.); harsh.ranawat@outlook.com (H.R.)
3 Department of Chemistry, Dibrugarh University, Dibrugarh, Assam 786004, India; surajitkonwer@dibru.ac.in
4 Institute of New Drug Development, China Medical University, No. 91, Hsueh-Shih Rd., Taichung 40402, Taiwan; ohnonancy@mail.cmu.edu.tw
5 Integrative Stem Cell Center, China Medical University Hospital, No. 2, Yude Rd., Taichung 40447, Taiwan
* Correspondence: zhuo0929@mail.cmu.edu.tw; Tel.: +886-4-2205-3366 (ext. 6508); Fax: +886-4-2207-1507
† These authors contributed equally to this work.

Academic Editors: Maria Elizabeth Tiritan, Madalena Pinto and Carla Sofia Garcia Fernandes
Received: 3 February 2019; Accepted: 10 March 2019; Published: 13 March 2019

Abstract: Chiral molecules are stereoselective with regard to specific biological functions. Enantiomers differ considerably in their physiological reactions with the human body. Safeguarding the quality and safety of drugs requires an efficient analytical platform by which to selectively probe chiral compounds to ensure the extraction of single enantiomers. Asymmetric synthesis is a mature approach to the production of single enantiomers; however, it is poorly suited to mass production and allows for only specific enantioselective reactions. Furthermore, it is too expensive and time-consuming for the evaluation of therapeutic drugs in the early stages of development. These limitations have prompted the development of surface-modified nanoparticles using amino acids, chiral organic ligands, or functional groups as chiral selectors applicable to a racemic mixture of chiral molecules. The fact that these combinations can be optimized in terms of sensitivity, specificity, and enantioselectivity makes them ideal for enantiomeric recognition and separation. In chiral resolution, molecules bond selectively to particle surfaces according to homochiral interactions, whereupon an enantiopure compound is extracted from the solution through a simple filtration process. In this review article, we discuss the fabrication of chiral nanoparticles and look at the ways their distinctive surface properties have been adopted in enantiomeric recognition and separation.

Keywords: chirality; racemic mixture; enantiomer; enantiomeric recognition; enantiomeric separation; surface-modified nanoparticle; chiral ligand

1. Introduction

Chirality is one of those fundamental properties of molecular systems which are ubiquitous in nature [1,2]. Notably, a number of important biological compounds including proteins, amino acids, nucleosides, sugars, and a number of hormones, which are the basic building blocks of life, are chiral and thus the chemistry of many fundamental biological (metabolic and regulatory) processes are directly controlled by such molecular systems [3,4]. Interestingly, enantiomers are mirror images of a chiral compound, e.g., chiral drugs, pesticides, herbicides, etc., which often exhibit profoundly different metabolic, pharmacological, therapeutic and/or toxicological properties [5]. Consequently, the two enantiomers of a chiral compound will react differently with the complementary receptor or enzyme molecule in a biological environment which is highly stereoselective or enantioselective [6]. This issue is particularly important for the pharmaceutical industry where many drugs currently in use are known to be chiral (about 56%) [4] out of which only about 25% are pure enantiomers. Besides, chirality is also important for other industries like agrochemical, food, petroleum, etc. [7–9]. For example, approximately 30–40% of the currently registered pesticides, insecticides and herbicides are chiral [6,8,10]. Most of these currently used chiral compounds are racemates or mixtures [11]. Importantly, while one enantiomer of a chiral drug or agrochemical product has desired effect (eutomer), the other may be inactive and/or cause adverse side effects in many cases (distomer) [3,6].

Given the tremendous importance of chirality in many biochemical processes, recent years witnessed enormous efforts among the researchers and manufacturing industries to prepare enantiopure compounds so that an administered chiral compound (drug or other chemical compounds) fits properly to the target binding site or receptor molecule. In this regard, the ideal solution would be the enantioselective synthesis of just one of the enantiomers (synthetic or chiral approach [9,12,13]) by using methods like isolation of natural compounds [14,15], fermentation [16], asymmetric synthesis [17], etc. Notably, the Nobel Prize in Chemistry for 2001 was awarded for the development of catalytic asymmetric synthesis with one half jointly to Knowles and Noyori "for their work on chirally catalyzed hydrogenation reactions" and the other half to Sharpless "for his work on chirally catalyzed oxidation reactions" [18]. Unfortunately very few enantiopure compounds can be obtained from natural sources. On the other hand, the use of asymmetric synthesis is limited due to the high catalyst cost and time-consuming synthesis procedure, albeit it is one of the most powerful methods to produce 100% enantiopure compounds [12,19,20].

As compared to chiral approach, the racemic approach involves recognition of racemates or mixtures and subsequent separation of the enantiomers, which is relatively cost-effective and presents lower level of difficulty [20]. This approach is based on a three-point interaction between the analyte and a chiral selector [12,21–23]. Briefly, a matched pair undergoes a three-point interaction whereas only one or two point interaction takes place in case of a mismatched pair, as shown in Figure 1. A number of chiral selectors, including proteins, cyclodextrin (CD), crown ethers, oligo- and polysaccharides, etc., have been developed till date. Unfortunately there is no universal chiral selector that can be efficiently applied for the recognition and/or separation of all types of chiral compounds [24].

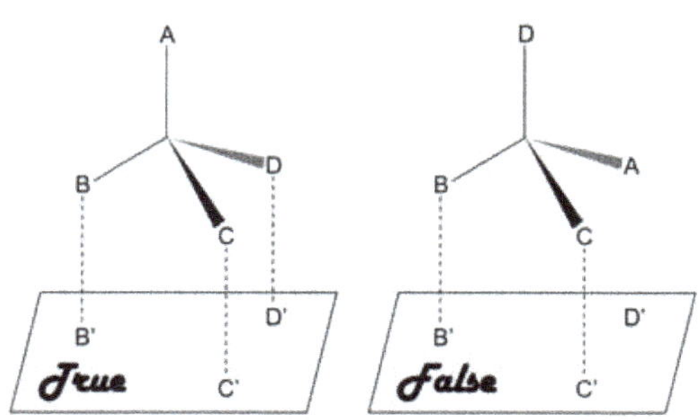

Figure 1. Three-point interaction model for describing the homochiral (true) and heterochiral (false) interactions.

A number of techniques have been developed during the recent past for recognition and/or separation of enantiomers obtained by using the racemic approach [11,25]. Most commonly used methods include chromatographic methods such as high-performance liquid chromatography (HPLC), liquid chromatography (LC), gas chromatography (GC), capillary electrochromatography (CEC), thin-layer chromatography (TLC), micellar chromatography (MC), supercritical fluid chromatography (SFC), and high-speed countercurrent chromatography (HSCCC) [26–36]. Other methods are capillary electrophoresis (CE) [29,30,37], crystallization resolution [38,39], liquid-liquid extraction (LLE) [12,40], membrane separation [41–43], kinetic resolution [13], etc. On the other hand, self-disproportionation of enantiomers (SDE) introduced by Soloshonok is another process which efficiently transforms an enantiomerically enriched compound (scalemic mixture, i.e., a mixture of enantiomers at a ratio other than 50:50 or 100:0) into completely racemic (enantio-depleted) and enantiomerically pure (enantioenriched) fractions [44–47]. Importantly, SDE occurs spontaneously whenever nonracemic compounds are subjected to any physiochemical processes, such as, precipitation, centrifugation, recrystallization, sublimation, force field, achiral chromatography, etc., under totally achiral conditions [48].

Each of these methods has their unique abilities for enantiomeric recognition, separation and quantification. Nevertheless, despite the availability of such advanced chiral resolution methods, it is still very difficult to separate the enantiomers with high efficiency because of the intrinsic difficulties associated with these methods, mainly due to the exactly same physiochemical properties of the enantiomer pairs [24,49,50]. In addition, most of them require using chiral columns in place of stationary phase and such processes are often expensive and cumbersome, and do not allow real time analysis. The advantages and limitations of the aforementioned methods that are widely used for chiral resolution are tabulated in Table 1.

Table 1. Comparison of currently existing chiral resolution methods. The table and the caption have been adapted with permission from [50], Copyright © 2008 The Royal Society of Chemistry. Note that the original Table is slightly modified and the method "Self-disproportionation of enantiomers" has been added.

Methods	Advantages	Disadvantages	Possible Scale
(a) Crystallization resolution			
(a1) Direct or preferential crystallization	Simplicity, low cost	Batch operation, resolving conglomerate	Small- and large-scale
(a2) Diastereomeric crystallization	Simplicity, wide applicability	Expensive, difficulty in finding appropriate resolving agents	Large-scale, industrial scale
(b) Kinetic resolution			
(b1) Chemical-mediated	High stability	Low efficiency	Preparative scale, large-scale
(b2) Enzyme-mediated	High resolving efficiency	Decreasing enzyme activity, narrow application range	Preparative scale, large-scale
(c) Chromatographic separation			
(c1) Supercritical fluid chromatography	Lower costs, [a] high efficiency, resolving most racemates	Low capacity,	Large-scale
(c2) Simulated moving bed chromatography	Continuous operation, [a] high efficiency, resolving most racemates	Low capacity,	Large-scale
(c3) Other chromatography	High efficiency, resolving most racemates	Low capacity, expensive, batch operation, slow and labor intensive	Analytical scale, preparative scale
(d) Membrane-based separation	Low cost, energy saving, high capacity, continuous operation and easy scale-up	Low number of transfer units per apparatus	Large-scale, industrial scale
(e) Self-disproportionation of enantiomers	Ubiquitous and spontaneous, simple, cost effective, and fully predictable (SDE via centrifugation), can be used for both liquid and crystalline Compounds (SDE via chromatography), all forms of liquid chromatography have the potential to give rise to SDE [44,48]	Does not occur with racemic compounds—instead it occurs only in case of partly enriched chiral compounds [51]	Analytical scale, preparative scale [44]

[a] Note: Advantages of (c1) and (c2) were obtained by comparing with high performance liquid chromatography.

Nanoparticles (NPs) represent an entirely new approach to chiral resolution. At present, this generally involves surface modification using chiral ligands; however, recent advances are making the recognition and separation of enantiomers far simpler. One particular achievement in enantiomeric recognition is colorimetric detection, which uses surface-modified NPs to convert recognition events into color changes observable to the naked eye or a UV-Vis spectrometer [52–62]. This makes it ideal for on-site chiral analysis and provides results instantaneously. The working principle is that interactions with specific enantiomers occur on the surface of metal NPs, which means that interactions can be monitored according to changes in surface plasmon resonance (SPR). Furthermore, chiral ligand-capped quantum dots (QDs) have also received considerable attention in this field, due to the size-dependent optical properties, bright chemiluminescence, and excellent chemical stability [63–65]. Beside the applications of magnetic nanoparticles (MNPs) such as catalysts, targeted administration and magnetic resonance imaging (MRI) [66,67], researchers have even capped the surfaces of MNPs with chiral ligands to promote specific interactions with the target enantiomers. It may find potential use in the design of new magneto-chiroptical devices [68]. A wide variety of chiral NPs have been developed for enantiomeric recognition, many of which are examined in Section 2. In enantiomeric separation, surface-modified NPs, as chiral selectors, are exposed to a racemic mixture of chiral molecules to perform the selective adsorption of one enantiomer, leaving an excess of the other enantiomer in solution to be removed through multiple rounds of centrifugation. Following centrifugation, the target enantiomers co-precipitate with the chiral NPs, whereas their enantiomeric counterparts remain in the supernatant [52,53,69]. Note that the chiral ligands chosen for enantiomeric separation are susceptible to denaturation and renaturation manipulated by external perturbations such as temperature and magnetic field, making them switchable and reusable. The used nanomaterials and chiral ligands are discussed in Section 3.

2. Enantiomeric Recognition by Chiral Nanoparticles

The ability to recognize the molecular chirality of enantiomers is significantly important owing to their critical role in drug development and biochemistry. Current discrimination of enantiomers has remained a challenge due to lack of efficient methods. NP-based enantiomeric recognition and separation have been widely discussed and studied in the past decade. Studies reported chiral -modified nanomaterials for catalysis [70,71], chiral drug separation [71] and sensing [54–60,63,68,72–74]. In this section, we will focus on the enantiomeric recognition and the fabrication of most commonly used gold and silver materials for chiral NPs.

2.1. Gold-Based Nanomaterials

Gold-based nanomaterials are the most widely used material for chiral NPs due to its unique optical properties with different shape and size, easy for surface medication and highly biocompatibility [75–77]. In addition, gold nanomaterials can be finely tuned by varying the morphological and reaction characteristic to yield desired size and shape of gold nanoparticles (AuNPs) including nanosphere, nanorod, nanocage and nanoflower. The majority of fabrication of gold materials is through gold-sulphur (Au-S) interaction formed between gold surface and either thiol or dithiol group. The surface-assembly monolayer (SAM) of gold-sulphur bond composes a strong and stable linkage for gold materials and the surface-modified moieties such as proteins, antibodies or polymers. Thiol-gold bond fabrication strategy for AuNP was used in several studies [54,58,70,71,73]. Kang et al. [54] developed an electrochemical sensor for enantioselective recognition of 3,4-dihydroxyphenylalanine (DOPA) based on penicillamine-modified AuNPs (Pen-AuNPs). Chiroptical activity was observed in small AuNPs (~2 nm) which are protected by thiolated Pen. This study was based on the chiral interaction between electroactive DOPA as a target molecule and Pen-AuNPs as chiral NPs. Keshvari's group applied L-cysteine (Cys)-capped AuNPs as a colorimetric sensor for enantioselective detection of naproxen racemic mixtures [73]. The enantioselective and rapid aggregation of L-Cys-modified AuNPs in the presence of *R*-naproxen makes this design capable of visual chiral sensing. The aggregation/agglomeration morphology of AuNPs were successfully demonstrated though transmission electron microscope (TEM), UV-visible spectroscopy, zeta potential, and Fourier-transform infrared spectroscopy (FTIR) measurements.

AuNP bound with chiral molecules on their surface have been used in many studies for enantiomeric recognition. While most commonly employed surface molecules are L-Cys and its derivatives, owing to the ability of the sulfur group easily to bind with NP surface. Other chiral moieties like ionic liquids are now being investigated [78,79]. Ionic liquids are non-molecular ionic solvents where the ions are loosely coordinated, hence their low melting point. They are also called molten or fused salts, with distinct properties such as low vapour pressure, good thermal and chemical stability and fast diffusion and migration of ions. Ionic liquids have been employed successfully in asymmetric synthesis, LC, CE and GC studies. Huang et al. [60] synthesized AuNPs adsorbed with 1-ethyl-3-methylimidazole L-tartrate (EMIML-Tar) and 1-ethyl-3-methylimidazole Lactate (EMIML-Lac) as chiral molecules. EMIML-Tar-AuNPs recognized chiral samples more potently between L-tyrosine (Tyr) and D-Tyr samples, and were considered for the rest of the study. An exclusive red-to-purple color change was observed when L-Tyr was added to the EMIML-Tar-AuNP solution, unlike the D enantiomer. The absorption peaks had also broadened, and spectral differences were quantified using a plot of the extinction ratio (A650/A520) against logarithmic concentration of the Tyr sample solution, as shown in Figure 2. Extinction coefficient values were significantly higher for L-Tyr exposed samples, demonstrating the chiral potency of the EMIML-Tar-AuNP solution. CE analysis was used to characterize the chiral interactions with different amino acids, where it was observed that EMIML-Tar-AuNPs interacted with L-Tyr, tryptophan (Trp) and phenylalanine (Phe) in decreasing order of strength. Visual color changes and absorption spectra changes were confirmed by TEM analysis which showed exclusive aggregation of EMIM-Tar-AuNPs caused by L-Tyr. The proposed mechanism of selective binding was based on shorter distance of the amino group of L-Tyr to carboxyl group of EMIM-Tar-AuNP when comparing the enantiomers of Tyr and hydroxyl group of Tyr allowing for three-point contact with the NPs when comparing Tyr with Trp and Phe.

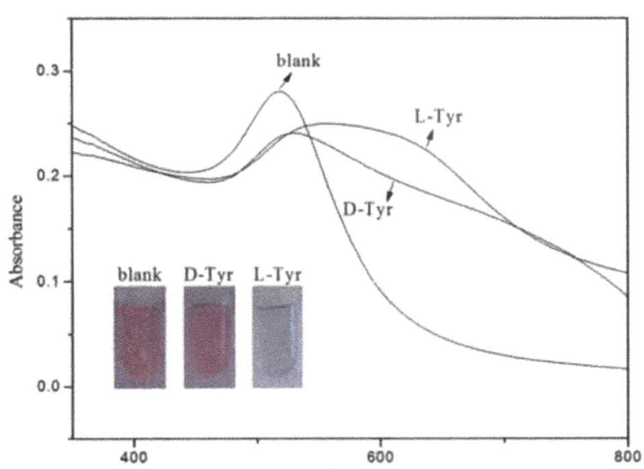

Figure 2. Photographs and UV-vis spectra of EMIML-Tar-AuNPs in the presence of D-Tyr or L-Tyr. Experimental conditions: 1.5 mL EMIML-Tar-AuNPs added with 0.5 mL 1.25 mmol/L D-Tyr or L-Tyr. The figure and the caption have been adapted with permission from [60], Copyright © 2016 OSA.

When AuNPs are used to discriminate between enantiomer they mostly require surface modification with chiral ligands that enable chiral recognition. Unlike nanospheres, gold nanorods (AuNRs) can take part in enantiomeric recognition independent of surface modifications as they exhibit chirality of their own. They differ from conventional NPs by being slightly elongated in structure and possessing two SPR bands, one for transverse and the other for longitudinal direction of NR. They are also more sensitive to their micro-environment. Therefore, another simple, sensitive, cheap, and easy to operate enantiorecognition method was developed by Wang's group [58]. Naked AuNR was employed as colorimetric probes for visual recognition of glutamine (Gln) enantiomers. The inherent chirality of AuNRs only aggregated in the presence of D-Gln, thereby resulting in appreciable blue-gray color changes of AuNR solution. On the other hand, no color changes of AuNR solution in the presence of L-Gln. D-Gln also caused a significant decrease in absorbance at 620 nm (longitudinal SPR band). The experiment was also modified and tried to detect enantiomeric excess and was successful in doing so. The experiment was also performed to test the enantiomeric recognition of other amino acids. The results did show some difference for the different enantiomers of these α-amino acids. However, this may be improved by changing the experimental conditions that cater to the specific amino acid being used. Cysteine however showed aggregation of the AuNRs when either of the enantiomers was present. The aggregations could be observed by visualizing the solution using a TEM.

Amino acids play a key role in cellular metabolism and protein composition. All amino acids except glycine exhibit chirality. Contemporary techniques of chirality-dependent separation of these amino acids, such as HPLC, CE and GC are expensive and require complicated upstream treatments. For this purpose, Song et al. [56] devised an inexpensive and easy colorimetric probe for visual enantiomeric recognition of right-handed and left-handed amino acid. The synthesis strategy for AuNP was $NaBH_4$ reduction method. L-Tartaric acid (TA) was found to be structure-similar to citric acid, which was the commonly used as both a reducing and capping agent for AuNP synthesis. Instead of three carboxyl group (-COOH) as citric acid, L-TA consists of two carboxyl groups, and is adsorbed onto the gold surface through the carboxyl group and provides steric barriers for repulsion. L-TA was also regarded as a chiral selector to separate various enantiomers, including many amino acids. The principle of the proposed technique (Figure 3) is that the change of dispersed AuNP solution to an aggregated state, caused by selective binding of enantiomers, is demonstrated by a red to blue shift. Enantiomer induced calorimetric changes for all 19 α-amino acids at 0.1 mM were observed, whose both L and D forms induced AuNP aggregation. Unlike the other 19 α-amino acids, Cys is the only molecule with a thiol-group, the sulfur atom of which can bind to the surface of AuNP causing the blue shift. On assessing optical rotation of L-TA-capped AuNPs and the amino acids, it was found that all caused a positive rotation, in accordance to the general observation of stronger homochiral bindings relative to heterochiral interactions. Hence, left-handed amino acids, which do not cause aggregation, can be discriminated from right-handed forms. For this, histidine (His) was analyzed further as a sample to prove this model. Solutions of L-His and D-His were analyzed against blank L-TA-capped AuNP solution. A visual change in the color of the solution was observed exclusively in L-His sample, measured as a reduction in the UV-Vis absorption spectra at 520 nm and emergence of a new 700 nm peak. TEM and dynamic light scattering (DLS) measurements established that the aggregation was exclusive to the L-His sample and the D-His sample did not affect the dispersity of L-TA-capped AuNPs in solution. The ratio of absorbance at 700 nm and 520 nm (A700/A520) was studied against different concentrations and enantiomer compositions to enable quantitative analysis. Using this technique for L-His, the limit of detection was 0.015 mM, the binding constant of His enantiomers with L-TA-capped AuNPs being 40.82 and 0.23 of L and D types, respectively. Mechanism of interaction between the right-handed amino acids and L-TA-capped AuNPs was proposed to be through carboxylic, hydroxyl and amino groups by hydrogen bonds. This was supported with FTIR spectral data. Owing to high stability of the synthesized NPs for a considerable amount of time and in a pH range of 3.0–9.0, along with the ability to discriminate enantiomers of 18 amino acids with cheap spectrophotometers as the chief measurement device, this technique can potentially be used for high-throughput applications.

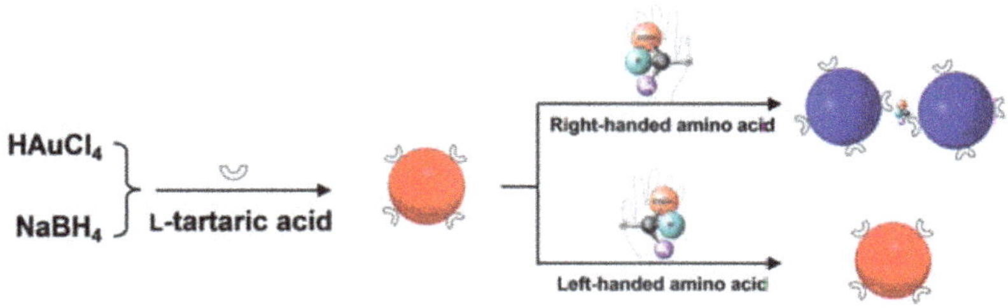

Figure 3. Schematic illustration of visual chiral recognition of right-handed and left-handed amino acid using L-tartaric acid-capped AuNPs as colorimetric probes. The figure and the caption have been adapted with permission from [56], Copyright © 2016 The Royal Society of Chemistry.

Additionally, Zhou et al. [57] established a carbon dots-gold nanoparticle (C-dots@AuNP) complex for chiral discrimination of glucose enantiomers according to colorimetric and fluorescence dual-mode signals. Cysteine was selected as a precursor to generate sulfhydryl decorated-C-dots, which is responsible for the formation of the C-dots@AuNP complex based on the strong tendency of the sulphur element to conjugate onto the surface of AuNPs. H_2O_2 is produced as a result of the enzymatic action of glucose oxidase (Gox) and this by-product helps formation of AuNPs by facilitating the reduction chloroauric acid. As shown in Figure 4, in the presence of carbon dots decorated with sulfhydryl groups (derived from Cys), the AuNPs form a complex—C-dots@AuNP. The carbon dots used show an absorption peak at around 350 nm and at an excitation wavelength of 340 nm, they fluoresce with maximum emission at 424 nm. The intensity of this emission is quenched due to the AuNPs when the C-dots@AuNPs complexes are formed. Thus fluorescence spectra in this case can be used to tell D- and L-glucose apart. Such a method is much simpler and faster when compared to using chiral ligands to modify the surface of NPs. The difference can also be seen visually since in the presence of D-glucose the solution turns reddish while it remains colorless when L-glucose is used. TEM images of glucose oxidase catalytic reaction solution show that C-dots cluster around AuNPs when D-glucose is used whereas no AuNPs are formed in the first place when L-glucose is used and C-dots are seen in a monodisperse state.

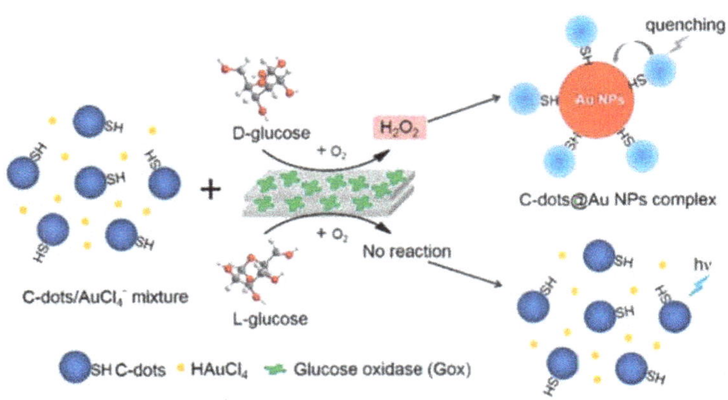

Figure 4. Schematic representation for the generation of C-dots@AuNP complex trigged by the stereoselective enzymatic reaction for the discrimination of glucose enantiomers. The figure and the caption have been adapted with permission from [57], Copyright © 2018 The Royal Society of Chemistry.

2.2. *Silver-Based Nanomaterials*

Silver-based nanomaterials are other widely used materials for enantiomeric recognition, separation and sensing. Silver nanomaterials possess unique chemical and physical properties like higher extinction coefficient, excellent biocompatibility and strong SPR absorption. This phenomenon depends on the size, shape and inter-particle distances of metal NPs and the surrounding environment. In addition, similar to AuNPs, AgNPs were found to have the inherent chirality. Those properties make AgNPs suitable for the fabrication of chiral sensors. Sun's group designed a sulfonated-substituted zinc tetraphenylporphyrin (ZnTPPS)-modified AgNP as colorimetric sensor for chiral detection of L-arginine (Arg) and His [55]. ZnTPPS was selected to fabricate the AgNPs for not only providing stability of NPs but also introducing enantiomeric recognition for L-Arg and L-His via Zn binding. ZnTPPS-AgNPs were found to be coordinately bound to N-H of L-Arg leading to formation of AgNP clusters (L-Arg-ZnTPPS-AgNPs) which were then characterized by TEM and UV-vis spectroscopy. In the presence of L-Arg, a color change from light yellow to yellow can be observed accompanied by a decrease in absorbance at 400 nm that arises from the SPR absorption of dispersed AgNPs. L-Arg induces aggregation of the ZnTPPS-AgNPs to form AgNP clusters, L-Arg-ZnTPPS-AgNPs. This was confirmed by TEM. Methionine (Met), His, Phe, Tyr and glutamic acid were tested using these modified AgNPs but only Met and His showed an obvious color change and His was chosen for further investigation. It was found that only L-His induced a color change from yellow brown to red accompanied by a significant decrease in the absorbance at 400 nm (Figure 5). A new band at 550 nm was observed and the absorbance ratio (R) at the two wavelengths were determined and was found to be enhanced only in the presence of L-His. D-His on the other hand had none of these effects, thereby proving the good chiral selectivity of these AgNPs. L-His induced further aggregation of the modified AgNPs and caused them to form larger agglomerates and this was clearly seen using TEM. In circular dichroism spectra, these NPs showed a slight chiral signal in the presence of D-His whereas L-His had a much stronger one. To satisfy the three-point contact model for enantiomeric recognition, they proposed a possible interaction mode between L-His and the AgNPs clusters via the carboxylic, amino and imidazole groups through electrostatic and hydrogen bond interactions. In a quantitative analysis study, they found that at very low concentrations of His, the NPs showed low chiral selectivity. Hence it showed in the end that this method was effective for enantiomeric recognition of His and Met.

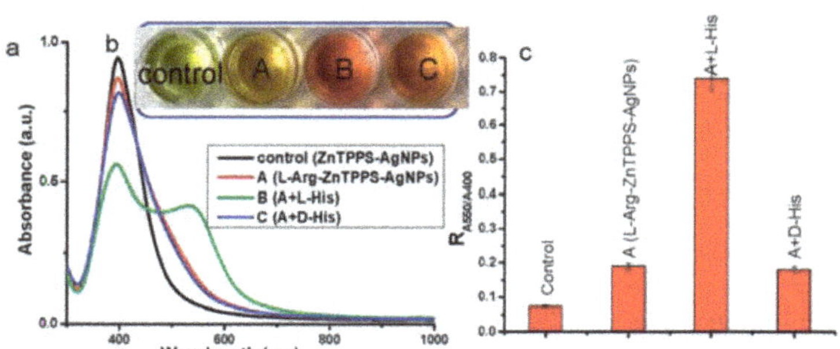

Figure 5. Schematic representation The UV-vis spectra (**a**), photographic images (**b**) and the value of R (A550/A400) (**c**) of silver nanoparticle cluster solution in the presence of D- or L-His. It is operated as follows: 0.5 mL of 1 mM solution of D- or L-His was added to 1.5 mL L-Arg-ZnTPPS-AgNPs solution and mixed for 10 min before measuring, respectively. The figure and the caption have been adapted with permission from [55], Copyright © 2012 The Royal Society of Chemistry.

Marzieh et al. [61] used chitosan-capped silver nanoparticles (CS-AgNPs) for enantiomeric recognition of the essential amino acid, Trp. They scanned optical cells containing the NPs and D- or L-Trp and the color values of each optical cell were then analyzed. The NPs were characterized using FTIR spectra, TEM, X-ray diffractometer (XRD) and UV-vis spectroscopy after their synthesis. When given concentrations of the Trp enantiomers were added separately to a solution of CS-AgNPs, UV-vis spectra were recorded after 30 min. A control without Trp was also tested in the same manner. While scanning for the color of the solution at selected areas of the scanned image for RGB analysis, the values were averaged. MATLAB software was used for assessing different types of colors and obtaining the CMYK values. Differences between CMYK values of the blank control were compared to those of solutions containing either of the Trp enantiomers. The maximum absorbance of the AgNPs was observed at 404 nm and represented the characteristic SPR band of the AgNPs. TEM images of the CS-AgNPs showed that they were uniformly distributed in aqueous solution with an average diameter of about 15 nm and their concentration was calculated using Beer's law. In the presence of L-Trp at sufficiently detectable concentrations, a color change from a yellowish to brown was observed and the CS-AgNPs were found to be aggregated when observed using TEM. No such observations were detected when D-Trp was used. Experiments were also performed by varying various experimental conditions to find the optimal settings of the experiment. They also found that enantiomeric composition of Trp could be determined from corresponding scanometry and spectrophotometric linear calibration curves. The possible interaction mechanism between the CS-AgNPs and L-Trp was proposed to be the hydrogen bond formation between -OH groups of surface of CS-AgNPs and the amine and carboxyl groups of L-Trp.

AgNPs exhibit a distance-relevant color and higher extinction coefficient than that of AuNPs of the same size, thereby making them more reliable as color reporting agents for colorimetric sensor design. Zhang et al. [52] used uridine 5′-trisphosphate (UTP)-capped AgNPs for chiral detection of Cys enantiomers. It was found that in the presence of D-Cys an appreciable color change from yellow to red took place but L-Cys showed no such change, as shown in Figure 6. Moreover, when these UTP-capped AgNPs were added to a racemic solution of Cys, they interacted specifically with one enantiomer leaving an excess of the other after centrifugation thus enabling enantiomeric separation. When UV-vis spectroscopy was performed, the peak at 400 nm was seen to undergo a red shift to 520 nm in the presence of D-Cys due to the SPR of AgNPs. Aggregation of the NPs also took place resulting in the decrease of the SPR band at 400 nm. This band was consequently broadened around 450-600 nm. No such effects were reported when L-Cys was used. The absorbance ratio at 520 nm and 400 nm relates to the quantities of dispersed and aggregated AgNPs and thus this was chosen to evaluate the performance of the AgNPs.

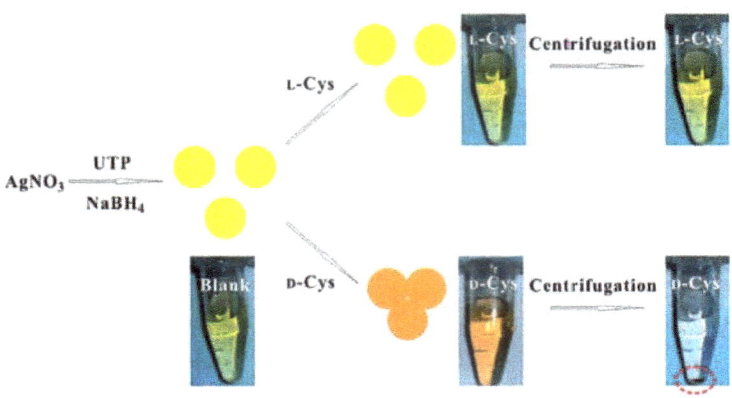

Figure 6. Colorimetric Discrimination of L- and D-Cys Using UTP-Capped AgNPs. The figure and the caption have been adapted with permission from [52], Copyright © 2011 American Chemical Society.

This ratio resulted in significant differences when D- and L-Cys were used. The ratio was found to increase with increasing concentration of D-Cys from 0.1 to 100 micromolars. No such event took place in case of L-Cys except when its concentration was above 100 micromolars where it may cause the AgNPs to lose stability and aggregate, but data suggested that aggregation caused by D-Cys was much more sensitive by at least two times that of L-Cys. UTP analogs were also tested in their role as stabilizing agents on the surface of the AgNPs. When adenosine 5′-triphosphate (ATP) was used, there was negligible discrimination between the two enantiomeric forms of Cys but gradual decrease in the 400 nm and increase in the 520 nm was observed along with the same color change as before. This proved that ATP coated AgNPs could be used to quantitatively detect Cys but not discriminate between the enantiomeric forms. Furthermore, Tashkhourian et al. [62] established AgNP-based colorimetric sensor to discriminate *R*-citalopram (Cit). Citalopram is a selective serotonin reuptake inhibitor used as an antidepressant drug. *S*-Cit was found 150 more potent than racemic Cit or *R*-Cit. Therefore, a simple and reliable detection system for chiral Cit is ponderable in pharmaceutical field.

Gold and silver nanoparticle-based enantiomeric recognition have been widely discussed and studied in the past decade. However, Bruckner et al. [80] demonstrated the use of polysaccharide-ester based amino acid enantiomers using gas chromatography-mass spectrometry (GCMS) in analysis of 24 h urine and blood sera. The experimental procedure involved GC of *N(O)*-pentafluoropropionyl (PFP)-amino acid-(2)-propyl esters on *N*-propionyl-L-valine-*tert*-butylamide polysiloxane, and subsequent filtration by selected ion monitoring mass spectrometry (SIM-MS). The PFP-(2)-propyl esters were synthesized using a mixture of 14 D,L-amino acids and γ-aminobutyric acid (Gaba). The highest amounts of chiral amino acids were determined as D-Serine and D-Alanine, both in urine samples and blood sera, with blood sera in relatively much lower amounts. Samples from fasting patients revealed decreased levels of D-amino acids, while time dependent analysis showed a continuous excretion in the urine.

3. Enantiomeric Separation by Chiral Nanoparticles

Applications of NPs in enantiomeric separation have shown promising results by enhancing the resolution, processing time and efficiency of the conventional separation methods [81]. The use of NPs in enantiomeric separation dates back to 1989 when Wallingford and Ewing reported the first application of polymer NPs (mentioned as monomolecular particles) in CE [82–84]. Since then a varieties of NPs have been used for efficient separation of enantiomers. Table 2 summarizes some of the recent applications of nanostructures in enantioseparation science.

Table 2. Some representative examples in which nanomaterials have been used for the separation of chiral compounds [24,85,86].

Type of Nanostructured Material		Method of Separation/Characterization	Chiral Selector/Template	Nanostructure Dimensions	Analytes/Analysed Compounds	Ref.
Metallic nanoparticles	AgNP	CE	β-CD	AgNPs were of the size of ca. 21 nm	1-phenyl-1-propanol, 1-phenyl-2-propanol, and 2-phenyl-1-propanol	[87]
		Colorimetry	Nucleotide-capped AgNPs	-	D,L-Cys	[52]
	AuNP	Optical polarimetry	D,L-Cys-AuNPs	Average diameter of AuNPs: ~5 nm	Propylene oxide	[88]
		Colorimetry	*N*-acetyl-L-Cys-capped AuNPs as chiral candidate	Size range from 6 to 8 nm	D,L-Tyr	[53]
		Optical polarimetry	Tetrahexahedral (THH, 24-sided) AuNPs modified with D- or L-Cys was used as chiral separator	Shape: rod-like; diameter: ~40 nm; length: ~100 nm	Propranolol	[89]
		CE	Streptomycin-modified gold nanoparticles (ST-AuNPs)	Particle size of AuNPs and ST-AuNPs was 53.1 nm and 79.2 nm, respectively	Adrenergic compounds: adrenaline, noradrenaline and isoprenaline	[86]
		Centrifugation	Functional nucleic acids-modified AuNPs	Diameter of AuNPs (for best separation efficiency): 55nm	D,L-Trp	[69]
		Pseudostationary phase-CEC	Thiolated β-CD-modified AuNPs	Average diameter: 9.5 ± 2.5 nm	Four amino acid enantiomers (D,L-Val, Leu, Glu and Asp) and three drug enantiomers (*R,S*-chlorpheniramine, zopiclone and carvedilol)	[90]
Metal Oxide nanoparticles	Iron Oxide (Fe_3O_4)	Direct separation using a magnet	(*R*)- and (*S*)-*N*-(2,2-dimethyl-4-pentanoyl)-proline-3,5-dimethylanilide	The average particle size of magnetic silica nanoparticles (MSNPs): 300 nm	*N*-(3,5-dinitrobenzoyl)-α-amino acid *N*-propylamides	[91]
		HPLC	Bovine serum albumin (BSA)	Mean diameter of Fe_3O_4: 400 nm; thickness of silica layer in $Fe_3O_4@SiO_2$: 60 nm	Trp, Phe and His	[66]
		HPLC	Carboxymethyl-β-CD	-	D,L-Trp, Phe and Tyr	[92]
		HPLC	BSA	Average size: 13.3 nm	Ibuprofen and ofloxacin	[93]
	Titanium dioxide (TiO_2)	HPLC	Cellulose tris-(3,5-dimethyl-phenylcarbamate)-coated TiO_2/SiO_2 chiral stationary phase (CSP)	Size of TiO_2/SiO_2 spheres: ~6 nm; pore diameter: ~7 nm	Eight basic indole ring derivative enantiomers	[94]
		CE	Tris-H_3PO_4 solution containing TiO_2 NPs as background electrolytes (BGEs)	-	β-adrenergic drugs (atenolol, eliprolol, clorprenaline, fenoterol, metoprolol, propranolol, and terbutaline) and clenbuterol	[95]

Table 2. *Cont.*

	Zirconiun dioxide (ZrO_2)	Separation using a magnet	Cellulose tris-(3,5-dimethylphenylcarbamate)	Average size: 340 nm	Basic β-blocker (β-antagonists) chiral drugs	[96]
Carbon nanostructures	Single-walled nanotubes (SWCNTs), multi-walled nanotubes (MWCNTs)	Electrokinetic chromatography (EKC)	SWCNTs and MWCNTs	SWCNT: diameters between 0.7 and 1.2 nm and lengths 2–20 mm; MWCNT: diameters between 6 and 20 nm and 1–5 mm length	(±)-ephedrine, (±)-norephedrine and (±)-*N*-methylephedrine	[97]
	SWCNTs	Microchip electrophoresis	BSA conjugated with the shortened carboxylic SWCNTs	-	Trp	[98]
		HPLC	CNT monolithic column coated with a pyrenyl derivative	Average diameter: 1 nm; length: < 10 nm	A series of 10 amino acids	[99]
		HPLC	SWCNTs in monolithic backbones	Average diameter: ~1 nm; length 1–10 μm	α- and β-blockers, antiinflammatory drugs, antifungal drugs, dopamine antagonists, norepinephrine-dopamine reuptake inhibitors, catecholamines, sedative hypnotics, diuretics, antihistaminics, anticancer drugs, and antiarrhythmic drugs	[100]
	MWCNTs	CE	β-CD	Interlayerspacing of 3.4 Å; typical diameter of 10–20 nm	Clenbuterol	[101]
		TLC	Hydroxypropyl-β-CD	Diameter: 10–20 nm; length: 2–20 μm	Clenbuterol	[102]
	Ionic liquid dispersed MWCNTs	EC	Chondroitin sulfate E	MWCNT (od: 10–20 nm, length 5–30 nm)	Racemic drugs (amlodopine, laudanosine, nefopam, citalopram, and propranolol)	[103]
	Carboxylated SWCNTs and MWCNTs	EKC	β-CD	Carboxylated SWCNTs: od 1~2 nm; Carboxylated MWCNTs: od 10~20 nm	Sulconazole, ketoconazole, citalopram hydrobromide, and nefopam hydrochloride	[104]
	Graphene	HPLC	Graphene nanosheets with tetracyanoethyle oxide (TCNEO) and (*S*)-(+)-2-pyrrolidinemethanol	-	Ibuprofen and thalidomide racemic mixtures	[105]
		TLC	D-TA-graphene	Thickness of graphene nanosheet: 2–3 nm	Racemic drugs (propranolol and ofloxacin)	[106]

Table 2. *Cont.*

	Graphene oxide (GO)	CEC	Methyl-β-CD	-	Anionic racemic drugs (naproxen,warfarin and pranoprofen)	[107]
		CEC	β-CD conjugated GO-magnetic nanocomposites (GO/Fe_3O_4 NCs)	Average size of about 8 nm	D,L-Trp	[108]
		Open-tubular capillary electrochromatography (OTCEC)	Bovine serum albumin-conjugated graphene oxide–magnetic nanocomposites $GO/Fe_3O_4/BSA$	-	Trp, threonine (Thr), and propranolol enantiomers	[109]
		HPLC	Reduced graphene oxide/silica gel (rGO/SiO_2)	Silica gel (particle size of 5 μm, pore size of 120 Å)	Benzene enriched enantiomers, ibuprofen, *trans*-stilbene oxide, 2-phenylcyclohexanone, praziquantel, propranolol, *R,S*-equol, ketoconazole, benzoin, and quinidine	[110]
		HPLC	Graphene oxide/poly(*N*-isopropyl-acrylamide-co-glycidyl methacrylate) (MGO/PNG-CD)	Diameter: ~80 nm; thickness: 8nm	D,L-Trp	[111]
Other nanoparticles	Polystyrene nanoparticles	CE	Hydroxypropyl (HP)-β-CD	Average diameter: 15 ± 5 nm	Propranolol	[112]
		Chromatographic technique	Sulfated β-CD	Average size of ethylene dimethacrylate-*N*-methacryloyl-L-His methyl ester NP: ~111.5 nm	Ofloxacin	[113]
	Mesoporous silica nanoparticles	Direct chiral separation, CE	Teicoplanin-conjugated mesoporous silica MNPs	Average diameter: ~600 nm; mean pore size: ~3.9 nm	D,L-Trp, Phe, D,L-Mandelic acid, (±)-1-Phenyl-1,2-ethanediol, and *N*-Benzoyl-D,L-alanine	[67]
		CEC	Cellulose tris-(3,5-dimethylphenyl-carbamate)	Particle size of ca. 600 nm and a pore size of ca. 3 nm	Tetrahydropalmatine and pindolol	[114]
		CEC	Pepsin		(±)-nefopam	[115]
		CE	Carboxymethyl-β-CD	Approximately 120 nm	Ephedrine and chlorpheniramine	[116]
		CE	BSA	Approximately 150 nm	Propranolol and Trp	[117]
	Metal-organic framework	HPLC	Chiral bridging ligand	-	2-butanol and 2-methyl-1-butanol HPLC	[118]
		GC	Chiral bridging ligand	-	Amino acid derivative	[119]

Notably, NPs represent a state of matter in the transition region between solids and molecular structures in the size range of 1~100 nm [103], where charge carriers are confined in one, two or all of the three dimensions resulting in quantum well structures (nanosheets), quantum wire structures (nanowires) or quantum dot structures (nanoparticles), respectively. Due to quantum confinement and large surface-to-volume ratio ($s/v > 1$), such structures exhibit unique size-dependent optical, electronic, magnetic, and catalytic properties [120]. In addition, nanostructures possess better chemical stability, low cytotoxicity, significant mechanical strength, and are easily modifiable in terms of their size, shape and surface properties [91,103]. Importantly, NPs enhance the performance of enantioseparation process due to their extremely large s/v ratio (e.g., nanospheres of 10 nm diameter can have a surface area as large as 600 m^2/cm^3 [121]) that maximizes the amount of binding sites [69,81]. In the following sections we summarize some of the recent applications of different types of nanostructures for the separation of enantiomers.

3.1. Metal Nanoparticles

Recent years have witnessed tremendous applications of metal NPs in diverse fields of research including material science, biotechnology, biomedical engineering, targeted drug delivery, environmental, etc. [122–124]. In addition, the potential of metal NPs in separation science has been already well recognized due to their easy synthesis procedure, large s/v ratio, controllable particle size, narrow size distribution, extraordinary biocompatibility, molecular detection properties, etc. [24,89,125].

In a pioneering work in 2005, Choi et al. [87] employed sulfonated β-CD as a chiral selector and Ag colloids as an additive in CE for the enantiomeric separation of arylalcohols (1-phenyl-1-propanol, 1-phenyl-2-propanol, and 2-phenyl-1-propanol). The group observed that the addition of Ag colloid to the running buffer improves the resolution significantly. Another simple, reliable and highly efficient colorimetric platform for enantiomeric separation and detection was proposed by Zhang et al. [52] by using UTP nucleotide-capped AgNPs. The enantioselective aggregation of AgNPs allowed the group achieving rapid colorimetric separation of D- and L-Cys.

AuNPs have also been applied extensively for enantiomeric separation. In 2005, Shao et al. [126] synthesized human serum albumin (HSA) immobilized gold nanotube membranes for the separation of warfarin enantiomers. Li et al. [127] used bovine serum albumin-gold nanoparticles (BSA-AuNPs) conjugate as chiral stationary phases (CSPs) in open-tubular capillary electrochromatography (OTCEC) for the enantiomeric separation of ephedrine and norephedrine isomers. Good resolutions of 1.18 and 2.15 were obtained for norephedrine and ephedrine isomers, respectively, within 250 s at an effective separation channel length of 36 mm. In another work, Yang et al. [90] employed thiolated β-CD-modified AuNPs as chiral selector in pseudostationary phase-CEC to separate four pairs of dinitrophenyl-labeled amino acid enantiomers and three pairs of drug enantiomers with enantioseparation resolution up to 4.7. The repeatability of this method has been improvised in a later work by immobilizing the thiolated β-CD-modified AuNPs onto the inner wall of a capillary to serve as stationary phase for enantioselective OTCEC separation [128]. Interestingly, Shukla et al. [88] reported a unique strategy for enantioselective adsorption of propylene oxide (PO) by using either D- or L-Cys-modified AuNPs. In particular, the group demonstrated the ability L-Cys (D-Cys)-modified AuNPs to selectively adsorb the (*R*)-propylene oxide ((*S*)-propylene oxide). Notably, this work was extended by using D- or L-Cys to functionalize tetrahexahedral (THH, 24-sided) AuNPs to separate a real chiral pharmaceutical-propranolol [89]. Recently, nucleic acid aptamer-functionalized AuNPs were used as chiral selector for the separation of racemic D,L-Trp. The method used centrifugation to separate the precipitate formed by the aptamer-specific enantiomer (L-Trp) bounded AuNPs (Figure 7) [69]. In another work, Su et al. [53] demonstrated the potential of laboratory synthesized chiral *N*-acetyl-L-Cys-capped AuNPs for high throughput enantiomeric separation of amino acid enantiomers via centrifugation. Recently, monolayer and multilayer AuNPs film capillary columns were fabricated through layer-by-layer self-assembly of AuNPs and their subsequent functionalization through self-assembly of thiolated β-CD in OTCEC [129]. It was observed that, as compared to monolayer AuNPs film capillary column, the three layer AuNPs film capillary column possess superior enantioseparation performance. Further, in 2018, Liu et al. [86] successfully added streptomycin-modified AuNPs in background electrolyte (BGE) solution in CE for the first time to separate a number of drug racemates, such as, adrenaline hydrochloride, noradrenaline bitartrate, and isoprenaline hydrochloride.

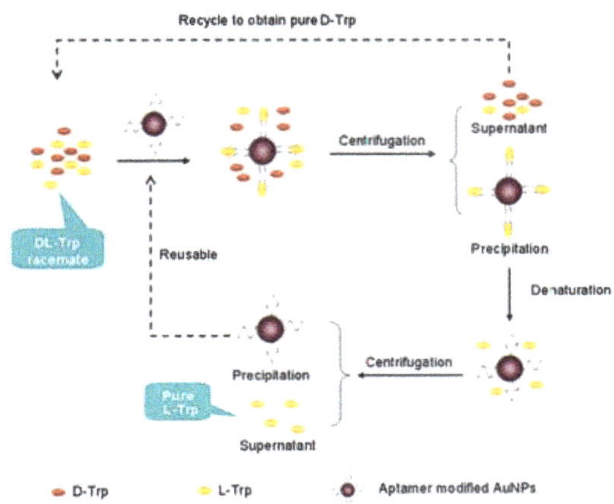

Figure 7. The schematic of the enantioseparation of D,L-Trp based on functional nucleic acids modified Au nanoparticles. The figure and the caption have been adapted with permission from [69], Copyright © 2013 Wiley.

3.2. Metal Oxide Nanoparticles

In recent years, metal oxide nanoparticles such as magnetite (Fe_3O_4), titania (TiO_2), Zirconia (ZrO_2), etc., have become crucial in enantiomeric separations, catalysis, sensing devices, cell labeling, drug delivery, and biomedical applications [101,130–132]. Notably, nanoscale Fe_3O_4 is one of the important phases of iron oxide and has been extensively used in the field of enantiomeric separation due to the possibility of their easy manipulation even at a lower magnetic field [85,133]. In a seminal work, Choi et al. [91] prepared magnetic silica nanoparticles, MSNPs (spherical SiO_2 NPs containing Fe_3O_4) of average size 300 nm in order to separate enantiomers of *N*-(3,5-dinitrobenzoyl)-α-amino acid *N*-propylamides. Interestingly, the group named the enantioseparation process as "enantioselective fishing" since the MSNPs tagged with an appropriate chiral selector readily formed complexes with one of the enantiomers which was subsequently separated simply by using a magnet. Enantiomeric separation of amino acids were also performed by employing water-soluble β-CD-modified Fe_3O_4 NPs fabricated by using a simple and convenient chemical route [134]. The group cleverly utilized a strategy to chirally select amino acids by β-CD and subsequently use Fe_3O_4 NPs as magnetic separators. In another work, enantiomeric separation of aromatic amino acids (D,L-Trp, Phe and Tyr) has also been achieved by using carboxymethyl-β-cyclodextrin (CM-β-CD)-functionalized MSNPs [92]. In the meantime, Arslan et al. [132] developed a novel enantioselective sorption approach based on Fe_3O_4 NPs functionalized with three different β-CDs. The group established that the novel CD-grafted Fe_3O_4 NPs significantly enhance their enantioseparation capabilities towards a set of chiral carboxylic acid molecules. Notably, Sayin et al. [135] grafted *p-tert*-butylcalix[8]arene derivative (C[8]-C4-COOH) onto Fe_3O_4 NPs and used both C[8]-C4-COOH and Fe_3O_4 NPs for the encapsulation of lipase in order to achieve enhanced catalysis and enantioselective resolution of racemic naproxen methyl ester. Chiral MNPs based on BSA and Fe_3O_4 NPs have been used recently to separate chiral drugs (ibuprofen and ofloxacin) [93] and amino acids (Figure 8) [66]. Most recently, chiral core/shell structured

MNPs were fabricated by using poly(*N*-isopropylacrylamide-co-glycidyl methacrylate)-β-cyclodextrin (PNG-CD) to form smart polymer brushes (shell) onto polydopamine coated Fe_3O_4 MNPs (core) [136]. The smart NPs (Fe_3O_4@PDA@PNG-CD), fabricated in this work, showed excellent magnetic separability where the β-CD units acted efficiently as the chiral selector to selectively recognize and bind the desired enantiomer forming host–guest inclusion complexes.

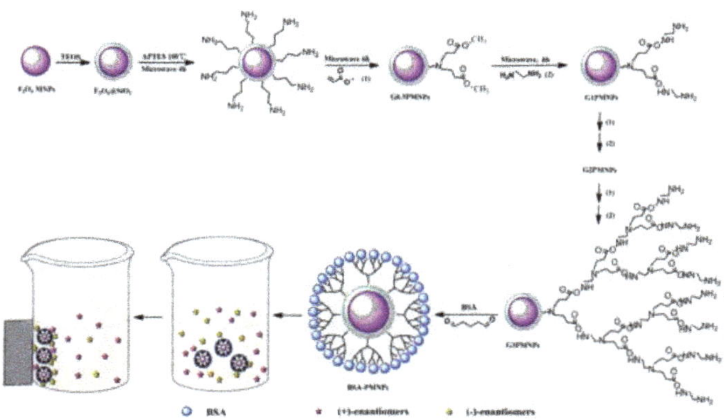

Figure 8. Preparation of BSA–PMNPs and the direct separation of the racemates. The figure and the caption have been adapted with permission from [66], Copyright © 2013 The Royal Society of Chemistry.

On the other hand, TiO_2 and ZrO_2, have also been well recognized as a stationary phase due to their favorable physiochemical properties [137–140]. The applicability of TiO_2 NPs in enantiomeric separation has been demonstrated by first coating it onto the micron sized silica spheres [94]. The TiO_2/SiO_2 particles were further coated with cellulose tris-(3,5-dimethylphenylcarbamate) (CDMPC) to prepare the CSP for the separation of eight indole ring derivative enantiomers. Notably, the use of TiO_2 NPs as additive in CE for the simultaneous separation of eight β-adrenergic drugs has also been reported [96]. In another work, a ZrO_2 based HPLC packing material ZrO_2/SiO_2 was prepared by multilayer coating of ZrO_2 NPs on the surfaces of silica spheres using layer-by-layer self-assembly technique [141]. The group analyzed the potential of ZrO_2/SiO_2 material for enantioseparation applications. Moreover, in another work, Kumar et al. [96] reported the fabrication of Fe_3O_4@ZrO_2 microspheres (Fe_3O_4 magnetic core covered by a ZrO_2 shell) of average size 340 nm, functionalized with CDMPC, for the separation of racemic chiral drugs. Interestingly, the excellent recyclability of the synthesized chiral ZrO_2 magnetic microspheres is quite promising for their use in the multiple enantiomeric separations.

3.3. Carbon-Based Nanomaterials

Importantly, among different varieties nano-engineered materials used for enantiomeric separation, carbon nanostructures have also attracted much attention due to their high mechanical strength, good chemical stability, high elasticity, and high thermal conductivity [142,143]. So far, a number of allotropic carbonaceous nanomaterials, such as, carbon nanotubes (CNTs), graphene and graphene oxides (GO) have been successfully applied in this area.

3.3.1. Carbon Nanotubes

Notably, CNTs can be considered as graphite sheets (sp^2 hybridised carbon atoms) wrapped up in the form of a cylinder that are usually capped by a fullerene-like structure with diameters ranging from few Å [144–146] to tens of a nanometer [102,147] and length up to a few micrometer (may extend to centimetres in certain occasions) [103,148,149]. CNTs were first discovered by Iijima [150] and have been one of the most researched materials of 21st century [151]. CNTs can be broadly classified into two types: multi-walled carbon nanotubes (MWCNTs) comprised of more than one coaxially rolled up graphite sheets and single-walled carbon nanotubes (SWCNTs). Due to their unique physicochemical properties and large chemically active surface area, which have been thoroughly discussed in several review articles [152–155], both MWCNTS and SWCNTs are emerging as one of the most promising candidates for numerous potential applications [156–158]. In the context of enantiomeric separation, CNTs are capable of improving the speed, selectivity, stability, and efficiency in chiral chromatographic separation [24,159]. In an exciting work, Ahmed et al. [100] demonstrated the creation of a CSP with good enantioselectivity by encapsulating small amount of SWCNTs into polymer monolithic backbones. Using enantioselective nano-HPLC separation, the group successfully separated twelve classes of pharmaceutical racemates. Under optimum concentration conditions, the group achieved satisfactory repeatability and observed that (6,5) SWCNTs displayed higher enantioselectivity as compared to the (7,6) SWCNTs. Notwithstanding the enormous potentials of CNTs to be used as sorptive material for enantiomeric separation, there are very few reports where such non-functionalized CNTs have been used for the separation of chiral compounds through affinity chromatography. This might be due to the reason that CNTs, especially chiral CNTs, alone may not be an effective adsorbent for enantiomeric separation [160,161].

Nevertheless under strong and specific chemical conditions the surfaces CNTs can be modified/functionalized to enhance their solubility and other intrinsic properties which render them CSPs or pseudostationary phases [24,97]. There are a number of efforts in the recent past where enantiomeric separation was successfully performed by the modification of CNTs with chiral selectors such as β-CD. For example, Na et al. [101] demonstrated a new strategy for enantiomeric separation of clenbuterol by CE using modified CNT as chiral selector in 2004. The group reported the formation of a large surface area platform by the NPs modified with β-CD that served as a pseudostationary chiral phase for the enhanced separation of enantiomers. In a similar work to separate the same kind of enantiomers, Yu et al. [102] reported a method to functionalize MWCNTs with hydroxypropyl-β-cyclodextrin (HP-β-CD) in order to use as a stationary phase additive in TLC, as shown in Figure 9. Recently, carbogenic NPs, carboxylated SWCNTs and carboxylated MWCNTs were used to modify enantioseparation systems to form pseudostationary phases with dextrin as chiral selectors in electrokinetic chromatography (EKC) [104]. Particularly the group established that the introduction of NPs into the buffer significantly could enhance the resolution of several drug enantiomers namely sulconazole, ketoconazole, citalopram hydrobromide, and nefopam hydrochloride. Under optimized dextrin concentration, buffer pH and buffer concentration resolution time of 15 min with resolution values in the range 1.41–4.52 were achieved in this work. In another work, a novel chiral selector was developed by conjugating BSA which is a type of protein with carboxylic SWCNTs [98]. The efficient enantioseparation ability of the BSA-SWCNT conjugate was illustrated by the successful separation of Trp enantiomers within 70 s with a resolution factor of 1.35 and separation length of 32 mm obtained by using poly(methyl methacrylate) based microchip electrophoresis. The potential of concentrated surfactant-coated MWCNTs as an alternative to other chiral selectors like CDs were also successfully demonstrated by using partial filling of the capillary in EKC at different experimental conditions such as pH, addition of organic modifiers, potential and injection time [97].

Recently, Zhao et al. [162] allowed SWCNTs to bind with the inner wall of a capillary column in GC in order to enhance the enantiomeric separation by achieving larger surface for the chiral ionic liquid stationary phase, which in turn increased the interaction between the stationary phases and the analytes. They prepared two capillary columns for the chromatography process: column A containing only the chiral ionic liquid and column B containing a mixture of the chiral ionic liquid and SWCNTs. Interestingly out of 12 model racemates, column B successfully separated eight chiral compounds as compared to the column A which separated only four of such compounds. In another work, Guillaume et al. [99] successfully developed a simple approach by using HPLC for the enantiomeric separation of amino acids by functionalizing SWCNTs with pyrenyl derivative of a chiral aminoglycoside called neomycin A (PNA). Magnetization of CNTs also showed promising results in the separation of enantiomers. Tarigh et al. [163] used the chiral selector L-threonine (Thr), anchored to the surface of magnetic MWCNTs, to successfully separate D,L-mandelic acid within a time period of 10 min.

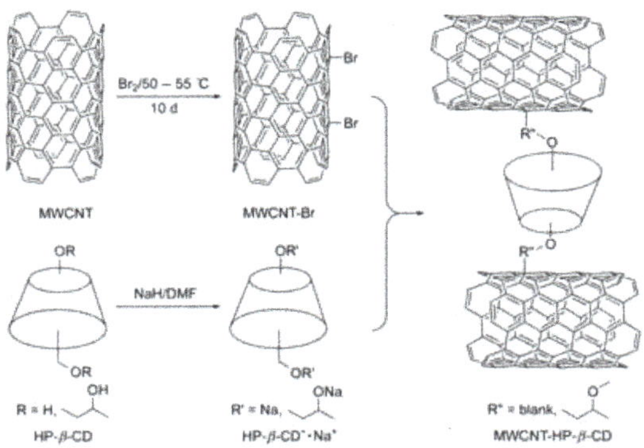

Figure 9. Schematic presentation of preparation of HP-β-CD modified MWCNTs. The figure and the caption have been adapted with permission from [102], Copyright © 2011 Wiley.

3.3.2. Graphene and Graphene Oxide Nanomaterials

Graphene is, on the other hand, a two-dimensional (2D) sheet of sp^2 hybridized carbon atoms with single atom thickness. In principle, graphene can be regarded as the basic building block of fullerenes (0D structure), carbon nanotubes (1D structure) and graphite (3D structure) [164,165]. It is the thinnest yet the strongest known material which exhibit exceptional optical, electronic, thermal, and adsorptive properties [166,167]. The profound impact of graphene in the field of material science has been well recognized by the Nobel Prize in Physics for 2010 awarded jointly to Geim and Novoselov of The University of Manchester, UK for their groundbreaking experiments related to graphene [168]. On the other hand, oxidization of graphite leads to the formation of graphene oxide (GO) decorated with various oxygen containing functionalities such as epoxide, carbonyl, carboxyl, and hydroxyl groups [169]. GO is attracting enormous attention due to its easy synthesis procedure, high yield and satisfactory dispersibility in organic solvents [170].

Notably, graphene and GO exhibit considerable potential in separation science [165] and has been used by several research groups for the enantiomeric separation of chiral molecules. For instance, Tu et al. [106] reported the graphene assisted resolution of two racemic drugs propranolol and ofloxacin using pure D-(−)-TA as the chiral selector in TLC. Interestingly, computational simulations using density functional theory also showed the applicability of nanoporous graphene, when functionalized by a chiral bouncer molecule [171]. In another exciting work, Candelaria et al. [105] demonstrated the fabrication of new type of chemically stable, versatile and cost-effective graphene-based CSPs for enantiomeric separation using LC. The group employed functionalized mesoporous 3D graphene nanosheets for the enantiomeric separation of pharmaceutical grade racemic mixtures of model ibuprofen and thalidomide.

In case of GO, Liang and his coworkers, in a series of publications, employed OTCEC for the separation of different types of enantiomers by using β-CD conjugated GO-magnetic nanocomposites (GO/Fe_3O_4 NCs) [108], BSA-conjugated GO-magnetic nanocomposites (GO/Fe_3O_4) [172] and BSA conjugated polydopamine-GO nanocomposites (PDA/GO/BSA) [109] as the stationary phases. Exploiting the large surface area, excellent biocompatibility of graphene and rough surface morphology of GO, the group used the OTCEC microdevices to successfully separate Trp, Thr and propranolol enantiomers with reasonably good resolution factors. Notably, as compared to other similar techniques, the group reported better resolution factor for the enantiomeric separation of D,L-Trp by using the novel PDA/GO/BSA stationary phase. Li et al. [170] reported a facile and efficient strategy to chirally functionalize GO with optically active helical polyacetylene chains. The group successfully used the GO hybrid as a chiral inducer for the enantioselective crystallization of alanine enantiomers where L-alanine was induced to crystallize in the form of rodlike crystals. There are also examples where GO coated fused silica capillaries were applied for the enantiomeric separation of the ephedrine–pseudoephedrine (E-PE) and -methylphenethylamine (-Me-PEA) isomers by using CE method [173]. On the other hand, GO-polymer coated fused-silica capillary columns were used to improve the enantiomeric separation of three anionic racemic drugs (naproxen, warfarin and pranoprofen) by using CEC with methyl-β-cyclodextrin (Me-β-CD) as chiral selector [107]. Using molecular modeling with AutoDock, the group studied the mechanism of GO-modification effect on the enantioseparation efficiency of the CEC system. In another work, Hong et al. [174] reported for the first time the development of GO-based affinity capillary silica monolith with HSA or pepsin as chiral selector for enantiomeric separation and proteolysis by using CEC. It was observed that as compared to affinity monoliths without GO, HSA-modified affinity capillary silica monoliths (HSA-GO-EDA@CSM) possess better enantiorecognition ability. Recently, Li et al. [110] reported better HPLC enantiomeric separation of benzene-enriched enantiomers by using a novel CSP based on GO. In this method, GO was first covalently coupled to silica gel microspheres which was subsequently reduced with hydrazine to form reduced graphene oxide@silica gel (rGO@SiO_2). Finally, the surfaces of rGO@SiO_2 were physically coated with cellulose tris-(3,5-dimethylphenylcarbamate) to prepare the CSP. Most recently, a smart multifunctional graphene oxide nanocomposite was prepared which showed exceptionally good selectivity, thermosensitivity and magnetic separability for the identification and enantiomeric separation of Trp enantiomers [111]. The synthesized nanocomposite was composed of GO nanosheets, immobilized superparamagnetic Fe_3O_4 nanoparticles, poly(*N*-isopropylacrylamide-co-glycidyl methacrylate) (PNG) chains and β-CD which was fabricated through a combination of surface-initiated atom transfer radical polymerization (SI-ATRP) and a ring-opening reaction, as shown in Figure 10.

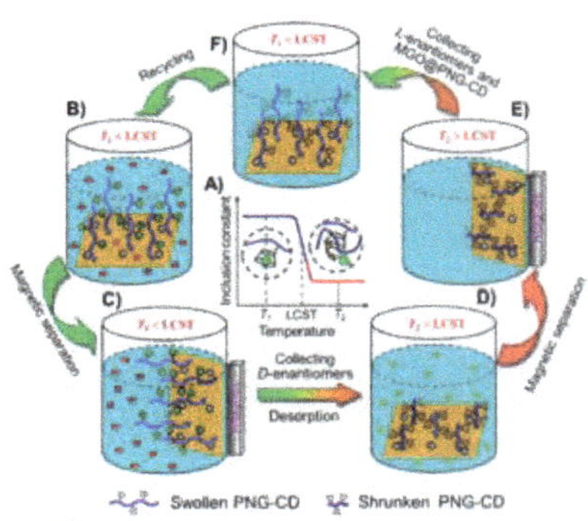

Figure 10. Schematic illustration of the thermosensitive chiral recognition and enantioseparation of AAs enantiomers by MGO@PNG-CD. (A) As the temperature is below the LCST of PNIPAM chains (T1), the β-CD units in PNG-CD can selectively recognize and accommodate L-enantiomers to form stable host-guest inclusion complexes, while as the temperature is above the LCST of PNIPAM chains (T2), the loaded L-enantiomers can desorb from the β-CD cavities automatically because of the reduction in the inclusion constants of β-CD/L-enantiomers complexes; (B) The MGO@PNG-CD is added in the AAs enantiomeric solution at T1 and the PNIPAM chains are swollen, and then L-enantiomers are recognized by β-CD; (C) Through an external magnetic field L-enantiomers are loaded on the MGO@PNG-CD, and D-enantiomers are remained in the enantiomeric solution for subsequent separation; (D) while the operating temperature is above the LCST of PNIPAM (T2), the PNIPAM chains occur to collapse, and the loaded L-enantiomers are released, and the MGO@PNG-CD is recycled; (E and F) the regenerated MGO@PNG-CD is recovered using a magnet and reused. The figure and the caption have been adapted with permission from [111], Copyright © 2018 American Chemical Society.

4. Conclusions

Enantiomeric recognition and separation is one of the most challenging tasks, particularly in the fields of contemporary pharmaceutical science, agrochemical science, material science, and many other rapidly expanding areas of research. Notably, the advent of modern high-throughput experimentation (HTE) technology has enabled the research and industrial laboratories to produce a large number of samples within a very short period of time. In comparison, the analytical methods (e.g., HPLC, GC, CE, etc.) which are generally employed for screening the reaction yield, enantiomeric purity, stability to racemization, enantiomeric excess (ee), and concentration of chiral compounds are relatively expensive, time consuming, laborious and produces solvent waste [175–177]. In this context, chiral sensing methods, capable of performing real time analysis of mixtures of enantiomers by utilizing inexpensive instrumentation with almost zero waste of expensive reagents, have significant scientific and industrial relevance and, therefore, are gaining increasing attention [177]. To date, a number of methods including electrochemical sensors, gravimetric-mass sensors-resonators, electrical sensors and chiroptical/spectroscopic sensors have been developed for fast and accurate differentiation of enantiomer [178]. Table 3 summarizes some of the recently used chiral sensing methods. Importantly, there are also examples where NPs have been used for the purpose of chiral sensing. For example, Tsourkas et al. [179] reported a enantioselective immunosensor by utilizing dextran-coated MNPs as magnetic relaxation switch that decreased the T2 relaxation time of water by 100 ms. The group successfully detected 0.1 μM of D-Phe in the presence of 10 mM of L-Phe (99.998% ee) by using NMR measurement of the T2 parameter. In another work, Wang et al. [58] reported a visual sensing platform based on AuNPs for visual recognition of Gln enantiomers. Silver nanoparticles (~400 nm) have been used to enhance the circular dichroism by two orders of magnitude [180]. Moreover, enantioselective sensors based on microcantilevers (MCs) with nanostructured (roughened) gold surfaces on one side showed promising results for the label-free stereoselective detection of α-amino acids [181]. Further, use of molecular imprinted polymer (MIP) nanowires/nanofibers showed immense potential to enhance the sensitivity of chiral sensors by allowing more binding sites (or cavities) due to the high surface area of such nanostructures [182,183]. Thus, integration of nanoscience and nanotechnology has a lot of potential to enhance the molecular sensing and signal transduction, which are the basic processes of chiral sensing [184]. A more detailed analysis of the chiral sensing methods can be found in several recent reviews [177,184,185] and are not discussed here.

In this review, we mainly focus on the application of nanostructured materials to amplify chiral recognition and separation. Notably, a wide array of nanostructures get benefited from the rapid development of nanotechnology are synthesized for the characterization and purification of racemates. The surfaces of NPs coated with chiral ligands (chiral NPs) have proved the potentialities for enantiomeric recognition based on the specific interactions between analytes and target enantiomers. Chiral NPs have paved a new route for enantiomeric recognition and surpassed the current cumbersome methods in this field, offering a relatively simple examination platform as discussed in Section 2. Notably, chiral NPs accommodate various nanocomposites to manipulate the molecular adsorption and aggregation, colloidal assembly, and to characterize chemical dynamics at particle surfaces as a useful modality for the understanding of enantioselective mechanism and future pharmaceutical applications. These surface effects also raise questions concerning the means by which biologically active compounds interact with chiral molecules, particularly with regard to enantioselective mechanisms at the nanoscale.

Table 3. Representative list of chiral sensing methods. The table derives from Manoli et al. [177].

	Sensors	Sub-Categories/Types	Advantages	Disadvantages	Related References
(a)	Electrochemical Sensors	Potentiometric sensors, voltammetric sensors	High sensitivity, simple operation, rapid detection, low cost, miniature size, low power requirements [186]	Poor durability, need for a reference electrode [177]	[187–190]
(b)	Gravimetric-Mass Sensors	Quartz crystal microbalance (QCM) devices based on: CDs, molecular imprinted polymers, biological recognition elements, etc.	Capability to measure sub-nanogram level changes, possibility of real-time condensed phase measurements, long time stability	Resolution degradation due to multi resonance modes of the cantilever, limited performance due to degraded quality factor and resolution in liquid medium [191], electrochemical QCM can only be used for studying electroplated, evaporated, or sputtered materials [185]	[192–194]
(c)	Electrical Sensors	Chemiresistors, organic field effect transistors, chemocapacitors	Ease of fabrication and simplicity in instrumentation, cost effective, large selection of materials and flexible	Low Thermal stability and low chemical stability (oxidation)	[195–198]
(d)	Optical sensors	SPR sensors, fluorescence spectroscopy, circular dichroism/optical rotation probes	Speed of detection, simplicity in the measurement procedure	Low sensitivity and poor tolerance to impurities [58,199]	[176,180,200–203]

On the other hand, separation of racemic compounds is accomplished by selecting the desired enantiomer by introducing chiral selectors of compatible size and structure to the separation system as a part of the CSP or as chiral additive in the mobile phases (CAMP) [9,204]. Notably, there still remain three major issues on the use of chiral selectors for the detection and/or separation of enantiomers. These are: (a) the broad spectrum of applications, (b) the cost of the selector and (c) the productivity of the separation (gram of pure enantiomers/kg of CSP/day). In this regard, it is noteworthy to mention that α-, β- and γ-CDs (oligosaccharides) and their derivatives seem to represent a kind of universal chiral selector not only due to their broad-spectrum of chiral selectivity and relatively low cost but also their applicability in most of the separation techniques, such as, TLC, CEC, HPLC, CE, crystallization, etc. In addition, modified polysaccharides, proline derivatives, Pirkle type chiral selectors are also finding widespread applications in both analytical and preparative scale separations [9,205,206]. Recently, the most important and widely used CSPs for enantio-recognition/separation mechanisms were discussed in several reviews [9,21,207–211]. Importantly, economics of the separation process including the selection of the CSP/CAMP and the separation technique depends mainly on the stage of production. For example, in early stage small scale laboratory productions, finding an optimum separation method which is capable of producing large amount of desired enantiomers within a suitable time period becomes more important than the cost of the process. For preparative scale productions, however, both cost of the process as well as its productivity plays an important role [212,213]. Importantly, during the transition from analytical to preparative scale productions, conventional separation techniques (e.g., batch chromatography) are often replaced by high throughput large scale production techniques like stimulated moving bed (SMB) technology [214]. Further, in preparative scale separations the cost of production may significantly vary depending on the type of CSP/CAMP used and the technique in which the separation process is conducted. Thus the primary challenge is to find an optimum (cost effective and most productive) separation process, especially by using all possible combinations of more efficient CSPs/CAMPs [213,215]. In this regard, the use of nanoparticles and nanocomposites in separation processes has the potential to circumvent such challenges. In Section 3, we described the state of the art in the application of nanostructured materials in the field of enantiomeric separation. Arguably the nanostructured materials will continue to play a vital role in the separation of chiral molecules, especially due to their extraordinary capacity to enhance the enantioseparation ability of conventional techniques.

To summarize, we note that enantioseparation science has completed a phenomenal journey of around 170 years since its first demonstration in 1848 by Louis Pasteur [216]. At the present time, combination of conventional enantioseparation techniques with the recent advances in nanoscience and nanotechnology is proving to be quite synergetic by producing exciting results. Despite all these developments, the major challenge that still remains is the development of a more flexible, efficient and cost-effective chiral resolution technique and/or chiral selector to efficiently separate the numerous newer and newer chiral compounds, especially chiral drugs that are continuously being introduced. Nevertheless, without forgetting the enormous research efforts being devoted in this area during the last few years, we anticipate that the near future shall witness smarter techniques with more universal abilities for enantiomeric sensing, recognition, and separation.

Author Contributions: A.G., N.M., S.K., H.R., N.-T.C., and G.-Y.Z. wrote and editing the manuscript, G.-Y.Z. conceived the work and organized the manuscript.

Funding: This work was funded by Ministry of Science and Technology, Taiwan, 107-2112-M-039-001.

Acknowledgments: We thank Arbaaz Sait and Shih-Ting Lin for the help on editing the manuscript.

Conflicts of Interest: The authors declare no conflict of interest.

References

1. Lough, W.J.; Wainer, I.W. *Chirality in Natural and Applied Science*, 1st ed.; Blackwell Science: Hoboken, NJ, USA, 2002.
2. Barron, L.D. Chirality and Life. *Space. Sci. Rev.* **2008**, *135*, 187–201. [CrossRef]
3. Sekhon, B.S. Enantioseparation of chiral drugs—An overview. *Int. J. PharmTech. Res.* **2010**, *2*, 1584–1594.
4. Nguyen, L.A.; He, H.; Pham-Huy, C. Chiral drugs: An overview. *Int. J. Biomed. Sci.* **2006**, *2*, 85. [PubMed]
5. Rukhlenko, I.D.; Tepliakov, N.V.; Baimuratov, A.S.; Andronaki, S.A.; Gun'ko, Y.K.; Baranov, A.V.; Fedorov, A.V. Completely chiral optical force for enantioseparation. *Sci. Rep.* **2016**, *6*, 36884. [CrossRef] [PubMed]
6. Liu, W.; Tang, M. Enantioselective activity and toxicity of chiral herbicides. In *Herbicides-Mechanisms and Mode of Action*; IntechOpen: London, UK, 2011; pp. 63–80.
7. Tombo, G.M.R.; Belluš, D. Chirality and crop protection. *Angew. Chem.* **1991**, *30*, 1193–1215. [CrossRef]
8. Wang, M.; Zhang, Q.; Cong, L.; Yin, W.; Wang, M. Enantioselective degradation of metalaxyl in cucumber, cabbage, spinach and pakchoi. *Chemosphere* **2014**, *95*, 241–246. [CrossRef]
9. Maier, N.M.; Franco, P.; Lindner, W. Separation of enantiomers: Needs, challenges, perspectives. *J. Chromatogr. A* **2001**, *906*, 3–33. [CrossRef]
10. Ye, J.; Zhao, M.; Niu, L.; Liu, W. Enantioselective environmental toxicology of chiral pesticides. *Chem. Res. Toxicol.* **2015**, *28*, 325–338. [CrossRef]
11. Ganapathy, S. *Chiral Separation Techniques: A Practical Approach*; Wiley: Hoboken, NJ, USA, 2008.
12. Schuur, B.; Verkuijl, B.J.; Minnaard, A.J.; de Vries, J.G.; Heeres, H.J.; Feringa, B.L. Chiral separation by enantioselective liquid-liquid extraction. *Org. Biomol. Chem.* **2011**, *9*, 36–51. [CrossRef]
13. Lorenz, H.; Seidel-Morgenstern, A. Processes to separate enantiomers. *Angew. Chem. Int. Ed. Engl.* **2014**, *53*, 1218–1250. [CrossRef]
14. Lahlou, M. The success of natural products in drug discovery. *Pharmacol. Pharm.* **2013**, *4*, 17–31. [CrossRef]
15. Das, P.; Srivastav, A.K. Phytochemical extraction and characterization of the leaves of Andrographispaniculata for its anti-bacterial, anti-oxidant, anti-pyretic and anti-diabetic activity. *Int. J. Innov. Res. Sci. Eng. Technol.* **2014**, *3*, 15176–15184. [CrossRef]

16. Cascaval, D.; Oniscu, C.; Galaction, A.I. Selective separation of amino acids by reactive extraction. *Biochem. Eng. J.* **2001**, *7*, 171–176. [CrossRef]
17. Shin, J.S.; Kim, B.G. Asymmetric synthesis of chiral amines with ω-transaminase. *Biotechnol. Bioeng.* **1999**, *65*, 206–211. [CrossRef]
18. Noyori, R. Asymmetric catalysis: Science and opportunities (Nobel lecture). *Angew. Chem. Int. Ed. Engl.* **2002**, *41*, 2008–2022. [CrossRef]
19. Paik, P.; Gedanken, A.; Mastai, Y. Enantioselective separation using chiral mesoporous spherical silica prepared by templating of chiral block copolymers. *ACS Appl. Mater. Interfaces* **2009**, *1*, 1834–1842. [CrossRef]
20. Francotte, E.R. Enantioselective chromatography as a powerful alternative for the preparation of drug enantiomers. *J. Chromatogr. A* **2001**, *906*, 379–397. [CrossRef]
21. Berthod, A. Chiral recognition mechanisms in enantiomers separations: A general view. In *Chiral Recognition in Separation Methods*, 1st ed.; Springer: Berlin, Germany, 2010; pp. 1–32.
22. Dalgliesh, C.E. 756. The optical resolution of aromatic amino-acids on paper chromatograms. *J. Chem. Soc. (Resumed)* **1952**, *137*, 3940–3942. [CrossRef]
23. Mane, S. Racemic drug resolution: A comprehensive guide. *Anal. Methods* **2016**, *8*, 7567–7586. [CrossRef]
24. Chang, C.; Wang, X.; Bai, Y.; Liu, H. Applications of nanomaterials in enantioseparation and related techniques. *Trends. Analyt. Chem.* **2012**, *39*, 195–206. [CrossRef]
25. Izake, E.L. Chiral discrimination and enantioselective analysis of drugs: An overview. *J. Pharm. Sci.* **2007**, *96*, 1659–1676. [CrossRef] [PubMed]
26. Chankvetadze, B. Recent developments on polysaccharide-based chiral stationary phases for liquid-phase separation of enantiomers. *J. Chromatogr. A* **2012**, *1269*, 26–51. [CrossRef]
27. Ali, I.; Aboul-Enein, H.Y. *Chiral Separations by Liquid Chromatography and Related Technologies*, 1st ed.; CRC Press: Boca Raton, FL, USA, 2003.
28. Scriba, G.K.E. *Chiral Separations: Methods and Protocols*, 2nd ed.; Humana Press: New York, NY, USA, 2016.
29. Gübitz, G.; Schmid, M.G. Chiral separation by chromatographic and electromigration techniques. A Review. *Biopharm. Drug. Dispos.* **2001**, *22*, 291–336. [CrossRef]
30. Ward, T.J.; Ward, K.D. Chiral Separations: A Review of Current Topics and Trends. *Anal. Chem.* **2012**, *84*, 626–635. [CrossRef]
31. Stalcup, A.M. Chiral Separations. *Annu. Rev. Anal. Chem.* **2010**, *3*, 341–363. [CrossRef] [PubMed]
32. Drašar, P.; Moravcova, J. Recent advances in analysis of Chinese medical plants and traditional medicines. *J. Chromatogr. B* **2004**, *812*, 3–21. [CrossRef] [PubMed]
33. Okamoto, Y.; Kawashima, M.; Hatada, K. Chromatographic resolution: XI. Controlled chiral recognition of cellulose triphenylcarbamate derivatives supported on silica gel. *J. Chromatogr. A* **1986**, *363*, 173–186. [CrossRef]
34. Ikai, T.; Yamamoto, C.; Kamigaito, M.; Okamoto, Y. Immobilized Polysaccharide-Based Chiral Stationary Phases for HPLC. *Polym. J.* **2006**, *38*, 91–108. [CrossRef]
35. Francotte, E.R. Enantioselective Chromatography: An Essential and Versatile Tool for the Analytical and Preparative Separation of Enantiomers. *CHIMIA* **1997**, *51*, 717–725.
36. Francotte, E.R. Polysaccharide Derivatives as Unique Chiral Selectors for Enantioselective Chromatography. *CHIMIA* **2017**, *71*, 430–450. [CrossRef]
37. Gübitz, G.; Schmid, M.G. Recent advances in chiral separation principles in capillary electrophoresis and capillary electrochromatography. *Electrophoresis* **2004**, *25*, 3981–3996. [CrossRef] [PubMed]
38. Wang, Y.; Chen, A. Crystallization-based separation of enantiomers. In *Stereoselective Synthesis of Drugs and Natural Products*, 1st ed.; Wiley: Hoboken, NJ, USA, 2013; pp. 1–20.
39. Lorenz, H.; Polenske, D.; Seidel-Morgenstern, A. Application of preferential crystallization to resolve racemic compounds in a hybrid process. *Chirality* **2006**, *18*, 828–840. [CrossRef] [PubMed]
40. Holbach, A.; Godde, J.; Mahendrarajah, R.; Kockmann, N. Enantioseparation of chiral aromatic acids in process intensified liquid–liquid extraction columns. *AIChE J.* **2015**, *61*, 266–276. [CrossRef]
41. Wang, Y.; Hu, Y.; Xu, J.; Luo, G.; Dai, Y. Immobilization of lipase with a special microstructure in composite hydrophilic CA/hydrophobic PTFE membrane for the chiral separation of racemic ibuprofen. *J. Memb. Sci.* **2007**, *293*, 133–141. [CrossRef]

42. Flores-López, L.Z.; Caloca, J.; Rogel-Hernández, E.; Espinoza-Gomez, H. Development of an enantioselective membrane from cellulose acetate propionate/cellulose acetate, for the separation of trans-stilbene oxide. *Cellulose* **2014**, *21*, 1987–1995. [CrossRef]
43. Higuchi, A.; Higuchi, Y.; Furuta, K.; Yoon, B.O.; Hara, M.; Maniwa, S.; Saitoh, M.; Sanui, K. Chiral separation of phenylalanine by ultrafiltration through immobilized DNA membranes. *J. Memb. Sci.* **2003**, *221*, 207–218. [CrossRef]
44. Han, J.; Kitagawa, O.; Wzorek, A.; Klika, K.D.; Soloshonok, V.A. The self-disproportionation of enantiomers (SDE): A menace or an opportunity? *Chem. Sci.* **2018**, *9*, 1718–1739. [CrossRef]
45. Jianlin, H.; Donna, J.N.; Alexander, E.S.; Vadim, A.S. Self-Disproportionation of Enantiomers via Sublimation; New and Truly Green Dimension in Optical Purification. *Curr. Org. Synth.* **2011**, *8*, 310–317.
46. Soloshonok, V.A.; Roussel, C.; Kitagawa, O.; Sorochinsky, A.E. Self-disproportionation of enantiomers via achiral chromatography: A warning and an extra dimension in optical purifications. *Chem. Soc. Rev.* **2012**, *41*, 4180–4188. [CrossRef]
47. Sorochinsky, A.E.; Aceña, J.L.; Soloshonok, V.A. Self-Disproportionation of Enantiomers of Chiral, Non-Racemic Fluoroorganic Compounds: Role of Fluorine as Enabling Element. *Synthesis* **2013**, *45*, 141–152. [CrossRef]
48. Han, J.; Wzorek, A.; Soloshonok, V.A.; Klika, K.D. The self-disproportionation of enantiomers (SDE): The effect of scaling down, potential problems versus prospective applications, possible new occurrences, and unrealized opportunities? *Electrophoresis* **2019**. [CrossRef]
49. Armstrong, D.W.; Chang, C.D.; Li, W.Y. Relevance of enantiomeric separations in food and beverage analyses. *J. Agric. Food. Chem.* **1990**, *38*, 1674–1677. [CrossRef]
50. Xie, R.; Chu, L.Y.; Deng, J.G. Membranes and membrane processes for chiral resolution. *Chem. Soc. Rev.* **2008**, *37*, 1243–1263. [CrossRef]
51. Gawley, R.E.; Aubé, J. Preface. In *Principles of Asymmetric Synthesis*, 2nd ed.; Elsevier: Amsterdam, The Netherlands, 2012.
52. Zhang, M.; Ye, B.-C. Colorimetric Chiral Recognition of Enantiomers Using the Nucleotide-Capped Silver Nanoparticles. *Anal. Chem.* **2011**, *83*, 1504–1509. [CrossRef]
53. Su, H.; Zheng, Q.; Li, H. Colorimetric detection and separation of chiral tyrosine based on *N*-acetyl-L-cysteine modified gold nanoparticles. *J. Mater. Chem.* **2012**, *22*, 6546–6548. [CrossRef]
54. Kang, Y.J.; Oh, J.W.; Kim, Y.R.; Kim, J.S.; Kim, H. Chiral gold nanoparticle-based electrochemical sensor for enantioselective recognition of 3,4-dihydroxyphenylalanine. *Chem. Commun.* **2010**, *46*, 5665–5667. [CrossRef]
55. Sun, Y.; Zhang, L.; Li, H. Chiral colorimetric recognition of amino acids based on silver nanoparticle clusters. *New J. Chem.* **2012**, *36*, 1442. [CrossRef]
56. Song, G.; Zhou, F.; Xu, C.; Li, B. A universal strategy for visual chiral recognition of alpha-amino acids with L-tartaric acid-capped gold nanoparticles as colorimetric probes. *Analyst* **2016**, *141*, 1257–1265. [CrossRef]
57. Zhou, J.; Duan, J.; Zhang, X.E.; Wang, Q.; Men, D. A chiral responsive carbon dots-gold nanoparticle complex mediated by hydrogen peroxide independent of surface modification with chiral ligands. *Nanoscale* **2018**, *10*, 18606–18612. [CrossRef]
58. Wang, Y.; Zhou, X.; Xu, C.; Jin, Y.; Li, B. Gold Nanorods as Visual Sensing Platform for Chiral Recognition with Naked Eyes. *Sci. Rep.* **2018**, *8*, 5296. [CrossRef]
59. Gautier, C.; Bürgi, T. Chiral *N*-Isobutyryl-cysteine Protected Gold Nanoparticles: Preparation, Size Selection, and Optical Activity in the UV−vis and Infrared. *J. Am. Chem. Soc.* **2006**, *128*, 11079–11087. [CrossRef]
60. Huang, L.; Chen, Y.-T.; Li, Y.-X.; Yu, L.-S. Application of Chiral Ionic Liquid-Modified Gold Nanoparticles in the Chiral Recognition of Amino Acid Enantiomers. *Appl. Spectrosc.* **2016**, *70*, 1649–1654. [CrossRef]
61. Jafari, M.; Tashkhourian, J.; Absalan, G. Chiral recognition of tryptophan enantiomers using chitosan-capped silver nanoparticles: Scanometry and spectrophotometry approaches. *Talanta* **2018**, *178*, 870–878. [CrossRef]
62. Tashkhourian, J.; Afsharinejad, M. A novel colorimetric sensor for sensitive determination of R-citalopram based on the plasmonic properties of silver nanoparticles. *New J. Chem.* **2017**, *41*, 13881–13888. [CrossRef]
63. Shahrajabian, M.; Ghasemi, F.; Hormozi-Nezhad, M.R. Nanoparticle-based Chemiluminescence for Chiral Discrimination of Thiol-Containing Amino Acids. *Sci. Rep.* **2018**, *8*, 14011. [CrossRef]
64. Carrillo-Carrión, C.; Cárdenas, S.; Simonet, B.M.; Valcárcel, M. Selective Quantification of Carnitine Enantiomers Using Chiral Cysteine-Capped CdSe(ZnS) Quantum Dots. *Anal. Chem.* **2009**, *81*, 4730–4733. [CrossRef]

65. Han, C.; Li, H. Chiral Recognition of Amino Acids Based on Cyclodextrin-Capped Quantum Dots. *Small* **2008**, *4*, 1344–1350. [CrossRef]
66. Wang, Y.; Su, P.; Wang, S.; Wu, J.; Huang, J.; Yang, Y. Dendrimer modified magnetic nanoparticles for immobilized BSA: A novel chiral magnetic nano-selector for direct separation of racemates. *J. Mater. Chem. B* **2013**, *1*, 5028–5035. [CrossRef]
67. Wu, J.; Su, P.; Huang, J.; Wang, S.; Yang, Y. Synthesis of teicoplanin-modified hybrid magnetic mesoporous silica nanoparticles and their application in chiral separation of racemic compounds. *J. Colloid Interface Sci.* **2013**, *399*, 107–114. [CrossRef]
68. Visheratina, A.K.; Purcell-Milton, F.; Serrano-García, R.; Kuznetsova, V.A.; Orlova, A.O.; Fedorov, A.V.; Baranov, A.V.; Gun'ko, Y.K. Chiral recognition of optically active $CoFe_2O_4$ magnetic nanoparticles by CdSe/CdS quantum dots stabilised with chiral ligands. *J. Mater. Chem. C* **2017**, *5*, 1692–1698. [CrossRef]
69. Huang, R.; Wang, D.; Liu, S.; Guo, L.; Wang, F.; Lin, Z.; Qiu, B.; Chen, G. Preparative separation of enantiomers based on functional nucleic acids modified gold nanoparticles. *Chirality* **2013**, *25*, 751–756. [CrossRef] [PubMed]
70. Oila, M.J.; Koskinen, A.M.P. Chirally modified gold nanoparticles: Nanostructured chiral ligands for catalysis. *ARKIVOC* **2006**, *xv*, 76–83.
71. Roy, S.; Pericas, M.A. Functionalized nanoparticles as catalysts for enantioselective processes. *Org. Biomol. Chem.* **2009**, *7*, 2669–2677. [CrossRef] [PubMed]
72. Farrag, M.; Tschurl, M.; Heiz, U. Chiral Gold and Silver Nanoclusters: Preparation, Size Selection, and Chiroptical Properties. *Chem. Mater.* **2013**, *25*, 862–870. [CrossRef]
73. Keshvari, F.; Bahram, M.; Farshid, A.A. Gold nanoparticles biofunctionalized (grafted) with chiral amino acids: A practical approach to determining the enantiomeric percentage of racemic mixtures. *Anal. Methods* **2015**, *7*, 4560–4567. [CrossRef]
74. Preiss, L.C.; Wagner, M.; Mastai, Y.; Landfester, K.; Munoz-Espi, R. Amino-Acid-Based Polymerizable Surfactants for the Synthesis of Chiral Nanoparticles. *Macromol. Rapid Commun.* **2016**, *37*, 1421–1426. [CrossRef] [PubMed]
75. Shah, M.; Badwaik, V.; Kherde, Y.; Waghwani, H.K.; Modi, T.; Aguilar, Z.P.; Rodgers, H.; Hamilton, W.; Marutharaj, T.; Webb, C.; et al. Gold nanoparticles: Various methods of synthesis and antibacterial applications. *Front Biosci.* **2014**, *19*, 1320–1344. [CrossRef]
76. Shah, M.; Badwaik, V.D.; Dakshinamurthy, R. Biological applications of gold nanoparticles. *J. Nanosci. Nanotechnol.* **2014**, *14*, 344–362. [CrossRef] [PubMed]
77. Schulz, F.; Tober, S.; Lange, H. Size-Dependent Phase Transfer Functionalization of Gold Nanoparticles To Promote Well-Ordered Self-Assembly. *Langmuir* **2017**, *33*, 14437–14444. [CrossRef] [PubMed]
78. Fei, Z.; Geldbach, T.J.; Zhao, D.; Dyson, P.J. From Dysfunction to Bis-function: On the Design and Applications of Functionalised Ionic Liquids. *Chem. Eur. J.* **2006**, *12*, 2122–2130. [CrossRef] [PubMed]
79. Rodriguez-Liviano, S.; Nuñez, N.O.; Rivera-Fernández, S.; de la Fuente, J.M.; Ocaña, M. Ionic Liquid Mediated Synthesis and Surface Modification of Multifunctional Mesoporous Eu:GdF_3 Nanoparticles for Biomedical Applications. *Langmuir* **2013**, *29*, 3411–3418. [CrossRef]
80. Brückner, H.; Schieber, A. Determination of amino acid enantiomers in human urine and blood serum by gas chromatography–mass spectrometry. *Biomed. Chromatogr.* **2001**, *15*, 166–172. [CrossRef] [PubMed]
81. González-Curbelo, M.Á.; Varela-Martínez, D.A.; Socas-Rodríguez, B.; Hernández-Borges, J. Recent applications of nanomaterials in capillary electrophoresis. *Electrophoresis* **2017**, *38*, 2431–2446. [CrossRef] [PubMed]
82. Wallingford, R.A.; Ewing, A.G. Capillary Electrophoresis. *Adv. Chromatogr.* **1989**, *29*, 1–76. [PubMed]
83. Duan, L.P.; Ding, G.S.; Tang, A.N. Preparation of chitosan-modified silica nanoparticles and their applications in the separation of auxins by capillary electrophoresis. *J. Sep. Sci.* **2015**, *38*, 3976–3982. [CrossRef] [PubMed]
84. Nilsson, C.; Nilsson, S. Nanoparticle-based pseudostationary phases in capillary electrochromatography. *Electrophoresis* **2006**, *27*, 76–83. [CrossRef]
85. Deng, X.; Li, W.; Ding, G.; Xue, T.; Chen, X. Synthesis and Applications of Functionalized Magnetic Nanomaterials in Enantioseparation. *Sep. Purif. Rev.* **2019**, *48*, 14–29. [CrossRef]
86. Liu, C.; Zhang, J.; Zhang, X.; Zhao, L.; Li, S. Enantiomeric separation of adrenaline, noradrenaline, and isoprenaline by capillary electrophoresis using streptomycin-modified gold nanoparticles. *Microchim. Acta* **2018**, *185*, 227. [CrossRef]

87. Choi, S.H.; Noh, H.J.; Lee, K.P. Chiral separation of arylalcohols by capillary electrophoresis using sulfonated β-cyclodextrin and Ag colloids as additives. *Bull. Korean Chem. Soc.* **2005**, *26*, 1549–1554.
88. Shukla, N.; Bartel, M.A.; Gellman, A.J. Enantioselective separation on chiral Au nanoparticles. *J. Am. Chem. Soc.* **2010**, *132*, 8575–8580. [CrossRef]
89. Shukla, N.; Yang, D.; Gellman, A.J. Enantiomeric separations of chiral pharmaceuticals using chirally modified tetrahexahedral Au nanoparticles. *Surf. Sci.* **2016**, *648*, 29–34. [CrossRef]
90. Yang, L.; Chen, C.; Liu, X.; Shi, J.; Wang, G.; Zhu, L.; Guo, L.; Glennon, J.D.; Scully, N.M.; Doherty, B.E. Use of cyclodextrin-modified gold nanoparticles for enantioseparations of drugs and amino acids based on pseudostationary phase-capillary electrochromatography. *Electrophoresis* **2010**, *31*, 1697–1705. [CrossRef]
91. Choi, H.J.; Hyun, M.H. Separation of enantiomers with magnetic silica nanoparticles modified by a chiral selector: Enantioselective fishing. *Chem. Commun.* **2009**, *42*, 6454–6456. [CrossRef]
92. Ghosh, S.; Fang, T.H.; Uddin, M.S.; Hidajat, K. Enantioselective separation of chiral aromatic amino acids with surface functionalized magnetic nanoparticles. *Coll. Surf. B Biointerfaces* **2013**, *105*, 267–277. [CrossRef]
93. Fu, Y.; Huang, T.; Chen, B.; Shen, J.; Duan, X.; Zhang, J.; Li, W. Enantioselective resolution of chiral drugs using BSA functionalized magnetic nanoparticles. *Sep. Purif. Technol.* **2013**, *107*, 11–18. [CrossRef]
94. Ge, J.; Zhou, W.; Zhao, L.; Shi, Y.P. Basic Indole Ring Enantiomer Separation on Cellulose Tris (3,5-dimethylphenylcarbamate) Coated TiO_2/SiO_2 Chiral Stationary Phase. *Anal. Lett.* **2007**, *40*, 2515–2523. [CrossRef]
95. Zhou, S.; Wang, Y.; De Beer, T.; Baeyens, W.R.; Fei, G.T.; Dilinuer, M.; Ouyang, J. Simultaneous separation of eight β-adrenergic drugs using titanium dioxide nanoparticles as additive in capillary electrophoresis. *Electrophoresis* **2008**, *29*, 2321–2329. [CrossRef]
96. Kumar, A.P.; Kim, J.H.; Thanh, T.D.; Lee, Y.I. Chiral zirconia magnetic microspheres as a new recyclable selector for the discrimination of racemic drugs. *J. Mater. Chem. B* **2013**, *1*, 4909–4915. [CrossRef]
97. Moliner-Martínez, Y.; Cárdenas, S.; Valcárcel, M. Evaluation of carbon nanostructures as chiral selectors for direct enantiomeric separation of ephedrines by EKC. *Electrophoresis* **2007**, *28*, 2573–2579. [CrossRef]
98. Weng, X.; Bi, H.; Liu, B.; Kong, J. On-chip chiral separation based on bovine serum albumin-conjugated carbon nanotubes as stationary phase in a microchannel. *Electrophoresis* **2006**, *27*, 3129–3135. [CrossRef]
99. Guillaume, Y.C.; André, C. Fast enantioseparation by hplc on a modified carbon nanotube monolithic stationary phase with a pyrenyl aminoglycoside derivative. *Talanta* **2013**, *115*, 418–421. [CrossRef]
100. Ahmed, M.; Yajadda, M.M.A.; Han, Z.J.; Su, D.; Wang, G.; Ostrikov, K.K.; Ghanem, A. Single-walled carbon nanotube-based polymer monoliths for the enantioselective nano-liquid chromatographic separation of racemic pharmaceuticals. *J. Chromatogr. A* **2014**, *1360*, 100–109. [CrossRef]
101. Na, N.; Hu, Y.; Ouyang, J.; Baeyens, W.R.; Delanghe, J.R.; Taes, Y.E.; Xie, M.; Chen, H.; Yang, Y. On the use of dispersed nanoparticles modified with single layer β-cyclodextrin as chiral selecor to enhance enantioseparation of clenbuterol with capillary electrophoresis. *Talanta* **2006**, *69*, 866–872. [CrossRef]
102. Yu, J.; Huang, D.; Huang, K.; Hong, Y. Preparation of Hydroxypropyl-β-cyclodextrin Cross-linked Multi-walled Carbon Nanotubes and Their Application in Enantioseparation of Clenbuterol. *Chin. J. Chem.* **2011**, *29*, 893–897. [CrossRef]
103. Zhang, Q.; Du, Y.; Du, S. Evaluation of ionic liquids-coated carbon nanotubes modified chiral separation system with chondroitin sulfate E as chiral selector in capillary electrophoresis. *J. Chromatogr. A* **2014**, *1339*, 185–191. [CrossRef]
104. Hua, X.; Du, Y.; Chen, J.; Xu, G.; Yu, T.; Zhang, Q. Evaluation of the enantioselectivity of carbon nanoparticles-modified chiral separation systems using dextrin as chiral selector by capillary electrokinetic chromatography. *Electrophoresis* **2013**, *34*, 1901–1907. [CrossRef]
105. Candelaria, L.; Frolova, L.V.; Kowalski, B.M.; Artyushkova, K.; Serov, A.; Kalugin, N.G. Surface-modified three-dimensional graphene nanosheets as a stationary phase for chromatographic separation of chiral drugs. *Sci. Rep.* **2018**, *8*, 147. [CrossRef]
106. Tu, F.Y.; Yu, L.Y.; Yu, J.G.; Chen, X.Q.; Fu, Q.; Jiao, F.P.; Peng, Z.G.; Zhang, T. Graphene as tunable stationary phase additive for enantioseparation. *Nano* **2013**, *8*, 1350069. [CrossRef]
107. Liu, Z.; Du, Y.; Feng, Z. Enantioseparation of drugs by capillary electrochromatography using a stationary phase covalently modified with graphene oxide. *Microchim. Acta* **2017**, *184*, 583–593. [CrossRef]

108. Liang, R.P.; Liu, C.M.; Meng, X.Y.; Wang, J.W.; Qiu, J.D. A novel open-tubular capillary electrochromatography using β-cyclodextrin functionalized graphene oxide-magnetic nanocomposites as tunable stationary phase. *J. Chromatogr. A* **2012**, *1266*, 95–102. [CrossRef]
109. Liang, R.P.; Wang, X.N.; Liu, C.M.; Meng, X.Y.; Qiu, J.D. Facile preparation of protein stationary phase based on polydopamine/graphene oxide platform for chip-based open tubular capillary electrochromatography enantioseparation. *J. Chromatogr. A* **2014**, *1323*, 135–142. [CrossRef] [PubMed]
110. Li, Y.; Li, Q.; Zhu, N.; Gao, Z.; Ma, Y. Cellulose type chiral stationary phase based on reduced graphene oxide@ silica gel for the enantiomer separation of chiral compounds. *Chirality* **2018**, *30*, 996–1004. [CrossRef] [PubMed]
111. Yang, X.R.; Song, X.D.; Zhu, H.Y.; Cheng, C.J.; Yu, H.R.; Zhang, H.H. Novel Smart Polymer-Brush-Modified Magnetic Graphene Oxide for Highly Efficient Chiral Recognition and Enantioseparation of Tryptophan Enantiomers. *ACS Appl. Bio. Mater.* **2018**, *1*, 1074–1083. [CrossRef]
112. Na, N.; Hu, Y.; Ouyang, J.; Baeyens, W.R.; Delanghe, J.R.; De Beer, T. Use of polystyrene nanoparticles to enhance enantiomeric separation of propranolol by capillary electrophoresis with Hp-beta-CD as chiral selector. *Anal. Chim. Acta* **2004**, *527*, 139–147. [CrossRef]
113. Aydoğan, C.; Karakoç, V.; Yılmaz, F.; Shaikh, H.; Denizli, A. Enantioseparation of Ofloxacin by Ligand Exchange Capillary Electrophoresis Using L-Histidine Modified Nanoparticles as Chiral Ligand. *Hacet. J. Biol. Chem.* **2013**, *41*, 29–36.
114. Dong, X.; Wu, R.A.; Dong, J.; Wu, M.; Zhu, Y.; Zou, H. A mesoporous silica nanoparticles immobilized open-tubular capillary column with a coating of cellulose tris (3,5-dimethylphenyl-carbamate) for enantioseparation in CEC. *Electrophoresis* **2008**, *29*, 3933–3940. [CrossRef]
115. Xu, S.; Mo, R.; Jin, C.; Cui, X.; Bai, R.; Ji, Y. Mesoporous silica nanoparticles incorporated hybrid monolithic stationary phase immobilized with pepsin for enantioseparation by capillary electrochromatography. *J. Pharm. Biomed. Anal.* **2017**, *140*, 190–198. [CrossRef]
116. Gong, Z.S.; Duan, L.P.; Tang, A.N. Amino-functionalized silica nanoparticles for improved enantiomeric separation in capillary electrophoresis using carboxymethyl-β-cyclodextrin (CM-β-CD) as a chiral selector. *Microchim. Acta* **2015**, *182*, 1297–1304. [CrossRef]
117. Li, W.; Ding, G.S.; Tang, A.N. Enantiomer separation of propranolol and tryptophan using bovine serum albumin functionalized silica nanoparticles as adsorbents. *RSC Adv.* **2015**, *5*, 93850–93857. [CrossRef]
118. Padmanaban, M.; Müller, P.; Lieder, C.; Gedrich, K.; Grünker, R.; Bon, V.; Senkovska, I.; Baumgärtner, S.; Opelt, S.; Paasch, S.; et al. Application of a chiral metal–organic framework in enantioselective separation. *Chem. Commun.* **2011**, *47*, 12089–12091. [CrossRef]
119. Xie, S.M.; Zhang, Z.J.; Wang, Z.Y.; Yuan, L.M. Chiral metal–organic frameworks for high-resolution gas chromatographic separations. *J. Am. Chem. Soc.* **2011**, *133*, 11892–11895. [CrossRef]
120. Sattler, K.D. *Handbook of Nanophysics: Nanoparticles and Quantum Dots*, 1st ed.; CRC Press: Boca Raton, FL, USA, 2016.
121. Hu, A.; Yee, G.T.; Lin, W. Magnetically recoverable chiral catalysts immobilized on magnetite nanoparticles for asymmetric hydrogenation of aromatic ketones. *J. Am. Chem. Soc.* **2005**, *127*, 12486–12487. [CrossRef]
122. Mei, W.; Wu, Q. Applications of Metal Nanoparticles in Medicine/Metal Nanoparticles as Anticancer Agents. In *Metal Nanoparticles: Synthesis and Applications in Pharmaceutical Sciences*; Wiley: Hoboken, NJ, USA, 2018; pp. 169–190.
123. Mody, V.V.; Siwale, R.; Singh, A.; Mody, H.R. Introduction to metallic nanoparticles. *J. Pharm. Bioallied Sci.* **2010**, *2*, 282. [CrossRef]
124. Khan, A.K.; Rashid, R.; Murtaza, G.; Zahra, A. Gold nanoparticles: Synthesis and applications in drug delivery. *Trop. J. Pharm. Res.* **2014**, *13*, 1169–1177. [CrossRef]
125. Tang, S.; Guo, Y.; Xiong, C.; Liu, S.; Liu, X.; Jiang, S. Nanoparticle-based monoliths for chromatographic separations. *Analyst* **2014**, *139*, 4103–4117. [CrossRef]
126. Shao, P.; Ji, G.; Chen, P. Gold nanotube membranes: Preparation, characterization and application for enantioseparation. *J. Memb. Sci.* **2005**, *255*, 1–11. [CrossRef]
127. Li, H.F.; Zeng, H.; Chen, Z.; Lin, J.M. Chip-based enantioselective open-tubular capillary electrochromatography using bovine serum albumin-gold nanoparticle conjugates as the stationary phase. *Electrophoresis* **2009**, *30*, 1022–1029. [CrossRef]

128. Li, M.; Liu, X.; Jiang, F.; Guo, L.; Yang, L. Enantioselective open-tubular capillary electrochromatography using cyclodextrin-modified gold nanoparticles as stationary phase. *J. Chromatogr. A* **2011**, *1218*, 3725–3729. [CrossRef]
129. Fang, L.L.; Wang, P.; Wen, X.L.; Guo, X.; Yu, J.; Guo, X.J. Layer-by-layer self-assembly of gold nanoparticles/thiols β-cyclodextrin coating as the stationary phase for enhanced chiral differentiation in open tubular capillary electrochromatography. *Talanta* **2017**, *167*, 158–165. [CrossRef]
130. Luo, X.; Morrin, A.; Killard, A.J.; Smyth, M.R. Application of nanoparticles in electrochemical sensors and biosensors. *Electroanalysis* **2006**, *18*, 319–326. [CrossRef]
131. Corti, M.; Lascialfari, A.; Micotti, E.; Castellano, A.; Donativi, M.; Quarta, A.; Cozzoli, P.D.; Manna, L.; Pellegrino, T.; Sangregorio, C. Magnetic properties of novel superparamagnetic MRI contrast agents based on colloidal nanocrystals. *J. Magn. Magn. Mater.* **2008**, *320*, e320–e323. [CrossRef]
132. Arslan, M.; Sayin, S.; Yilmaz, M. Enantioselective sorption of some chiral carboxylic acids by various cyclodextrin-grafted iron oxide magnetic nanoparticles. *Tetrahedron Asymmetry* **2013**, *24*, 982–989. [CrossRef]
133. Caruntu, D.; Caruntu, G.; O'Connor, C.J. Magnetic properties of variable-sized Fe_3O_4 nanoparticles synthesized from non-aqueous homogeneous solutions of polyols. *J. Phys. D* **2007**, *40*, 5801. [CrossRef]
134. Chen, X.; Rao, J.; Wang, J.; Gooding, J.J.; Zou, G.; Zhang, Q. A facile enantioseparation for amino acids enantiomers using β-cyclodextrins functionalized Fe_3O_4 nanospheres. *Chem. Commun.* **2011**, *47*, 10317–10319. [CrossRef] [PubMed]
135. Sayin, S.; Akoz, E.; Yilmaz, M. Enhanced catalysis and enantioselective resolution of racemic naproxen methyl ester by lipase encapsulated within iron oxide nanoparticles coated with calix [8] arene valeric acid complexes. *Org. Biomol. Chem.* **2014**, *12*, 6634–6642. [CrossRef] [PubMed]
136. Zhu, H.Y.; Song, X.D.; Yang, X.R.; Cheng, C.J.; Yu, H.R.; Zhang, H.H. Smart chiral magnetic nanoparticles for highly efficient enantioseparation of tryptophan enantiomers. *J. Mater. Sci.* **2019**, *54*, 2960–2974. [CrossRef]
137. Winkler, J.; Marmé, S. Titania as a sorbent in normal-phase liquid chromatography. *J. Chromatogr. A* **2000**, *888*, 51–62. [CrossRef]
138. Yu, J.C.; Qu, F.; Lin, J.; Lam, H.; Chen, Z. Ion chromatographic separation of anions and cations on a titania packed column. *J. Liq. Chromatogr. Relat. Technol.* **2001**, *24*, 367–380.
139. Kumar, A.P.; Park, J.H. Chiral separation of basic compounds on a cellulose 3, 5-dimethylphenylcarbamate-coated zirconia monolithin basic eluents by capillary electrochromatography. *J. Chromatogr. A* **2011**, *1218*, 6548–6553. [CrossRef] [PubMed]
140. Hong, J.S.; Park, J.H. Chiral separation of basic compounds on sulfated β-cyclodextrin-coated zirconia monolith by capillary electrochromatography. *Bull. Korean Chem. Soc.* **2013**, *34*, 1809–1813. [CrossRef]
141. Dun, H.; Zhang, W.; Wei, Y.; Xiuqing, S.; Li, Y.; Chen, L. Layer-by-layer self-assembly of multilayer zirconia nanoparticles on silica spheres for HPLC packings. *Anal. Chem.* **2004**, *76*, 5016–5023. [CrossRef] [PubMed]
142. Zarei, M.; Zarei, M.; Ghasemabadi, M. Nanoparticle improved separations: From capillary to slab gel electrophoresis. *Trends Anal. Chem.* **2017**, *86*, 56–74. [CrossRef]
143. Valcárcel, M.; Cárdenas, S.; Simonet, B.M.; Moliner-Martínez, Y.; Lucena, R. Carbon nanostructures as sorbent materials in analytical processes. *Trends Anal. Chem.* **2008**, *27*, 34–43. [CrossRef]
144. Qin, L.C.; Zhao, X.; Hirahara, K.; Miyamoto, Y.; Ando, Y.; Iijima, S. Materials science: The smallest carbon nanotube. *Nature* **2000**, *408*, 50. [CrossRef] [PubMed]
145. Wang, N.; Tang, Z.K.; Li, G.D.; Chen, J.S. Materials science: Single-walled 4 Å carbon nanotube arrays. *Nature* **2000**, *408*, 50–51. [CrossRef] [PubMed]
146. Zhao, X.; Liu, Y.; Inoue, S.; Suzuki, T.; Jones, R.O.; Ando, Y. Smallest carbon nanotube is 3 Å in diameter. *Phys. Rev. Lett.* **2004**, *92*, 125502. [CrossRef] [PubMed]
147. Ren, Z.F.; Huang, Z.P.; Xu, J.W.; Wang, J.H.; Bush, P.; Siegal, M.P.; Provencio, P.N. Synthesis of large arrays of well-aligned carbon nanotubes on glass. *Science* **1998**, *282*, 1105–1107. [CrossRef] [PubMed]
148. Zhang, R.; Zhang, Y.; Zhang, Q.; Xie, H.; Qian, W.; Wei, F. Growth of half-meter long carbon nanotubes based on Schulz–Flory distribution. *ACS Nano* **2013**, *6*, 6156–6161. [CrossRef] [PubMed]
149. Zhang, J.; Albelda, M.T.; Liu, Y.; Canary, J.W. Chirality: The Pharmacological, Biological, and Chemical Consequences of Molecular Asymmetry. *Chirality* **2005**, *17*, 404–420. [CrossRef] [PubMed]
150. Iijima, S. Helical microtubules of graphitic carbon. *Nature* **1991**, *354*, 56–58. [CrossRef]
151. Ying, L.S.; bin Mohd Salleh, M.A.; Mohamed Yusoff, H.B.; Rashid, S.B.A.; Razak, J.B.A. Continuous production of carbon nanotubes—A review. *J. Ind. Eng. Chem.* **2011**, *17*, 367–376. [CrossRef]

152. Herrera-Herrera, A.V.; González-Curbelo, M.Á.; Hernández-Borges, J.; Rodríguez-Delgado, M.Á. Carbon nanotubes applications in separation science: A review. *Anal. Chim. Acta* **2012**, *734*, 1–30. [CrossRef]
153. Baughman, R.H.; Zakhidov, A.A.; De Heer, W.A. Carbon nanotubes–the route toward applications. *Science* **2002**, *297*, 787–792. [CrossRef]
154. Ren, X.; Chen, C.; Nagatsu, M.; Wang, X. Carbon nanotubes as adsorbents in environmental pollution management: A review. *Chem. Eng. J.* **2011**, *170*, 395–410. [CrossRef]
155. Coleman, J.N.; Khan, U.; Blau, W.J.; Gun'ko, Y.K. Small but strong: A review of the mechanical properties of carbon nanotube-polymer composites. *Carbon* **2006**, *44*, 1624–1652. [CrossRef]
156. Breuer, O.; Sundararaj, U. Big returns from small fibers: A review of polymer/carbon nanotube composites. *Polym. Compos.* **2004**, *25*, 630–645. [CrossRef]
157. Harrison, B.S.; Atala, A. Carbon nanotube applications for tissue engineering. *Biomaterials* **2007**, *28*, 344–353. [CrossRef]
158. Song, J.; Wang, F.; Yang, X.; Ning, B.; Harp, M.G.; Culp, S.H.; Hu, S.; Huang, P.; Nie, L.; Chen, J.; et al. Gold nanoparticle coated carbon nanotube ring with enhanced raman scattering and photothermal conversion property for theranostic applications. *J. Am. Chem. Soc.* **2016**, *138*, 7005–7015. [CrossRef]
159. Hemasa, A.L.; Naumovski, N.; Maher, W.A.; Ghanem, A. Application of carbon nanotubes in chiral and achiral separations of pharmaceuticals, biologics and chemicals. *Nanomaterials* **2017**, *7*, 186. [CrossRef]
160. Power, T.D.; Skoulidas, A.I.; Sholl, D.S. Can chiral single walled carbon nanotubes be used as enantiospecific adsorbents? *J. Am. Chem. Soc.* **2002**, *124*, 1858–1859. [CrossRef]
161. Suárez, B.; Simonet, B.M.; Cárdenas, S.; Valcárcel, M. Surfactant-coated single-walled carbon nanotubes as a novel pseudostationary phase in capillary EKC. *Electrophoresis* **2007**, *28*, 1714–1722. [CrossRef]
162. Zhao, L.; Ai, P.; Duan, A.-H.; Yuan, L.-M. Single-walled carbon nanotubes for improved enantioseparations on a chiral ionic liquid stationary phase in GC. *Anal. Bioanal. Chem.* **2011**, *399*, 143–147. [CrossRef]
163. Tarigh, G.D.; Shemirani, F. In situ immobilization of a general resolving agent on the magnetic multi-wall carbon nanotube for the direct enantioenrichment of DL-mandelic acid. *Talanta* **2015**, *144*, 899–907. [CrossRef]
164. Wang, Y.; Li, Z.; Wang, J.; Li, J.; Lin, Y. Graphene and graphene oxide: Biofunctionalization and applications in biotechnology. *Trends Biotechnol.* **2011**, *29*, 205–212. [CrossRef]
165. Liang, X.; Hou, X.; Chan, J.H.; Guo, Y.; Hilder, E.F. The application of graphene-based materials as chromatographic stationary phases. *Trends Anal. Chem.* **2018**, *98*, 149–160. [CrossRef]
166. Geim, A.K. Graphene: Status and prospects. *Science* **2009**, *324*, 1530–1534. [CrossRef]
167. Dai, J.F.; Wang, G.J.; Ma, L.; Wu, C.K. Surface properties of graphene: Relationship to graphene-polymer composites. *Rev. Adv. Mater. Sci.* **2015**, *40*, 60–71.
168. The 2010 Nobel Prize in Physics—Press Release. Available online: https://www.nobelprize.org/uploads/2018/06/press-9.pdf (accessed on 17 January 2019).
169. Ray, S.C. Application and uses of graphene oxide and reduced graphene oxide. In *Applications of Graphene and Graphene-Oxide Based Nanomaterials*, 1st ed.; Elsevier: Amsterdam, The Netherlands, 2015; pp. 39–55.
170. Li, W.; Liang, J.; Yang, W.; Deng, J. Chiral functionalization of graphene oxide by optically active helical-substituted polyacetylene chains and its application in enantioselective crystallization. *ACS Appl. Mater. Interfaces* **2014**, *6*, 9790–9798. [CrossRef]
171. Hauser, A.W.; Mardirossian, N.; Panetier, J.A.; Head-Gordon, M.; Bell, A.T.; Schwerdtfeger, P. Functionalized graphene as a gatekeeper for chiral molecules: An alternative concept for chiral separation. *Angew. Chem. Int. Ed.* **2014**, *53*, 9957–9960. [CrossRef]
172. Liang, R.P.; Meng, X.Y.; Liu, C.M.; Wang, J.W.; Qiu, J.D. Enantiomeric separation by open-tubular capillary electrochromatography using bovine-serum-albumin-conjugated graphene oxide–magnetic nanocomposites as stationary phase. *Microfluid. Nanofluid.* **2014**, *16*, 195–206. [CrossRef]
173. Ye, N.; Li, J.; Xie, Y.; Liu, C. Graphene oxide coated capillary for chiral separation by CE. *Electrophoresis* **2013**, *34*, 841–845. [CrossRef] [PubMed]
174. Hong, T.; Chen, X.; Xu, Y.; Cui, X.; Bai, R.; Jin, C.; Li, R.; Ji, Y. Preparation of graphene oxide-modified affinity capillary monoliths based on three types of amino donor for chiral separation and proteolysis. *J. Chromatogr. A* **2016**, *1456*, 249–256. [CrossRef] [PubMed]
175. De los Santos, Z.A.; Ding, R.; Wolf, C. Quantitative chirality sensing of amines and amino alcohols via Schiff base formation with a stereodynamic UV/CD probe. *Org. Biomol. Chem.* **2016**, *14*, 1934–1939. [CrossRef] [PubMed]

176. Mei, X.; Wolf, C. Enantioselective Sensing of Chiral Carboxylic Acids. *J. Am. Chem. Soc.* **2004**, *126*, 14736–14737. [CrossRef] [PubMed]
177. Manoli, K.; Magliulo, M.; Torsi, L. Chiral Sensor Devices for Differentiation of Enantiomers. In *Differentiation of Enantiomers II*, 1st ed.; Springer: Berlin, Germany, 2013; pp. 133–176.
178. De los Santos, Z.A.; Yusin, G.; Wolf, C. Enantioselective sensing of carboxylic acids with a bis(urea)oligo(phenylene)ethynylene foldamer. *Tetrahedron* **2019**, *75*, 1504–1509. [CrossRef]
179. Tsourkas, A.; Hofstetter, O.; Hofstetter, H.; Weissleder, R.; Josephson, L. Magnetic Relaxation Switch Immunosensors Detect Enantiomeric Impurities. *Angew. Chem.* **2004**, *43*, 2395–2399. [CrossRef] [PubMed]
180. Lieberman, I.; Shemer, G.; Fried, T.; Kosower, E.M.; Markovich, G. Plasmon-Resonance-Enhanced Absorption and Circular Dichroism. *Angew. Chem.* **2008**, *47*, 4855–4857. [CrossRef]
181. Dutta, P.; Tipple, C.A.; Lavrik, N.; Datskos, P.G.; Hofstetter, H.; Hofstetter, O.; Sepaniak, M.J. Enantioselective Sensors Based on Antibody-Mediated Nanomechanics. *Anal. Chem.* **2003**, *75*, 2342–2348. [CrossRef]
182. Prasad, B.B.; Madhuri, R.; Tiwari, M.P.; Sharma, P.S. Enantioselective recognition of d- and l-tryptophan by imprinted polymer-carbon composite fiber sensor. *Talanta* **2010**, *81*, 187–196. [CrossRef]
183. Huang, J.; Wei, Z.; Chen, J. Molecular imprinted polypyrrole nanowires for chiral amino acid recognition. *Sens. Actuat. B Chem.* **2008**, *134*, 573–578. [CrossRef]
184. Ariga, K.; Richards, G.J.; Ishihara, S.; Izawa, H.; Hill, J.P. Intelligent Chiral Sensing Based on Supramolecular and Interfacial Concepts. *Sensors* **2010**, *10*, 6796–6820. [CrossRef]
185. Vashist, S.K.; Vashist, P. Recent Advances in Quartz Crystal Microbalance-Based Sensors. *J. Sens.* **2011**, *2011*, 571405. [CrossRef]
186. Gou, H.; He, J.; Mo, Z.; Wei, X.; Hu, R.; Wang, Y.; Guo, R. A Highly Effective Electrochemical Chiral Sensor of Tryptophan Enantiomers Based on Covalently Functionalize Reduced Graphene Oxide with L-Lysine. *J. Electrochem. Soc.* **2016**, *163*, B272–B279. [CrossRef]
187. Basozabal, I.; Gómez-Caballero, A.; Unceta, N.; Aranzazu Goicolea, M.; Barrio, R.J. Voltammetric sensors with chiral recognition capability: The use of a chiral inducing agent in polyaniline electrochemical synthesis for the specific recognition of the enantiomers of the pesticide dinoseb. *Electrochim. Acta* **2011**, *58*, 729–735. [CrossRef]
188. Chen, Y.; Chen, L.; Bi, R.; Xu, L.; Liu, Y. A potentiometric chiral sensor for l-Phenylalanine based on crosslinked polymethylacrylic acid–polycarbazole hybrid molecularly imprinted polymer. *Anal. Chim. Acta* **2012**, *754*, 83–90. [CrossRef]
189. Lahav, M.; Kharitonov, A.B.; Willner, I. Imprinting of Chiral Molecular Recognition Sites in Thin TiO_2 Films Associated with Field-Effect Transistors: Novel Functionalized Devices for Chiroselective and Chirospecific Analyses. *Chem. Eur. J.* **2001**, *7*, 3992–3997. [CrossRef]
190. Chen, Q.; Zhou, J.; Han, Q.; Wang, Y.; Fu, Y. A new chiral electrochemical sensor for the enantioselective recognition of penicillamine enantiomers. *J. Solid State Electrochem.* **2012**, *16*, 2481–2485. [CrossRef]
191. Bayraktar, E.; Eroglu, D.; Ciftlik, A.T.; Kulah, H. A MEMS based gravimetric resonator for mass sensing applications. In Proceedings of the 2011 IEEE 24th International Conference on Micro Electro Mechanical Systems, Cancun, Mexico, 23–27 January 2011; pp. 817–820.
192. Stefan, R.-I.; van Staden, J.F.; Aboul-Enein, H.Y. Amperometric biosensors/sequential injection analysis system for simultaneous determination of S- and R-captopril. *Biosens. Bioelectron.* **2000**, *15*, 1–5. [CrossRef]
193. Ng, S.-C.; Sun, T.; Chan, H.S.O. Durable chiral sensor based on quartz crystal microbalance using self-assembled monolayer of permethylated β-cyclodextrin. *Macromol. Symp.* **2003**, *192*, 171–182. [CrossRef]
194. Luo, M.L.; Zhang, W.G.; Zhang, S.; Fan, J.; Su, W.C.; Yin, X. Self-Assembly and Chiral Recognition of Quartz Crystal Microbalance Chiral Sensor. *Chirality* **2009**, *22*, 411–415. [CrossRef]
195. De Lacy Costello, B.P.J.; Ratcliffe, N.M.; Sivanand, P.S. The synthesis of novel 3-substituted pyrrole monomers possessing chiral side groups: A study of their chemical polymerisation and the assessment of their chiral discrimination properties. *Synth. Met.* **2003**, *139*, 43–55. [CrossRef]
196. Torsi, L.; Farinola, G.M.; Marinelli, F.; Tanese, M.C.; Omar, O.H.; Valli, L.; Babudri, F.; Palmisano, F.; Zambonin, P.G.; Naso, F. A sensitivity-enhanced field-effect chiral sensor. *Nat. Mater.* **2008**, *7*, 412–417. [CrossRef]
197. Kurzawski, P.; Bogdanski, A.; Schurig, V.; Wimmer, R.; Hierlemann, A. Opposite Signs of Capacitive Microsensor Signals upon Exposure to the Enantiomers of Methyl Propionate Compounds. *Angew. Chem.* **2008**, *47*, 913–916. [CrossRef]

198. Zhang, S.; Ding, J.; Liu, Y.; Kong, J.; Hofstetter, O. Development of a Highly Enantioselective Capacitive Immunosensor for the Detection of α-Amino Acids. *Anal. Chem.* **2006**, *78*, 7592–7596. [CrossRef]
199. Zhao, Y.; Swager, T.M. Simultaneous Chirality Sensing of Multiple Amines by 19F NMR. *J. Am. Chem. Soc.* **2015**, *137*, 3221–3224. [CrossRef]
200. Ben-Amram, Y.; Riskin, M.; Willner, I. Selective and enantioselective analysis of mono- and disaccharides using surface plasmon resonance spectroscopy and imprinted boronic acid-functionalized Au nanoparticle composites. *Analyst* **2010**, *135*, 2952–2959. [CrossRef]
201. Bentley, K.W.; Wolf, C. Comprehensive Chirality Sensing: Development of Stereodynamic Probes with a Dual (Chir)optical Response. *J. Org. Chem.* **2014**, *79*, 6517–6531. [CrossRef]
202. De los Santos, Z.A.; Joyce, L.A.; Sherer, E.C.; Welch, C.J.; Wolf, C. Optical Chirality Sensing with a Stereodynamic Aluminum Biphenolate Probe. *J. Org. Chem.* **2018**. [CrossRef]
203. De los Santos, Z.A.; Legaux, N.M.; Wolf, C. Chirality sensing with stereodynamic copper(I) complexes. *Chirality* **2017**, *29*, 663–669. [CrossRef]
204. Drăghici, C.; Chirila, E.; Sica, M. Enantioselectivity of Chiral Pesticides in the Environment. In *Environmental Security Assessment and Management of Obsolete Pesticides in Southeast Europe*, 1st ed.; Springer: Berlin, Germany, 2013; pp. 91–102.
205. Ward, T.J.; Ward, K. Recent Progress in Chiral Stationary Phase Development and Current Chiral Applications. *LC-GC Eur.* **2012**, *30*, 43–45.
206. Ward, T.J.; Ward, K. Recent Progress in Chiral Stationary Phase Development and Current Chiral Applications. *LC-GC Eur.* **2014**, *32*, 20–23.
207. Padró, J.M.; Keunchkarian, S. State-of-the-art and recent developments of immobilized polysaccharide-based chiral stationary phases for enantioseparations by high-performance liquid chromatography (2013–2017). *Microchem. J.* **2018**, *140*, 142–157. [CrossRef]
208. Mitchell, C.R.; Armstrong, D.W. Cyclodextrin-Based Chiral Stationary Phases for Liquid Chromatography. In *Chiral Separations: Methods and Protocols*; Gübitz, G., Schmid, M.G., Eds.; Humana Press: New York, NY, USA, 2004; pp. 61–112.
209. Zhu, Q.; Scriba, G.K.E. Advances in the Use of Cyclodextrins as Chiral Selectors in Capillary Electrokinetic Chromatography: Fundamentals and Applications. *Chromatographia* **2016**, *79*, 1403–1435. [CrossRef]
210. Berthod, A. Chiral Recognition Mechanisms. *Anal. Chem.* **2006**, *78*, 2093–2099. [CrossRef]
211. Cavazzini, A.; Pasti, L.; Massi, A.; Marchetti, N.; Dondi, F. Recent applications in chiral high performance liquid chromatography: A review. *Anal. Chim. Acta* **2011**, *706*, 205–222. [CrossRef]
212. Dingenen, J.; Kinkel, J. Preparative chromatographic resolution of racemates on chiral stationary phases on laboratory and production scales by closed-loop recycling chromatography. *J. Chromatogr. A* **1994**, *666*, 627–650. [CrossRef]
213. Cox, G.B. Preparative chiral separations: From laboratory scale to production. *Chromatogr. Today* **2009**, *4*, 4–7.
214. Pais, L.S.; Loureiro, J.; Rodrigues, A. Chiral separation by SMB chromatography. *Sep. Purif. Technol.* **2000**, *20*, 67–77. [CrossRef]
215. Fernandes, C.; Tiritan, M.; Pinto, M. Chiral Separation in Preparative Scale: A Brief Overview of Membranes as Tools for Enantiomeric Separation. *Symmetry* **2017**, *9*, 206. [CrossRef]
216. Flack, H. Louis Pasteur's discovery of molecular chirality and spontaneous resolution in 1848, together with a complete review of his crystallographic and chemical work. *Acta Crystallogr. A* **2009**, *65*, 371–389. [CrossRef]

molecules

MDPI

Article

Probing the Structural Determinants of Amino Acid Recognition: X-Ray Studies of Crystalline Ditopic Host-Guest Complexes of the Positively Charged Amino Acids, Arg, Lys, and His with a Cavitand Molecule

Giovanna Brancatelli [1], Enrico Dalcanale [2], Roberta Pinalli [2,*] and Silvano Geremia [1,*]

1 Centre of Excellence in Biocrystallography, Department of Chemical and Pharmaceutical Sciences, University of Trieste, 34127 Trieste, Italy; giovanna_brancatelli@hotmail.com

2 Department of Chemistry, Life Sciences and Environmental Sustainability, University of Parma, and INSTM, UdR Parma, Parco Area delle Scienze 17/A, 43124 Parma, Italy; enrico.dalcanale@unipr.it

* Correspondence: roberta.pinalli@unipr.it (R.P.); sgeremia@units.it (S.G.); Tel.: +39-0521-905464 (R.P.); +39-040-558-3936 (S.G.)

Academic Editors: Maria Elizabeth Tiritan, Madalena Pinto and Carla Sofia Garcia Fernandes
Received: 19 November 2018; Accepted: 18 December 2018; Published: 19 December 2018

Abstract: Crystallization of tetraphosphonate cavitand Tiiii[H, CH_3, CH_3] in the presence of positively charged amino acids, namely arginine, lysine, or histidine, afforded host-guest complex structures. The X-ray structure determination revealed that in all three structures, the fully protonated form of the amino acid is ditopically complexed by two tetraphosphonate cavitand molecules. Guanidinium, ammonium, and imidazolium cationic groups of the amino acid side chain are hosted in the cavity of a phosphonate receptor, and are held in place by specific hydrogen bonding interactions with the P=O groups of the cavitand molecule. In all three structures, the positively charged α-ammonium groups form H-bonds with the P=O groups, and with a water molecule hosted in the cavity of a second tetraphosphonate molecule. Furthermore, water-assisted dimerization was observed for the cavitand/histidine ditopic complex. In this 4:2 supramolecular complex, a bridged water molecule is held by two carboxylic acid groups of the dimerized amino acid. The structural information obtained on the geometrical constrains necessary for the possible encapsulation of the amino acids are important for the rational design of devices for analytical and medical applications.

Keywords: amino acid; phosphonate cavitand; molecular recognition; X-ray structure; host-guest chemistry; lysine; arginine; histidine

---

1. Introduction

L-Arginine (Arg), L-lysine (Lys), and L-histidine (His) belong to the class of 20 proteinogenic amino acids. Arg is classified as a conditionally essential amino acid for adults [1], and it serves as a precursor for methylguanidoacetic acid (creatine), which plays an essential role in the energy metabolism of muscle, nerve, and testis, and accounts for Arg catabolism of (4-aminobutyl)guanidine (agmatine) by decarboxylation. Arg is involved as a substrate in the urea cycle, and in nitric oxide production, the removal of ammonia from the body, cell division, and functioning of the immune system [2]. Lysine is an essential amino acid that plays an important role in the solvation, structure, and activity of proteins [3]. Arg and Lys are mostly exposed to the globular protein surface, and they are involved in the electrostatic interactions in protein-protein and protein-DNA complexes. Furthermore, through salt bridge formation with the complementary Glu and Asp residues, they contribute to the protein stability [4,5]. Arg and Lys side chain methylations are fundamental in post-translational modifications of histones, essential in epigenetic regulation. Methylations occur primarily at Lys and Arg residues on the histone backbone [6], affecting transcriptional regulation [7], assembly of heterochromatin, and cell cycle progression [8].

His is a conditionally essential amino acid, and plays an important role in the growth and repair of tissues, in the control of the metal elements transmission in biological bases [9], and in protein interactions [10]. Histidine is present in the naturally occurring histidine-rich protein II (HRP2), a common target for rapid diagnostic tests for malaria [11]

The unique structures of the three proteinogenic basic amino acids, Arg, Lys, and His, having long and positively charged side chains, make them interesting targets among the twenty amino acids for the molecular recognition of a specific side chain in a peptide [10,12].

In the literature, there are many examples reporting the use of synthetic macrocycles for Arg and Lys amino acid recognition [12–15], while the use of molecular receptors for histidine recognition is less explored [16].

Cavitands are a class of abiotic macromolecular receptors based on a resorcinarene scaffold having enforced cavities of molecular dimensions [17,18]. In the design of cavitands, the choice of the bridging groups connecting the phenolic hydroxyls of the resorcinarene scaffold determines shape, rigidity, dimensions, and complexation properties of the resulting cavity [19]. The selection can be made based on the particular class of analytes to be detected. Besides shape complementary, the selective recognition of a guest by a cavitand host requires the presence of specific weak interactions such as hydrogen bonding [20], π-π stacking [21], CH-π [22], and cation-π interactions [23]. The degree of sophistication achieved in controlling weak host–guest interactions in cavitands is such that it allows the rational design of synthetic receptors according to the analyte to be detected. The different functionalization of the cavity upper rim leads to the synthesis of cavitands presenting remarkable molecular recognition properties towards different guests, like aromatic and halogenated hydrocarbons [24–28], short chain alcohols [29,30], and *N*-methylammonium salts [31]. The ability of the cavitands to selectively recognize analytes can be exploited in solution [31], and at the gas-solid [25,26] and solid-liquid interface [28,30]. In particular, tetraphosphonate cavitands, named Tiiii [32], were successfully employed in the molecular recognition of biological relevant *N*-methyl ammonium salts, like sarcosine in urine [33,34]. The origin of Tiiii selectivity toward these species can be attributed to the presence of three different interaction modes: (i) $N^{+}\cdots O{=}P$ cation-dipole interactions; (ii) cation-π interactions of the $^{+}N–CH_3$ group with the π basic cavity; and (iii) two simultaneous hydrogen bonds between two adjacent P=O bridges and the two nitrogen protons. Recently, the ability of Tiiii in selectively complex different amino acids, both in solution and in the solid state, was demonstrated [35]. The work focused on the interaction of Tiiii with the ammonium/*N*-methyl ammonium group of 13 amino acids, neglecting the complexation of amino acids with biologically relevant substituents like guanidinium (Arg) and imidazole (His). These side chains are relevant targets for protein camouflage, since molecular receptors that recognize protein surfaces are important tools for recognition and activity modulation of proteins [36–40]. Sulfonatocalix[4]arene and cucurbiturils turned out to be the preferred synthetic receptors for protein surface recognition [41–43].

A recent work of Paton and co-workers describes the cation-π interactions of neutral aromatic ligands, using benzene as an archetype, with the cationic amino acid residues Arg and Lys via ab initio calculations, symmetry-adapted perturbation theory (SAPT), and a systematic meta-analysis of all available Protein Data Bank (PDB) X-ray structures [44]. The Lys-arene interaction is predicted to be weakened by polar surroundings to the point that it has a negligible effect on an aromatic ligand binding mode. By contrast, the cation-π interaction made by Arg residues with aromatic ligands is more robust to changes in the surrounding environment.

Here, we report the crystal structures of three complexes between the Tiiii cavitand as host and Arg, Lys, and His as guests. The purpose of the study is to determine which are the interactions responsible for amino acid recognition in the three cases, and to identify the role of the amino acids side chain in the complexation event.

2. Results and Discussion

Within a systematic crystallization program to assess the complexation properties of tetraphosphonate cavitands towards amino acids in the solid state [36], we obtained crystals of the fully protonated form of proteinogenic positively charged amino acids, such as Arg, Lys, and His. Crystallization trials were performed using the vapor diffusion method with sitting drops technique in Linbro multi-well plates containing trifluoroethanol (TFE) as solvent and PEG300 as precipitant (see Material and Methods). Within the class of tetraphosphonate cavitand, Tiiii[H, CH_3, CH_3], a compact synthetic receptor without substituents at its lower rim (Scheme 1), was chosen for its great tendency to crystallize. Then, accurate crystal structures of the Arg, Lys, and His amino acids complexed with the tetraphosphonate cavitand were determined by single crystal X-ray diffraction using synchrotron radiation as X-ray source, with crystals quenched at cryogenic temperatures (see Material and Methods).

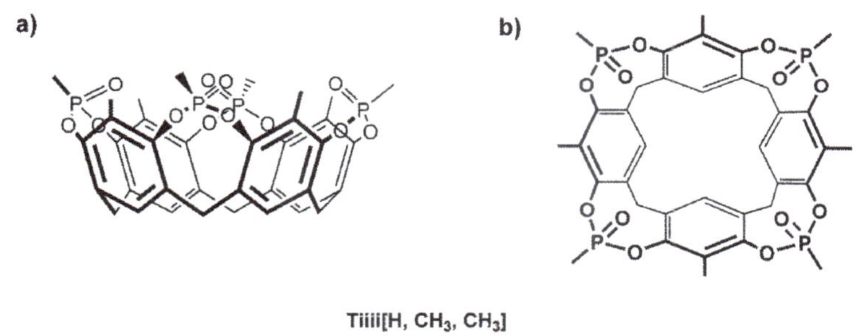

Scheme 1. Molecular sketch of the (**a**) side and (**b**) top view of cavitand Tiiii[H, CH_3, CH_3].

The analysis of the diffraction patterns revealed that the single crystals obtained by co-crystallization of the cavitand in the presence of the fully protonated amino acids were monoclinic ($P2_1$ space group) for the Arg and His complexes, while the Tiiii/Lys complex crystallized in the orthorhombic $P2_12_12$ space group. The asymmetric unit of the 2 Tiiii[H, CH_3, CH_3]•Arg·2HCl and 2 Tiiii[H, CH_3, CH_3]•Lys·2HCl crystal structures is composed of two host and one guest molecule, forming a 2:1 host-guest ditopic complex (Figure 1a,b), while, in the 2 Tiiii[H, CH_3, CH_3]•His·2HCl crystal structure, four crystallographically independent cavitands and two amino acids molecules form a sort of dimeric 2:1 host-guest ditopic complex (Figure 1c).

In all three structures, a chloride ion was modeled at the lower rim of the cavitand, forming weak C–H···Cl^- interactions with the aromatic CH fragments (green spheres in Figure 1).

The relevant geometric parameters describing the host-guest interactions are reported in Table 1. The X-ray structure determination revealed two binding geometries for the side chain fragments of arginine, lysine, and histidine. Guanidinium, ammonium, and imidazolium cationic groups of the side chain are hosted in the cavity of the phosphonate receptor, and are held in place by specific hydrogen bonding interactions with the P=O groups of a cavitand molecule (Figures 2 and 3).

The guanidinium group of Arg was found to be statistically disordered over two equally populated positions (conformation I and II). In conformation I, the side chain of Arg interacts with all four P=O groups of Tiiii (Table 1 and Figure 2a,b), and its terminal atoms are inserted below the main plane, defined by the oxygen atoms of the P=O groups. The planar guanidinium group forms a dihedral angle of 40° with the mean plane of the four oxygen atoms of the P=O groups of Tiiii host. In conformation II, the planar guanidinium group is more perpendicular with respect to the oxygen atoms' plane (dihedral angle of 63°), and forms only two strong H-bonds with two opposite P=O groups (Table 1 and Figure 2c,d). In this case, a NH_2 group is located in the center of the cavity of the tetraphosphonate host, and it is slightly more deeply inserted in the cavity with respect to the other conformation (Table 1 and Figure 2b,d).

The ammonium group of Lys, located above the P=O plane, interacts and forms strong H-bonds with two adjacent P=O groups of Tiiii (Table 1 and Figure 3a,b) and with a water molecule inserted in the Tiiii cavity. This internal water molecule forms two H-bonds with the remaining oxygen atoms of the P=O groups (Table 1 and Figure 3a,b).

Table 1. Relevant geometric parameters (Å) describing the host-guest interactions. Tiiii(1) and Tiiii(2) cavitands form H-bonds with the α-ammonium group and with the side chain of the amino acid, respectively. The out of plane (oop) of the more inserted non-H atoms of the amino acid (or water molecule) is calculated with respect to the mean plane, defined by the four oxygen atoms of the cavitand P=O groups. A negative sign is used for atoms below the plane and inserted in the cavity of Tiiii. His(1) and His(2) correspond to the two crystallographic independent molecules of the dimeric 2 Tiiii[H, CH_3, CH_3]•His·2HCl complex. Values in brackets correspond to the II conformation of Arg.

Tiiii(1) H-Bonds	Arg	Lys	His(1)	His(2)
$N–O_2$	2.71.	2.71.	2.81.	2.71.
$N–O_3$	2.80	2.83	2.77	2.78
$N–O_W$	2.74	2.72	2.64	2.66
$O_W–O_1$	2.90	2.87	2.76	2.81
$O_W–O_4$	2.83	2.90	2.73	2.87
oop N	0.70	0.55	0.89	0.89
oop O_w	−0.65	−0.86	0.36	0.13
Tiiii(2) H-Bonds	**Arg**	**Lys**	**His(1)**	**His(2)**
O_1	$N_{\eta2}$ 2.93 (2.70)	N_ζ 2.84	$N_{\delta1}$ 2.70	$N_{\delta1}$ 2.69
O_2	$N_{\eta1}$ 2.92 (3.47)	N_ζ 2.76		
O_3	N_ε 2.76	O_W 2.68	$N_{\varepsilon2}$ 2.71	$N_{\delta1}$ 2.67
O_4	$N_{\eta2}$ 2.73 (3.45)	O_w 2.71		
O_w		N_ζ 2.72		
oop	$N_{\eta1}$ −0.54 (−0.64)	N_ζ 0.72	$C_{\varepsilon1}$ −0.47	$C_{\varepsilon1}$ −0.38
oop		O_w 0.29		

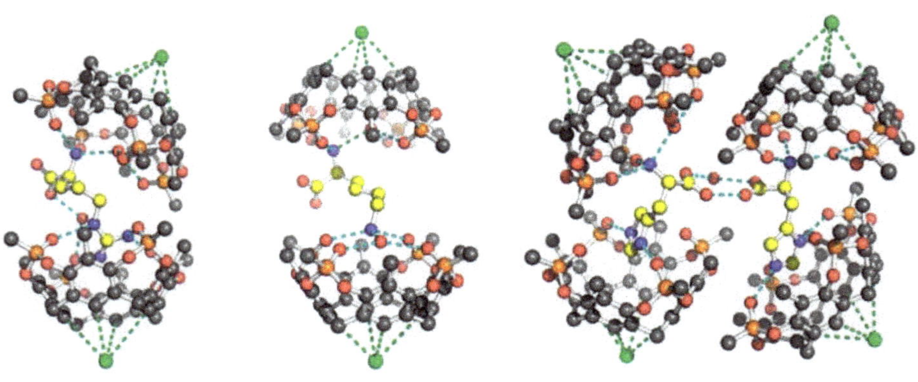

(a) (b) (c)

Figure 1. Ball and stick models of the crystallographic host-guest complexes. CPK colors with the amino acid guests evidenced by yellow carbon atoms. Hydrogen atoms have been omitted for clarity. The H-bonds are represented by cyan dashed lines. The chloride anions (green spheres) form weak C–H···Cl^- interactions (green dashed lines) with the lower rim of the cavitand (**a**) 2 Tiiii[H, CH_3, CH_3]•Arg·2HCl 2:1 ditopic complex (**b**) 2 Tiiii[H, CH_3, CH_3]•Lys·2HCl 2:1 ditopic complex (**c**) dimeric 2 Tiiii[H, CH_3, CH_3]•His·2HCl 2:1 ditopic complex.

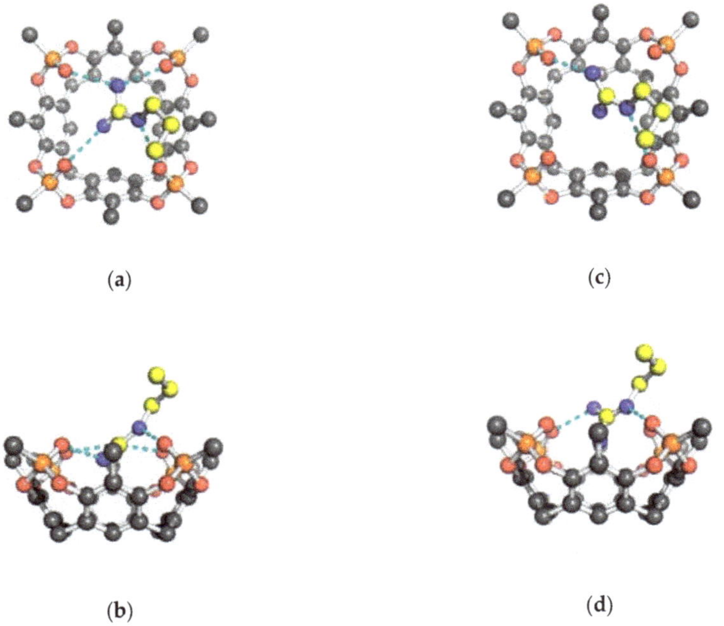

Figure 2. Orthogonal views of the H-bond interactions (cyan dashed lines) between the side chain of Arg amino acid (yellow carbons) and tetraphosphonate cavitand Tiiii (grey carbons) (**a**,**b**) conformation I; (**c**,**d**) conformation II. Other atoms color code: P, orange; N, blue; O, red. Hydrogen atoms have been omitted for clarity.

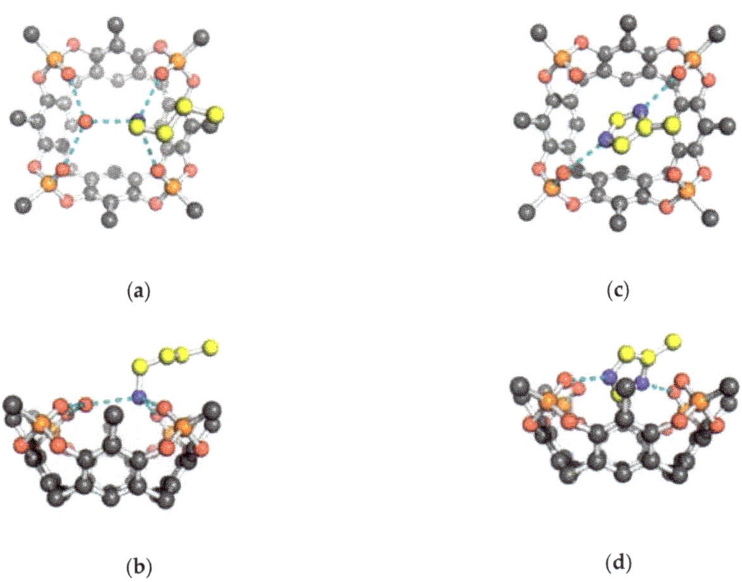

Figure 3. Orthogonal views of the H-bond interactions (cyan dashed lines) between the amino acid side chain (yellow carbons) and tetraphosphonate cavitand (black carbons) (**a**,**b**) 2 Tiiii[H, CH_3, CH_3]• Lys 2HCl complex; (**c**,**d**) 2 Tiiii[H, CH_3, CH_3]• His 2HCl complex. Hydrogen atoms have been omitted for clarity.

In the His crystal, both crystallographically independent 2:1 host-guest complexes have a very similar side chain interaction with the cavitand (Figure 1c). The imidazolium cationic group of the His amino acid is inserted with its $C_{\varepsilon 1}$ atom into the center of the cavity, and it forms two strong H-bonds with two opposite P=O groups (Table 1 and Figure 3c,d). The dihedral angle between the imidazolium group and the oxygen atoms plane of Tiiii is 57° and 67°, in the two crystallographically independent complexes, respectively. It is interesting to note that, while the interaction of the Lys side chain is assisted by a buried water molecule, no extra water molecules are presented in the cavity of Tiiii hosting the Arg or His side chains. This is probably due to the fact that the guanidinium and imidazolium groups are able to form H-bonds with two opposite P=O groups (see Figures 2c and 3c). In such arrangements, a water molecule does not have enough space inside the cavity and any possibility to form double H-bonds with two vicinal P=O groups, as in the case of the complexes with the ammonium group (Figures 3a and 4a). On the other hand, Lys side chain has a higher charge density compared with the Arg and His side chains, in which the conjugation between double bond and nitrogen lone pairs delocalizes the positive charge. Consequently, desolvation of the planar guanidinium and imidazolium groups should be more favorable with respect to the ammonium group of Lys amino acid.

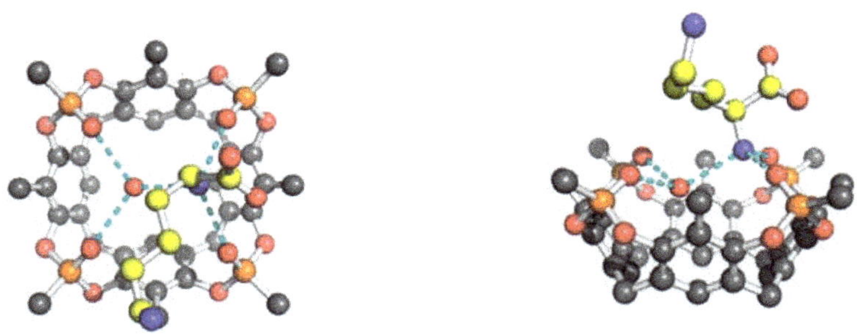

(**a**) (**b**)

Figure 4. Orthogonal views of the H-bond interactions (cyan dashed lines) between the α-ammonium group of Lys amino acid (yellow carbons) and Tiiii (gray carbons) mediated by a water molecule (red sphere). Hydrogen atoms have been omitted for clarity. (**a**) top view (**b**) side view.

In all three structures, the fully protonated form of the amino acid is ditopically complexed (Figure 1). The positively charged α-ammonium groups form H-bonds with the P=O groups, and with a water molecule hosted in the cavity of a second tetraphosphonate molecule. The nitrogen atom is above the O mean plane, defined by the oxygen atoms of the P=O units, while the water molecule is located below this plane (Figure 4). More specifically, in the Lys complex, this water molecule is much more inserted in the cavity with respect to the analogous water molecule on the opposing cavitand that hosts the ζ-ammonium group (1.15 Å, see Table 1). This different behavior can be associated with the presence, in the α ammonium group complexation, of a methylene β carbon that covers the cavity (Figure 4a). On the other hand, the complexation of the linear side chain of Lys leaves the water molecule free to adopt a different position in the hydrophobic cavity (Figure 3a).

This amino acid/cavitand interaction mediated by a water molecule has been observed in all three complexes, namely with Arg, Lys, and His, and it involves the constant part of the α-amino acid. This "nonspecific" amino acid H-bond interaction was already reported in previously determined host-guest complexes involving the hydrophobic amino acids: Ala, Val, Leu, and Ile [35].

A particular H-bond interaction between the carboxylic acid group of Arg and a P=O group of cavitand mediated by a water molecule has been observed (Figure 5a). This interaction, noted only in 2 Tiiii[H, CH_3, CH_3]•Arg 2HCl complex, links the two cavitands of the 2:1 complex by an H-bond network that involves the α-ammonium and α-carboxylic acid groups of the guest.

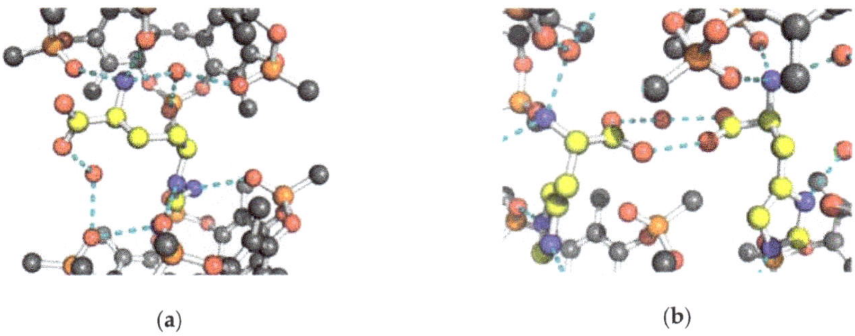

Figure 5. H-bond interactions between the amino acid (yellow carbons) and tetraphosphonate cavitand (gray carbons) (**a**) H-bond interaction between the carboxylic acid group of Arg amino acid and Tiiii mediated by a water molecule (red sphere); (**b**) dimerization of the 2 Tiiii[H, CH_3, CH_3]•His 2HCl complex mediated by a water molecule (red sphere). Hydrogen atoms have been omitted for clarity.

An interesting difference within the positively charge proteinogenic amino acids is the formation, in the solid state, of water-assisted dimerization of the cavitand/histidine 2:1 complex. In this structure, a bridged water molecule is held by two carboxylic acid groups of the dimerized amino acid (Figure 5b). From the topological point of view, this dimer shows that the couple of 2:1 complexes are mutually oriented in such a way that the side chain of the guest amino acid has the same direction and orientation (Figure 1c).

The cavitand units of the 2:1 complexes are in a typical cone conformation with the amino acid guest, connecting the two bowls, oriented slantwise to allow for optimal host-guest interaction. Despite the formation of 2:1 complexes with similar α-ammonium/Tiiii interaction and analogue side chain complexation, the crystal structures of the three amino acid/cavitand complexes show a quite different spatial arrangement of the two cavitands.

The overall shape of the 2 Tiiii[H, CH_3, CH_3]•Arg 2HCl complex is quasi-capsular, with the Arg amino acid captured between the two almost-aligned cavitands. In particular, the two bowls, facing each other, hold the Arg amino acid in a clamp-like arrangement. The dihedral angle between the mean planes of the four oxygen atoms of the P=O groups of two cavitands, representing the opening angle of the clamp, is 36° (Figure 6a).

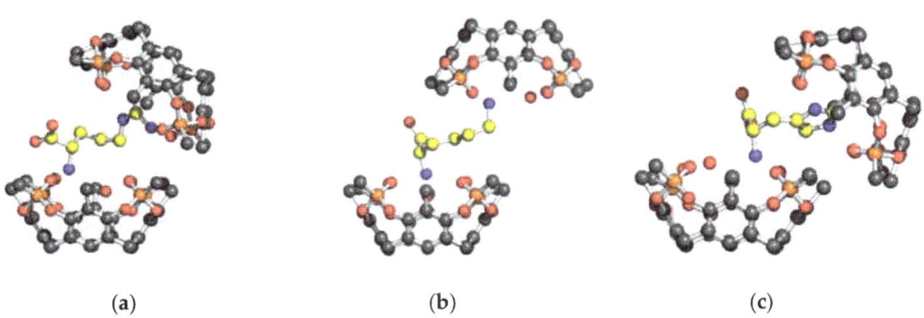

Figure 6. Ball and stick models of the Tiiii/amino acid 2:1 complexes view parallel to the mean planes of the four oxygen atoms of the P=O groups of Tiiii hosts (**a**) 2 Tiiii[H, CH_3, CH_3]•Arg 2HCl 2:1 ditopic complex (**b**) 2 Tiiii[H, CH_3, CH_3]•Lys 2HCl 2:1 ditopic complex (**c**) 2 Tiiii[H, CH_3, CH_3]•His 2HCl 2:1 ditopic complex. Hydrogen atoms have been omitted for clarity.

In the 2 Tiiii[H, CH_3, CH_3]•Lys 2HCl complex, the two cavitand bowls are almost perfectly anti-parallel, while they are offset by about a half cavitand (Figure 6b). This misalignment significantly opens the clamp formed by the two host molecules.

In the 2 Tiiii[H, CH_3, CH_3]•His 2HCl complex, the two cavitand bowls are both shifted laterally, by more than half cavitand, and rotated by about 48° (dihedral angle between the P=O planes), resulting in a more open clamp with respect to the other two cases (Figure 6c).

With respect to the ditopic behavior of positively charge amino acids, an unexpected result obtained in this crystal structure analysis is the trend observed in the opening of the clamp of the 2:1 complex. The trend observed, His > Lys > Arg, is in reverse order with respect to the length of amino acid side chain Arg > Lys > His. This reverse trend can be attributed to dimer formation, in the case of the 2 Tiiii[H, CH_3, CH_3]•His 2HCl complex (Figure 1c), and to the different behavior between Arg and Lys residues in the insertion of the side chain into the cavitand (Table 1 and Figures 2b and 3b).

3. Materials and Methods

3.1. Co-Crystallization Experiments

For the co-crystallization experiments, the phosphonate cavitand Tiiii[H, CH_3, CH_3], with the P=O groups pointing inward the cavity, was synthesized as previously described [45], and 2,2,2-trifluoroethanol (TFE) and amino acids (L-arginine, L-lysine, and L-histidine) were purchased from Sigma-Aldrich (St. Louis, MO, USA), and used as supplied.

Crystals of host-guest complexes containing amino acids were obtained by co-crystallization microscale experiments with the sitting drop vapor diffusion technique. The co-crystallization solution was prepared by adding, to a TFE solution (30 mM) of cavitand Tiiii[H, CH_3, CH_3], a solution containing the desired amino acid. The solutions of amino acids were prepared in 1 M HCl (see Table S1 for experimental details). As a general procedure, 4 μL of the solution containing the cavitand and the amino acid were set on a microbridge, and then 4 μL of the reservoir solution were added to the drop. The drop was left to equilibrate against 1 mL of reservoir solution, containing the organic polymer polyethylene glycol (PEG) 300 in the range between 5 and 50% (*v/v*) as precipitant agent. Crystals were left to grow at the temperatures of 4 °C. After 1 month, good quality crystals suitable for X-ray diffraction analysis were obtained.

3.2. Crystal Structure Determination

Data collections were carried out at the Macromolecular crystallography XRD1 beamline of the Elettra synchrotron (Trieste, Italy), by employing the rotating crystal method and the cryo-cooling technique. Routinely, the crystal was mounted in a loop and flash frozen at 100 K with liquid nitrogen without adding further cryoprotectant, thanks to the presence of PEG 300 in the mother liquor. Diffraction data of Arg, Lys, and His/cavitand complexes were indexed and integrated using the XDS package [46]. Scaling was carried out with AIMLESS, for the dataset collected from crystals of Arg, Lys, and His/cavitand supramolecular complexes.

All structures were solved by direct methods using SIR2011 [47]. Non-hydrogen atoms at full occupancy, or with population higher than 0.5, were anisotropically refined (H atoms at the calculated positions) by full-matrix least-squares methods on F^2 using SHELXL-13. Restraints on the geometrical and thermal parameters of the disordered solvent molecules (DFIX, DELU, ISOR) were introduced during the last refinement cycles. Several TFE co-crystallized solvent molecules were found in the asymmetric units for all the structures. In all the structures, electronic density that could be attributed to highly disordered solvent molecules was detected. The contribute of the disordered solvent to the overall scattering was removed through the SQUEEZE function of PLATON software [48]. A detailed discussion of the refinement for each structure is provided below.

3.2.1. Structure Refinement of 2 Tiiii[H, CH_3, CH_3]•Arg 2HCl Complex

In the asymmetric unit of the complex of Tiiii[H, CH_3, CH_3] with doubly protonated Arg, a dimeric supramolecular host-guest complex was found together with 2.9 TFE solvent molecules and 3.7 water molecules. The guanidinium fragment of arginine was found disordered over two positions refined with equal occupancy. The residual electron density of 155 electrons/cell, found in the voids of the crystal (corresponding to 9% of the cell volume), was attributed to about 1.5 highly disordered CF_3CH_2OH molecules in asymmetric units. A refinement using reflections modified by the SQUEEZE procedure [48] behaved well, and the *R*-factor was reduced from 9.1% to 7.5%.

3.2.2. Structure Refinement of 2 Tiiii[H, CH_3, CH_3]•Lys 2HCl Complex

In the asymmetric unit of the crystal of the complex of Tiiii[H, CH_3, CH_3] with doubly protonated Lys, a dimeric host-guest complex was detected together with 0.8 TFE solvent molecules and 2.8 water molecules. The residual electron density of 1097 electrons/cell, found in the voids of the crystal (corresponding to 28% of the cell volume), was attributed to a disordered chlorine ion and about 5.3 CF_3CH_2OH molecules in asymmetric units. A refinement using reflections modified by the SQUEEZE procedure [48] behaved well, and the *R*-factor was reduced from 14.3% to 6.8%.

3.2.3. Structure Refinement of 2 Tiiii[H, CH_3, CH_3]•His 2HCl Complex

In the asymmetric unit of the crystal of the complex 2 Tiiii[H, CH_3, CH_3] with doubly protonated His, two crystallographically independent dimeric host-guest complexes were detected together with 7.82 TFE solvent molecules and four water molecules. The residual electron density of 191.7 electrons/cell, found in the voids of the crystal (corresponding to 7% of the cell volume), was attributed to about 2.1 CF_3CH_2OH molecules in asymmetric units. A refinement using reflections modified by the SQUEEZE procedure behaved well, and the *R*-factor was reduced from 7.4% to 6.8%.

Crystal data and refinement details are reported in Table S2. CCDC 1869186 (Tiiii/His), 1869187 (Tiiii/Arg), and 1876524 (Tiiii/Lys) contain the supplementary crystallographic data for this paper. These data can be obtained free of charge via http://www.ccdc.cam.ac.uk/conts/retrieving.html (or from the CCDC, 12 Union Road, Cambridge CB2 1EZ, UK; Fax: +44-1223-336033; E-mail: deposit@ccdc.cam.ac.uk)

4. Conclusions

Molecular recognition of amino acids is a topic of particular interest because it involves several important applicative aspects, such as advanced therapeutic approaches towards target proteins, as well as sensor applications for biochemical analysis and immobilization techniques for protein purification and/or characterization [38,49]. Within a systematic crystallization program, to assess the complexation properties of a cavitand molecule towards amino acids [35], we obtained crystals of the fully protonated form of the three proteinogenic positively charged amino acids, Arg, Lys, and His complexed with the tetraphosphonate cavitand Tiiii[H, CH_3, CH_3] in cone conformation. The X-ray structure revealed that, in all three complexes, the fully protonated form of these amino acids has a ditopic behavior. Each amino acid is complexed by two tetraphosphonate cavitand molecules. The side chain of the amino acid is hosted by a cavitand molecule from one side, while the α-ammonium group interacts with a second cavitand molecule on the opposite side. Then, an overall 2:1 host-guest complex is assembled. In particular, the guanidinium, ammonium, and imidazolium cationic groups of the side chain of the amino acids, Arg, Lys, and His, are respectively hosted in the cavity of a phosphonate receptor, and they are held in place by specific hydrogen bonding interactions with the P=O groups of the cavitand molecule. The positively charged α-ammonium group of the three amino acids forms H-bonds with two P=O groups, and with a water molecule hosted in the cavity of a second tetraphosphonate molecule. The chloride counter ion has a crucial role in the stabilization of this complex because it has been detected at the bottom of the lower rim of both cavitands, just on the opposite site of the cavity. The chloride ions form weak C–H···Cl^- interactions with the aromatic CH fragments of both Tiiii units, balancing the two positive charges of the guest. It is interesting to note that, in the solid state, the cavitand/histidine 2:1 complex forms a dimer through the carboxylic acid group of the amino acid. This dimerization, with formation of a supramolecular 4:2 host-guest complex, is assisted by a crucial bridged water molecule. This phenomenon has been observed only for this specific amino acid complex. The analysis of the relative disposition of the cavitands in the 2:1 host-guest complexes has evidenced an unexpected ditopic behavior of the proteinogenic positively charged amino acids. In particular, the trend observed in the opening of the clamp formed by the two cavitands, His > Lys > Arg, is inverted with respect to the trend of the length of amino acid side chain Arg > Lys > His. The bulkiest guest, Arg, is almost completely encapsulated by two phosphonate cavitands, while the shorter Lys and the even shorter His gradually show a more open structure of the complex. The overall architecture observed in these complexes provide support for the rational design of ditopic cavitands as molecular clamps for encapsulation of proteinogenic positively charged amino acids. [50,51].

Supplementary Materials: The following are available online. Table S1: Experimental details for the co-crystallization experiments of the host-guest complexes of cavitand Tiiii[H, CH_3, CH_3] with the selected amino acids., Table S2: Crystal data and structure refinement details for the host-guest complexes.

Author Contributions: Conceptualization, R.P. and S.G.; Methodology, E.D.; Formal Analysis, G.B.; Investigation, G.B., R.P., E.D., S.G.; Data Curation, G.B.; Writing—Original Draft Preparation, R.P., S.G.; Writing—Review and Editing, E.D.; Funding Acquisition, S.G.

Funding: The University of Trieste is acknowledged for financial support, FRA2016.

Acknowledgments: We thank the Elettra Synchrotron (Trieste, Italy) and the staff of the XRD1 beamline for their technical assistance.

Conflicts of Interest: The authors declare no conflict of interest.

References

1. Tapiero, H.; Mathé, G.; Couvreur, P.; Tew, K.D. Arginine. *Biomed. Pharmacother.* **2002**, *56*, 439–445. [CrossRef]
2. Stechmiller, J.K.; Childress, B.; Cowan, L. Arginine Supplementation and Wound Healing. *Nutr. Clin. Pract.* **2005**, *20*, 52–61. [CrossRef] [PubMed]
3. Vila, J.A.; Ripoll, D.R.; Villegas, M.E.; Vorobjev, Y.N.; Scheraga, H.A. Role of Hydrophobicity and Solvent-Mediated Charge-Charge Interactions in Stabilizing α-Helices. *BioPhys. J.* **1998**, *75*, 2637–2646. [CrossRef]
4. Strickler, S.S.; Gribenko, A.V.; Gribenko, A.V.; Keiffer, T.R.; Tomlinson, J.; Reihle, T.; Loladze, V.V.; Makhatadze, G.I. Protein Stability and Surface Electrostatics: A Charged Relationship. *Biochemistry* **2006**, *45*, 2761–2766. [CrossRef]
5. Borders, C.L., Jr.; Broadwater, J.A.; Bekeny, P.A.; Salmon, J.E.; Lee, A.S.; Eldridge, A.M.; Pett, V.B. A structural role for arginine in proteins: Multiple hydrogen bonds to backbone carbonyl oxygens. *Protein Sci.* **1994**, *3*, 541–548. [CrossRef]
6. Tsukada, Y.; Fang, J.; Erdjument-Bromage, H.; Warren, M.E.; Borchers, C.H.; Tempst, P.; Zhang, Y. Histone demethylation by a family of JmjC domain-containing proteins. *Nature* **2006**, *439*, 811–816. [CrossRef]
7. Stewart, M.D.; Li, J.; Wong, J. Relationship between Histone H3 Lysine 9 Methylation, Transcription Repression, and Heterochromatin Protein 1 Recruitment. *J. Mol. Cell. Biol.* **2005**, *25*, 2525–2538. [CrossRef] [PubMed]
8. Nakayama, J.; Rice, J.C.; Strahl, B.D.; Allis, C.D.; Grewal, S.I. Role of Histone H3 Lysine 9 Methylation in Epigenetic Control of Heterochromatin Assembly. *Science* **2001**, *292*, 110–113. [CrossRef] [PubMed]
9. Chen, G.N.; Wu, X.P.; Duan, J.P.; Chen, H.Q. A study on electrochemistry of histidine and its metabolites based on the diazo coupling reaction. *Talanta* **1999**, *49*, 319–330.
10. Liao, S.-M.; Du, Q.-S.; Meng, J.-Z.; Pang, Z.-W.; Huang, R.-B. The multiple roles of histidine in protein interactions. *Chem. Cent. J.* **2013**, *7*, 44. [CrossRef]

11. Beadle, C.; Long, G.W.; Weiss, W.R.; McElroy, P.D.; Maret, S.M.; Oloo, A.J.; Hoffman, S.L. Diagnosis of malaria by detection of Plasmodium falciparum HRP-2 antigen with a rapid dipstick antigen-capture assay. *Lancet* **1994**, *343*, 564–568. [CrossRef]
12. Chen, H.; Gu, L.; Yin, Y.; Koh, K.; Lee, J. Molecular Recognition of Arginine by Supramolecular Complexation with Calixarene Crown Ether Based on Surface Plasmon Resonance. *Int. J. Mol. Sci.* **2011**, *12*, 2315–2324. [CrossRef] [PubMed]
13. Fokkens, M.; Schrader, T.; Klärner, F.-G. A Molecular Tweezer for Lysine and Arginine. *J. Am. Chem. Soc.* **2005**, *127*, 14415–14421. [CrossRef] [PubMed]
14. Ngola, S.M.; Kearney, P.C.; Mecozzi, S.; Russell, K.; Dougherty, D.A. A Selective Receptor for Arginine Derivatives in Aqueous Media. Energetic Consequences of Salt Bridges that Are Highly Exposed to Water. *J. Am. Chem. Soc.* **1999**, *121*, 1192–1201. [CrossRef]
15. Julian, R.R.; Akin, M.; May, J.A.; Stoltz, B.M.; Beauchamp, J.L. Molecular recognition of arginine in small peptides by supramolecular complexation with dibenzo-30-crown-10 ether. *Int. J. Mass Spectrom.* **2002**, *220*, 87–96. [CrossRef]
16. Chen, Z.; Liu, J.; Han, Y.; Zhu, L. A novel histidine assay using tetraphenylporphyrin manganese (III) chloride as a molecular recognition probe by resonance light scattering technique. *Anal. Chim. Acta* **2006**, *570*, 109–115. [CrossRef]
17. Cram, D.J.; Cram, J.M. *Container Molecules and Their Guests*; Stoddart, J.F., Ed.; The Royal Society of Chemistry: Cambridge, UK, 1994.
18. Pinalli, R.; Pedrini, A.; Dalcanale, E. Cavitands. In *Comprehensive Supramolecular Chemistry II*; Atwood, J.L., Ed.; Elsevier: Amsterdam, The Netherlands, 2017; pp. 87–115. ISBN 9780128031995.
19. Pinalli, R.; Dalcanale, E.; Ugozzoli, F.; Massera, C. Resorcinarene-based cavitands as building blocks for crystal engineering. *CrystEngComm* **2016**, *18*, 5788–5802. [CrossRef]
20. Rebek, J., Jr. Molecular Recognition with Model Systems. *Angew. Chem. Int. Ed. Engl.* **1990**, *29*, 245–255. [CrossRef]
21. Hunter, C.H.; Lawson, K.R.; Perkins, J.; Urch, C.J. Aromatic interactions. *J. Chem. Soc., Perkin Trans.* **2001**, 2, 651–669. [CrossRef]
22. Nishio, M.; Hirota, M.; Umezawa, Y. *The CH-π Interactions: Evidence, Nature and Consequences*; Wiley-VCH: New York, NY, USA, 1998.
23. Dougherty, D.A. The Cation-π Interaction. *Acc. Chem. Res.* **2013**, *46*, 885–893. [CrossRef]
24. Bianchi, F.; Mattarozzi, M.; Betti, P.; Bisceglie, F.; Careri, M.; Mangia, A.; Sidisky, L.; Ongarato, S.; Dalcanale, E. Innovative Cavitand-Based Sol-Gel Coatings for the Environmental Monitoring of Benzene and Chlorobenzenes via Solid-Phase Microextraction. *Anal. Chem.* **2008**, *80*, 6423–6430. [CrossRef]
25. Tudisco, C.; Fragalà, M.E.; Giuffrida, A.E.; Bertani, F.; Pinalli, R.; Dalcanale, E.; Compagnini, G.; Condorelli, G.G. Hierarchical Route for the Fabrication of Cavitand-Modified Nanostructured ZnO Fibers for Volatile Organic Compound Detection. *J. Phys. Chem. C* **2016**, *120*, 12611–12617. [CrossRef]
26. Trzciński, J.W.; Pinalli, R.; Riboni, N.; Pedrini, A.; Bianchi, F.; Zampolli, S.; Elmi, I.; Massera, C.; Ugozzoli, F.; Dalcanale, E. In Search of the Ultimate Benzene Sensor: The EtQxBox Solution. *ACS Sensors* **2017**, *2*, 590–598. [CrossRef] [PubMed]
27. Pinalli, R.; Pedrini, A.; Dalcanale, E. Environmental gas sensing with cavitands. *Chem Eur. J.* **2018**, *24*, 1010–1019. [CrossRef]
28. Giannetto, M.; Pedrini, A.; Fortunati, S.; Brando, D.; Milano, S.; Massera, C.; Tatti, R.; Verucchi, R.; Careri, M.; Dalcanale, E.; et al. Sensing of halogenated aromatic hydrocarbons in water with a cavitand coated piezoelectric device. *Sens. Actuators B Chem.* **2018**, *276*, 340–348. [CrossRef]
29. Maffei, F.; Betti, P.; Genovese, D.; Montalti, M.; Prodi, L.; De Zorzi, R.; Geremia, S.; Dalcanale, E. Highly Selective Chemical Vapor Sensing by Molecular Recognition: Specific Detection of C_1–C_4 Alcohols with a Fluorescent Phosphonate Cavitand. *Angew. Chem. Int. Ed.* **2011**, *50*, 4654–4657. [CrossRef]
30. Melegari, M.; Massera, C.; Pinalli, R.; Yebeutchou, R.M.; Dalcanale, E. Supramolecular sensing of short chain alcohols with mixed-bridged thio-phosphonate cavitands. *Sens. Actuators B Chem.* **2013**, *179*, 74–80. [CrossRef]
31. Dionisio, M.; Oliviero, G.; Menozzi, D.; Federici, S.; Yebeutchou, R.M.; Schmidtchen, F.P.; Dalcanale, E.; Bergese, P. Nanomechanical Recognition of *N*-Methylammonium Salts. *J. Am. Chem. Soc.* **2012**, *134*, 2392–2398. [CrossRef]
32. Pinalli, R.; Suman, M.; Dalcanale, E. Cavitands at work: From molecular recognition to supramolecular sensors. *Eur. J. Org. Chem.* **2004**, *3*, 451–462. [CrossRef]

33. Biavardi, E.; Tudisco, C.; Maffei, F.; Motta, A.; Massera, C.; Condorelli, G.G.; Dalcanale, E. Exclusive recognition of sarcosine in water and urine by a cavitand-functionalized silicon surface. *Proc. Natl. Acad. Sci. USA* **2012**, *109*, 2263–2268. [CrossRef]
34. Valenti, G.; Rampazzo, E.; Biavardi, E.; Villani, E.; Fracasso, G.; Marcaccio, M.; Bertani, F.; Ramarli, D.; Dalcanale, E.; Paolucci, F.; et al. An electrochemiluminescence-supramolecular approach to sarcosine detection for early diagnosis of prostate cancer. *Faraday Discuss.* **2015**, *185*, 299–309. [CrossRef] [PubMed]
35. Pinalli, R.; Brancatelli, G.; Pedrini, A.; Menozzi, D.; Hernández, D.; Ballester, P.; Geremia, S.; Dalcanale, E. The origin of selectivity in the complexation of *N*-methyl amino acids by tetraphosphonate cavitands. *J. Am. Chem. Soc.* **2016**, *138*, 8569–8580. [CrossRef]
36. Liu, Y.; Perez, L.; Mettry, M.; Easley, C.J.; Hooley, R.J.; Zhong, W. Self-Aggregating Deep Cavitand Acts as a Fluorescence Displacement Sensor for Lysine Methylation. *J. Am. Chem. Soc.* **2016**, *138*, 10746–10749. [CrossRef]
37. Liu, Y.; Perez, L.; Gill, A.D.; Mettry, M.; Li, L.; Wang, Y.; Hooley, R.J.; Zhong, W. Site-Selective Sensing of Histone Methylation Enzyme Activity via an Arrayed Supramolecular Tandem Assay. *J. Am. Chem. Soc.* **2017**, *139*, 10964–10967. [CrossRef]
38. Pinalli, R.; Pedrini, A.; Dalcanale, E. Biochemical Sensing with cavitands. *Chem. Soc. Rev.* **2018**, *47*, 7006–7026. [CrossRef] [PubMed]
39. McGovern, R.E.; Fernandes, H.; Khan, A.R.; Power, N.P.; Crowley, P.B. Protein camouflage in cytochrome c-calixarene complexes. *Nat. Chem.* **2012**, *4*, 527–533. [CrossRef]
40. Rennie, M.L.; Fox, G.C.; Pérez, J.; Crowley, P.B. Auto-regulated Protein Assembly on a Supramolecular Scaffold. *Angew. Chem. Int. Ed.* **2018**, *57*, 13764–13769. [CrossRef]
41. McGovern, R.E.; McCarthy, A.A.; Crowley, P.B. Protein assembly mediated by sulfonatocalix[4]arene. *Chem. Commun.* **2014**, *50*, 10412–10415. [CrossRef]
42. Chinai, J.M.; Taylor, A.B.; Ryno, L.M.; Hargreaves, N.D.; Morris, C.A.; Hart, P.J.; Urbach, A.R. Molecular recognition of insulin by a synthetic receptor. *J. Am. Chem. Soc.* **2011**, *133*, 8810–8813. [CrossRef]
43. Guagnini, F.; Antonik, P.M.; Rennie, M.L.; O'Byrne, P.; Khan, A.R.; Pinalli, R.; Dalcanale, E.; Crowley, P.B. Cucurbit[7]uril-Dimethyllysine Recognition in a Model Protein. *Angew. Chem. Int. Ed.* **2018**, *57*, 7126–7130. [CrossRef]
44. Kumar, K.; Woo, S.M.; Siu, T.; Cortopassi, W.A.; Duarte, F.; Paton, R.S. Cation–π interactions in protein–ligand binding: Theory and data-mining reveal different roles for lysine and arginine. *Chem. Sci.* **2018**, *9*, 2655–2665. [CrossRef] [PubMed]
45. Melegari, M.; Suman, M.; Pirondini, L.; Moiani, D.; Massera, C.; Ugozzoli, F.; Kalenius, E.; Vainiotalo, P.; Mulatier, J.-C.; Dutasta, J.-P.; et al. Supramolecular Sensing with Phosphonate Cavitands. *Chem. Eur. J.* **2008**, *14*, 5772–5779. [CrossRef] [PubMed]
46. Kabsch, W. XDS. *Acta Cryst.* **2010**, *D66*, 125–132. [CrossRef] [PubMed]
47. Burla, M.C.; Caliandro, R.; Camalli, M.; Carrozzini, B.; Cascarano, G.L.; Giacovazzo, C.; Mallamo, M.; Mazzone, A.; Polidori, G.; Spagna, R. SIR2011: A new package for crystal structure determination and refinement. *J. Appl. Cryst.* **2012**, *45*, 357–361. [CrossRef]
48. Spek, A.L. PLATON SQUEEZE: A tool for the calculation of the disordered solvent contribution to the calculated structure factors. *Acta Cryst.* **2015**, *C71*, 9–18. [CrossRef]
49. Van Dun, S.; Ottmann, C.; Milroy, L.C.; Brunsveld, L. Supramolecular Chemistry Targeting Proteins. *J. Am. Chem. Soc.* **2017**, *139*, 13960–13968. [CrossRef] [PubMed]
50. Menozzi, E.; Busi, M.; Ramingo, R.; Campagnolo, M.; Geremia, S.; Dalcanale, E. Design and Self-Assembly of Ditopic and Tetratopic Cavitand Complexes. *Chem. Eur. J.* **2005**, *11*, 3136–3148. [CrossRef]
51. Busi, M.; Cantadori, B.; Boccini, F.; De Zorzi, R.; Geremia, S.; Dalcanale, E. Molecular Recognition with Ditopic Cavitand Re Complexes. *Eur. J. Org. Chem.* **2011**, *2011*, 2629–2642. [CrossRef]

Sample Availability: Samples of the compound Tiiii[H, CH_3, CH_3] is available from the authors.

molecules

MDPI

Review

Stereoselective Multicomponent Reactions in the Synthesis or Transformations of Epoxides and Aziridines †

Allan Ribeiro da Silva [1], Deborah Araujo dos Santos [1,2], Marcio Weber Paixão [1] and Arlene Gonçalves Corrêa [1,*]

[1] Centre of Excellence for Research in Sustainable Chemistry, Department of Chemistry, Federal University of São Carlos, 13565-905 São Carlos, SP, Brazil; allan.silva.bsb@hotmail.com (A.R.d.S.); deborah.araujo89@gmail.com (D.A.d.S.); mwpaixao@ufscar.br (M.W.P.)

[2] São Carlos Institute of Chemistry, University of São Paulo, 13563-120 São Carlos, SP, Brazil

* Correspondence: agcorrea@ufscar.br; Tel.: +55-16-33518082

† Dedicated to Professor Madalena Pinto on the occasion of her 70th anniversary.

Academic Editors: Maria Elizabeth Tiritan, Madalena Pinto and Carla Sofia Garcia Fernandes
Received: 16 January 2019; Accepted: 7 February 2019; Published: 11 February 2019

Abstract: Small ring heterocycles, such as epoxides and aziridines, are present in several natural products and are also highly versatile building blocks, frequently involved in the synthesis of numerous bioactive products and pharmaceuticals. Because of the potential for increased efficiency and selectivity, along with the advantages of environmentally benign synthetic procedures, multicomponent reactions (MCRs) have been explored in the synthesis and ring opening of these heterocyclic units. In this review, the recent advances in MCRs involving the synthesis and applications of epoxides and aziridines to the preparation of other heterocycles are discussed emphasizing the stereoselectivity of the reactions.

Keywords: multicomponent reactions; epoxides; aziridine; asymmetric synthesis; green synthesis

1. Introduction

Heterocyclic compounds are very important due to their biological and/or pharmacological properties. In particular, the small ring heterocycles, such as epoxides and aziridines, are present in several bioactive natural products [1]. Among the epoxides, the following examples of enzymatic inhibitors deserve mention: scyphostatin, which was isolated from a mycelial extract of *Trichopeziza mollissima* by Nara et al. in 1999, is a neutral sphingomyelinase inhibitor [2]; and E-64, a promiscuous irreversible cysteine protease inhibitor that is broadly reactive toward the papain family, was isolated from *Aspergillus japonicus* in 1978 by Hanada et al. [3]. Moreover, 3,4-epoxy-6,9-heneicosadiene was identified as a sex pheromone component of the eucalyptus brown-looper *Thyrinteina arnobia*, which is considered an important pest in Brazilian native plants [4].

Concerning the aziridine rings, we should emphasize the antitumor antibiotics mitomycin C, isolated from cultures of *Streptomyces caespitosus*, that has been used in combination chemotherapy for a variety of tumors and as an antifibrotic agent for several surgical procedures [5]; and azinomycin B, a remarkable natural product containing both epoxide and aziridine rings in its structure, originally isolated from *Streptomyces sahachiroi* (Figure 1) [6].

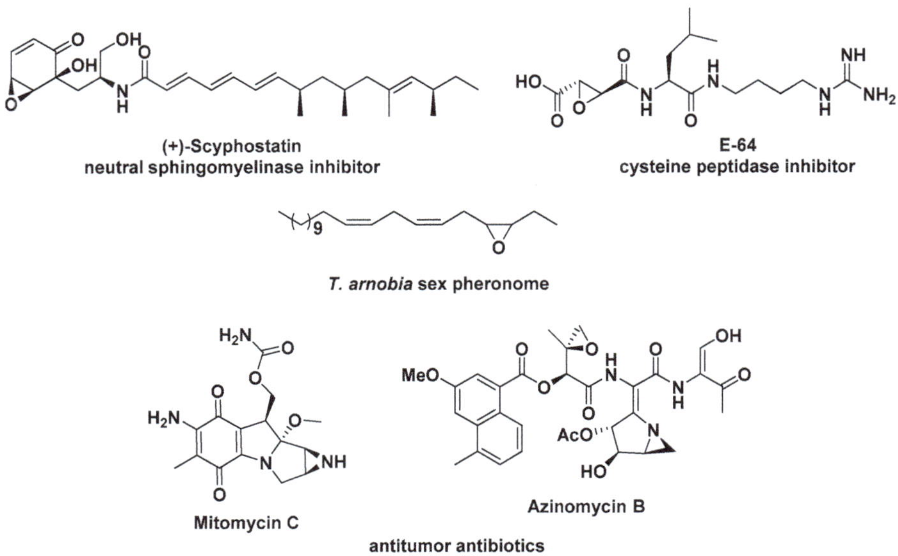

Figure 1. Examples of natural products possessing epoxide and aziridine rings.

The three-membered rings are also highly versatile building blocks, frequently involved in the synthesis of natural compounds and incorporated into the molecular scaffold of numerous bioactive products and pharmaceuticals [7,8].

Several methodologies have been already reported for the stereoselective synthesis of epoxides and aziridines from alkenes catalyzed by transition metals, such as the seminal work by Professor Sharpless using Ti(O*i*Pr)$_4$ and diethyl tartrate [9], the Jacobsen–Katsuki epoxidation employing Mn(III)–salen complexes [10,11], and also Rh(II)-aziridination [12,13]. The asymmetric synthesis of epoxides, cyclopropanes, and aziridines using organocatalysis was reviewed by Meninno and Lattanzi [14]. Moreover, a recent review covered the synthesis of natural and/or biologically relevant three-membered heterocycles and derivatives based on asymmetric epoxidation, aziridination, azirination, and thiirination reactions [15]. The use of three-membered rings as intermediates to the synthesis of biologically active compounds has also been extensively explored [16–18].

The development of efficient, stereoselective, and environmentally benign synthetic protocols has become extremely important in the last several decades. An effective approach to achieve these goals is to carry out a multistep reaction or synthesis in a one-pot procedure, since several synthetic transformations and bond-forming steps can be carried out in a single pot, while circumventing several purification procedures at the same time. This strategy, named "pot economy" by Professor Hayashi, minimizes chemical waste generation, saves time, and simplifies practical aspects [19].

An outstanding example by Smith III and Kim described the use of silyl 1,3-dithianes and epoxides in a solvent-controlled Brook rearrangement to access differentially protected monosilyl 1,5-diol moieties with precise stereocontrols [20]. This chemistry was further explored with the use of *N*-Ts aziridines, as the second electrophile to access protected 1,5-amino alcohols, which could be exploited as advanced intermediates for the construction of 3,5-disubstituted indolizidine rings. According to the authors, this transformation probably occurs via an intramolecular alkylation of the 1,5-amino alcohols, where the nitrogen would act as a nucleophile, attacking electrophilic carbons bearing activated oxygen substituents.

This strategy was applied in the total synthesis of (−)-indolizidine 223AB having epoxide (+)-**2** and aziridine (−)-**4** as building blocks (Scheme 1). The key step for the synthesis of chiral epoxide (+)-**2** involved a Jacobsen's hydrolytic kinetic resolution (HKR) employing the (*R*,*R*)-Salen

Co(III) catalyst [21]. Aziridine (−)-**4** was prepared from D-norvaline in three steps [22]. Then, lithiation of dithiane, followed by addition of epoxide (+)-**2**, warming to −25 °C over a period of 1 h, stirring for an additional 4 h at −25 °C, and then addition of aziridine (−)-**4** in Et_2O containing hexamethylphosphoramide (HMPA, 0.65 equivalents) to promote the solvent-controlled Brook rearrangement furnished (−)-**5** in 56% isolated yield. Removal of the *t*-butylsilyl groups (TBS) using tetrabutylammonium fluoride (TBAF, 95% yield), bismesylation followed by treatment with potassium carbonate in MeOH for 3 h, and then addition of excess sodium amalgam (5%) directly to the reaction mixture furnished the desired product in excellent yield (95%). Finally, reductive removal of the dithiane with Raney Ni (69% yield) completed the synthesis of (−)-indolizidine 223AB.

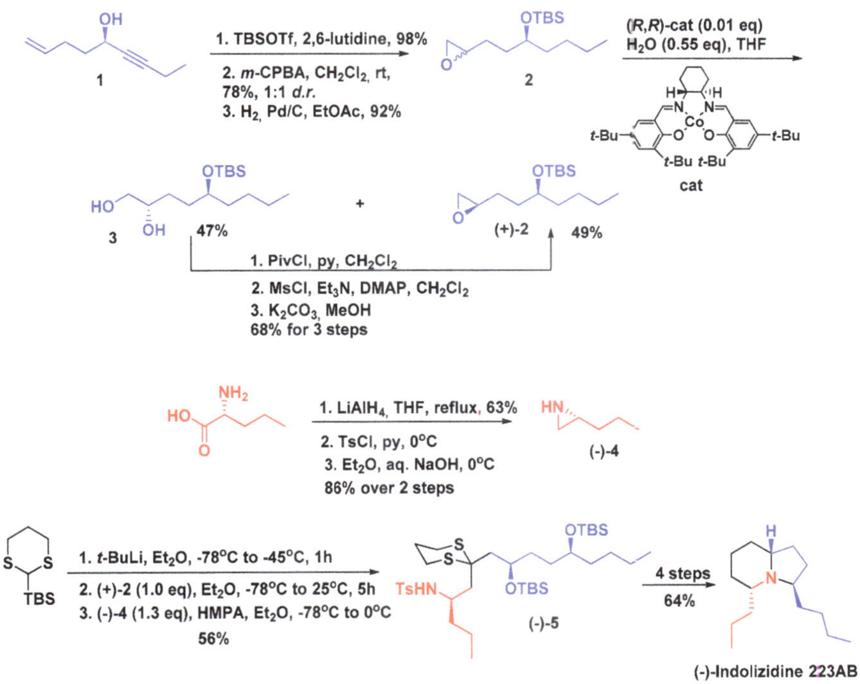

Scheme 1. Total synthesis of (−)-indolizidine 223AB.

The multicomponent and domino reactions have demonstrated a remarkable impact on the synthesis of complex products, with biologically or pharmacological active products among these [23]. The multicomponent reactions (MCRs) involve three or more reactants that are introduced concurrently and form a single product which contains the essential parts of the starting materials [24]. Such reactions allow the straightforward synthesis of intricate molecules in a one-pot fashion without the isolation and purification of intermediates, therefore leading to lower costs, time, and energy consumption. Additionally, MCRs are modular and convergent in nature and also an important source of molecular diversity [25]. Because of the potential for increased efficiency and selectivity, along with the advantages of environmentally benign synthetic procedures and catalyst reusability, association of heterogeneous catalysis and MCRs have been explored [26,27]. Recently, three-component reactions of amines, epoxides, and carbon dioxide have emerged as a powerful strategy for the synthesis of organic carbamates [28].

Despite the intensive research in the field of asymmetric synthesis in the last several decades, the asymmetric MCRs are still a challenge, especially those based on isocyanides (IMCRs) [29]. In this context, several groups have investigated asymmetric approaches to IMCRs, and recently very elegant solutions were found for the catalytic enantioselective version of some Passerini- or Ugi-type reactions [30,31]. Thus, in this review, the recent advances in MCRs involving the synthesis and applications of epoxides and aziridines is discussed with emphasis on the stereoselectivity of the reactions.

2. Epoxides

The epoxides are considered versatile starting materials in organic synthesis. The inherent polarity coupled with ring strain makes epoxides susceptible to various reactions with different reagents (nucleophiles, electrophiles, acids, and bases) to produce new useful functional groups.

2.1. Epoxide Synthesis

Amino epoxides are important building blocks in the synthesis of many natural products and analogues. Roy and coworkers demonstrated a transition-metal-free methodology for the synthesis of multisubstituted α-amino epoxides via a three-component reaction (3-CR), in which 2-(trimethylsilyl)aryl triflate (**6**) was employed to generate the aryne species in situ that reacted with *N*-substituted aziridines (**7**) and aldehydes (**8**) (Scheme 2). The authors proposed that the addition of aryne to the aziridine substituted with an electron withdrawing group (EWG) generated the zwitterionic species **A**; then, the intramolecular proton transfer provided a strained aziridine ylide **B** which added to the carbonyl component, and the resulting alkoxide **C** promoted the aziridine ring opening and further epoxide closure. This elegant approach furnished the desired epoxides in good yields and diastereoselectivity with good tolerance to different functional groups [32].

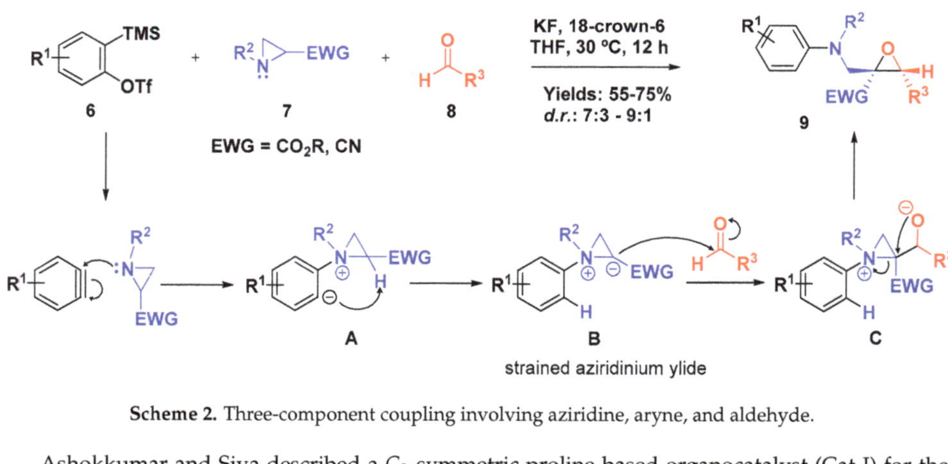

Scheme 2. Three-component coupling involving aziridine, aryne, and aldehyde.

Ashokkumar and Siva described a C_3-symmetric proline-based organocatalyst (Cat I) for the synthesis of α,β-epoxy ketones (**12**) (Scheme 3) via a directly oxidative coupling involving styrenes (**10**) and aldehydes (**11**) in the presence of *tert*-butyl hydroperoxide (TBHP) as oxidant. The authors proposed that the reaction mechanism might follow a radical pathway in which the catalyst plays a crucial role to provide good enantioselectivity to the product. In general, the reaction presented a good tolerance for different substituents, including a less active aliphatic aldehyde [33].

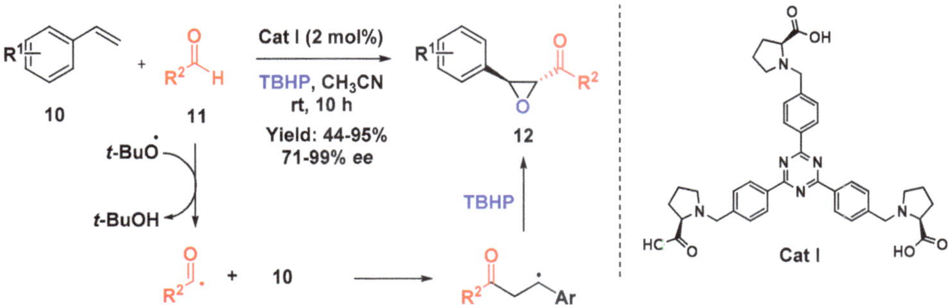

Scheme 3. Proline-based chiral C_3-symmetric organocatalyst for the synthesis of epoxide derivatives through a domino reaction.

The use of asymmetric 3-substituted 2,3-epoxy-aldehydes in multicomponent reactions might be a challenge since those species can easily undergo epimerization under either acidic or basic conditions. To overcome this issue, our group designed the one-pot synthesis of epoxy-α-acyloxycarboxamides (**14**) from asymmetric organocatalyzed epoxidation of α,β-unsaturated aldehydes **13** followed by the Passerini 3-CR (Scheme 4). For the asymmetric epoxidation a new diarylprolinol silyl ether catalyst (Cat II) was developed, which enabled the use of green solvents such as ethanol/water mixture providing the desired epoxy-aldehydes with good yields and enantioselectivity. Then, the other components of Passerini-3CR (carboxylic acid and isocyanide) were added and after 24 h, highly functionalized epoxy α-acyloxycarboxamides **14** were obtained in good yields and stereoselectivity [34]. Those adducts were submitted to inhibition assays against cathepsins K, V, and L, in which the **LSPN423** was the most potent and selective against cathepsins L. Further investigations demonstrated that **LSPN423** is a tight binding uncompetitive inhibitor with an inhibition constant (K_i) of 1.33 μM [35].

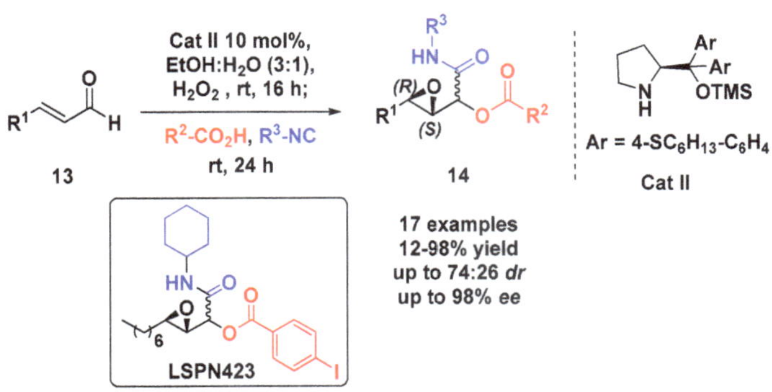

Scheme 4. Tandem asymmetric organocatalytic epoxidation/Passerini reaction.

2.2. Epoxide Ring Opening

Vignatti and Luis-Barrera recently described the synthesis of chiral 1,2,2′-aminodialcohols **18** via a three-component approach in which occurred the ring opening of two distinct epoxides **16** and **17** by a single aniline component (**15**) under solvent-free conditions. This was catalyzed by squaramide moieties present in a metal–organic framework (Sq_IRMOF-16) consisting of a zinc metal cluster (Zn_4O) bridged by dicarboxylate linkers, forming a three-dimensional mesopore system (Scheme 5). The construction of the catalyst was planned so that the catalytic sites were available in all three dimensions, and also so that the pores were large enough to host the intermediates generated during the reaction. The authors demonstrated that MOF pores decorated with squaramide moieties were more effective than only the squaramide as catalyst in the proposed conditions. Enantiomerically pure epoxides were employed and the asymmetry was retained in the products. From the three-component reaction, they could obtain either homo-disubstituted ($R^2 = R^3$) or hetero-disubstituted amino diols ($R^2 \neq R^3$) by controlling the addition of substrates. The polarity and size of the epoxide substituents directly affected the yield: bulky hydrophobic chains, such as $–C_{11}H_{23}$ and $–C_{10}H_{21}$, required longer reaction times (3–4 days) and provided the products in moderate yields (43 and 56%, respectively). On the other hand, an increase in substituent polarity lead to an increase of both rate and yield, even when bulky groups, such as $–CH_2OTBDMS$, were used [36].

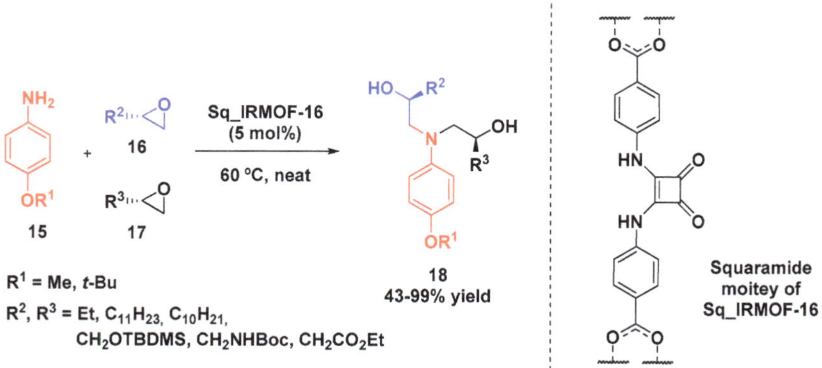

Scheme 5. Epoxide ring-opening multicomponent reaction catalyzed by Sq_IRMOF-16.

The synthesis of multisubstituted asymmetric 1,3-oxazolidines **22** was reported by Hong et al. in a three-component reaction between anilines (**19**), mono-substituted epoxides (**21**), and ethyl glyoxalate (**20**) (Scheme 6). To obtain the enantiomerically enriched product, they used a Ti(IV) complex with the chiral ligand (*S*)-BINOL in a 1:2 ratio as catalyst, with a small amount of trifluoroacetic acid (TFA). Since only satisfactory yields were observed—around 50%—the authors further investigated the possibility that a kinetic resolution might be occurring.

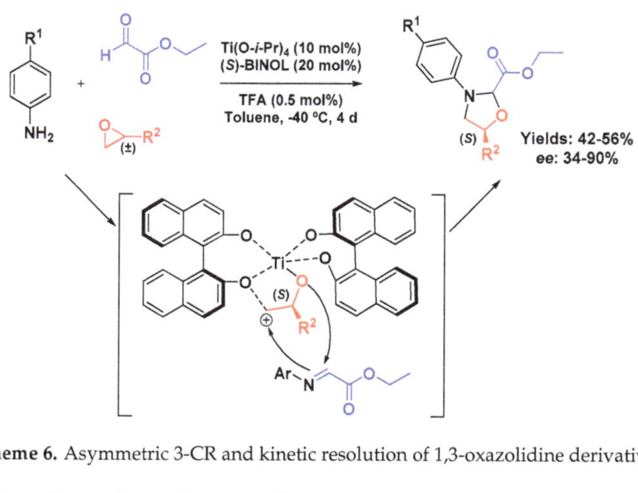

Scheme 6. Asymmetric 3-CR and kinetic resolution of 1,3-oxazolidine derivatives.

Moreover, from the analysis of unreacted epoxide, they could detect that the substrate was no longer a racemate. Additional experiments have shown that only *S*-epoxide was consumed, confirming the kinetic resolution hypothesis. They also could observe the regiospecificity of the reaction, since from all the experiments—even at 10 mmol scale—only the C-4 isomer was detected; consequently, no C-5 isomer was found, which corroborates the proposed mechanism where the complex of Ti(IV) with (*S*)-BINOL promoted the epoxide ring opening (only *S*-isomer) that immediately reacts with the imine from aniline and ethyl glyoxalate, giving the desired product [37].

Murray and coworkers described an anion relay strategy to the three-component linchpin coupling involving a tandem epoxide opening/Cu-catalyzed allylation, where two C–C bonds and one new stereogenic center were formed (Scheme 7). By using enantiopure epoxides **23** the authors observed excellent diastereoselectivity to the *anti*-isomer (up to 20:1). The $[B(pin)]_2$–C(H)Li was employed as a bifunctional linchpin [38] so it could react with two electrophiles: first, promoting the epoxide opening with poor diastereoselectivity (intermediate **A**), and second, the subsequent deborylative transmetalation that generated the organocopper **B**, which reacted with the allyl bromide **C**, affording the desired products. According to mechanistic studies the step of formation of the organocopper species determines the *anti*-selectivity of the product. The 1,3-hydroxy-homoallylboronates **26** obtained were proven to be very useful for further functionalization, such as the stereoselective synthesis of a key intermediate of the alkaloid (+)-allo-sedamine; this methodology was also successfully extended to a sequential four-step process to obtain 1,3-polyol motifs [39].

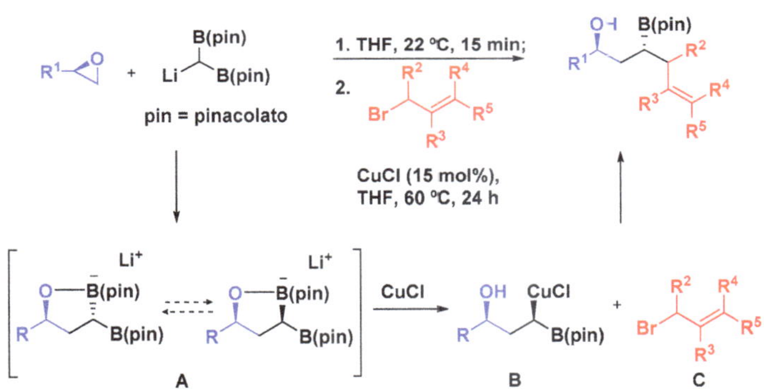

Scheme 7. Copper-catalyzed three-component linchpin coupling.

In 2010, Yeung and coworkers first described a new catalyst-free electrophilic MCR involving olefins, cyclic ethers, primary amines, and *N*-bromo-succinimide (NBS) as a bromide source (Scheme 8) [40]. The initial studies used diverse cyclic ethers **28**. According to experimental observations, the authors proposed the MCR pathways in which NBS promoted the bromination of the olefin (**27**), then the cationic intermediate underwent ring opening by nucleophilic addition of the cyclic ether; in addition, the intermediate **30** was captured by the sulphonamide ($NsNH_2$), furnishing the desired product **31**. The use of ethylene oxide as the cyclic ether component enabled subsequent reactions to provide biologically active morpholine derivatives, such as the (±)-phenmetrazine and (±)-phendimetrazine, two norepinephrine–dopamine releasing agents.

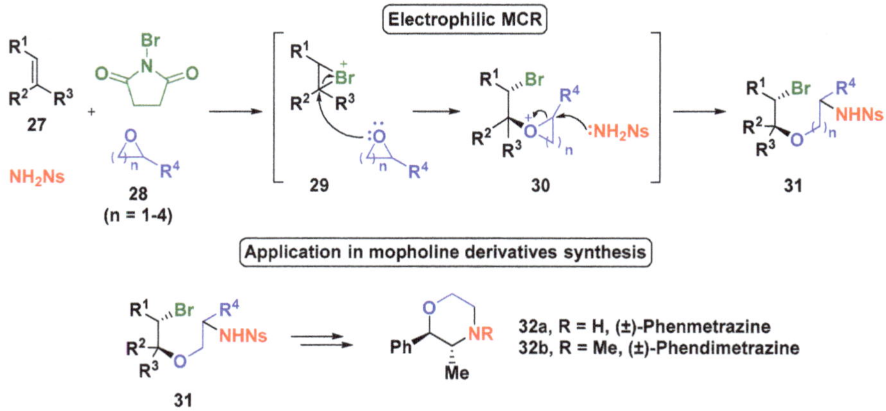

Scheme 8. One-pot electrophilic aminoalkoxylation reaction.

In further investigations concerning electrophilic MCR employing epoxides, Zhou et al. observed that monosubstituted epoxides could be opened via Markovnikov or anti-Markovnikov pathways giving products with low regioselectivity (Scheme 9) [41]. Then, studies concerning the electronic effect of the epoxide substituents over the ring opening selectivity demonstrated a preference for the anti-Markovnikov product when epichlorohydrin was used. As previously described, the treatment of the multicomponent product **34** with K_2CO_3 in MeCN afforded the corresponding morpholines **35** in good yields. The attempt to perform both steps in one pot was successful, as well as the use of enantiopure (*R*)-epichlorohydrin in order to provide the asymmetric morpholine **37**; in this case, not only was the configuration of epoxide maintained, but the racemic olefin **36** was desymmetrized, affording a single stereoisomer with excellent enantiomeric excess.

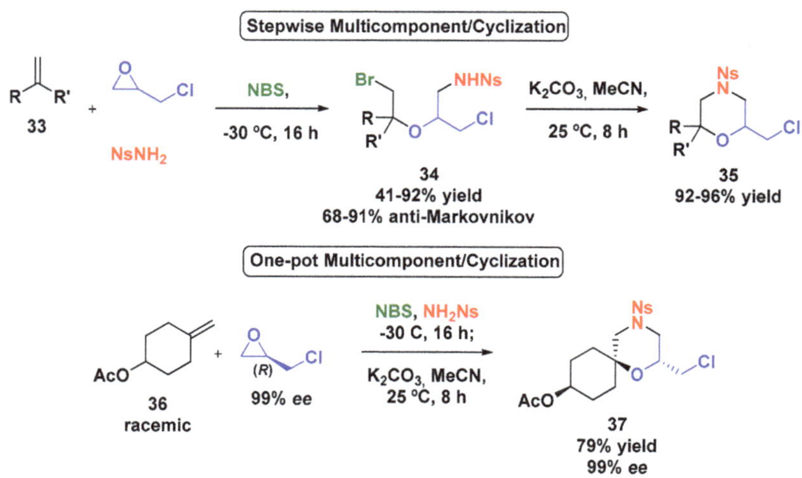

Scheme 9. Synthesis of trisubstituted morpholines through an electrophilic multicomponent reaction using epichlorohydrin, olefin, nosyl amide, and *N*-bromosuccinimide.

In order to demonstrate the availability of the electrophilic MCR using epichlorohydrin in the synthesis of bioactive morpholines, Zhou and Yeung reported the formal synthesis of reboxetine, a norepinephrine reuptake inhibitor, and also the enantioselective synthesis of a carnitine acetyltransferase inhibitor (Scheme 10). The main strategy for the synthesis of both products built upon the previous methodology described by Yeung and employed tert-butyldimethylsilyl ethers (OTBS) or hydrogen as substituent to **38** aim to obtain the morpholine skeleton [41], followed by the substitution of chloride to introduce a suitable functionality, as well as removal of the nosyl group [42].

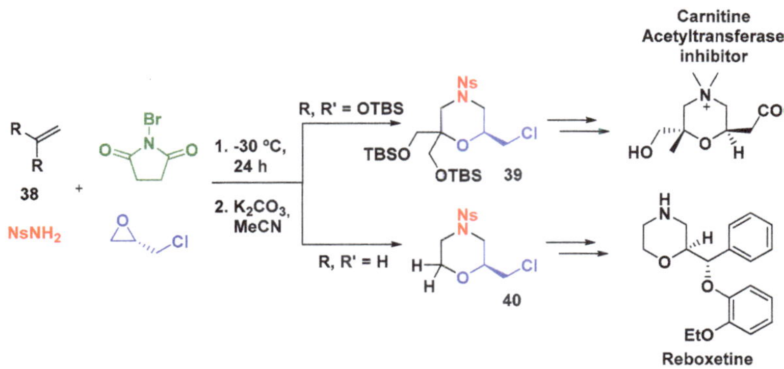

Scheme 10. Formal synthesis of Carnitine acetyltransferase and Reboxetine.

Coates et al. proposed a 3-CR with epoxides, carbon monoxide, and isocyanates based on previous mechanistic studies regarding the carbonylation of epoxides to form β-lactones catalyzed by the $[(salph)Al(THF)_2]^+[Co(CO)_4]^-$ complex, in which the intermediate **42** could be trapped by an electrophile such as isocyanates (Scheme 11). This 3-CR gave 1,3-oxazinane-2,4-diones (OD), a versatile intermediate in the synthesis of α,β-unsaturated carbonyl systems, β-ketoesters, and β-hydroxy-acid, among other functionalities. The substrate scope revealed that the electrophilicity of the isocyanate was crucial to the efficiency of the MCR, aryl isocyanates with electron-withdrawing substituents afforded the products with excellent yields and chemoselectivity; on the other hand, in the reaction with aryl isocyanates bearing electron-donating substituents, the products were obtained in lower yields and chemoselectivity, since β-lactones side products were observed. Therefore, the addition of intermediate **42** to isocyanate might be the rate-determining step. Further mechanistic studies gave additional information concerning reaction pathways: (1) β-lactone cannot be converted to OD; thus, β-lactone is not an intermediate in the OD synthesis; (2) NMR experiments with labeled OD prepared from ^{13}CO confirmed the proposed mechanism, in which the CO was incorporated adjacent to the β-carbon; and (3) 1,2-disubstituted epoxides afforded OD with inversion of configuration in the β position to the ring oxygen and retention in the α position, e.g., *cis*-epoxide gave *trans*-OD and *trans*-epoxide gave *cis*-OD [43].

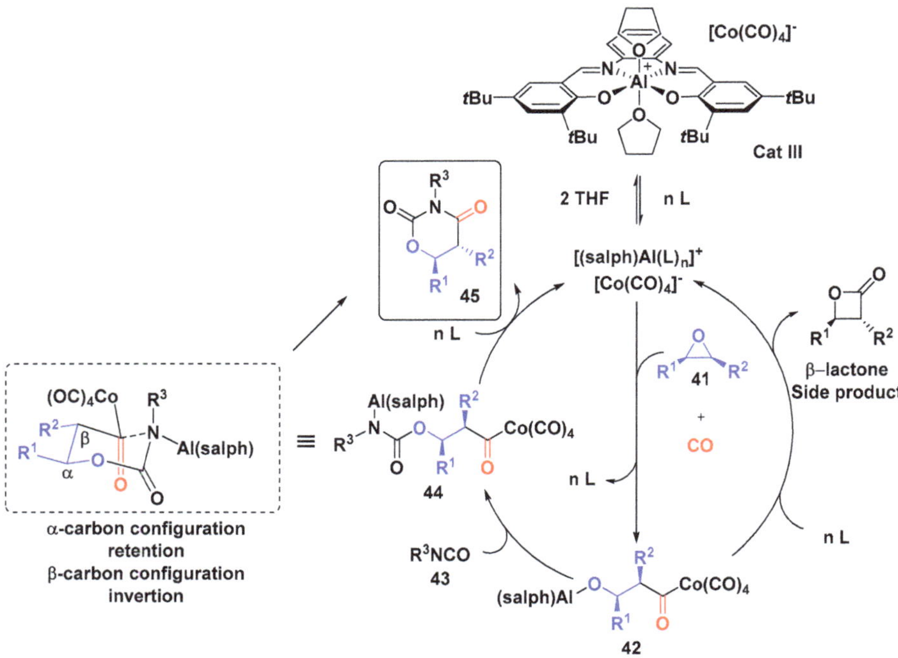

Scheme 11. Three-component synthesis of 1,3-oxazinane-2,4-diones catalyzed by the $[(salph)Al(THF)_2]^+[Co(CO)_4]^-$ complex.

In 2014, Zhang et al. and Liu and Sun reported simultaneously a cobalt-catalyzed multicomponent reaction involving epoxides (**46**), imines (**47**), and carbon monoxide (Scheme 12) [44,45]. Interestingly, both groups aimed at the copolymerization of imines with CO, using the epoxide to generate acylcobalt species as a catalyst; however, the major product obtained was 1,3-oxazinan-4-ones (**48**), rather than the desired polypeptide. Since it was the first one-pot methodology described for the synthesis of this class of compounds, the authors further investigated this 3-CR. In general, alkyl-substituted epoxides provided the 2,6-disubstituted product due to the selective ring-opening at the less hindered carbon.

On the other hand, Liu and Sun observed the opposite regioselectivity for styrene oxide (2,5-disubstituted product), which was attributed to higher stability of the benzylic carbocation. The chairlike conformation of the six-membered cyclic transition state model can be used to rationalize the diastereoselectivity observed when monosubstituted epoxides were employed. They performed in situ IR experiments in order to obtain information concerning the reaction mechanism. The first account was that $[Ph_3SiCo(CO)_4]$ pre-catalyst was converted to $[HCo(CO)_4]$—the catalytic species—from β-hydrogen elimination of oxirane. To prove this hypothesis, they employed the $[HCo(CO)_4]$ generated in situ from the alcoholysis of $[Ph_3SiCo(CO)_4]$ in methanol and the reaction proceeded at a higher rate. The IR experiments also indicated that imine addiction to the acylcobalt species is the rate-determining step of the reaction [40].

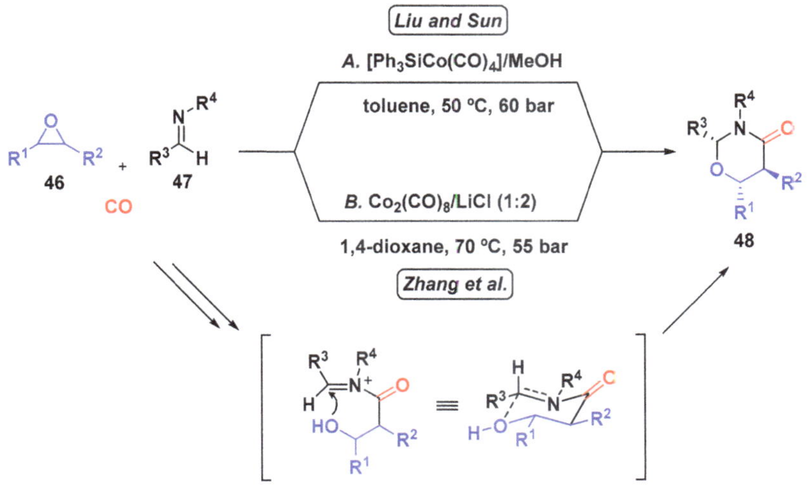

Scheme 12. Cobalt-catalyzed 3-CR for the synthesis of 1,3-oxazinan-4-ones.

In the context of multicomponent reactions involving the transition metals, undoubtedly, the Catellani reaction is one of the most known. This reaction concerns an ipso functionalization of an aryl iodide and, concomitantly, a C–H activation in the ortho position catalyzed by palladium and promoted synergistically by 2-nobornene species. Recently, Zhang's group developed an elegant method of C–H activation in a Catellani reaction employing epoxides, and, posteriorly, aziridine, as the alkylating agent for this transformation [46]. The salient features of the strategy include its broad substrate scope and its high atom economy, since an epoxide (or aziridine) can be incorporated into the product in its entirely, without the need for any sacrificial group.

The developed protocol begins with the mixture of an aryl iodide **49**, an alkylating reagent (epoxide) **50**, and electron-deficient olefin as terminating reagent using $Pd(OAc)_2$ as catalyst, XPhos as the ligand, the potassium salt of 5-norbornene-2-carboxylic acid as mediator, and *N*-methyl-2-pyrrolidone (NMP) as solvent at 60 °C (Scheme 13). After 12 h, the product was obtained in moderate to excellent yields (25–95%), and a total of 45 examples were prepared. Among these examples were the aryl iodide, the electron withdraw olefin, and a large range of epoxides including natural product derivatives and macrocyles. Chiral epoxides were also tested and a regioselective ring opening was observed for all products with no decrease of ee. Moreover, the products of the reaction are poised to undergo oxa-Michael reactions, thus allowing expedient access to isochroman scaffolds **52**.

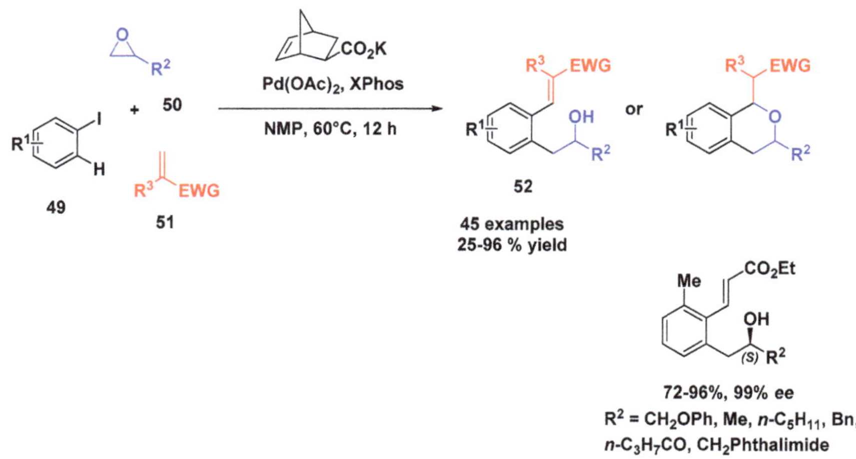

Scheme 13. Palladium-catalyzed Catellani reaction using epoxides.

3. Aziridines

Aziridines are a powerful synthetic building block widely used in the synthesis of different nitrogen-containing derivatives [47]. Due to their feasibility as synthetic precursors and building blocks, in this section we describe recent strategies for the aziridination and transformation of this heterocycle in multicomponent protocols.

3.1. Synthesis of Aziridines

There are many methods for the synthesis of aziridines and, currently, most of them involve diastereo- or enantioselective strategies to achieve this three-membered ring. However, few papers have reported the synthesis of this ring using three or more components.

In 2016, Xu and coworkers described a three-component reaction for the preparation of α-phosphonyloxy-β-amino ester derivatives through an [1,2]-phospha-Brook rearrangement followed by addition of an imine substrate as an electrophile (Mannich coupling) (Scheme 14) [48]. In this MCR, using *N*-tosyl or *N*-4-methoxyphenylsulphonyl as a protecting group, 18 examples of α-phosphonyloxy-β-amino esters **55** were obtained in high yields and at >20:1 *dr* for most entries. Using diphenylphosphinyl as a protecting group, a different behavior was observed. This new protecting group increases the nucleophilicity of nitrogen, therefore facilitating the intramolecular elimination of the diethoxyphosphate group. This modification affords the aziridines **56** in good to excellent yields (52%–99%) and diastereoselectivity (>20:1 *dr*).

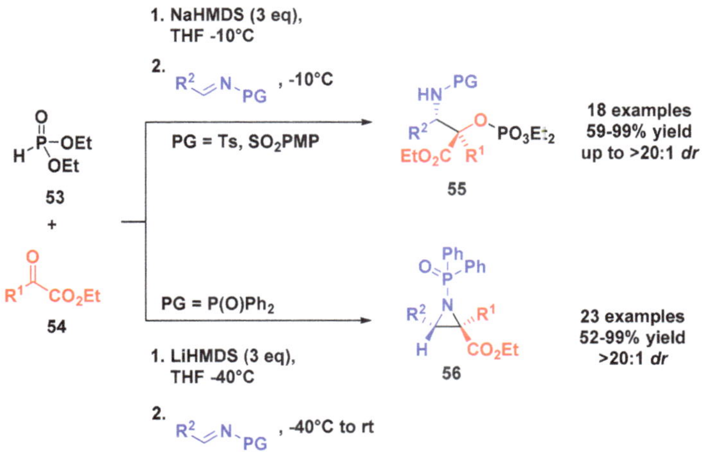

Scheme 14. Three-component coupling of diethyl phosphite, α-ketoesters, and *N*-protected imines.

Concerning enantioselective protocols, aziridination from an MCR perspective has been scarcely described [49,50]. The first examples for this transformation were described in 1999 and consisted of stoichiometric reactions, whereas in 2009, organocatalyzed multicomponent aziridination reactions appeared in the literature [51,52]. Following Akiyama's strategy [53], Bew and coworkers reported a multicomponent asymmetric Brønsted-acid-catalyzed aza-Darzens reaction for the synthesis of *N*-aryl-*cis*-aziridine carboxylate esters [54]. Alkyl diazo acetates and aromatic or hetero aromatic aldehydes were employed to afford 10 examples in good yields (61%–98%) and with mostly >90% ee; however, this score was dependent on the use of ortho-*tert*-butoxy aniline at −60°C to ensure the performance of this transformation.

Regardless of Bew's reports, Wulff and coworkers had been contributing to the development of a multicomponent catalytic asymmetric aziridination reaction. The reported procedure starts with the addition of diazo ester compounds to *N*-protected aldimine catalyzed by a chiral Brønsted acid [55] (Scheme 15). This transformation usually affords *cis*-aziridine-2-carboxylate **59** in good yields and enantiomeric excess. This transformation begins with a facile method for in situ preparation of BOROX catalyst derivative from (*S*)-VAPOL or (*S*)-VANOL and three equivalents of $B(OPh_3)$. When the BOROX catalyst and protonated imine are close enough, an ion pair is formed and the diazo ester compound attacks, delivering the *cis*-aziridine.

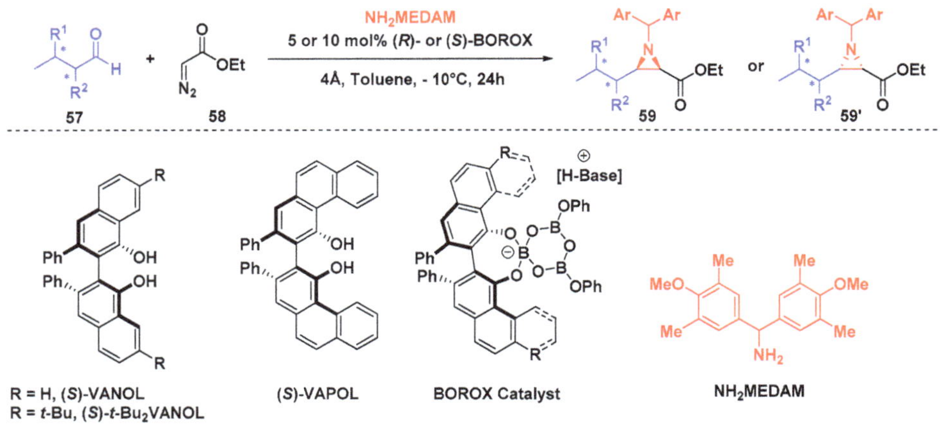

Scheme 15. Multicomponent aziridination by BOROX catalysis.

Later, Wulff and coworkers evaluated the catalytic asymmetric *cis*-aziridination reaction from chiral aldehydes and the transition states using (*R*)- or (*S*)-BOROX. The reaction uses NH_2MEDAM and α- or β-substituted chiral aldehydes at −10 °C and 5–10 mol % BOROX catalyst using (*R*)- or (*S*)-VAPOL, VANOL, and tBu_2VANOL as ligands. A variety of different substrates reveal a strong dependence on the match between the substrate and the catalyst to achieve good yields and enantioselectivity.

Nevertheless, previous reports suggest that aziridination reactions using diazo acetamides instead of diazo esters afford *trans*-aziridine **62** [56–58]—the change in the stereochemical outcome could be rationalized by the reversal in the ordering of both the substrates **60** and **61** H-bonding to the BOROX catalyst, which causes a flip in the face selectivity of bond formation to the imine carbon [59]. Based on this information, Wulff and coworkers optimized the multicomponent *trans*-aziridination reaction for aryl and alkyl aldehydes **60** exploring the BUDAM protecting group on nitrogen (Scheme 16) [60]. With reaction conditions very similar to those employed before, the authors achieved excellent results for aromatic aldehydes, whereas alkyl aldehydes were highly dependent of the substrate, catalyst, and the time required for imine bond formation.

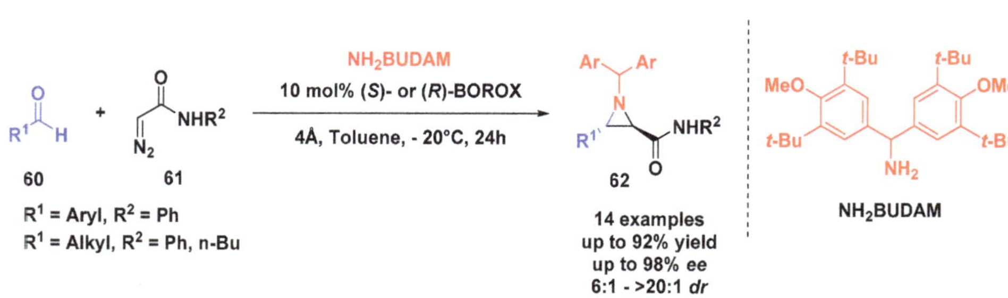

Scheme 16. Enantioselective multicomponent *trans*-aziridination of imines with diazo acetamides.

3.2. Aziridine Ring Opening

Over recent years, the aziridine moiety has been used in several methods involving ring opening strategies allowing access to domino chemical transformations in a one-pot procedure or without the purification/isolation of intermediates.

Based on the susceptibility of C–N bond cleavage in aziridine, Majee and coworkers reported the synthesis of 1,3-oxazolidines under neat conditions [61]. Based on previous work [62], the authors described a practical and simple multicomponent method to achieve this heterocycle. The protocol consists of a mixture of aziridine, formaldehyde, and formic acid in 1:1:1 ratio at 100 °C under neat conditions, and 1,3-oxazolidines were obtained in good yields (Scheme 17). The proposed mechanism begins with nucleophilic attack of nitrogen from the aziridine ring to the protonated formaldehyde. The oxygen atom adds to the benzylic position, opening the aziridine ring and closing the final five-membered oxazolidine.

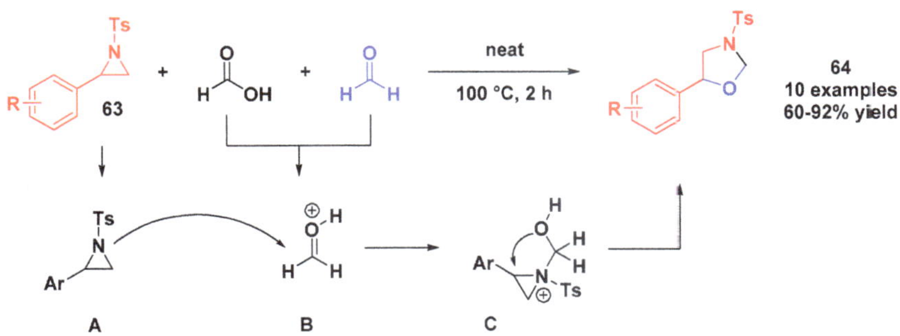

Scheme 17. Protocol for the synthesis of 1,3-oxazolidines described by Majee and coworkers.

Larionov and coworkers developed a 3-CR employing aziridine, aryne, and acetonitrile as the nucleophile [63]. Different research groups have since extended this concept to other nucleophiles such as fluoride [64], phenols, carboxylic acids [65], water [66], and aldehyde as the electrophile (Scheme 18) as shown before in Section 2.1.

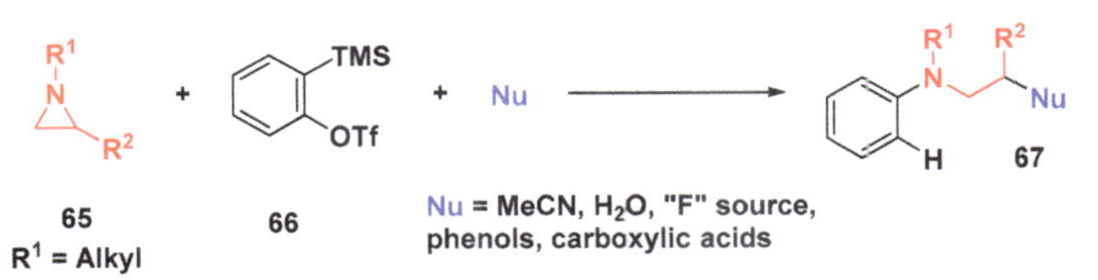

Scheme 18. Aryne multicomponent coupling and use of aziridines as nucleophilic triggers.

Wani et al. developed a domino ring-opening cyclization (DROC) of activated aziridines and epoxides with nitrones via dual catalysis using $LiClO_4/Bu_4NBF_4$ "on water" furnishing the corresponding 1,2,4-oxadiazinanes and 1,4,2-dioxazinanes, respectively. The authors also explored a model reaction in a multicomponent stepwise fashion, with benzaldehyde, *N*-methyl hydroxyl-amine hydrochloride, and 2-phenyl-*N*-tosylaziridine as substrates, obtaining the desired product 1,2,4-oxadiazinane **69** in good yield (78%) as a single diastereoisomer (Scheme 19) [67]. According to the proposed mechanism, the catalysts promoted the regioselective aziridine ring opening by nitrone followed by cyclization affording the *cis*-disubstituted product.

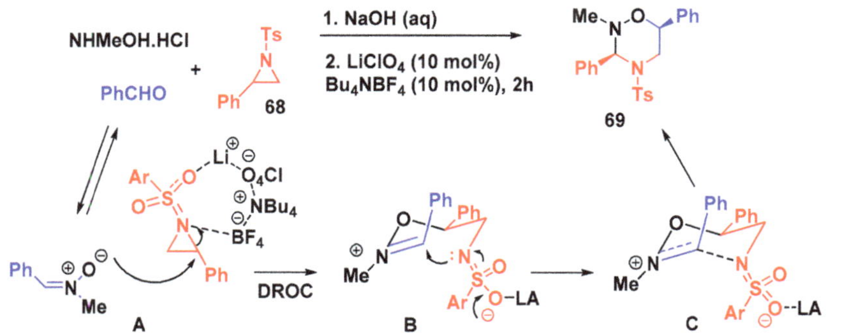

Scheme 19. Multicomponent approach for the synthesis of 1,2,4-oxadiazinane.

Another example of DROC of activated aziridines was published by Zhou and Yeung, who developed an electrophilic halogen-induced domino reaction for the synthesis of pyrrolidine, piperidine, or azepanes. The experimental protocol starts with an enantiopure olefinic aziridine, and *N*-bromosuccinimide (NBS) and $NsNH_2$ as the halogenation agent and nucleophilic partner, respectively. For all cases, the aziridine **70** furnished good yields and stereoselectivity for the enantiopure substituted azepane, and no piperidine was detected [68].

From further investigations concerning electrophilic DROC employing ethenylaziridine (n = 1) [69], the same authors observed that an aziridinium ion intermediate formed in a shorter carbon chain could be opened by NH_2Ns at either a terminal or internal position from aziridine to give pyrrolidine or piperidine, respectively (Scheme 20). The preference for the formation of pyrrolidine was attributed to the substitution having taken place at the less hindered terminal carbon position.

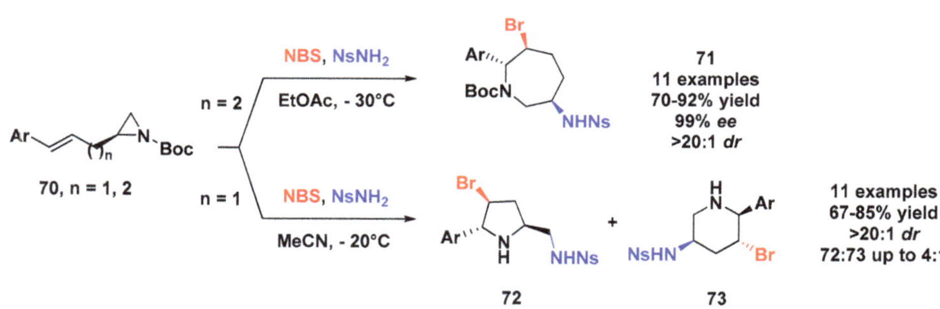

Scheme 20. NBS-induced aminocyclization–aziridine ring expansion cascade of **70**.

Another example of aziridine opening reaction was described by Zhou and coworkers [70]. Inspired by the discovery that the epoxides could act as alkylating reagents for the Catellani reaction, the authors explored the same behavior for aziridine aiming for the synthesis of tetrahydroisoquinolines (Scheme 21). Using Pd/tri-2-furanylphosphine (Pd/TFP) as a catalyst, a norbornene mediator, K_2CO_3 in MeCN as solvent at 70 °C, the authors evaluated the aryl iodine, the *N*-protected aziridine, and electron-deficient olefin. Using this methodology, 42 examples were prepared in a mild, chemo- and regioselective ring opening, including an enantiopure aziridine. This protocol provided a versatile way to access tetrahydroisoquinolines.

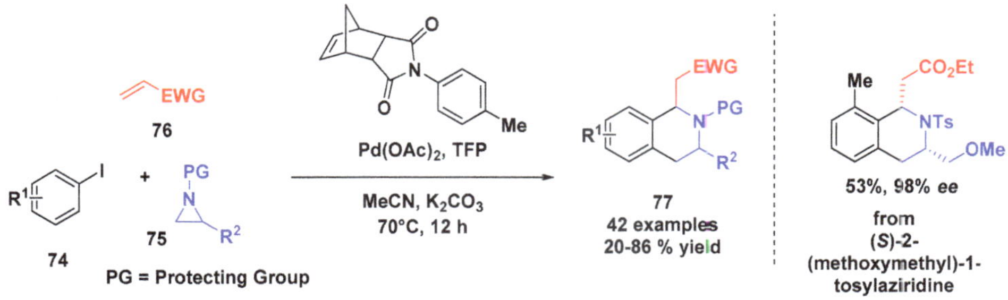

Scheme 21. Palladium-catalyzed Catellani reaction for synthesis of tetrahydroisoquinolines.

Beyond the last report, few papers are available on the one-pot multicomponent synthesis of tetrahydroisoquinolines. Based on the ready access and good reactivity of *N*-sulfonylaziridine, Xing and coworkers reported a stereoselective three-component synthesis of 1,4-disubstituted tetrahydroisoquinolines that provided a good choice for convergent synthesis of this core skeleton [71]. The 3-CR of aziridines, arenes, and aldehydes was performed with BF3OEt$_2$ and anhydrous $MgSO_4$ as an additive under dichloroethane (DCE) at 60 °C (Scheme 22). In most cases, *cis*-products were obtained in moderate yields with good regio- and diastereoselectivities. In the proposed mechanism, the Lewis acid promoted the ring opening of *N*-tosyl aziridine via a Fridel–Crafts substitution. Next, the imine sulfonyl compound further underwent Lewis-acid-catalyzed Pictet–Spengler condensation in a cascade fashion, leading to *cis*-1,4-disubstituted tetrahydroisoquinolines.

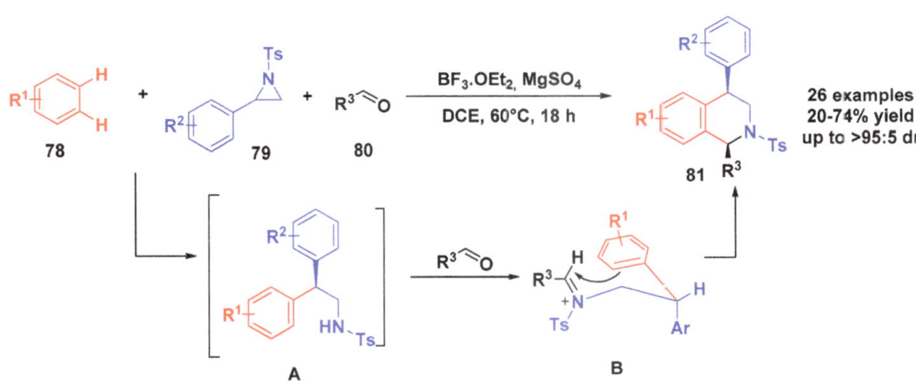

Scheme 22. Tandem three-component reactions between aziridines, arenes, and aldehydes.

Yudin's group expounded aziridine synthesis or functionalization, exploring the aziridine aldehyde dimer transformation in different ways, including diverse protocols involving MCRs. Enantiomeric enriched aziridine aldehyde dimers can be obtained from natural α-amino acids in five steps [72]. Recently, our research group described an alternative synthetic route to these dimers starting from α,β-unsaturated aldehydes through an asymmetric organocatalytic aziridination followed by Boc deprotection, under mild conditions and in two steps [73].

The aziridine aldehyde dimer has been employed as a building block to access, for example, a wide range of functionalized diamine compounds from a Petasis borono–Mannich process (Scheme 23). Using excess of morpholine, boronic acid, and aziridine aldehyde dimer under hexafluoroisopropanol (HFIP), the diamine **84** was obtained with high *syn* diastereoselectivity [74]. Another example was reported in 2017, in which the 1,2-aziridinyl propargylic amines (**86**) were obtained with good yield and stereoselectivity through a zinc-catalyzed multicomponent reaction under trifluoroethanol (TFE) at room temperature [75].

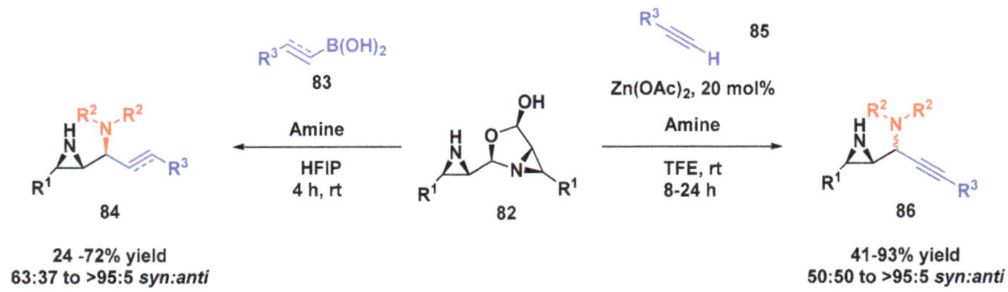

Scheme 23. Example of aziridine aldehyde dimer as an inducing group for building blocks synthesis.

Furthermore, in 2010, Yudin's group reported a *trans*-diastereoselective cyclization of amino acids and peptides using a disrupted Ugi reaction (as it was called by the authors) [76]. This protocol has as its main aspect the reversibly autoprotected aziridine aldehyde dimer **87** when applied in an Ugi reaction. This characteristic is related to iminium ion formation prior to the selectivity-determining isocyanide addition; then, the exocyclic aziridine intercepts the carbonyl group of the mixed anhydride, which undergoes solvolysis. The disrupted Ugi protocol afforded piperazinone **90** from three components; however, as showed in 2014 [77], there is a substantial difference in reactivity between secondary and primary amino acids in this kind of Ugi reaction (Scheme 24). The authors showed that the relative stereochemistry is controlled by both the amino acid and the aziridine aldehyde dimer under TFE. In the case of the chiral primary amino acid, this transformation was selective for the *trans*-substituted products **90** while the chiral secondary or protected amino acids afforded *cis*-products. Besides this, a diverse range of functionalized isocyanides were screened in the disrupted Ugi reaction to obtain chiral piperazinones in high stereoselectivities [78].

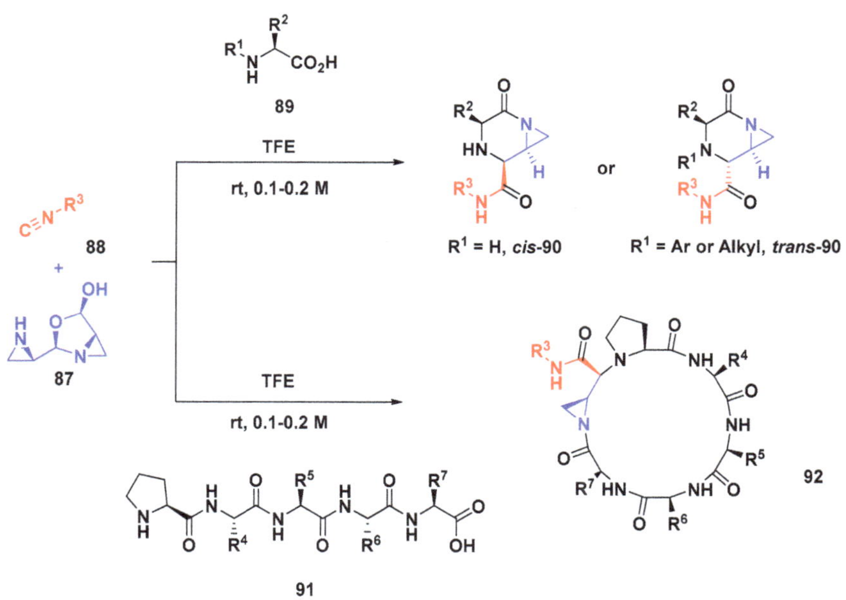

Scheme 24. Disrupted Ugi reaction with aziridine aldehyde dimers, isocyanides, and amino acids (or peptides).

A computational study concerning the mechanism pathway, including the factors contributing to stereochemistry induction, was also reported by Yudin [79]. The same protocol was also extended subsequently to evaluate the multicomponent reactivity of linear peptides **91** towards peptide macrocycles **92** from disrupted Ugi reactions [80,81].

4. Conclusions

In conclusion, we have shown in this review that stereoselective MCRs can be efficiently employed in the synthesis of epoxides and aziridines, as well as in their transformation via ring opening on more functionalized compounds including other heterocycles, with potential pharmacological properties.

However, considering the green context of the MCRs, in several reports that were discussed in this work, quite toxic solvents were still employed, such as, for example, HMPA, DCE and THF. Although there is a long path to be travelled, alternative solvents that have low toxicity, are easy to recycle, and are inert should be considered. Furthermore, it is expected that investigations focusing on asymmetric catalysis will continue to grow in the near future, on the road to more sustainable chemistry.

Author Contributions: Conceptualization: A.G.C.; writing-original draft preparation: A.G.C., D.A.d.S. and A.R.d.S.; writing-review & editing: A.G.C., D.A.d.S. and A.R.d.S.; visualization A.G.C., D.A.d.S., A.R.d.S. and M.W.P.; supervision: A.G.C.

Funding: The authors gratefully acknowledge FAPESP (grants 2014/50249-8, 2017/23867-0 and 2018/09961-7), GlaxoSmithKline, CAPES (Finance Code 001), and CNPq (grants 302634/2016-9 and 429748/2018-3) for funding and fellowships.

Conflicts of Interest: The authors declare no conflict of interest.

References

1. *Aziridines and Epoxides in Organic Synthesis*; Yudin, A.K. (Ed.) John Wiley & Sons, Inc.: Hoboken, NJ, USA, 2006. [CrossRef]
2. Nara, F.; Tanakaa, M.; Hosoya, T.; Suzuki-Konagai, K.; Ogita, T. Scyphostatin, a Neutral Sphingomyelinase Inhibitor from a Discomycete, *Trichopeziza Mollissima*: Taxonomy of the Producing Organism, Fermentation, Isolation, and Physico-Chemical Properties. *J. Antibiot.* **1999**, *52*, 525–530. [CrossRef] [PubMed]
3. Barrett, A.J.; Kembhavi, A.A.; Brown, M.A.; Kirschke, H.; Knight, C.G.; Tamai, M.; Hanada, K. L-*Trans*-Epoxysuccinyl-Leucylamido(4-Guanidino)Butane (E-64) and Its Analogues as Inhibitors of Cysteine Proteinases Including Cathepsins B, H and L. *Biochem. J.* **1982**, *201*, 189–198. [CrossRef] [PubMed]
4. Moreira, J.A.; Neppe, T.; De Paiva, M.M.; Deobald, A.M.; Batista-Pereira, L.G.; Paixão, M.W.; Corrêa, A.G. Studies towards the Identification of the Sex Pheromone of *Thyrinteina Arnobia*. *J. Braz. Chem. Soc.* **2013**, *24*, 1933–1941. [CrossRef]
5. Paz, M.M.; Zhang, X.; Lu, J.; Holmgren, A. A New Mechanism of Action for the Anticancer Drug Mitomycin C: Mechanism-Based Inhibition of Thioredoxin Reductase. *Chem. Res. Toxicol.* **2012**, *25*, 1502–1511. [CrossRef] [PubMed]
6. Corre, C.; Lowden, P.A.S. The First Biosynthetic Studies of the Azinomycins: Acetate Incorporation into Azinomycin B. *Chem. Commun.* **2004**, *8*, 990–991. [CrossRef] [PubMed]
7. Lapinsky, D.J. Three-Membered Ring Systems. In *Progress in Heterocyclic Chemistry*; Gribble, G.W., Joule, J.A., Eds.; Elsevier: Oxford, UK, 2015; Volume 27, pp. 61–85. ISBN 9780081000243.
8. Rotstein, B.H.; Zaretsky, S.; Rai, V.; Yudin, A.K. Small Heterocycles in Multicomponent Reactions. *Chem. Rev.* **2014**, *114*, 8323–8359. [CrossRef]
9. Katsuki, T.; Sharpless, K.B. The First Practical Method for Asymmetric Epoxidation. *J. Am. Chem. Soc.* **1980**, *102*, 5974–5976. [CrossRef]
10. Zhang, W.; Loebach, J.L.; Wilson, S.R.; Jacobsen, E.N. Enantioselective Epoxidation of Unfunctionalized Olefins Catalyzed by Salen Manganese Complexes. *J. Am. Chem. Soc.* **1990**, *112*, 2801–2803. [CrossRef]
11. Irie, R.; Noda, K.; Ito, Y.; Matsumoto, N.; Katsuki, T. Catalytic Asymmetric Epoxidation of Unfunctionalized Olefins. *Tetrahedron Lett.* **1990**, *31*, 7345–7348. [CrossRef]
12. Jat, J.L.; Paudyal, M.P.; Gao, H.Y.; Xu, Q.L.; Yousufuddin, M.; Devarajan, D.; Ess, D.H.; Kurti, L.; Falck, J.R. Direct Stereospecific Synthesis of Unprotected N-H and N-Me Aziridines from Olefins. *Science* **2014**, *343*, 61–65. [CrossRef]

13. Ma, Z.; Zhou, Z.; Kürti, L. Direct and Stereospecific Synthesis of N-H and N-Alkyl Aziridines from Unactivated Olefins Using Hydroxylamine-O-Sulfonic Acids. *Angew. Chem. Int. Ed.* **2017**, *56*, 9886–9890. [CrossRef] [PubMed]
14. Meninno, S.; Lattanzi, A. Asymmetric Organocatalytic Journey into the World of Three-Membered Rings. *Catal. Today* **2017**, *285*, 39–48. [CrossRef]
15. Helene, P. Synthesis of Natural and/or Biologically Relevant Three-Membered Heterocycles and Derivatives through Asymmetric Epoxidation, Aziridination, Azirination and Thiirination. *Lett. Org. Chem.* **2018**, *15*, 171–182. [CrossRef]
16. Wang, C. Electrophilic Ring Opening of Small Heterocycles. *Synthesis* **2017**, *49*, 5307–5319. [CrossRef]
17. Ahmad, S.; Zahoor, A.F.; Naqvi, S.A.R.; Akash, M. Recent Trends in Ring Opening of Epoxides with Sulfur Nucleophiles. *Mol. Divers.* **2018**, *22*, 191–205. [CrossRef] [PubMed]
18. Akhtar, R.; Naqvi, S.A.R.; Zahoor, A.F.; Saleem, S. Nucleophilic Ring Opening Reactions of Aziridines. *Mol. Divers.* **2018**, *22*, 447–501. [CrossRef] [PubMed]
19. Hayashi, Y. Pot Economy and One-Pot Synthesis. *Chem. Sci.* **2016**, *7*, 866–880. [CrossRef] [PubMed]
20. Smith, A.B.; Kim, D.-S. A General, Convergent Strategy for the Construction of Indolizidine Alkaloids: Total Syntheses of (−)-Indolizidine 223AB and Alkaloid (−)-205B. *J. Org. Chem.* **2006**, *71*, 2547–2557. [CrossRef]
21. Schaus, S.E.; Brandes, B.D.; Larrow, J.F.; Tokunaga, M.; Hansen, K.B.; Gould, A.E.; Furrow, M.E.; Jacobsen, E.N. Highly Selective Hydrolytic Kinetic Resolution of Terminal Epoxides Catalyzed by Chiral (Salen)CoIII Complexes. Practical Synthesis of Enantioenriched Terminal Epoxides and 1,2-Diols. *J. Am. Chem. Soc.* **2002**, *124*, 1307–1315. [CrossRef]
22. Oka, T.; Yasusa, T.; Ando, T.; Watanabe, M.; Yoneda, F.; Ishida, T.; Knoll, J. Enantioselective Synthesis and Absolute Configuration of (−)-1-(Benzofuran-2-yl)-2-Propylaminopentane, ((−)-BPAP), a Highly Potent and Selective Catecholaminergic Activity Enhancer. *Bioorg. Med. Chem.* **2001**, *9*, 1213–1219. [CrossRef]
23. Rivera, D.G.; Paixão, M.W. Interplay between Organocatalysis and Multicomponent Reactions in Stereoselective Synthesis. In *Stereochemistry and Global Connectivity: The Legacy of Ernest L. Eliel Volume 2*; ACS Symposium Series; American Chemical Society: Washington, DC, USA, 2017; pp. 4–49. [CrossRef]
24. Cioc, R.C.; Ruijter, E.; Orru, R.V.A. Multicomponent Reactions: Advanced Tools for Sustainable Organic Synthesis. *Green Chem.* **2014**, *16*, 2958–2975. [CrossRef]
25. Das, D. Multicomponent Reactions in Organic Synthesis Using Copper-Based Nanocatalysts. *ChemistrySelect* **2016**, *1*, 1959–1980. [CrossRef]
26. Climent, M.J.; Corma, A.; Iborra, S. Homogeneous and Heterogeneous Catalysts for Multicomponent Reactions. *RSC Adv.* **2012**, *2*, 16–58. [CrossRef]
27. Lima, C.G.S.; Moreira, N.M.; Paixão, M.W.; Corrêa, A.G. Heterogenous Green Catalysis: Application of Zeolites on Multicomponent Reactions. *Curr. Opin. Green Sustain. Chem.* **2019**, *15*, 7–12. [CrossRef]
28. Hosseinian, A.; Ahmadi, S.; Mohammadi, R.; Monfared, A.; Rahmani, Z. Three-component reaction of amines, epoxides, and carbon dioxide: A straightforward route to organic carbamates. *J. CO2 Util.* **2018**, *27*, 381–389. [CrossRef]
29. Wang, Q.; Wang, D.-X.; Wang, M.-X.; Zhu, J. Still Unconquered: Enantioselective Passerini and Ugi Multicomponent Reactions. *Acc. Chem. Res.* **2018**, *51*, 1290–1300. [CrossRef]
30. Riva, R. Enantioselective Four-Component Ugi Reactions. *Science* **2018**, *361*, 1072–1073. [CrossRef]
31. Zhang, J.; Yu, P.; Li, S.-Y.; Sun, H.; Xiang, S.-H.; Wang, J. (Joelle); Houk, K.N.; Tan, B. Asymmetric Phosphoric Acid–catalyzed Four-Component Ugi Reaction. *Science* **2018**, *361*, 8707. [CrossRef]
32. Roy, T.; Thangaraj, M.; Gonnade, R.G.; Biju, A.T. Synthesis of Functionalized Amino Epoxides by a Three-Component Coupling Involving Aziridines, Arynes and Aldehydes. *Chem. Commun.* **2016**, *52*, 9044–9047. [CrossRef]
33. Ashokkumar, V.; Siva, A. One-Pot Synthesis of α,β-Epoxy Ketones through Domino Reaction between Alkenes and Aldehydes Catalyzed by Proline Based Chiral Organocatalysts. *Org. Biomol. Chem.* **2017**, *15*, 2551–2561. [CrossRef]
34. Deobald, A.M.; Corrêa, A.G.; Rivera, D.G.; Paixão, M.W. Organocatalytic Asymmetric Epoxidation and Tandem Epoxidation/Passerini Reaction under Eco-Friendly Reaction Conditions. *Org. Biomol. Chem.* **2012**, *10*, 7681. [CrossRef] [PubMed]

35. Santos, D.A.; Deobald, A.M.; Cornelio, V.E.; Ávila, R.M.D.; Cornea, R.C.; Bernasconi, G.C.R.; Paixão, M.W.; Vieira, P.C.; Corrêa, A.G. Asymmetric synthesis and evaluation of epoxy-α-acyloxycarboxamides as selective inhibitors of cathepsin L. *Bioorg. Med. Chem.* **2017**, *25*, 4620–4627. [CrossRef] [PubMed]
36. Vignatti, C.; Luis-Barrera, J.; Guillerm, V.; Imaz, I.; Mas-Ballesté, R.; Alemán, J.; Maspoch, D. Squaramide-IRMOF-16 Analogue for Catalysis of Solvent-Free, Epoxide Ring-Opening Tandem and Multicomponent Reactions. *ChemCatChem* **2018**, *10*, 3995–3998. [CrossRef]
37. Hong, X.-W.; Zhou, Y.-Q.; Bai, C.-B.; Wang, N.-X.; Xing, Y.; Zhang, W.; Wang, Y.-J.; Lan, X.-W.; Xie, Y.; Li, Y.-H. Asymmetric Synthesis of 1,3-Oxazolidine Derivatives with Multi-Component Reaction and Research of Kinetic Resolution. *Molecules.* **2015**, *20*, 17208–17220. [CrossRef] [PubMed]
38. Smith, A.B.; Duffey, M.O. Bifunctional molecular linchpins: A three-component coupling protocol employing 2-bromoallyltrimethylsilane. *Synlett.* **2004**, *2004*, 1363–1366. [CrossRef]
39. Murray, S.A.; Liang, M.Z.; Meek, S.J. Stereoselective Tandem Bis-Electrophile Couplings of Diborylmethane. *J. Am. Chem. Soc.* **2017**, *139*, 14061–14064. [CrossRef]
40. Zhou, L.; Tan, C.K.; Zhou, J.; Yeung, Y.Y. Facile, efficient, and catalyst-free electrophilic aminoalkoxylation of olefins: Scope and application. *J. Am. Chem. Soc.* **2010**, *132*, 10245–10247. [CrossRef]
41. Zhou, J.; Zhou, L.; Yeung, Y.Y. Multicomponent approach in the synthesis of 2,2,6-trisubstituted morpholine derivatives. *Org. Lett.* **2012**, *14*, 5250–5253. [CrossRef]
42. Zhou, J.; Yeung, Y.Y. Synthesis of reboxetine intermediate and carnitine acetyltransferase inhibitor via NBS-induced electrophilic multicomponent reaction. *J. Org. Chem.* **2014**, *79*, 4644–4649. [CrossRef]
43. Church, T.L.; Byrne, C.M.; Lobkovsky, E.B.; Coates, G.W. A new multicomponent reaction catalyzed by a [Lewis acid]$^+$[$Co(CO)_4$]$^-$ catalyst: Stereospecific synthesis of 1,3-oxazinane-2,4-diones from epoxides, isocyanates, and CO. *J. Am. Chem. Soc.* **2007**, *129*, 8156–8162. [CrossRef]
44. Zhang, Y.; Ji, J.; Zhang, X.; Lin, S.; Pan, Q.; Jia, L. Cobalt-catalyzed cyclization of carbon monoxide, imine, and epoxide. *Org. Lett.* **2014**, *16*, 2130–2133. [CrossRef] [PubMed]
45. Liu, L.; Sun, H. [$HCo(CO)_4$]-catalyzed three-component cycloaddition of epoxides, imines, and carbon monoxide: Facile construction of 1,3-oxazinan-4-ones. *Angew. Chem. Int. Ed.* **2014**, *53*, 9865–9869. [CrossRef] [PubMed]
46. Cheng, H.G.; Wu, C.; Chen, H.; Chen, R.; Qian, G.; Geng, Z.; Wei, Q.; Xia, Y.; Zhang, J.; Zhang, Y.; et al. Epoxides as Alkylating Reagents for the Catellani Reaction. *Angew. Chem. Int. Ed.* **2018**, 3444–3448. [CrossRef] [PubMed]
47. Singh, G.S.; Sudheesh, S.; Keroletswe, N. Recent Applications of Aziridine Ring Expansion Reactions in Heterocyclic Synthesis. *ARKIVOC* **2018**, 50–113. [CrossRef]
48. Jiang, J.; Liu, H.; Lu, C.-D.; Xu, Y.-J. Diethyl Phosphite Initiated Coupling of α-Ketoesters with Imines for Synthesis of α-Phosphonyloxy-β-Amino Acid Derivatives and Aziridine-2-Carboxylates. *Org. Lett.* **2016**, *18*, 880–883. [CrossRef] [PubMed]
49. Minakata, S.; Ando, T.; Nishimura, M.; Ryu, I.; Komatsu, M. Novel Asymmetric and Stereospecific Aziridination of Alkenes with a Chiral Nitridomanganese Complex. *Angew. Chem. Int. Ed.* **1999**, *37*, 3392–3394. [CrossRef]
50. Albrecht, Ł.; Jiang, H.; Jørgensen, K.A. A Simple Recipe for Sophisticated Cocktails: Organocatalytic One-Pot Reactions—Concept, Nomenclature, and Future Perspectives. *Angew. Chem. Int. Ed.* **2011**, *50*, 8492–8509. [CrossRef] [PubMed]
51. Gupta, A.K.; Mukherjee, M.; Wulff, W.D. Multicomponent Catalytic Asymmetric Aziridination of Aldehydes. *Org. Lett.* **2011**, *13*, 5866–5869. [CrossRef]
52. Gupta, A.K.; Mukherjee, M.; Hu, G.; Wulff, W.D. BOROX Catalysis: Self-Assembled Amino-BOROX and Imino-BOROX Chiral Brønsted Acids in a Five Component Catalyst Assembly/Catalytic Asymmetric Aziridination. *J. Org. Chem.* **2012**, *77*, 7932–7944. [CrossRef]
53. Akiyama, T.; Suzuki, T.; Mori, K. Enantioselective Aza-Darzens Reaction Catalyzed by A Chiral Phosphoric Acid. *Org. Lett.* **2009**, *11*, 2445–2447. [CrossRef]
54. Bew, S.P.; Liddle, J.; Hughes, D.L.; Pesce, P.; Thurston, S.M. Chiral Brønsted Acid-Catalyzed Asymmetric Synthesis of N-Aryl-Cis-Aziridine Carboxylate Esters. *Angew. Chem. Int. Ed.* **2017**, *56*, 5322–5326. [CrossRef] [PubMed]
55. Mukherjee, M.; Zhou, Y.; Dai, Y.; Gupta, A.L.K.; Pulgam, V.R.; Staples, R.J.; Wulff, W.D. Catalyst-Controlled Multicomponent Aziridination of Chiral Aldehydes. *Chem. Eur. J.* **2017**, *23*, 2552–2556. [CrossRef] [PubMed]

56. Zeng, X.; Zeng, X.; Xu, Z.; Lu, M.; Zhong, G. Highly Efficient Asymmetric Trans-Selective Aziridination of Diazoacetamides and N-Boc-Imines Catalyzed by Chiral Brønsted Acids. *Org. Lett.* **2009**, *11*, 3036–3039. [CrossRef] [PubMed]
57. Desai, A.A.; Wulff, W.D. Controlled Diastereo- and Enantioselection in a Catalytic Asymmetric Aziridination. *J. Am. Chem. Soc.* **2010**, *132*, 13100–13103. [CrossRef] [PubMed]
58. Hashimoto, T.; Uchiyama, N.; Maruoka, K. Trans-Selective Asymmetric Aziridination of Diazoacetamides and N-Boc Imines Catalyzed by Axially Chiral Dicarboxylic Acid. *J. Am. Chem. Soc.* **2008**, *130*, 14380–14381. [CrossRef] [PubMed]
59. Vetticatt, M.J.; Desai, A.A.; Wulff, W.D. How the Binding of Substrates to a Chiral Polyborate Counterion Governs Diastereoselection in an Aziridination Reaction: H-Bonds in Equipoise. *J. Am. Chem. Soc.* **2010**, *132*, 13104–13107. [CrossRef] [PubMed]
60. Zhou, Y.; Gupta, A.K.; Mukherjee, M.; Zheng, L.; Wulff, W.D. Multicomponent Catalytic Asymmetric Synthesis of Trans-Aziridines. *J. Org. Chem.* **2017**, *82*, 13121–13140. [CrossRef] [PubMed]
61. Chakraborty Ghosal, N.; De, A.; Mahato, S.; Santra, S.; Zyryanov, G.V.; Majee, A. An Efficient Synthesis of Oxazolidines by Tandem Ring-Opening/Closing Reaction of Ts-Aziridine Using Formic Acid. *ChemistrySelect* **2018**, *3*, 10509–10514. [CrossRef]
62. Ghosal, N.C.; Santra, S.; Zyryanov, G.V.; Hajra, A.; Majee, A. Conversion of Aziridines to Oxazolidines through Geminal Difunctionalization of Vinyl Arenes or by Tandem Ring-Opening/Closing Reaction of Aziridine Itself. *Tetrahedron Lett.* **2016**, *57*, 3551–3555. [CrossRef]
63. Stephens, D.; Zhang, Y.; Cormier, M.; Chavez, G.; Arman, H.; Larionov, O.V. Three-Component Reaction of Small-Ring Cyclic Amines with Arynes and Acetonitrile. *Chem. Commun.* **2013**, *49*, 6558–6560. [CrossRef]
64. Tang, C.-Y.; Wang, G.; Yang, X.-Y.; Wu, X.-Y.; Sha, F. One-Pot Synthesis of α-Fluoro-β-Amino Acid and Indole Spiro-Derivatives via CN Bond Cleavage/Formation. *Tetrahedron Lett.* **2014**, *55*, 6447–6450. [CrossRef]
65. Roy, T.; Bhojgude, S.S.; Kaicharla, T.; Thangaraj, M.; Garai, B.; Biju, A.T. Employing Carboxylic Acids in Aryne Multicomponent Coupling Triggered by Aziridines/Azetidines. *Org. Chem. Front.* **2016**, *3*, 71–76. [CrossRef]
66. Roy, T.; Baviskar, D.R.; Biju, A.T. Synthesis of N-Aryl β-Amino Alcohols by Trifluoroacetic Acid Promoted Multicomponent Coupling of Aziridines, Arynes, and Water. *J. Org. Chem.* **2015**, *80*, 11131–11137. [CrossRef]
67. Wani, I.A.; Sayyada, M.; Ghorai, M.K. Domino ring-opening cyclization (DROC) of activated aziridines and epoxides with nitrones via dual-catalysis "on water". *Chem. Commun.* **2017**, *53*, 4386–4389. [CrossRef]
68. Zhou, J.; Yeung, Y.-Y. N-Bromosuccinimide-Induced Aminocyclization–Aziridine Ring-Expansion Cascade: An Asymmetric and Highly Stereoselective Approach toward the Synthesis of Azepane. *Org. Lett.* **2014**, *16*, 2134–2137. [CrossRef] [PubMed]
69. Zhou, J.; Yeung, Y.-Y. Diastereoselective Synthesis of Functionalized Pyrrolidines through N-Bromosuccinimide-Induced Aziridine Ring Expansion Cascade of Cinnamylaziridine. *Org. Biomol. Chem.* **2014**, *12*, 7482–7485. [CrossRef]
70. Qian, G.; Bai, M.; Gao, S.; Chen, H.; Zhou, S.; Cheng, H.-G.; Yan, W.; Zhou, Q. Modular One-Step Three-Component Synthesis of Tetrahydroisoquinolines Using a Catellani Strategy. *Angew. Chem. Int. Ed.* **2018**, *57*, 10980–10984. [CrossRef]
71. Xing, S.; Ren, J.; Wang, K.; Cui, H.; Li, W.; Yan, H. Lewis Acid Promoted Three-Component Reactions of Aziridines, Arenes and Aldehydes: An Efficient and Diastereoselective Synthesis of Cis-1,4-Disubstituted Tetrahydroisoquinolines. *Tetrahedron* **2015**, *71*, 6290–6299. [CrossRef]
72. Assem, N.; Hili, R.; He, Z.; Kasahara, T.; Inman, B.L.; Decker, S.; Yudin, A.K. Role of Reversible Dimerization in Reactions of Amphoteric Aziridine Aldehydes. *J. Org. Chem.* **2012**, *77*, 5613–5623. [CrossRef]
73. Monteiro, J.L.; Moreira, N.M.; dos Santos, D.A.; Paixão, M.W.; Corrêa, A.G. Step economy strategy for the synthesis of amphoteric aminoaldehydes, key intermediates for reduced hydantoins. *Pure Appl. Chem.* **2018**, *90*, 121–132. [CrossRef]
74. Liew, S.K.; He, Z.; St. Denis, J.D.; Yudin, A.K. Stereocontrolled Synthesis of 1,2- and 1,3-Diamine Building Blocks from Aziridine Aldehyde Dimers. *J. Org. Chem.* **2013**, *78*, 11637–11645. [CrossRef] [PubMed]
75. St. Denis, J.D.; Liew, S.K.; Scully, C.C.G.; Yudin, A.K. Activation of Alkynylzinc Reagents by a Hemiaminal-Driven Catalytic Microenvironment. *Eur. J. Org. Chem.* **2016**, *2017*, 419–423. [CrossRef]
76. Hili, R.; Rai, V.; Yudin, A.K. Macrocyclization of Linear Peptides Enabled by Amphoteric Molecules. *J. Am. Chem. Soc.* **2010**, *132*, 2889–2891. [CrossRef] [PubMed]

77. Zaretsky, S.; Adachi, S.; Rotstein, B.H.; Hickey, J.L.; Scully, C.C.G.; St. Denis, J.D.; Courtemanche, R.; Yu, J.C.Y.; Chung, B.K.W.; Yudin, A.K. Stereocontrolled Disruption of the Ugi Reaction toward the Production of Chiral Piperazinones: Substrate Scope and Process Development. *J. Org. Chem.* **2014**, *79*, 9948–9957. [CrossRef] [PubMed]
78. Heine, N.B.; Kaldas, S.J.; Belding, L.; Shmatova, O.; Dudding, T.; Nenajdenko, V.G.; Studer, A.; Yudin, A.K. Synthesis of Chiral Piperazinones Using Amphoteric Aziridine Aldehyde Dimers and Functionalized Isocyanides. *J. Org. Chem.* **2016**, *81*, 5209–5216. [CrossRef] [PubMed]
79. Belding, L.; Zaretsky, S.; Rotstein, B.H.; Yudin, A.K.; Dudding, T. Shifting the Energy Landscape of Multicomponent Reactions Using Aziridine Aldehyde Dimers: A Mechanistic Study. *J. Org. Chem.* **2014**, *79*, 9465–9471. [CrossRef]
80. Zaretsky, S.; Hickey, J.L.; Tan, J.; Pichugin, D.; St. Denis, M.A.; Ler, S.; Chung, B.K.W.; Scully, C.C.G.; Yudin, A.K. Mechanistic Investigation of Aziridine Aldehyde-Driven Peptide Macrocyclization: The Imidoanhydride Pathway. *Chem. Sci.* **2015**, *6*, 5446–5455. [CrossRef] [PubMed]
81. Treder, A.P.; Hickey, J.L.; Tremblay, M.-C.J.; Zaretsky, S.; Scully, C.C.G.; Mancuso, J.; Doucet, A.; Yudin, A.K.; Marsault, E. Solid-Phase Parallel Synthesis of Functionalised Medium-to-Large Cyclic Peptidomimetics through Three-Component Coupling Driven by Aziridine Aldehyde Dimers. *Chem. Eur. J.* **2015**, *21*, 9249–9255. [CrossRef]

molecules

Article

Hemi-Synthesis of Chiral Imine, Benzimidazole and Benzodiazepines from Essential Oil of *Ammodaucus leucotrichus* subsp. *leucotrichus*

Farid Chebrouk [1,2], Khodir Madani [1,*], Brahim Cherfaoui [2,5], Leila Boukenna [2,5], Mónica Válega [3], Ricardo F. Mendes [4], Filipe A. A. Paz [4], Khaldoun Bachari [2], Oualid Talhi [2,3,*] and Artur M. S. Silva [3,*]

1 Laboratory of Biomathematics, Biochemistry, Biophysics and Scientometrics (L3BS), Faculty of Nature and Life Sciences (FSNV), University of Bejaia, 06000 Bejaia, Algeria; chebroukfarid@yahoo.fr
2 Research Center Scientific and Technical in Analyzes Physico-Chimiques CRAPC, BP384, Bou-Ismail, 42004 Tipaza, Algeria; cherfaoui.br@gmail.com (B.C.); lboukenna15@gmail.com (L.B.); bachari2000@yahoo.fr (K.B.)
3 QOPNA and LAQV-REQUIMTE, Department of Chemistry, University of Aveiro, 3810-193 Aveiro, Portugal; mvalega@ua.pt
4 CICECO—Aveiro Institute of Materials, Department of Chemistry, University of Aveiro, 3810-193 Aveiro, Portugal; rfmendes@ua.pt (R.F.M.); filipe.paz@ua.pt (F.A.A.P.)
5 Laboratory of Applied Organic Chemistry, Faculty of Chemistry, University of Science and Technology Houari Boumediene, BP 32, Alia Bab-Ezzouar, 16111 Algiers, Algeria
* Correspondence: khodir.madani@univ-bejaia.dz (K.M.); oualid.talhi@ua.pt (O.T.); artur.silva@ua.pt (A.M.S.S.); Tel.: +213-24-325-779 (O.T.); +351-234-370-714 (A.M.S.S.)

Academic Editor: Maria Elizabeth Tiritan
Received: 27 January 2019; Accepted: 6 March 2019; Published: 10 March 2019

Abstract: The hemi-synthesis of chiral imine, benzimidazole and benzodiazepine structures is reported by the condensation of (*S*)-(−)-perillaldehyde, the major phytochemical of *Ammodaucus leucotrichus* subsp. *leucotrichus* essential oil, with different amine derivatives of 2,3-diaminomaleonitrile, *o*-phenylenediamine and 3-[(2-aminoaryl)amino]dimedone. The reaction proceeds in situ at ambient temperature without prior isolation of the natural (*S*)-(−)-perillaldehyde. Final products precipitate in the ethanolic reaction medium. 2D NMR and single-crystal X-ray diffraction studies were used to unequivocally characterize the structures in solution and in the solid state, respectively. Chiral HPLC analysis confirms the formation of unique enantiomers and diastereomeric mixtures.

Keywords: (*S*)-(−)-perillaldehyde; *Ammodaucus leucotrichus*; amines; hemi-synthesis; essential oil; 2D NMR; single-crystal X-ray diffraction; chiral-HPLC

1. Introduction

Over the past decades, there has been a growing transition in drugs from natural materials, natural products, or their simple derivatives to more potent natural-mimicking synthetic prototypes [1]. Essential oils' components still play a major role in this combinatorial chemistry and are considered promising sources of stereospecific structures. They can indeed serve as substrates in hemi-synthesis, leading to new active molecules. For instance, (+)-carotol, the major constituent of carrot seed essential oil, was used as a starting material in the hemi-synthesis of ten cytotoxic hydroindene-derived chiral synthons [2]. A novel series of essential oils oriented chiral esters with high insecticidal activity have been synthesized based on the scaffold of the natural pyrethrin [3]. *Cymbopogon schoenanthus* essential oil is mainly composed of piperitone (68.2%), which was used for in situ preparation of three anti-parasitic carbasones [4].

Ammodaucus leucotrichus Cosson & Durieu subsp. *leucotrichus* belongs to the *Apiaceae* (*Umbelliferae*) plant family. It is an endemic species growing in the Saharan and sub-Saharan countries of North Africa, extending up to Egypt and tropical Africa [5]. Generally, it grows in wadis with granite boulders (temporary water streams) [6]. This plant plays an important role in traditional medicine in North African countries. It is used as a remedy for cardiac diseases [7], rheumatism, asthma and stomach diseases [8,9]. The major components in the essential oil of *A. leucotrichus* from Morocco were reported to be perillaldehyde (63.6%) and limonene (26.8%) [5]. A similar phytochemical profile is reported for the essential oil of *A. leucotrichus* growing wild at different areas in the Southern Algerian Sahara, characterized by perillaldehyde (37.5–84.4%) and limonene (7.0%–29.2%) [10–12]. (*S*)-(−)-Perillaldehyde laevorotatory form is found most abundantly in essential oils of other plants such as the woody shrub *Conyza newii* [13]. This aldehyde is extensively used as a food additive for flavouring and especially as an ingredient in perfumes [14], but it also exhibits antioxidant, antidepressant and other biological activities [15].

In general, aldehydes are most commonly used in the synthesis of a large range of interesting molecules, such as imines and benzodiazepines. Many preparations of various imines by the condensation of different aldehydes with aliphatic or aromatic amines are reported in the literature [16,17]. This kind of molecule is known to possess significant biological properties, such as antitumor, insecticidal, antibacterial, anti-tuberculosis, antimicrobial and anticonvulsant activities [18]. Benzodiazepines represent a well-known class of therapeutics displaying anticonvulsant, anti-inflammatory, analgesic and anti-depressive effects [19]. A literature survey reveals that a fair amount of work has been published in the stereospecific synthesis of benzodiazepine derivatives by the condensation of aldehydes and several heterocyclic compounds [20,21]. In particular, a simple and fast method has recently been described by us for the elaboration of novel benzodiazepine structures [22].

Following this last method, we propose herein the hemi-synthesis of novel chiral imine, benzimidazole and benzodiazepine structures from the essential oil of *A. leucotrichus* by condensation of its major constituent (*S*)-(−)-perillaldehyde (81.0%) with different amine substrates, such as 2,3-diaminomaleonitrile, *o*-phenylenediamine and 3-[(2-aminoaryl)amino]dimedone. The GC-MS analysis of our *A. leucotrichus* essential oil allowed the identification of perillaldehyde (81.0%) and limonene (12.4%) as major constituents (see Supplementary Materials). The combination of amines with the enantiopure perillaldehyde may lead to functional synergy and render the final chiral molecules with improved biological properties, since it is well-known that the biological effect is intrinsically related to one of the enantiomers. In addition, the toxicological, pharmacokinetic behaviour and metabolism of the enantiomers may also be different [23].

2. Results and Discussion

2.1. Chemistry and Mechanism

Perillaldehyde, the major constituent of the *A. leucotrichus* essential oil, reacts in situ with the aforementioned amines without the need for prior purification or isolation from the oil matrix. The condensation reaction proceeds under similar operating conditions as previously published [22], by employing ethanol as a solvent at ambient temperature and under catalyst-free conditions over a period of 12 h. The reaction is easily worked up by filtration of the formed solid showing a high purity of products **1–4**, isolated in moderate-to-good yields (47%–82%) after washing with ethanol, ultimately avoiding any further chromatographic purifications (Scheme 1).

The final products **1–4** are chiral because the condensation reactions are carried out over the aldehyde group, ultimately maintaining the configuration of the asymmetric centre of (*S*)-perillaldehyde. This has direct implications in the performed crystallographic studies as detailed below. The reaction mechanisms follow the classical imine condensation to afford compound **1**, 2-amino-3-{[(*E*)-[(*S*)-4-(prop-1-en-2-yl) cyclohex-1-en-1-yl)methylene]amino}maleonitrile, obtained from the reaction of the main constituent of the *A. leucotrichus* essential oil with 2,3-diaminomaleonitrile (Scheme 2). In a similar way, *o*-phenylenediamine condenses on perillaldehyde to afford the corresponding imine as an intermediate. The latter undergoes a second amine attack, giving rise to an imidazole ring closure after spontaneous dehydrogenation at ambient temperature, therefore generating (*S*)-2-[4-(prop-1-en-2-yl)cyclohex-1-en-1-yl]-1*H*-benzo[*d*]imidazole **2** via a tandem process (Scheme 2). In common reaction cases, the imidazole ring closure occurs at higher temperatures (heating at reflux) [24]. However, our experimental protocols were strategically oriented to perform hemi-synthetic transformations at room temperature, which allows a rapid precipitation/crystallization of the final chiral product upon formation. Under these relatively low temperature conditions, compound **1** could not cyclize to imidazole, while the benzimidazole ring closure in compound **2** was possible due to chemical stabilization of the mentioned heterocyclic system. The last experiment was performed using 3-[(2-aminoaryl)amino]dimedone which, according to our previous study [22], originates 3,3-dimethyl-11-[(*S*)-4-(prop-1-en-2-yl)cyclohex-1-en-1-yl]-2,3,4,5,10,11-hexahydro-1*H*-dibenzo[*b*,*e*][1,4]diazepin-1-ones **3** and **4** via imine condensation and diazepine ring closure that creates a new asymmetric centre over the diazepine ring (Scheme 2).

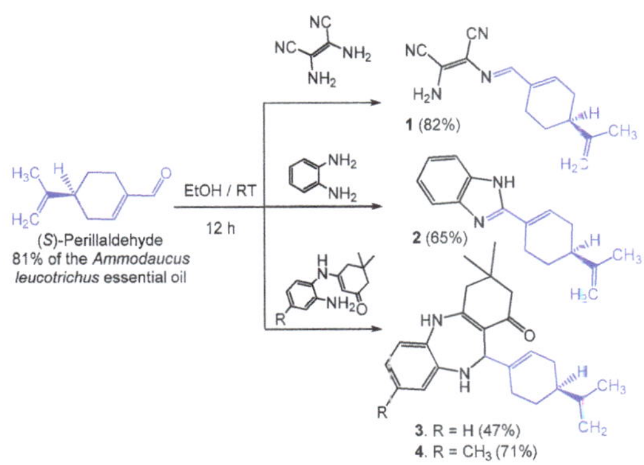

Scheme 1. In situ hemi-synthesis of imine **1**, benzimidazole **2** and benzodiazepines **3–4** from the main constituent of the *Ammodaucus leucotrichus* essential oil.

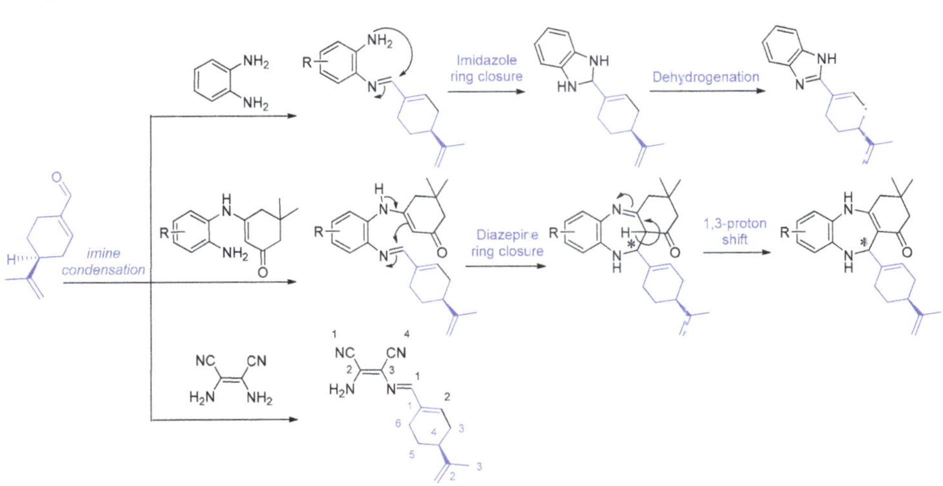

Scheme 2. Mechanistic pathway for the hemi-synthesis of imine **1**, benzimidazole **2** and benzodiazepine **3–4** structures form essential oil of *A. leucotrichus* in situ.

2.2. Chiral HPLC and Stereochemistry

(*S*)-(−)-Perillaldehyde is found in essential oils in an optically active laevorotatory form due to the presence of an asymmetric carbon [13]. This carbon was chemically resistant to our amine nucleophilic attacks on perillaldehyde where chemical alterations were mainly brought to the aldehyde function to produce compounds **1–4** (including imine, imidazole and 1,4-diazepines) and maintaining the asymmetric carbon unchanged. This fact was further studied, however, by chiral-HPLC analysis performed on imine **1**, imidazole **2** and 1,4-benzodiazepine **3**, where the presence of a sole chromatographic peak indicated the existence of one enantiomer as noted in the case of compounds **1** and **2** (Figure 1, see Supplementary Materials). Structurally, 1,4-benzodiazepines **3** and **4** are composed of the asymmetric perillaldehyde moiety attached to the new connecting asymmetric carbon which belongs to the diazepine ring. Stereochemical and mechanistic considerations suggest the presence of a 1:1 ratio of two diastereomers (11-*R*, 4′-*S*- and 11-*S*, 4′-*S*-isomers) (Scheme 2). This was proved with the help of chiral-HPLC, showing the presence of two chromatographic peaks with 1:1 integrating areas, ultimately supporting the formation of a diastereomeric mixture of 1,4-benzodiazepine **3** (see Supplementary Materials). According to chiral chromatographic profiles of **1–3**, our hemi-synthetic process was efficient without any side products being formed, showing high purity of the unique enantiomeric forms of compounds **1** and **2**, along with diastereomeric mixtures of 1,4-benzodiazepines **3** and **4**.

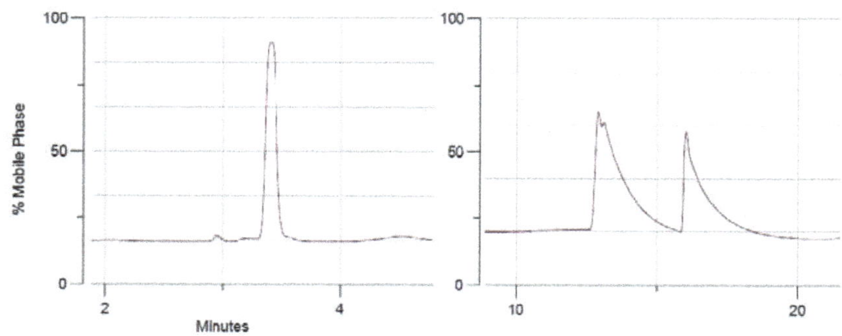

Figure 1. Chiral-HPLC separation of compound **1** (pure enantiomer—**left**) and compound **3** (diasteromeric mixture—**right**).

2.3. Nuclear Magnetic Resonance Spectroscopy

The structures of **1–4** were characterized by 1D and 2D NMR spectroscopy (see Supplementary Materials). The ^{1}H-NMR spectrum of compounds **1–4** suggested the presence of the asymmetric perillalkyl unit, which is mainly characterized by aliphatic protons of methylene groups appearing as multiplets between δ_H 1.0 and 3.0 ppm with their positions confirmed based on DEPT-135, HSQC and HMBC experiments (Figure 2). Vinylic protons of the perillalkyl moiety were attributed around 4.40–5.00 ppm (for the extra-cyclic protons) and 5.00–6.70 ppm (for the intra-cyclic ones). Using DEPT-135 and HSQC experiment, all the protonated carbons were assigned, especially those from δ_C 20 to 40 ppm due to aliphatic methyl and methylene groups of perillalkyl radical (see Supplementary Materials).

The structures of **1–4** were further confirmed based on the HMBC experiments. The main HMBC connectivities observed over the perillalkyl unit are depicted in Figure 2. Aromatic and vinylic quaternary carbons were localized via HMBC correlations with their neighbouring protons, especially in the case of 1,4-benzodiazepines **3** and **4**, where the proton H-11 was shown to establish four different connectivities with the surrounding quaternary carbons, along with the carbonyl C-1 (Figure 2). A deep examination of both ^{1}H and ^{13}C-NMR spectra of 1,4-benzodiazepines **3** and **4** revealed the presence of duplicated signals with a 1:1 ratio according to proton integration, which is clear evidence of the presence of diastereomeric mixtures (see Supplementary Materials).

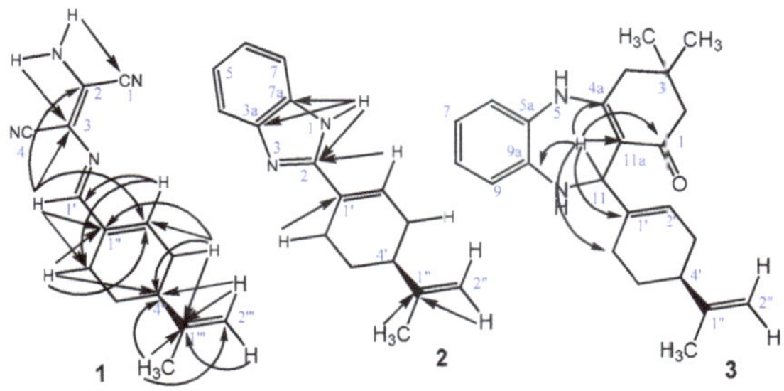

Figure 2. HMBC correlations of compound **1–3**.

2.4. Single-Crystal X-ray Diffraction

Single-crystal X-ray diffraction studies were further used to determine the spatial arrangement of the new compounds **1–4** (Figures 3 and 4). Crystals of compounds **1–4** were directly obtained from the reaction batches after filtration and washing with ethanol. This is noteworthy because use of the natural (*S*)-(−)-perillaldehyde as a starting chiral reagent means that all the resulting structures crystallized in chiral and non-centrosymmetric space groups. The absence of strongly diffracting elements in all structures means that the absolute configurations of asymmetric carbons could not be determined from the X-ray studies, but they were nevertheless unveiled from the chiral HPLC studies performed.

Compounds **1** and **2** only have a single asymmetric carbon centre belonging to the parent perillalkyl unit, exhibiting the (*S*)-configuration (Figure 3). While **1** crystallized in a non-centrosymmetric monoclinic space group $P2_1$ with the asymmetric unit having only one molecule, **2** crystallized in the non-centrosymmetric orthorhombic space group $P2_12_12_1$, with two molecules composing the asymmetric unit. These two molecules had the same configuration for the asymmetric carbon centre with the inequivalence arising from simple rotations of the various molecular units composing the molecule. The presence of donor and acceptor units capable of engaging in strong hydrogen bonds was well evidenced in the crystal packing of both **1** and **2**. In **1**, the two nitrile groups and the amine moiety were engaged in strong intermolecular N–H···N hydrogen bonds, forming a $\mathbf{R}^3{}_3(14)$ graph set ($d_{N\cdots N}$ = 3.012(3)–3.028(3) Å and <(NHN) = 147–168°), ultimately dictating the way the molecules closely pack in the solid state. In **2**, there was a single N–H···N interaction connecting the two crystallographically independent molecular units, leading to the formation of a $\mathbf{C}^2{}_2(8)$ graph set motif ($d_{N\cdots N}$ = 2.853(4) –2.854(4) Å and <(NHN) = 163°) [25].

Compounds **3** and **4** both crystallized in the non-centrosymmetric monoclinic *I*2 space group, with the asymmetric units being composed of a pair of (*S*,*S*)- and (*R*,*S*)-diastereomers (in Figure 4 only the (*S*,*S*)-diastereomers are depicted for the two compounds). According to our previous mechanistic descriptions, the diazepine cyclisation leads to the formation of an extra asymmetric carbon (C-11), with the one from perillaldehyde (C-4′) always maintaining its original (*S*)-configuration. We further note that chiral HPLC clearly showed the presence of a mixture of two diastereomers, which agrees with the crystallographic studies performed. The configuration of all the stereocentres was determined to be (11-*R*,4′-*S*) and (11-*S*,4′-*S*). These results are also in agreement with the aforementioned 2D-NMR investigations.

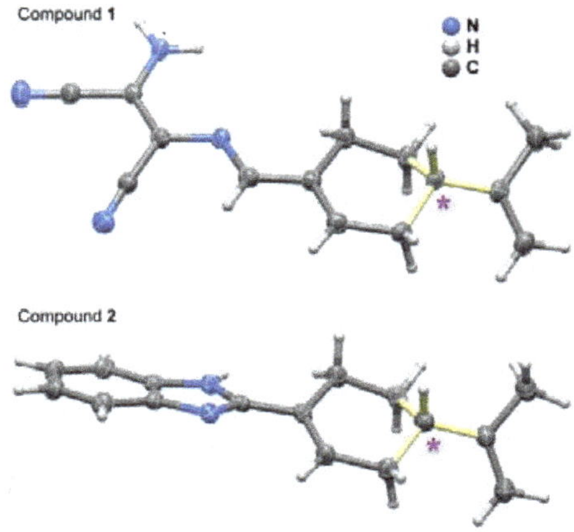

Figure 3. Schematic representation of molecular units present in the crystal structures of compounds **1** and **2** (note: the second identical molecule present in the asymmetric unit of **2** was omitted for clarity). Asymmetric carbon centres are depicted by an asterisk and the bonds in light yellow. Non-hydrogen atoms are represented as thermal ellipsoids drawn at the 50% probability level and hydrogen atoms as small spheres with arbitrary radii.

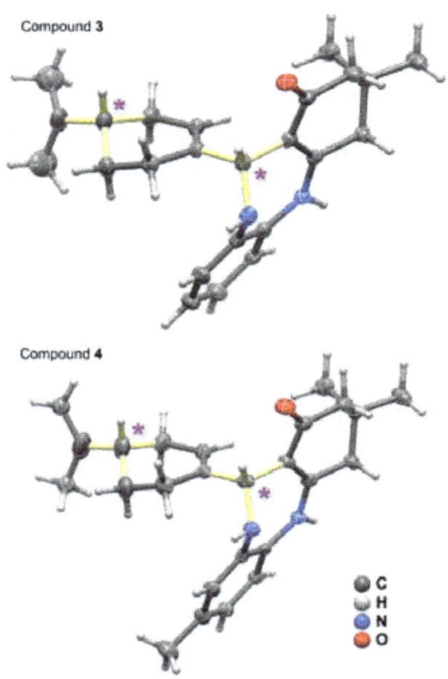

Figure 4. Schematic representation of the (*S*,*S*)-diastereomers molecular units present in the crystal structures of compounds **3** and **4** (note: the (*S*,*R*)-diastereomer molecules present in the asymmetric units were omitted for clarity). Asymmetric carbon centres are depicted by an asterisk and the bonds are in light yellow. Non-hydrogen atoms are represented as thermal ellipsoids drawn at the 50% probability level and hydrogen atoms as small spheres with arbitrary radii.

3. Materials and Methods

3.1. General Remarks

Melting points were measured with a Büchi Melting Point B-540 apparatus (BÜCHI Labortechnik AG, Flawil, Switzerland). NMR spectra were recorded on a Bruker AVANCE 500 spectrometer (Bruker, Wissembourg, France, 400 MHz for ^{1}H and 100 MHz for ^{13}C), in DMSO-d_6 as solvent. Chemical shifts (δ) were reported in ppm and coupling constants (*J*) in Hz and the internal standard was TMS. Unequivocal ^{13}C assignments were made with the aid of 2D gHSQC and gHMBC experiments (delays for one-bond and long-range $J_{C/H}$ couplings were optimized for 145 and 7 Hz, respectively). High-resolution mass spectra (ESI$^+$-HRMS) were measured with a micrOTOF-Q 98 spectrometer (Bruker Daltonics, Hamburg, Germany). GC-MS analysis of *A. leucotrichus* subsp. *leucotrichus* essential oil was performed on a Hewlett-Packard computerized system comprising a 6890 gas chromatograph coupled with a 5973A mass spectrometer (Agilent Technologies, Santa Clara, CA, United States). All chemicals and solvents were purchased from commercial sources and were used as received. Chiral-HPLC separation of compounds **1–3** was performed on chiral stationary phase at 25 °C using a CHIRALPAK® IA (Chiral Technologies Europe, Illkirch, France, amylose-tris(3,5-dimethylphenylcarbamate) immobilized on 5 μm silica-gel, 250 mm × 4.6 mm ID). The mobile phase used was hexane/acetone (isocratic mode, 50:50 (*v*/*v*)) at a flow rate of 1.0 mL/min. The UV detector was set at 220 nm (Figure S21 in the Supplementary Information). An injection of 20 μL of 1.0 g/L concentrated samples of dissolved compounds **1–3** was used in the mobile phase.

3.2. Plant Material and Extraction Procedure

The aerial parts of *A. leucotrichus* were collected from Ghardaia (Septentrional Algerian Sahara) in April 2017. They were identified by the botanists of the National Agronomic Institute in El-Harrach Algeria. Air-dried fruits of *A. leucotrichus* were submitted to water distillation with a Clevenger-type apparatus (Merck KGaA, Darmstadt, Germany) for 3 h. The final product yielded 2.23% of deep-blue liquid oil. The obtained essential oil was dried over anhydrous sodium sulphate and, after filtration, it was stored at +4 °C.

3.3. General Procedure for the Hemi-Synthesis of Compounds **1–4**

Amine derivatives (1.2 mmol, 2,3-diaminomaleonitrile 129.6 mg; *o*-phenylenediamine 129.6 mg; 3-[(2-aminoaryl)amino]dimedone 276 mg; CH_3-3-[(2-aminoaryl)amino]dimedone 292.8 mg) were added to an ethanolic solution (15 mL) of *A. leucotrichus* essential oil (1 mmol of perillaldehyde, calculated on the basis of 81% of mass, approximately 0.2 mL or 185 mg of the crude oil), under constant agitation at ambient temperature, and the reaction was left for 12 h. Crystals of compounds **1–4** were gradually formed in the reaction medium. Precipitates of **1–4** were filtered off and washed with ethanol after the required time (as monitored by TLC).

(S)-2-Amino-3-{[(E)-(4-(prop-1-en-2-yl)cyclohex-1-en-1-yl)methylene]amino}maleonitrile (**1**). $C_{14}H_{16}N_4$. Brown crystals; yield: 197 mg (82%); m.p. 184–186 °C. ^{1}H-NMR (400 MHz, DMSO-d_6): δ 1.30–1.50 (m, 1H, H-5″), 1.73 (s, 3H, 1‴-CH$_3$), 1.77–1.89 (m, 1H, H-5″), 2.07–2.27 (m, 3H, H-4″, H-3″, H-6″), 2.39 (dd, *J* 14.9, 4.0, Hz, 1H, H-3″), 2.56–2.70 (m, 1H, H-6″), 4.66–4.79 (m, 2H, H-2‴), 6.56 (d, *J* 4.7 Hz, 1H, H-2″), 7.48 (s, 2H, 2-NH2), 7.85 (s, 1H, H-1′) ppm. ^{13}C-NMR (100 MHz, DMSO-d_6): δ 21.0 (1‴-CH3), 23.7 (C-6″), 26.8 (C-5″), 31.8 (C-3″), 40.7 (C-4″), 103.8 (C-3), 109.7 (C-2‴), 114.2 (C-4), 114.9 (C-1), 126.2 (C-2), 137.9 (C-1″), 142.6 (C-2″), 149.1 (C-1‴), 158.4 (C-1′) ppm HRMS-ESI$^-$: *m*/*z* calcd. for $[C_{14}H_{16}N_4 + H]^+$: 241.1453; found: 241.1441.

(S)-2-[4-(Prop-1-en-2-yl)cyclohex-1-en-1-yl]-1H-benzo[d]imidazole (**2**). $C_{16}H_{18}N_2$. White crystals; yield: 155 mg (65%); m.p. 220–222 °C. ^{1}H-NMR (400 MHz, DMSO-d_6): δ 1.47–1.65 (m, 1H, H-5′), 1.77 (s, 3H, 1″-CH$_3$), 1.89–1.99 (m, 1H, H-5′), 2.07–2.49 (m, 4H, H-4′, 2 × H-3′, H-6′), 2.75–2.86 (m, 1H, H-6′), 4.55–5.00 (m, 2H, H-2″), 6.79 (d, *J* 4.0 Hz, 1H, H-2′), 7.07–7.20 (m, 2H, H-7, H-5), 7.42 (d, *J* 7.6 Hz, 1H,

H-4), 7.56 (d, *J* 7.6 Hz, 1H, H-6), 12.31 (s, 1H, 1-NH) ppm. ^{13}C-NMR (100 MHz, DMSO-d6): δ 21.1 (1″-CH_3), 25.8 (C-6′), 27.3 (C-5′), 30.9 (C-3′), 40.5 (C-4″), 109.6 (C-2″), 111.2 (C-4), 118.8 (C-6), 121.4 (C-7), 122.3 (C-5), 128.8 (C-1′), 129.2 (C-2′), 135.0 (C-7a), 143.9 (C-3a), 149.2 (C-1″), 158.4 (C-2) ppm. HRMS-ESI$^+$: *m/z* calcd. for $[C_{16}H_{18}N_2 + H]^+$: 239.1548; found: 239.1537.

*3,3-Dimethyl-11-[(S)-4-(prop-1-en-2-yl)cyclohex-1-en-1-yl]-2,3,4,5,10,11-hexahydro-1H-dibenzo[b,e][1,4] diazepin-1-one (**3**, mixture of two diastereomers).* $C_{24}H_{30}N_2O$. Yellowish crystals; yield: 170 mg (47%); m.p. 194–196 °C. ^{1}H-NMR (400 MHz, DMSO-d_6): δ 1.02, 1.03, 1.05 and 1.06 (4s, 6H, 3-CH_3, 2× diast.), 1.56 and 1.58 (2s, 3H, 1″-CH_3, 2 × diast.), 1.06–2.25 (m, 7H, H-3′, H-4′, H-5′, H-6′, 2× diast.), 2.02–2.19 (m, 2H, H-2, 2 × diast.), 2.49–2.53 (m, 2H, H-4, 2× diast.), 4.40–4.43 and 4.50–4.53 and 4.54–4.57 and 4.58–4.62 (4m, 2H, H-2″, 2× diast.), 4,81–4.89 (m, 1H, H-11), 5.07 and 5.09 (d, *J* 6.9 Hz 1H, H-2′, 2× diast.), 5.77 and 5.82 (2d, *J* 6.0 Hz, 1H 10-NH, 2× diast.), 6.61–6.77 (m, 3H, H-7, H-8, H-9), 6.90–6.92 and 6.92–6.93 (2m, 1H, H-6, 2× diast.), 8.530 and 8.533 (s, 1H, 5-NH) ppm. ^{13}C-NMR (100 MHz, DMSO-d_6): δ 20.9 and 21.2 (1″-CH_3, 2× diast.) 27.9 and 28.0 and 29.0 and 29.1 (3-CH_3, 2× diast.), 26.8 and 27.5 (C-6′, 2× diast.), 27.8 and 28.1 (C-5′,2× diast.), 30.3 and 30.5 (C-3′), 32.2 (C-3), 40.4 and 41.1 (C-4′, 2× diast.), 44.5 (C-4), 50.1 (C-2), 57.0 and 57.8 (C-11, 2× diast.), 109.1 and 109.2 (C-2″, 2× diast.), 110.2 and 110.4 (C-11a, 2× diast.), 119.84 and 119.86 (C-6), 120.17, 120.22, 120.26, 120.39, 120.51 and 120.73 (C-2′, C-7, C-9, 2× diast.), 122.7 and 122.8 (C-8, 2× diast.), 131.9 and 131.8 (C-5a, 2× diast.)), 138.3 and 138.7 (C-9a, 2× diast.)), 139.1 and 139.3 (C-1′, 2× diast.), 149.2 and 149.7 (C-1″, 2× diast.), 154.85 and 154.87 (C-4a, 2× diast.), 192.2 and 192.3 (C-1, C=O, 2× diast.) ppm. HRMS-ESI$^+$: *m/z* calcd. for $[C_{24}H_{30}N_2O + H]^+$: 363.2436; found: 363.2419.

*3,3,8-Trimethyl-11-[(S)-4-(prop-1-en-2-yl)cyclohex-1-en-1-yl]-2,3,4,5,10,11-hexahydro-1H-dibenzo[b,e][1,4] diazepin-1-one (**4**, mixture of two diastereomers).* $C_{25}H_{32}N_2O$. Yellowish crystals; yield: 267 mg (71%); m.p. 201–202 °C. ^{1}H-NMR (400 MHz, DMSO-d_6): δ 1.01, 1.02, 1.04 and 1.06 (4s, 6H, 3-CH_3, 2× diast.), 1.56 and 1.59 (2s, 3H, 1″-CH_3, 2× diast.), 1.06–2.16 (m, 7H, H-3′, H-4′, H-5′, H-6′, 2× diast.), 2.11 and 2.13 (2s, 3H, 8-CH_3, 2× diast.), 2.02–2.19 (m, 2H, H-2, 2× diast.), 2.49–2.51 (m, 2H, H-4, 2× diast.), 4.36–4.40 and 4.48–4.53 and 4.55–4.58 and 4.58–4.62 (4m, 2H, H-2″, 2× diast.), 4.78–4.87 (m, 1H, H-11), 5.06 and 5.09 (d, *J* 6.9 Hz 1H, H-2′, 2× diast.), 5.71 and 5.74 (2d, *J* 6.0 Hz, 1H 10-NH, 2× diast.), 6.41–6.59 (m, 2H, H-6, H-9), 6.81 (dd, *J* 2.8 and 8.4 Hz, 1H, H-7), 8.50 and 8.51 (2s, 1H, 5-NH, 2× diast.) ppm. ^{13}C-NMR (100 MHz, DMSO-d_6): δ 20.77 and 20.79 (8-CH3, 2× diast.), 20.9 and 21.3 (1″-CH_3, 2× diast.), 27.8 and 28.0 and 29.1 and 29.2 (3-CH_3, 2× diast.), 26.7 and 27.3 (C-6′, 2× diast.), 27.8 and 28.1 (C-5′,2× diast.), 30.2 and 30.5 (C-3′), 32.1 (C-3), 40.4 and 41.1 (C-4′, 2× diast.), 44.5 (C-4), 50.1 (C-2), 57.0 and 57.7 (C-11, 2× diast.), 109.1 and 109.2 (C-2″, 2× diast.), 109.9 and 110.0 (C-11a, 2× diast.), 120.1 (C-2′), 120.22 and 120.24 (C-7, 2× diast.), 120.50 and 120.52 (C-6, 2× diast.), 120.8 and 120.9 (C-9, 2× diast.), 129.3 and 129.4 (C-8, 2× diast.), 131.5 and 131.6 (C-5a, 2× diast.)), 138.5 and 138.9 (C-9a, 2× diast.)), 139.0 and 139.1 (C-1′, 2× diast.), 149.1 and 149.7 (C-1″, 2× diast.), 154.79 and 154.82 (C-4a, 2× diast.), 191.9 and 192.0 (C-1, C=O, 2× diast.) ppm. HRMS-ESI$^+$: *m/z* calcd. for $[C_{25}H_{32}N_2O + H]^+$: 377.2548; found: 377.2577.

4. Conclusions

In summary, we used the natural (*S*)-(−)-perillaldehyde of *A. leucotrichus* subsp. *leucotrichus* essential oil as a chiral reagent to stereospecifically prepare imine, imidazole and dibenzo[*b,e*][1,4] diazepin-1-ones, which is an efficient, simple and economic hemi-synthetic protocol towards pure chiral compounds. We proved that natural compounds, when present in high amount in essential oil or extracts, can be used as excellent starting materials to prepare new structures in hemi-synthetic processes without their prior isolation and purification. The reported method is advantageously applicable to generate chiral compounds, avoiding the use of expensive chiral catalysts and isolated natural compounds. An arsenal of analytical tools, including 2D NMR, single-crystal X-ray diffraction and chiral HPLC were helpful to identify pure enantiomers and some diastereomeric mixtures. Further hemi-synthetic studies are being conducted on the same and other essential oils' major compounds to extend the scope and determine the limits of the reported method. Biological screening and comparative studies are also foreseen.

Supplementary Materials: The Supplementary Materials are available online. Supplementary Information (ESI) available: [NMR spectroscopic data for all the reported compounds **1–4**; Chiral-HPLC analysis of compounds **1–3**; Specific rotation analysis for pure enantiomers of compounds **1** and **2**; Single-crystal X-ray diffraction data (CIF files and check CIF reports) and additional crystallographic details for compounds **1–4**.

Author Contributions: O.T. conceptualized the work and co-wrote the manuscript; F.C. performed the extraction and synthetic experimental work and wrote the original draft preparation; B.C. and L.B. participated in the synthetic experimental work; M.V. performed all the chromatographic experimental work; R.F.M. and F.A.A.P. were responsible for all X-ray diffraction work; K.M. and K.B. co-wrote the manuscript; A.M.S.S. co-conceptualized the work and co-wrote the manuscript.

Funding: Thanks are due to University of Aveiro and FCT/MEC for the financial support to the QOPNA research project (FCT UID/QUI/00062/2019), the project CICECO-Aveiro Institute of Materials, POCI-01-0145-FEDER-007679 (FCT Ref. UID /CTM /50011/2019), financed by national funds and when appropriate co-financed by FEDER under the PT2020 Partnership Agreement, and of the bilateral project PT-DZ/0005 and to the Portuguese NMR Network. We would like also to thank the General Directorate for Scientific Research and Technological Development—DGRSDT of Algeria and Agence Thématique de Recherche en Sciences et Technologie ATRST for approving the co-financed bilateral project PT-DZ/0005.

Conflicts of Interest: There are no conflicts of interest to declare.

References

1. Appendino, G.; Minassi, A.; Taglialatela-Scafati, O. Recreational drug discovery: natural products as lead structures for the synthesis of smart drugs. *Nat. Prod. Rep.* **2014**, *31*, 880–904. [CrossRef] [PubMed]
2. Bonikowski, R.; Kula, J.; Bujacz, A.; Wajs-Bonikowska, A.; Zaklos-Szyda, M.; Wysocki, S. Hydroindene-derived chiral synthons from carotol and their cytotoxicity. *Tetrahedron Asymm.* **2012**, *23*, 1038–1045. [CrossRef]
3. Huangyong, L.; Changshui, C.; Cao, X. Essential oils-oriented chiral esters as potential pesticides: Asymmetric syntheses, characterization and bio-evaluation. *Ind. Crops Prod.* **2015**, *76*, 432–436.
4. Sakirigui, A.; Gbaguidi, F.; Kasséhin, U.; JacquesPoupaert; Accrombessi, G.; Kotchoni, S. Structural and antitrypanosomal data of different carbasones of piperitone. *Data Brief* **2016**, *9*, 1039–1043. [CrossRef] [PubMed]
5. Velasco-Negueruela, A.; Pérez-Alonso, M.J.; Pérez de Paz, P.L.; Palá-Paúl, J.; Sanz, J. Analysis by gas chromatography–mass spectrometry of the volatiles from the fruits of *Ammodaucus leucotrichus* subsp. *leucotrichus* and subsp. *nanocarpus* grown in North Africa and the Canary Islands, respectively. *J. Chromatogr. A* **2006**, *1108*, 273–275. [CrossRef] [PubMed]
6. Anthelme, F.; Waziri Mato, M.; Maley, J. Elevation and local refuges ensure persistence of mountain specific vegetation in the Nigerien Sahara. *J. Arid Environ.* **2008**, *72*, 2232–2242. [CrossRef]
7. Jouad, H.; Haloui, M.; Rhiouani, H.; El Hilaly, J.; Eddouks, M. Ethnobotanical survey of medicinal plants used for the treatment of diabetes, cardiac and renal diseases in the North centre region of Morocco (Fez–Boulemane). *J. Ethnopharmacol.* **2001**, *77*, 175–182. [CrossRef]
8. Merzouki, A.; Ed-derfoufi, F.; Molero Mesa, J. Contribution to the knowledge of Rifian traditional medicine. II: Folk medicine in Ksar Lakbir district (NW Morocco). *Fitoterapia* **2000**, *71*, 278–307. [CrossRef]
9. Hammiche, V.; Maiza, K. Traditional medicine in Central Sahara: Pharmacopoeia of Tassili N'ajjer. *J. Ethnopharmacol.* **2006**, *105*, 358–367. [CrossRef] [PubMed]
10. Dahmane, D.; Dob, T.; Krimat, S.; Nouasri, A.; Metidji, H.; Ksouri, A. Chemical composition, antioxidant and antibacterial activities of the essential oils of medicinal plant *Ammodaucus leucotrichus* from Algeria. *J. Essent. Oil Res.* **2017**, *29*, 48–55. [CrossRef]
11. Khaldi, A.; Meddah, B.; Moussaoui, A.; Sonnet, P. Anti-mycotoxin Effect and Antifungal Properties of Essential Oil from *Ammodaucus leucotrichus Coss. & Dur.* on Aspergillus flavus and Aspergillus ochraceus. *J. Essent. Oil Bear.* **2017**, *20*, 36–44.
12. Abu Zarga, M.H.; Al-Jaber, H.I.; Baba Amer, Z.Y.; Sakhrib, L.; Al-Qudah, M.A.; Al-humaidi, J.Y.G.; Abaza, I.F.; Afifi, F.U. Chemical Composition, Antimicrobial and Antitumor Activities of Essential Oil of *Ammodaucus leucotrichus* Growing in Algeria. *JAPN* **2013**, *3*, 224–231.

13. Mayeku, W.P.; Omollo, N.I.; Odalo, O.J.; Hassanali, A. Chemical composition and mosquito repellency of essential oil of *Conyza newii* propagated in different geographical locations of Kenya. *Med. Vet. Entomol.* **2014**, *28*, 253–256. [CrossRef] [PubMed]
14. Wang, C.Y.; Wang, S.Y.; Chen, C. Increasing Antioxidant Activity and Reducing Decay of Blueberries by Essential Oils. *J. Agric. Food Chem.* **2008**, *56*, 3587–3592. [CrossRef] [PubMed]
15. Tian, J.; Zeng, X.; Lü, A.; Zhu, A.; Peng, X.; Wang, Y. Perillaldehyde, a potential preservative agent in foods: Assessment of antifungal activity against microbial spoilage of cherry tomatoes. *Food Sci. Technol.* **2015**, *60*, 63–70. [CrossRef]
16. Appelt, H.R.; Oliveira, J.S.; Santos, R.C.V.; Rodrigues, O.E.D.; Santos, M.Z.; Heck, E.F.; Rosa, L.C.R. Synthesis and Antimicrobial Activity of Carbohydrate Based Schiff Bases: Importance of Sugar Moiety. *Int. J. Carbohydr. Chem.* **2013**, *2013*. [CrossRef]
17. Kundu, A.; Shakil, N.A.; Saxena, D.B.; Pankaj; Kumar, J.; Walia, S. Microwave assisted solvent-free synthesis and biological activities of novel imines (Schiff bases). *J. Environ. Sci. Health* **2009**, *44*, 428–434. [CrossRef] [PubMed]
18. Sonnekar, V.S.; Jadhav, W.N.; Dake, D.S.; Pawar, R. Synthesis and antimicrobial and antifungal activities of novel bis-imine derivatives. *Res. J. Pharm. Biol. Chem. Sci.* **2013**, *4*, 1411–1418.
19. Missaoui, B.E.; Ouahrani, M.R.; Kouadri, Y.; Chebrouk, F.; Gherraf, N. Synthesis of Novel Heterocyclic Compounds Containing 1,5-Benzodiazepine. *Asian J. Chem.* **2015**, *27*, 2175–2177. [CrossRef]
20. Kolos, N.N.; Yurchenko, E.N.; Orlov, V.D.; Shishkina, S.V.; Shishkin, O.V. Investigation of the products of interaction of cyclic diketones with nitrogen-containing 1,4-binucleophiles. *Chem. Heterocycl. Compd.* **2004**, *40*, 1550–1559. [CrossRef]
21. Gupta, S.; Gupta, P.; Sachar, A.; Sharma, R.L. Facile and one pot synthetic routes for various novel, differently fused and promising heteropolycycles. *J. Heterocycl. Chem.* **2010**, *47*, 334–349.
22. Cherfaoui, B.; Lakhdari, H.; Bennamane, N.; Ameraoui, R.; Talhi, O.; Almeida Paz, F.A.; Bachari, K.; Kirsch, G.; Nejar-Bellara, K.; Silva, A.M.S. Dibenzo[b,e][1,4]diazepin-1-ones and their Ring-Opened Derivatives: Revisited Synthesis, 2D NMR and Crystal Structure. *Synlett* **2017**, *28*, 2247–2252.
23. Ling, I.; Podanyi, B.; Hamori, T.; Solyom, S. Asymmetric reduction of a carbon-nitrogen double bond: enantioselective synthesis of 4,5-dihydro-3H-2,3-benzodiazepines. *J. Chem. Soc. Perkin Trans. 1* **1995**, *11*, 1423–1427. [CrossRef]
24. Al-Azmi, A.; Elassara, A.A.; Booth, B.L. The chemistry of diaminomaleonitrile and its utility in heterocyclic synthesis. *Tetrahedron* **2003**, *59*, 2749–2763. [CrossRef]
25. Grell, J.; Bernstein, J.; Tinhofer, G. Graph-set analysis of hydrogen-bond patterns: some mathematical conceptsWork supported by grant No. I-0333-263.06/93 from the GIF, the German-Israeli Foundation for Scientific Research and Development. *Acta Crystallogr. Sect. B Struct. Sci.* **1999**, *55*, 1030–1043. [CrossRef]

Sample Availability: Samples of the compounds **1–4** are available from the authors.

molecules

MDPI

Article

Synthesis of New Proteomimetic Quinazolinone Alkaloids and Evaluation of Their Neuroprotective and Antitumor Effects

Solida Long [1], Diana I. S. P. Resende [1,2], Anake Kijjoa [2,3], Artur M. S. Silva [4], Ricardo Fernandes [5,6], Cristina P. R. Xavier [5,6], M. Helena Vasconcelos [5,6,7], Emília Sousa [1,2,*] and Madalena M. M. Pinto [1,2]

1 Laboratory of Organic and Pharmaceutical Chemistry (LQOF), Department of Chemical Sciences, Faculty of Pharmacy, University of Porto, Rua de Jorge Viterbo Ferreira, 228, 4050-313 Porto, Portugal; up201502099@ff.up.pt (S.L.); dresende@ff.up.pt (D.I.S.P.R.); madalena@ff.up.pt (M.M.M.P.)

2 Interdisciplinary Centre of Marine and Environmental Research (CIIMAR), 4450-208 Matosinhos, Portugal; ankijjoa@icbas.up.pt

3 ICBAS-Instituto de Ciências Biomédicas Abel Salazar, University of Porto, 4050-313 Porto, Portugal

4 Organic Chemistry and Natural Products Unit (QOPNA), Department of Chemistry, University of Aveiro, 3810-193 Aveiro, Portugal; artur.silva@ua.pt

5 Instituto de Investigação e Inovação em Saúde (i3S), University of Porto, 4200-135 Porto, Portugal; ricardojorgefernandes@gmail.com (R.F.); cristinax@ipatimup.pt (C.P.R.X.); hvasconcelos@ipatimup.pt (M.H.V.)

6 Cancer Drug Resistance Group, Institute of Molecular Pathology and Immunology of the University of Porto (IPATIMUP), 4200-135, Porto, Portugal

7 Laboratory of Microbiology, Department of Biological Sciences, Faculty of Pharmacy, University of Porto, 4050-313 Porto, Portugal

* Correspondence: esousa@ff.up.pt; Tel.: +351-2-2042-8689

Received: 8 January 2019; Accepted: 30 January 2019; Published: 1 February 2019

Abstract: New quinazolinone derivatives of the marine-derived alkaloids fiscalin B (**3**) and fumiquinazoline G (**1**), with neuroprotective and antitumor effects, were synthesized. Eleven quinazolinone-containing indole alkaloids were synthesized, proceeding the *anti* analogs via a one-pot method, and the *syn* analogs by the Mazurkiewicz-Ganesan approach. The neuroprotection capacity of these compounds on the rotenone-damage human neuroblastoma cell SH-SY5y was evaluated using the MTT assay. Compounds **1**, **3**, **5**, and **7** showed more than 25% protection. The antitumor activity was investigated using the sulforhodamine B assay and some compounds were tested on the non-malignant MCF-12A cells. Fumiquinazoline G (**1**) was the most potent compound, with GI_{50} values lower than 20 μM. Compounds **5**, **7**, and **11** were more active in all tumor cell lines when compared to their enantiomers. Compounds **5**, **7**, **10**, and **11** had very little effect in the viability of the non-malignant cells. Differences between enantiomeric pairs were also noted as being essential for these activities the *S*-configuration at C-4. These results reinforce the previously described activities of the fiscalin B (**3**) as substance P inhibitor and fumiquinazoline G (**1**) as antitumor agent showing potential as lead compounds for the development of drugs for treatment of neurodegenerative disorders and cancer, respectively.

Keywords: antitumor; neuroprotection; quinazolinones; fiscalin B; fumiquinazoline; enantioselectivity

1. Introduction

The pathophysiology of neurodegenerative diseases is poorly understood, and there are few therapeutic options, making neuroprotective drug discovery appealing for medicinal chemists.

Although cancer and neurodegeneration have very distinct pathological disorders, over recent years growing evidence indicates that they share common molecular pathways [1]. Furthermore, it is recognized that several drugs used in the treatment of neurodegenerative diseases display antitumor effects while some antitumor drugs are neuroprotective [2].

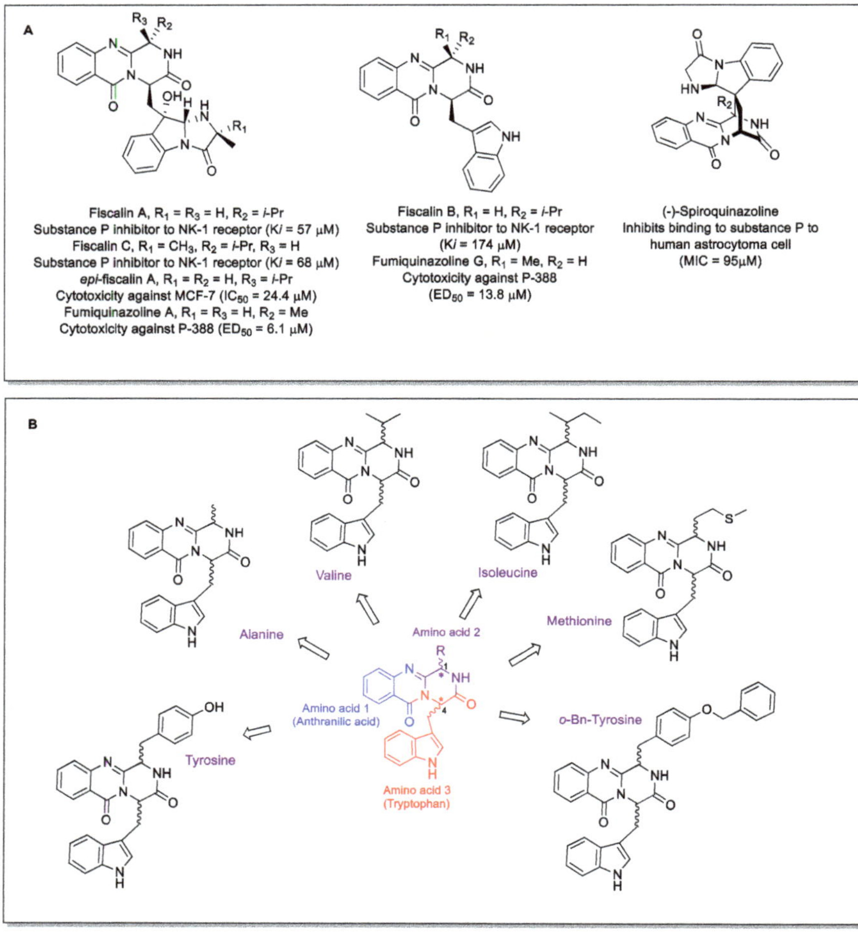

Figure 1. (**A**) Structure of natural quinazolinone-containing piperazine linked to an indole moiety such as substance P receptor antagonists and antitumor agents. (**B**) Proposed conformation constraint peptidomimetics synthetic quinazolinone alkaloids with different substituents at C-1.2.

Marine-derived indolylmethylpyrazinoquinazoline alkaloids, with a pyrazino[2,1-*b*]quinazoline-3,6-dione linked to an indole moiety (Figure 1), have attracted our attention due to their promising antitumor activities [3], with *epi*-fiscalin A [4], fumiquinazoline A [5–9], fumiquinazoline G [7], and versiquinazolines [10] as the most active analogs. Moreover, the response of fiscalins A-C [11] and (−)-spiroquinazoline [12] (Figure 1A) as substance P inhibitors was also reported as a novel neuroprotective therapy in the intrastriatal 6-hydroxydopamine model of early stage of Parkinson's disease (PD) [13]. It is well known that among compounds implicated in neurodegeneration, non-proteinogenic amino acids may cause significant collateral neurodegenerative damage [14]. Rodgers *et* al. [15] reported that proteomimetic L-tyrosine of L-DOPA is cytotoxic in vitro and capable of generating protein aggregation, whereas non-protein amino acid β-methylamino-L-alanine (BMAA) has been linked to neurological diseases such as amyotrophic lateral sclerosis (ALS) and PD since BMAA was detected in brain protein of LAS and PD patients. In the previous work, we have described syntheses of a series of fiscalin B derivatives, which showed weak to moderate antitumor activity against non-small cell lung cancer (NCI-H460) and colorectal adenocarcinoma (HCT-15) cell lines [16]. These findings led us to develop a small library of proteomimetic quinazolinone-derived compounds (Figure 1B) with different configurations at C-1 and C-4 to investigate their action on neurodegenerative disorders as well as to further explore their potential as tumor cell growth inhibitors, putting in evidence the influence of the stereochemistry of the derivatives.

2. Results

2.1. Chemistry

Two synthetic approaches were used to prepare the *syn* and *anti* enantiomers of quinazolinone alkaloids. The *syn* enantiomers **1** (fumiquinazoline G) and **2** were synthesized by the Mazurkiewicz-Ganesan procedure [17] (Scheme 1A) by coupling anthranilic acid (**i**) with D-tryptophan methyl ester (**ii**) for **1** or with L-tryptophan methyl ester (**vi**) for **2**, using 1,1,3,3-tetramethylaminium tetrafluoroborate (TBTU) in alkaline condition to obtain the dipeptide **iii** or **vii**. Then, the coupling of **iii** or **vii** with N-protected α-amino acid chloride in a two-phase Schotten-Baumann condition yielded a tripeptide (dehydrate β-keto amides) **v** or **ix**. The oxazole intermediates were obtained by adding the dehydrating agent, triphenylphosphine (Ph_3P), and I_2 to dehydrate β-keto amide **v** or **ix**, and N-deprotection by 20% piperidine afforded **1** and **2**. On the other hand, a highly effective and environmentally friendly approach using a microwave-assisted multicomponent polycondensation of amino acids was used to prepare a series of the *anti* enantiomers of pyrazinoquinazoline alkaloids [18], as described in our previous work [16]. This methodology was used to synthesize new derivatives of fiscalin B (**3**) and fumiquinazoline G (**1**), **4, 5, 6, 7**, and **8** (Scheme 1B). The *syn* isomer **9** was obtained along with **8,** and both were isolated by preparative thin layer chromatography (TLC). Diastereoisomers of **10** and **11** were obtained after deprotection of O-benzyl group from **8** and **9**, respectively, using boron trichloride, according to Okaya et al. [19] with a slight modification. Compound **12** was also synthesized using microwave irradiation from 3,5-dichloroanthranilic acid (**xiii**). The purity of the compounds was determined by a reversed-phase liquid chromatography (LC, C18, $MeOH:H_2O$; 60:40 or $CH_3CN:H_2O$; 50:50) and was found to be higher than 90%. The enantiomeric ratio (er) was determined by a chiral LC equipped with amylose tris-3,5-dimethylphenylcarbamate column, using hexane:EtOH (80:20) or (70:30) as a mobile phase.

The reaction carried out using microwave with high temperature resulted not only in low yields of the products in the range of 2.2 to 21.7%, but also with a high degree of epimerization (Scheme 1). Contrary to what has been found in our previous study [16] that the reaction under a microwave irradiation was regioselective and yielded only *anti* isomers, the synthesis of **8**, by a microwave irradiation, produced also its *syn* epimer, **9** [4-(benzyloxy)-1-methylbenzyl at C-1], with a 22% yield. This study suggested that microwave irradiation is beneficial for the synthesis of quinazolinone alkaloids with bulky substituents at C-1 which was previously reported as unsuccessful by Mazurkiewicz-Ganesan method [17]. However, this methodology failed for the synthesis of *syn* enantiomers as described in the experimental section for **4** and **6**. The *syn* enantiomers of **1** and **2** were synthesized by Mazurkiewicz-Ganesan approach [17] and gave moderate yields (37 and 26%, respectively). Compounds **10** and **11** were obtained by deprotection in good yields (30 and 69%, respectively).

Moreover, the methodology involving microwave irradiation was characterized by producing partial epimerizations. Surprisingly, **4** and **5**, with three stereogenic centers, gave a higher enantiomeric ratio (er) of 99%. Similar to the previous report for fiscalin B analogs [16], the multi-step approach gave a better yield and, in most cases, higher enantiomeric ratios due to milder conditions; nonetheless, the one-pot reaction is a faster alternative to provide *anti* enantiomers with diversity of substituents at C-1.

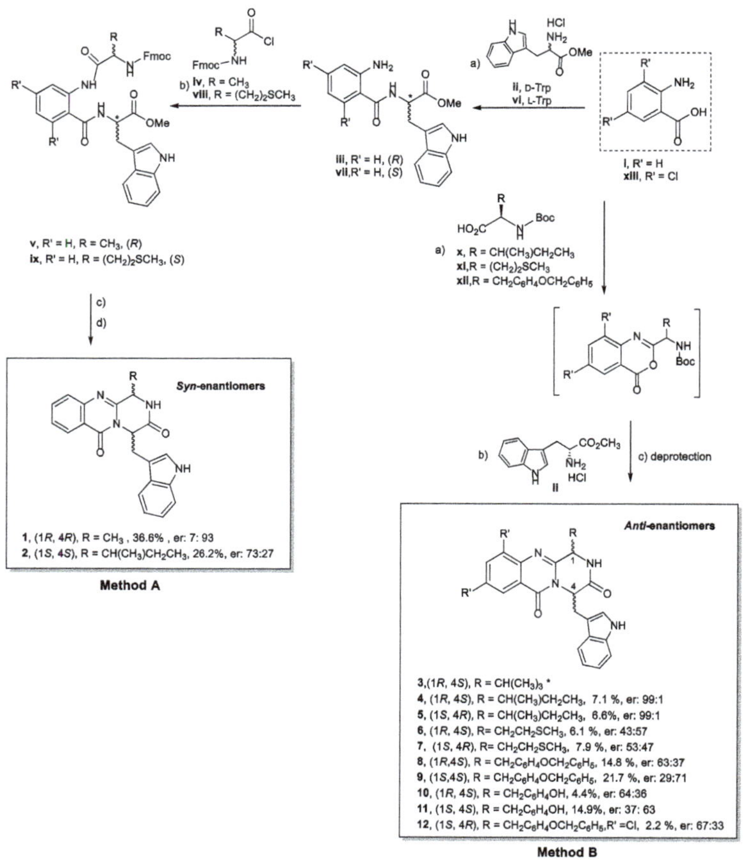

Scheme 1. (**A**) Mazurkiewicz-Ganesan approach for **1** and **2**. Reagents and conditions (**a**) CH_3CN, TBTU, Et_3N, rt, 5 h; (**b**) CH_2Cl_2/aq.Na_2CO_3, rt, 3 h; (**c**) dried CH_2Cl_2, Ph_3P, I_2, $EtN(i\text{-}Pr)_2$, rt, overnight; (**d**) piperidine in CH_2CH_2, rt, 12 min, then CH_3CN, DMAP, reflux 19 h. Fmoc = fluorenylmethyloxycarbonyl; DMAP = 4-(dimethylamino)pyridine, TBTU = 1,1,3,3-tetramethylaminium tetrafluoroborate. (**B**) One-pot synthesis of pyrazinoquinazolinone alkaloids **3**, **4**, **5**, **6**, **7**, **8**, **9**, **10**, **11**, and **12**. Reagents and conditions: (**a**) dried pyridine, $(PhO)_3P$, 55 °C, 16–24 h; (**b**) dried pyridine, $(PhO)_3P$, 220 °C, 1.5 min; (**c**) BCl_3, CH_2Cl_2, −78 °C for 10 min then 25 °C for 6 h; Boc = tert-butyloxycarbonyl, R′= H when no mentioned; er = enantiomeric ratio calculated from the peak area from chiral LC experiments (by using equation X × 100/Xn in which X is the peak area of each peak and Xn is the total peak area. * referred to previous work [16].

2.2. Structure Elucidation

A series of 1D and 2D NMR experiments and HRMS were used to confirm the structures of all the new compounds (Supplementary information, Figures S1–S57). The amide proton (H-2) appeared as a broad singlet (*brs*) in the *anti* isomers, i.e., **4**, **5**, **6**, **7**, **8**, **10**, and **12**, but as a doublet (*d*) in the *syn* isomers, i.e., **1**, **2**, **9**, and **11** (Supplementary information, Figures S1–S21). Each stereoisomer exhibited different chemical shift values for H-2, H-4, and H-1. For example, for isomers **2**, **4**, and **5**, with the *iso*butyl group at C-1, the *syn* and the *anti* isomers were distinguished by the chemical shift values of H-1 and H-1′; which were *ca.* δ_H 4.03 and 0.93, respectively for the *syn* isomer (**2**), and at *ca.* δ_H 2.80 and 2.3, respectively, for the *anti* isomers (**4** and **5**), due to the absence of the shielding effect by the aromatic ring of the indole moiety [20].

This assignment was also confirmed by HMBC correlations which distinguished between the isomers of **2**, **4**, and **5** by the presence or absence of correlations from H-2 to C-3, C-4, and C-14. In **2**, H-2 showed no correlation to C-3, C-4, and C-14 whereas in **4** and **5**, H-2 showed correlations to those three carbons. H-4 also showed correlations to different carbons among isomers. H-4 showed correlations to C-3 and C-4′ in **2**, but to C-3, C-14, C-4′, and C-5′ in **4**, while there were no such correlations observed in **5**. H-1 of the *syn* isomer of **2** showed correlation only to C-14 while in the *anti* isomer of **4** and **5**, it displayed correlations to C-14, C-1′ and C-3″ (Figure 2). In contrast, in **1**, H-2 and H-4 showed no correlations to carbons that are two or three bonds away (^2J or ^3J), and H-1 showed cross peaks to C-3 and C-14. Furthermore, H-8 and H-10 of **12** (with Cl at C-9 and C-11) appeared as two doublets (δ_H 8.26, J = 2.4 Hz and δ_H 7.85, J = 2.4 Hz, respectively) while in other compounds H-8 appeared as a double doublet and H-10 as double-double doublet. The NOESY spectrum for compound **6** with an *anti* configuration (*1R, 4S*) showed correlations from H-1 to H-1′ and H-3′ (methyl group attached to S atom) and H-2, while H-4 showed cross peak to H-4′ (Supplementary information, Figure S56). The NOESY spectrum for the *syn* configuration compound **2** (*1S, 4S*) exhibited correlations between H-1 and H-1′, H-2′, H-3″, and H-2 as well as from H-4 to H-4′. Also, correlations between H-4′ and H-1′ in the *syn* configuration compound **2** could be noted while these was absent in the *anti* configuration compound **6** (Supplementary information, Figure S57). These observations were similar to the previously described for fiscalin B isomers [16].

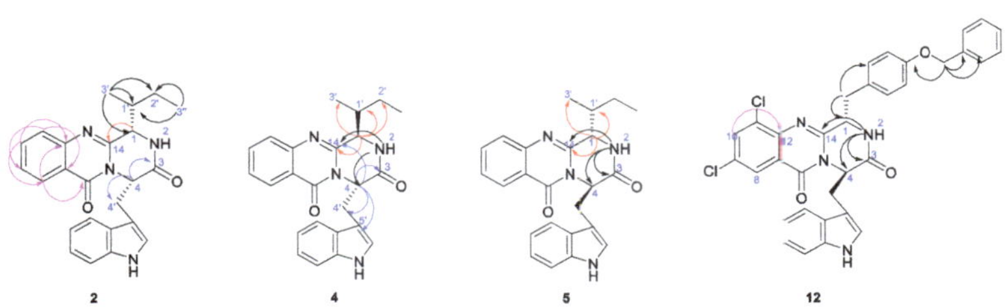

Figure 2. Key HMBC correlations for **2**, **4**, **5**, and **12**.

2.3. Neuroprotection Activity

The neuroprotection assay was performed on human neuroblastoma cell SH-SY5y treated with rotenone, a toxin that acts by interfering with the electron transport chain in mitochondria, inhibiting the transfer of electron from iron-sulfur centers in complex I to ubiquinone. This in turn interferes with NADH, therefore creating reactive oxygen species (ROS), which can damage DNA and other components, leading to cell death [21,22]. In animal experimentation, rotenone reproduces features of PD, including selective nigrostriatal dopaminergic degeneration and alpha-synuclein-positive cytoplasmic inclusions [23]. Furthermore, rotenone triggers mitochondrial impairment, oxidative damage, and cell death in neuronal culture, phenomena that are common in neurodegenerative diseases [24].

In this assay, the SH-SY5y cells were treated with 2 μM of rotenone for 24 h. The MTT assay, which assesses cell metabolic activity through the activity of NAD(P)H-dependent cellular oxidoreductase enzyme that reflect the number of viable cells [24], was used for quantifying the cell death. The cellular protection of **1**, **2**, **4**, **5**, **6**, **7**, **8**, **9**, **10**, **11**, and **12** against the toxin was determined by MTT assay and expressed as percentage referred to the cell treated with rotenone at 10 different concentrations to produce the dose-dependent curve. Synthetic fiscalin B (**3**), previously obtained by some of us [16] and was reported as a substance P antagonist [12], was also tested in this assay for comparison. Compounds considered as neuroprotective must have (i) more than 25% of protection, (ii) statistically significant difference, and (iii) protection is more than one dose [23,24].

Most of the compounds, at the highest concentration tested (100 μM), were found to increase toxicity, which ensures that the compounds were assayed at their maximal tolerated dose (MTD). Compounds **1**, **2**, **3**, **4**, **5**, **10**, **11**, and **12** showed high toxicity at 100 μM and some of them also at

50 μM. The compounds which showed neuroprotective activity in rotenone-treated in vitro model were **1**, **3**, **5**, and **7**. Additionally, some neuroprotective effect was also observed for **11** while **2**, **4**, **6**, **8**, **9**, **10**, and **12** did not display any neuroprotection (Figure 3). Compound **1**, **3**, **5**, and **7** showed more than 25% of protection of the cell death at least in one concentration. Compound **7** exhibited the best neuroprotective activity, with rotenone inhibitory of 47, 40, 42, 39, and 31% at the concentrations of 1.56, 3.13, 6.25, 12.5, and 25 μM, respectively.

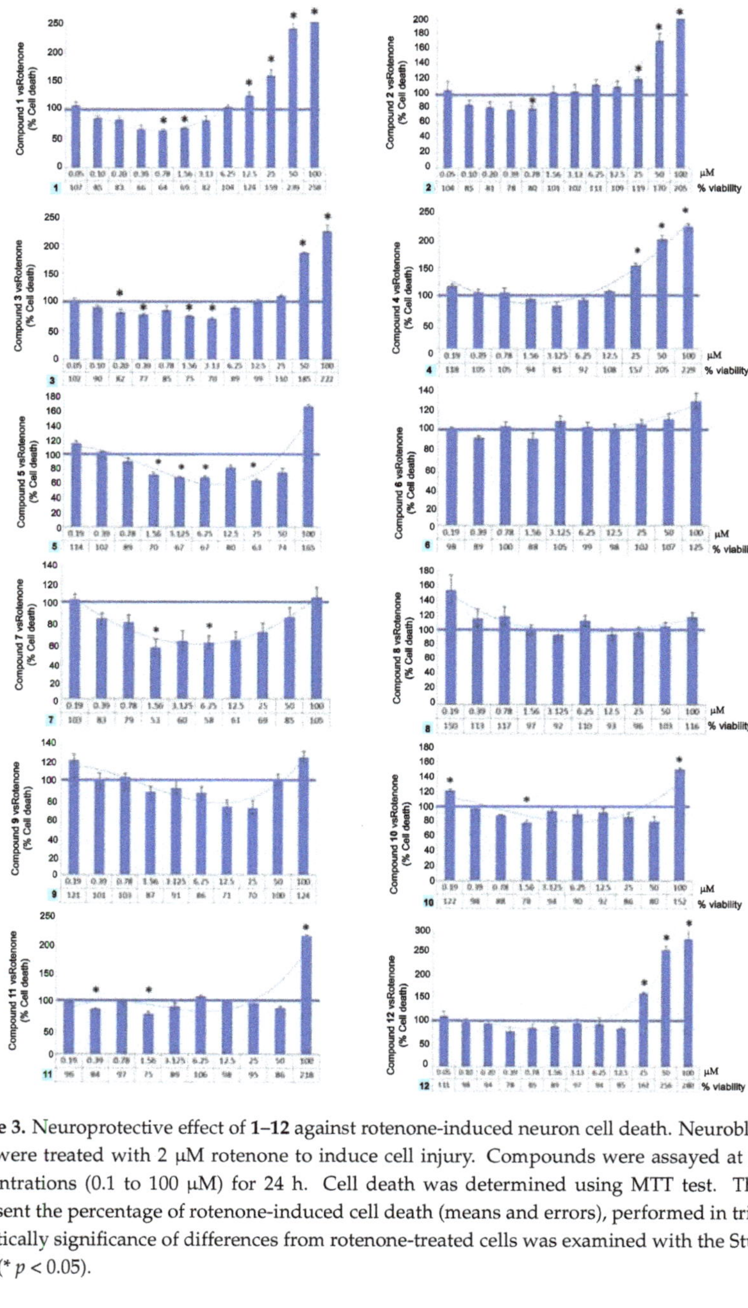

Figure 3. Neuroprotective effect of **1–12** against rotenone-induced neuron cell death. Neuroblastoma cells were treated with 2 μM rotenone to induce cell injury. Compounds were assayed at several concentrations (0.1 to 100 μM) for 24 h. Cell death was determined using MTT test. The data represent the percentage of rotenone-induced cell death (means and errors), performed in triplicate. Statistically significance of differences from rotenone-treated cells was examined with the Student's t-test (* $p < 0.05$).

2.4. Tumor Cell Growth Inhibitory Activity

Compounds **1–2, 4–8, 10** and **11** were tested for tumor cell growth inhibitory activity on three human tumor cell lines: NCI-H460 (non-small cell lung cancer), BxPC3 (human pancreatic adenocarcinoma), and PANC1 (human pancreatic adenocarcinoma) using the sulforhodamine B (SRB) colorimetric assay [16,25]. Cells were exposed to five concentrations of each compound (at a maximum concentration of 25, 150 or 200 μM, depending on the compound) for 48 h. Doxorubicin was used as a positive control for NCI-H460 cell line, and gemcitabine was used as a positive control for the BxPC3 and PANC1 cell lines. The antitumor activity was reported as GI_{50} concentration (drug concentration that inhibits the growth of cancer cells by 50%).

Compounds **2, 4–8, 10,** and **11** showed weak to moderate growth inhibitory, with GI_{50} ranking from 27.93 ± 0.8 to 151.07 ± 2.9 μM (Table 1). In general, compounds with the indolylmethyl substituent on C-4 whose configuration of C-4 is *R* showed better antitumor activity in all cell lines when compared to those with 4*S* configuration. This was evidenced by stronger antitumor activities of **1, 5,** and **7** than those of **2, 4,** and **6**. Only fumiquinazoline G (**1**) with *R*-configuration for both C-1 and C-4 showed strong growth inhibitory effect in all cancer cell lines tested (GI_{50} ranging from 7.62 ± 0.7 to 17.34 ± 1.7 μM). In addition, these results agree with those reported in our previous publication in which enantiomers with *R*-configuration at C-1 and C-4 showed better antitumor activity than enantiomers with *S*-configuration [16]. Unfortunately, **9** and **12** tested in this study could not be evaluated regarding their tumor cell growth inhibitory activity due to contamination of the compounds (data not shown).

Table 1. Growth inhibition (GI_{50}) concentration of **1–2, 4–8, 10,** and **11** against NCI-H460, BxPC3, and PANC1 human tumor cell lines.

Compounds	GI_{50} (μM)		
	NCI-H460	**BxPC3**	**PANC1**
1	7.62 ± 0.7	17.34 ± 1.7	10.06 ± 0.8
2	65.38 ± 4.0	104.77 ± 10.9	86.30 ± 11.1
4	69.26 ± 2.9	88.81 ± 9.6	78.17 ± 8.4
5	40.11 ± 5.0	57.44 ± 4.7	60.68 ± 4.8
6	151.07 ± 2.9	126.06 ± 21.5	112.44 ± 12.7
7	61.37 ± 2.3	99.10 ±5.8	76.65 ± 4.8
8	72.68 ± 6.2	115.42 ± 12.1	74.90 ± 7.2
10	38.00 ± 1.5	50.59 ± 2.3	34.13 ± 1.8
11	46.25 ± 5.8	29.87 ± 3.7	27.93 ± 0.8
Gemcitabine	-	0.20 ± 0.08	0.73 ± 0.22
Doxorubicin	0.0124 ± 0.0018	-	-

The GI_{50} concentrations (μM) were determined by the SRB assay and results are the mean of ± SEM of three independent experiments. Gemcitabine was use as a positive control for the BxPC3 and PANC1 cell lines, and doxorubicin as a positive control for the NCI-H460 cell line. (-) indicates not-determined.

2.5. Activity in Non-Tumor Cells

Compounds presenting the best neuroprotection and/or antitumor effects, namely **5, 7, 10,** and **11,** were also evaluated against the non-malignant MCF-12A human breast epithelial cells. For that, one concentration of each compound (corresponding to approximately the highest GI_{50} value obtained in the cell growth inhibitory activity assay) was tested and the percentage of cell growth inhibition was determined by the SRB assay. In this assay, the duration of the SRB assay had to be longer since non-malignant cells have a much slower growth rate than tumor cells. This longer duration of the assay also allowed to evaluate possible delayed effects of the compounds in these non-malignant cells. Therefore, the assay in the MCF-12A cells was performed following 7 days of treatment with compounds (48 h with compound incubation plus 5 days without the compounds). As shown in Table 2, all the tested compounds caused a small effect in the growth of these non-malignant cells, meaning that the cell growth inhibition detected in MCF-12A cells after 7 days of treatment was much lower than that detected in the tumor cell lines after 2 days of treatment (when tested at the same concentration).

Table 2. The percentage of cell growth inhibition (relative to the control) of **5**, **7**, **10**, and **11** in the non-malignant MCF-12A human breast epithelial cells.

Compounds	Concentration (μM) *	% Cell Growth Inhibition (Relative to the Control)
5	65	71.69 ± 7.9
7	100	89.56 ± 3.7
10	50	75.63 ± 4.7
11	50	71.15 ± 2.0

* These concentrations correspond to approximately the highest GI_{50} concentrations determined in the tumor cell lines tested (from Table 1). The values were determined by the SRB assay and results are the mean of ± SEM of three independent experiments.

3. Structural-Activity Relationship (SAR)

Structure-activity relationship analysis showed that the obtained results were consistent with data previously reported for the natural product fumiquinazoline G [26] and fiscalin B [11,27–29] and their derivatives [16] (Figure 4). Moreover, it was found that the configurations of C-1 and C-4 have strong influence on antitumor activity since fumiquinazoline G (**1**), with *R*-configurations at C-1 and C-4, showed the strongest antitumor effect against NCI-H460, BxPC3, and PANC1 cell lines. Comparing the enantiomeric pairs **4** and **5**, compound **5** with also *R*-configuration at C-4 and *S* at C-1 showed stronger antitumor effect. In contrast, their isomer, compound **2**, with *S*-configurations both at C-1 and C-4 were *S* exhibited the weakest inhibitory effect on all cell lines. In addition, the substituent at C-1 also affected the antitumor activity. Alkyl residues (isoleucine residues) in **2**, **4**, and **5** showed better antitumor activity than that found in compounds with a sulfur atom (methionine residues) such as **6** and **7**. Aromatic groups such as the tyrosine residue present in **10** and **11** produced good antitumor activity, with GI_{50} values ranging from 27.93 ± 0.8 to 50.59 ± 2.3 μM, but the substitution by benzyl groups as in the case of **8** caused a 2-fold decrease in the antitumor activity. Regarding neuroprotection capacity, SAR suggests that the *R*-configuration at C-4 is also important (when comparing compounds **6** and **7**, Figure 3); however, increasing the molecular weight of C-1 substituent has a negative effect in neuroprotection. In addition, this study confirmed that quinazolinone alkaloids which act as substance P inhibitors (i.e., fiscalin B, **3**), showed a potential as neuroprotective agents. Therefore, the studied fumiquinazoline-derived alkaloids showed promising antitumor and neuroprotection effects and deserves to be further explored.

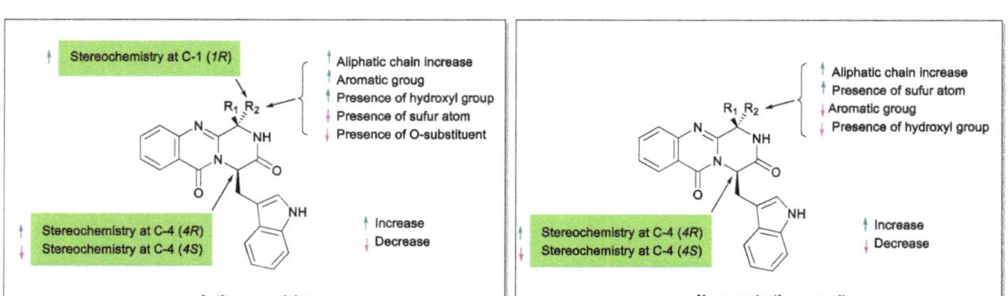

Figure 4. Structure-activity relationship studies of quinazolinone alkaloids **1–12** with antitumor and neuroprotective activities.

4. Materials and Methods

4.1. General Procedure

All reagents were from analytical grade. Dried pyridine and triphenylphosphite were purchased from Sigma (Sigma-Aldrich Co. Ltd., Gillingham, UK). Anthranilic acid (**i**) and Protected amino acids (**ii** and **vii**) were purchased from TCI (Tokyo Chemical Industry Co. Ltd., Chuo-ku, Tokyo, Japan). Column chromatography purifications were performed using flash silica Merck 60, 230–400 mesh (EMD Millipore corporation, Billerica, MA, USA) and preparative TLC was carried out on precoated plates Merck Kieselgel 60 F_{254} (EMD Millipore corporation, Billerica, MA, USA), spots were visualized with UV light (Vilber Lourmat, Marne-la-Vallée, France). Melting points were measured in a Köfler microscope and are uncorrected. Infrared spectra were recorded in a KBr microplate in a FTIR spectrometer Nicolet iS10 from Thermo Scientific (Waltham, MA, USA) with Smart OMNI-Transmission accessory (Software 188 OMNIC 8.3, Thermo Fisher Scientific Inc., Austin, TX, USA). ^{1}H and ^{13}C NMR spectra were recorded in $CDCl_3$ or DMSO-d_6 (Deutero GmbH, Kastellaun, Germany) at room temperature unless otherwise mentioned on Bruker AMC instrument (Bruker Biosciences Corporation, Billerica, MA, USA), operating at 300 MHz for ^{1}H and 75 MHz for ^{13}C). Carbons were assigned according to HSQC and or HMBC experiments. Optical rotation was measured at 25 °C using the ADP 410 polarimeter (Bellingham + Stanley Ltd., Royal Tunbridge Wells, Kent, UK), using the emission wavelength of sodium lamp, concentrations are given in g/100 mL. Qualitative GC-MS analyses were performed on a Trace GC 2000 Series ThermoQuest gas chromatography (Thermo Fisher Scientific Inc., Austin, TX, USA) equipped with ion-trap GCQ Plus ThermoQuest Finnigan mass detector (Thermo Fisher Scientific Inc.). Chromatographic separation was achieved using a capillary column (30 m × 0.25 mm × 0.25 µm, cross-linked 5% diphenyl and 95% dimethyl polysiloxane) from Thermo Scientific™ (Thermo Fisher Scientific Inc.) and high-purity helium C-60 as carrier gas. High resolution mass spectra (HRMS) were measured on a Bruker FTMS APEX III mass spectrometer (Bruker Corporation, Billerica, MA, USA) recorded as ESI (Electrospray) made in Centro de Apoio Cientifico e Tecnolóxico á Investigation (CACTI, University of Vigo, Pontevendra, Spain). The purity of synthesized compounds was determined by reversed-phase LC with diode array detector (DAD) using C18 column (Kimetex®, 2.6 EV0 C18 100 Å, 150 × 4.6 mm), and the mobile phase was methanol:water (60:40) or acetonitrile:water (50:50). Enantiomeric ratio was determined by chiral LC (LCMS-2010EV, Shimadzu, Lisbon, Portugal), employing a system equipped with a chiral column (Lux® 5 µm Amylose-1, 250 × 4.6 mm) and UV-detection at 254 nm, mobile phase was hexane:ethanol (80:20) and the flow rate was 0.5 mL/min. Compound **3** was obtained according to previous described method [17]. Neuroprotection studies were performed in Fundación Centro de Excelencia en Investigación de Medicamentos Innovadores en Andalucía, MEDINA (Granada, Spain).

4.2. General Conditions for the Synthesis of Compound (1R,4R)-4-((1H-indol-3-yl)methyl)-1-((R)-methyl)-1,2-dihydro-6H-pyrazino[2,1-b]quinazoline-3,6(4H)-dione (1)

To a mixture of anthranilic acid (**i**, 287 mg, 2.39 mmol) and TBTU (920 mg, 2.86 mmol, 1.2 equiv) in acetonitrile (20 mL) was added Et_3N (833 µL, 4.78 mmol, 2 equiv) and D-tryptophan methyl ester (**ii**, 521 mg, 2.39 mmol) at room temperature. After stirring for 5 h, the reaction mixture was concentrated under reduced pressure. The residue was dissolved in CH_2Cl_2 and washed with 1 M HCl, extracted with CH_2Cl_2 (3 × 100 mL), dried with Na_2SO_4, filtered, and concentrated. The residue was purified by flash chromatography (eluent 1% MeOH in CH_2Cl_2) to yield **iii** as a white solid. ^{1}H NMR and ^{13}C NMR referred to the previous work [16]. To a solution of **iii** (140 mg, 0.416 mmol) in dried CH_2Cl_2 (10 mL) *N*-Fmoc-D-alanine-Cl [30] (**iv**,182 mg, 0.5 mmol) was added. The mixture was stirred for 30 min, followed by the addition of aqueous Na_2CO_3 (1 M, 8 mL, 8 mmol). After continuous stirring for 3 h, the mixture was extracted with CH_2Cl_2 (4 × 100 mL), dried with Na_2SO_4, filtered, and concentrated. The residue was purified by flash chromatography (eluent: 5% MeOH in CH_2Cl_2) to give **v** (220.4 mg, 84.2%) as a white solid. ^{1}H NMR (300 MHz, $CDCl_3$) δ 11.48 (s, 1H), 8.58 (d, 1H,

J = 8.4 Hz), 8.13 (s, 1H), 7.76 (d, 2H, J = 7.5 Hz), 7.66 (d, 1H, J = 7.1 Hz), 7.59 (t, 1H, J = 7.4 Hz), 7.49-7.26 (m, 8H), 7.17 (t, 1H, J = 7.2 Hz), 7.06 (t, 1H, J = 7.6 Hz), 7.00 (t, 1H, J = 7.7 Hz), 6.97 (s, 1H), 6.71 (d, 1H, J = 7.6 Hz), 5.55 (d, 1H, J = 6.8), 5.03 (dt, 1H, J = 7.6, 5.3 Hz), 4.44 (m, 2H), 4.36 (1m, 1H), 4.26 (t, 1H, J = 7.0 Hz), 3.73 (s, 3H), 3.40 (dd, 1H, J = 15.3, 5.8 Hz), 3.34 (dd, 1H, J = 15.3, 5.3 Hz), 1.53 (d, 3H, J = 7.0 Hz) and ^{13}C NMR (75 MHz, $CDCl_3$) 172.7, 172.2, 168.8, 156.6, 144.2, 143.8, 141.4, 138.6, 136.4, 132.8, 127.8, 127.4, 127.2, 125.3, 123.4, 123.3, 122.1, 121.6, 120.9, 120.0, 119.5, 118.3, 111.7, 109.2, 67.3, 53.6, 52.7, 52.2, 47.3, 27.3, 18.4 (See in [31]). To a solution of **v** (183.2 mg, 0.278 mmol) in dried CH_2Cl_2 (20 mL) Ph_3P (365 mg, 1.39 mmol, 5 equiv), I_2 (345 mg, 1.36 mmol. 4.9 equiv), and *N*,*N*-diisopropylethylamine (489 µL, 2.81 mmol, 10 equiv) were added. The reaction mixture was stirred at room temperature for 5 h, quenched with aqueous Na_2CO_3, and extracted with CH_2Cl_2 (3 × 100 mL), dried with Na_2SO_4, filtered, and concentrated. Hexane was added to remove an excess of Ph_3P, the precipitate was filtered and was treated with CH_2Cl_2 (10 mL) and piperidine (2.5 mL, 20%) at room temperature for 20 min, followed by solvent evaporation to provide the solid which was triturated with hexane (1 × 200 mL), CH_2Cl_2/PhMe (1 × 200 mL), and hexane (1 × 200 mL). The vacuum-dried crude residue was dissolved in CH_3CN (10 mL) in the presence of DMAP (64 mg, 0.53 mmol) and refluxed for 19 h. The reaction mixture was purified by preparative TLC (EtOAc:MeOH:CH_2Cl_2, 50:2.5:47.5) to afford **1**. Yield: 22.4 mg, 21.11%; *er* = 7:93; mp: 105.9–106.5 °C; $[\alpha]_D^{30} = -117.64$ (c 0.034; $CHCl_3$); υ_{max} (KBr) 3406, 2924, 2852, 1678, 1473, 1329, and 1261 cm^{-1}; ^{1}H NMR (300 MHz, $CDCl_3$): δ 8. 38 (dd, 1H, J = 8.0 and 1.1 Hz, CH), 8.18 (br, 1H, NH-Trp), 7.78 (ddd, 1H, J = 8.5, 7.1, and 1.6 Hz, CH), 7.57 (d, 1H, J 7.4 Hz, CH), 7.54 (dd, 1H, J = 8.0 and 1.0 Hz, CH), 7.29 (d, 1H, J = 3.1 Hz, CH-Trp), 7.27 (d, 1H, J = 2.0 Hz, CH-Trp), 7.08 (ddd, 1H, J = 9.5, 7.0 and 0.9 Hz, CH-Trp), 6.83 (ddd, 1H, J = 8.7, 7.1 and 1.0 Hz, CH-Trp), 6.73 (d, 1H, J = 2.3 Hz, CH-Trp), 6.70 (d, 1H, J = 2.0 Hz, NH-amide), 5.54 (dd, 1H J = 5.2 and 3.6 Hz, CH*-Trp), 4.46 (qd, 1H, J = 6.9 and 2.8 Hz, CH*-ala), 3.78 (dd, 1H, J = 14.9 and 5.3 Hz, CH_2-Trp), 3.70 (dd, 1H, J = 14.9 and 3.4 Hz, CH_2-Trp), 0.58 (d, 3H, J = 7.0 Hz, CH_3-ala); ^{13}C NMR (75 MHz, $CDCl_3$): δ 167.3 (C=O), 161.0 (C=O), 151.2 (C=N), 147.2 (C), 135.7 (C-Trp), 134.9 (CH), 127.9 (C-Trp), 126.9 (CH), 126.8 (CH), 126.7 (CH), 123.7 (CH-Trp), 122.3 (CH-Trp), 120.1 (C), 119.9 (CH-Trp), 118.7 (CH-Trp), 111.1 (CH-Trp), 109.4 (C-Trp), 56.9 (CH*-Trp), 51.9 (CH*-ala), 27.1 (CH_2-Trp), 22.8 (CH_3-ala. (+)-HRMS-ESI *m/z* 359.1505 (M + H)$^+$, (calculated for $C_{21}H_{18}N_4O_2$, 358.1430).

4.3. General Conditions for the Synthesis of Compound (1S,4S)-4-((1H-indol-3-yl)methyl)-1-((S)-sec-butyl) -1,2-dihydro-6H-pyrazino[2,1-b]quinazoline-3,6(4H)-dione (**2**)

To a mixture of anthranilic acid (**i**, 287 mg, 2.39 mmol) and TBTU (920 mg, 2.86 mmol, 1.2 equiv) in acetonitrile (20 mL), Et_3N (833 µL, 4.78 mmol, 2 equiv) and L-tryptophan methyl ester (**vi**, 521 mg, 2.39 mmol) were added at room temperature. After stirring for 5 h, the reaction mixture was concentrated under reduced pressure. The residue was dissolved in CH_2Cl_2 and washed with 1 M HCl, extracted with CH_2Cl_2 (3 × 100 mL), dried with Na_2SO_4, filtered, and concentrated. The residue was purified by flash chromatography (eluent 1% MeOH in CH_2Cl_2) to yield **vii** as a white solid. ^{1}H NMR and ^{13}C NMR referred to the previous work [16]. To a solution of **vii** (304 mg, 0.901 mmol) in dried CH_2Cl_2 (30 mL), *N*-Fmoc-L-isoleucine-Cl [30] (**viii**, 395 mg, 1.08 mmol) was added. The mixture was stirred for 30 min, followed by addition of aqueous Na_2CO_3 (1 M, 16 mL, 16 mmol). After continuous stirring for 3 h, the mixture was extracted with CH_2Cl_2 (4 × 100 mL), dried with Na_2SO_4, filtered, and concentrated. The residue was purified by flash chromatography (eluent: 5% MeOH in CH_2Cl_2 to give **ix** (475.8 mg, 86.2%) as a white solid. ^{1}H NMR (300 MHz, $CDCl_3$): δ 11.43 (s, 1H), 8.59 (d, 1H, J = 8.3 Hz), 8.19 (s, 1H), 7.76 (d, 2H J = 7.4 Hz), 7.69-7.56 (m, 2H), 7.53-7.28 (m, 7H), 7.17 (t, 1H, J = 7.3 Hz), 7.06 (t, 1H, J = 7.4 Hz), 7.01 (d, 1H, J = 7.6 Hz), 6.96 (d, 1H, J = 3.5 Hz), 6.72 (d, 1H, J = 7.6 Hz), 5.55 (d, 1H, J = 8.4 Hz), 5.06 (dd, 1H, J = 12.6 and 5.2 Hz), 4.39 (dd, 2H, J = 16.4 and 9.1 Hz), 3.73 (s, 3H), 3.37 (m, 2H), 2.11-2.00 (m, 1H), 1.68-1.48 (m, 2H), 1.03 (d, 3H, J = 6.8 Hz), 0.96 (t, 3H, J = 7.3 Hz) and ^{13}C NMR (75 MHz, $CDCl_3$) 172.1, 170.2, 168.6, 144.1, 143.8, 141.3, 139.0, 136.1, 132.9, 127.7, 127.5, 127.1, 127.1, 126.9, 125.3, 125.2, 123.2, 122.8, 122.4, 121.4, 120.2, 120.0, 119.8, 118.5, 111.4, 109.7, 67.2, 61.3, 53.3, 52.6, 47.3, 37.9, 31.4, 27.3, 15.8, 11.7. To a solution of **ix** (291.5 mg, 0.432 mmol) in dried CH_2Cl_2 (20 mL) Ph_3P (565 mg, 2.16 mmol, 5 equiv), I_2 (448 mg, 2.12 mmol. 4.9 equiv), and *N,N*-diisopropylethylamine (753 µL, 4.32 mmol, 10 equiv) were added. The reaction mixture was stirred at room temperature for 5 h, quenched with aqueous Na_2CO_3, and extracted with CH_2Cl_2 (3 × 100 mL), dried with Na_2SO_4, filtered, and concentrated. Hexane was added to remove an excess of Ph_3P, the precipitate was filtered and was treated with CH_2Cl_2 (10 mL) and piperidine (2.5 mL, 20%) at room temperature for 20 min, followed by solvent evaporation to provide the solid which was triturated with hexane (1 × 200 mL), CH_2Cl_2/PhMe (1 × 200 mL), and hexane (1 × 200 mL). The vacuum-dried crude residue was dissolved in CH_3CN (10 mL) in the presence of DMAP (158 mg, 1.39 mmol) and refluxed for 19 h. The reaction mixture was purified by preparative TLC (EtOAc: MeOH: CH_2Cl_2, 50:2.5:47.5) to afford **2**. Yield: 36 mg, 36.2%; enantiomeric ratio (er) = 73:27; m.p: 181–183 °C; $[\alpha]_D^{30}$ = +346.40 (*c* 0.051; $CHCl_3$); υ_{max} (KBr) 3375, 3187, 2880, 1684, 1662, 1472, 1434, and 1261 1 cm^{-1}; ^1H NMR (300 MHz, $CDCl_3$): δ 8.38 (dd, 1H, *J* = 8.0 and 1.1 Hz, CH), 8.07 (br, 1H, NH-Trp), 7.79 (ddd, 1H, *J* = 8.5, 7.1, and 1.6 Hz, CH), 7.62 (dd, *J* = 8.2 and 0.5 Hz, CH), 7.54 (ddd, 1H, *J* = 8.2, 7.2, and 1.2 Hz, CH), 7.48 (d, 1H, *J* 7.9 Hz, CH-Trp), 7.28 (d, 1H, *J* = 8.5 Hz, CH-Trp), 7.12 (ddd, 1H, *J* = 8.1, 7.1 and 1.1 Hz, CH-Trp), 6.94 (ddd, 1H, *J* = 8.0, 7.1 and 1.0 Hz, CH-Trp), 6.87 (d, 1H, *J* = 2.3 Hz, CH-Trp), 6.55 (d, 1H, *J* = 3.1 Hz, NH-amide), 5.52 (dd, 1H, *J* = 6.4 and 3.6 Hz, CH*-Trp), 4 03 (dd, 1H, *J* = 8.0 and 3.5 Hz, CH-Ile), 3.83 (dd, 1H, *J* = 14.9 and 6.4 Hz, CH_2-Trp), 3.73 (dd, 1H, *J* = 14.8 and 3.5 Hz, CH_2-Trp), 0.99–0.85 (m, 1H, CH*-Ile), 0.85–0.69 (m, 2H, CH_2-Ile), 0.66 (d, 3H, *J* = 6.5 Hz, CH_3-Ile), 0.58 (t, 3H *J* = 7.1 Hz, CH_3-Ile); ^{13}C NMR (75 MHz, $CDCl_3$): δ 167.8 (C=O), 161.4 (C=O), 149.4 (C=N), 146.8 (C), 135.9 (C-Trp), 134.7 (CH), 127.9 (C-Trp), 127.1 (CH), 127.1 (CH), 126.8 (CH), 123.5 (CH-Trp), 122.3 (CH-Trp), 120.3 (C), 120.0 (CH-Trp), 119.0 (CH-Trp), 111.0 (CH-Trp), 110.2 (C-Trp), 60.8 (CH*-Ile), 57.6 (CH*-Trp), 40.8 (CH*-Ile), 27.1 (CH_2-Trp), 24.3 (CH_2-Ile), 15.3 (CH_3-Ile), 10.4 (CH_3-Ile); (+)-HREM-ESI *m/z* 401.1967 $(M + H)^+$, 423.1787 $(M + Na)^+$ (calculated for $C_{24}H_{25}N_4O_2$, 400.1899).

4.4. General Conditions for the Synthesis of Quinazolinone-3,6-(4H)-Diones Compound **4**, **5**, **6**, **7**, **8**, *and* **9**

In a closed vial anthranilic acid (**i**, 28 mg, 200 µmol), *N*-Boc-L-isoleucine (**x**, 44 mg, 200 µmol) for **4** and **5**, or *N*-Boc-L-methionine (**xi**, 46 mg, 200 µmol) for **6** and **7**, or *N*-Boc-*o*-Bn-Tyrosine (**xii**, 74 mg, 200 µmol) for **8** and **9**, and triphenylphosphite (63 µL, 220 µmol) were added along with 1 mL of dried pyridine. The vial was heated in heating block with stirring at 55 °C for 16–24 h. After cooling the mixture to room temperature, D-tryptophan methyl ester hydrochloride (**ii**) for **5**, and **7**, L-tryptophan methyl ester hydrochloride (**vi**) for **4**, **6**, **8** and **9** (51 mg, 200 µmol) was added, and the mixture was irradiated in the microwave at the constant temperature at 220 °C for 1.5 min. Reaction mixtures were prepared in the same conditions and treated in parallel. After removing the solvent with toluene, the crude product was purified by flash column chromatography using hexane:EtOAc (60:40) as a mobile phase. The preparative TLC was performed using CH_2Cl_2:Me_2CO (95:5) as mobile phase. The major compound appeared as a black spot with no fluorescence under the UV light (366 nm). The desirable compounds **4**, **5**, **6**, **7**, **8**, and **9** were collected as yellow solids. Before analysis, compounds were recrystallized from methanol.

(1R,4S)-4-((1H-indol-3-yl)methyl)-1-((S)-sec-butyl)-1,2-dihydro-6H-pyrazino[2,1-b]quinazoline-3,6(4H)- dione (4). Yield: 29.2 mg, 7.1%; *er* = 3:97; m.p: 220–221 °C; $[\alpha]_D^{30}$ = + 484.7 (*c* 0.037; $CHCl_3$); υ_{max} (KBr) 3373, 3059, 2880, 1684, 1662, 1472, 1434, and 1261 cm^{-1}; ^1H NMR (300 MHz, $CDCl_3$): δ 8. 38 (dd, 1H, *J* = 7.9 and 1.2 Hz, CH), 8.07 (br, 1H, NH-Trp), 7.78 (ddd, 1H, *J* = 8.4, 7.1, and 1.6 Hz, CH), 7.57 (d, *J* = 8.0 Hz, CH), 7.52 (d, 1H, *J* = 6.0 Hz, CH), 7.47 (d, 1H, *J* = 8.0 Hz, CH-Trp), 7.28 (d, 1H, *J* = 8.2 Hz, CH-Trp), 7.12 (t, 1H, *J* = 7.1 Hz, CH-Trp), 6.96 (t, 1H, *J* = 7.5 Hz, CH-Trp), 6.57 (d, 1H, *J* = 2.4 Hz, CH-Trp), 5.68 (dd, 1H, *J* = 5.2 and 2.7 Hz, CH*-Trp), 5.64 (s, 1H, NH-amide), 3.76 (dd, 1H, *J* = 14.8 and 2.7 Hz, CH_2-Trp), 3.63 (dd, 1H, *J* = 14.9 and 5.3 Hz, CH_2-Trp), 2.80 (d, 1H, *J* = 2.4 Hz, CH*-Ile), 2.36 (dt, 1H, *J* 14.9 and 7.5 Hz, CH*-Ile), 0.98 (m, 2H, CH_2-Ile), 0.88 (d, 3H, *J* = 6.5 Hz, CH_3-Ile), 0.64 (t, 3H *J* = 6.4 Hz, CH_3-Ile); ^{13}C NMR (75 MHz, $CDCl_3$): δ 169.5 (C=O), 160.9 (C=O), 150.7 (C=N), 147.1 (C), 136.1 (C-Trp), 134.7 (CH), 127.2 (CH), 127.1 (CH), 127.0 (C-Trp), 126.9 (CH), 123.6 (CH-Trp), 122.8 (CH-Trp), 120.2 (C), 120.1 (C-Trp), 118.8 (CH-Trp), 111.1 (CH-Trp), 109.4 (C-Trp), 56.8 (CH*-Trp), 55.1

(CH*-Ile), 35.8 (CH*-Ile), 27.4 (CH_2-Trp), 25.8 (CH_2-Ile), 13.2 (CH_3-Ile), 11.0 (CH_3-Ile; (+)-HRMS-ESI *m/z* 401.1964 $(M + H)^+$ (calculated for $C_{24}H_{25}N_4O_2$, 400.1899).

(1S,4R)-4-((1H-indol-3-yl)methyl)-1-((S)-sec-butyl)-1,2-dihydro-6H-pyrazino[2,1-b]quinazoline-3,6(4H)-dione (5). Yield: 27.2 mg, 6.6%; *er* = 3:97; m.p: 218–220 °C; $[\alpha]_D^{30} = -372.6$ (*c* 0.034; $CHCl_3$); υ_{max} (KBr) 3373, 3059, 2880, 1684, 1662, 1472, 1434, 1261 cm^{-1}; ^1H NMR (300 MHz, $CDCl_3$): δ 8. 37 (dd, 1H, *J* = 7.9 and 1.2 Hz, CH), 8.07 (br, 1H, NH-Trp), 7.78 (ddd, 1H, *J* = 8.4, 7.1, and 1.6 Hz, CH), 7.57 (d, *J* = 8.0 Hz, CH), 7.52 (d, 1H, *J* = 6.0 Hz, CH), 7.47 (d, 1H, *J* = 8.0 Hz, CH-Trp), 7.28 (d, 1H, *J* = 8.2 Hz, CH-Trp), 7.12 (t, 1H, *J* = 7.1 Hz, CH-Trp), 6.96 (t, 1H, *J* = 7.5 Hz, CH-Trp), 6.57 (d, 1H, *J* = 2.4 Hz, CH-indole), 5.68 (dd, 1H, *J* = 5.2 and 2.7 Hz, CH*-Trp), 5.52 (s, 1H, NH-amide), 3.76 (dd, 1H, *J* = 14.8 and 2.7 Hz, CH_2-Trp), 3.63 (dd, 1H, *J* = 14.9 and 5.3 Hz, CH_2-Trp), 2.80 (d, 1H, *J* = 2.4 Hz, CH*-Ile), 2.37 (dt, 1H, *J* = 14.9 and 7.5 Hz, CH*-Ile), 0.88 (m, 2H, *J* = 6.7 Hz, CH_2-Ile), 0.62 (d, 3H, *J* = 6.5 Hz, CH_3-Ile), 0.46 (t, 3H *J* = 6.4 Hz, CH_3-Ile); ^{13}C NMR (75 MHz, $CDCl_3$): δ 169.4 (C=O), 160.9 (C=O), 150.1 (C=N), 147.1 (C), 136.1 (C-Trp), 134.7 (CH), 127.2 (CH), 127.2 (CH), 127.0 (CH-Trp), 126.9 (CH), 123.6 (CH-Trp), 122.7 (CH-Trp), 120.2 (C), 120.1 (C-Trp), 118.8 (CH-Trp), 111.1 (CH-Trp), 109.4 (C-indol), 56.8 (CH*-Trp), 55.5 (CH*-Ile), 35.6 (CH*-Ile), 27.4 (CH_2-Trp), 25.9 (CH_2-Ile), 13.2 (CH_3-Ile), 11.0 (CH_3-Ile; (+)-HRMS-ESI *m/z* 401.1973 $(M + H)^+$ (calculated for $C_{24}H_{25}N_4O_2$, 400.1899).

(1R,4S)-4-((1H-indol-3-yl)methyl)-1-(2-(methylthio)ethyl)-1,2-dihydro-6H-pyrazino[2,1-b]quinazoline- 3,6(4H)-diones (**6**). Yield: 27 mg, 6.1%; m.p: 198–200.7 °C; $[\alpha]_D^{30} = +74.1$ (*c* 0.045; $CHCl_3$); υ_{max} (KBr) 3295, 3067, 2915, 1682, 1600, 1470, 770, and 697 cm^{-1}; ^1H NMR (300 MHz, $CDCl_3$): δ 8. 37 (dd, 1H, *J* = 8.0 and 1.2 Hz, CH), 8.07 (br, 1H, NH-Trp), 7.78 (ddd, 1H, *J* = 8.4, 7.0, and 1.5 Hz, CH), 7.57 (d, 1H, *J* = 2.5 Hz, CH), 7.53 (dd, 1H, *J* = 8.2 and 1.1 Hz, CH), 7.41 (d, 1H, *J* = 8.0 Hz, CH-Trp), 7.30 (d, 1H, *J* = 8.3 Hz, CH-Trp), 7.13 (t, 1H, *J* = 7.7 Hz, CH-Trp), 6.93 (t, 1H, *J* = 7.0 Hz, CH-Trp), 6.71 (d, 1H, *J* = 2.4 Hz, CH-Trp), 6.37 (s, 1H, NH-amide), 5.67 (dd, 1H, *J* = 5.2 and 3.0 Hz, CH*-Trp), 3.74 (dd, 1H, *J* = 15.0 and 3.0 Hz, CH_2-Trp), 3.65 (dd, 1H, *J* = 14.9 and 5.3 Hz, CH_2-Trp), 2.99 (dd, 1H, *J* = 6.6 and 3.6 Hz, CH*-Met), 2.49-2.13 (m, 4H, CH_2-Met), 1.96 (s, 3H, CH_3-Met); ^{13}C NMR (75 MHz, $CDCl_3$): δ 169.3 (C=O), 161.1 (C=O), 150.3 (C=N), 147.1 (C), 136.0 (C-Trp), 134.7 (CH), 127.2 (CH), 127.2 (C-Trp), 126.9 (CH), 123.5 (CH-Trp), 122.7 (CH-Trp), 120.2 (C), 120.2 (C-Trp), 118.7 (CH-Trp), 111.1 (CH-Trp), 109.5 (C-Trp), 57.2 (CH*Trp), 52.8 (CH*-Met), 30.7 (CH_2-S-Met), 29.7 (CH_2-Met), 27.2 (CH_2-Trp), 15.3 (CH_3-Met); HRMS-ESI m/z 419.1544 $(M + H)^+$ (calculate for $C_{23}H_{23}N_4O_2S$, 418.1463).

(1S,4R)-4-((1H-indol-3-yl)methyl)-1-(2-(methylthio)ethyl)-1,2-dihydro-6H-pyrazino[2,1-b]quinazoline-3,6(4H)-diones (7). Yield: 34.8 mg, 7.9%; *er* = 49: 51; m.p.: 197–200 °C; $[\alpha]_D^{30} = -56.9$ (*c* 0.041; $CHCl_3$); υ_{max} (KBr) 3290, 3058, 2918, 2854, 1684, 1670, 1602, 773, and 695 cm^{-1}; ^1H NMR (300 MHz, $CDCl_3$): δ 8. 37 (dd, 1H, *J* = 8.0 and 1.1 Hz, CH), 8.12 (br, 1H, NH-Trp), 7.78 (ddd, 1H, *J* = 8.4, 7.2, and 1.5 Hz, CH), 7.57 (d, *J* = 8.3 Hz, CH), 7.54 (d, 1H, *J* = 7.0 Hz, CH), 7.40 (d, 1H, *J* = 8.0 Hz, CH-Trp), 7.30 (d, 1H, *J* = 8.2 Hz, CH-Trp), 7.12 (t, 1H, *J* = 7.1 Hz, CH-Trp), 6.92(t, 1H, *J* = 7.0 Hz, CH-Trp), 6.71 (d, 1H, *J* = 2.4 Hz, CH-Trp), 5.67 (dd, 1H, *J* = 5.4 and 3.1 Hz, CH*-Trp), 6.54 (s, 1H, NH-amide), 3.73 (dd, 1H, *J* = 15.0 and 2.9 Hz, CH_2-Trp), 3.65 (dd, 1H, *J* = 15.1 and 5.5 Hz, CH_2-Trp), 3.01 (dd, 1H, *J* = 6.6 and 3.6 Hz, CH*-Met), 2.48-2.20 (m, 4H, CH_2-Met), 1.96 (s, 3H, CH_3-Met); ^{13}C NMR (75 MHz, $CDCl_3$): δ 169.5 (C=O), 161.0 (C=O), 150.6 (C=N), 146.9 (C), 136.1 (C-Trp), 134.7 (CH), 127.2 (C-Trp), 127.2 (CH), 126.9 (CH), 123.5 (CH-Trp), 122.6 (CH-Trp), 120.2 (C), 120.1 (C-Trp), 118.7 (CH-Trp), 111.2 (CH-Trp), 109.5 (C-Trp), 57.2 (CH*-Trp), 52.8 (CH*-Met), 30.7 (CH_2-S-Met), 29.7 (CH_2-Met), 27.2 (CH_2-Trp), 15.3 (CH_3-Met); (+)-HRMS-ESI *m/z* 419.1526 $(M + H)^+$, (calculated for $C_{23}H_{23}N_4O_2S$, 418.1463).

(1R,4S)-4-((1H-indol-3-yl)methyl)-1-(4-(benzyloxy)benzyl)-1,2-dihydro-6H-pyrazino[2,1-b]quinazoline-3,6(4H)-diones (8). Yield: 81.9 mg, 14.8%; *er* = 63:37; m.p.: 226.9–227.9 °C; $[\alpha]_D^{30} = +46.7$ (*c* 0.05; $CHCl_3$); υ_{max} (KBr) 3393, 3268, 2954, 1671, 1611, 1511, 1465, 1240, 772, and 697 cm^{-1}; ^1H NMR (300 MHz, $CDCl_3$): δ 8. 39 (dd, 1H, *J* = 8.0 and 1.2 Hz, CH), 8.05 (br, 1H, NH-Trp), 7.80 (ddd, 1H, *J* = 8.5, 7.2 and 1.5 Hz, CH), 7.62 (d, *J* = 8.0 Hz, CH), 7.56 (dt, 1H, *J* = 7.3, 7.7, and 1.1 Hz, CH), 7.45-7.39 (m, 5H, CH-Bz), 7.32 (d, 2H, *J* = 8.0 Hz, CH-Trp), 7.17 (t, 1H, *J* = 7.5 Hz, CH-Trp), 6.88 (t, 1H, *J* = 7.5 Hz, CH-Trp),

6.76 (d, 2H, J = 9.0 Hz, CH-Tyr), 6.61 (d, 1H, J = 3.0 Hz, CH-Trp), 6.39 (d, 2H, J = 8.5 Hz, CH-Tyr), 5.64 (dd, 1H, J = 5.2 and 2.7, CH*-Trp), 5.35 (s, 1H, NH-amide), 5.05 (s, 2H, CH_2-Bz), 3.75 (dd, 1H, J = 14.9 and 2.6 Hz, CH_2-Trp), 3.67 (dd, 1H, J = 15.0 and 5.3 Hz, CH_2-Trp), 3.52 (dd, 1H, J = 14.7 and 3.6 Hz, CH_2-Tyr), 2.89 (dd, J = 11.1 and 3.6 Hz, CH*-Tyr), 2.46 (dd, 1H, J = 14.7 and 11.2 Hz, CH_2-Tyr); ^{13}C NMR (75 MHz, $CDCl_3$): δ 169.2 (C=O), 160.7 (C=O), 157.9 (C-Tyr), 151.0 (C=N), 147.0 (C), 137.0 (C-Trp), 136.1(C-Bz), 134.8 (CH), 128.7 (CH-Tyr (2)), 128.6 (CH-Bz (2)), 128.0 (C-Tyr), 127.4 (CH-Bz), 127.3 (CH-Trp), 127.2 (CH), 127.1 (CH-Bz (2)), 127.0 (CH), 126.9 (CH), 123.8 (CH-Trp), 122.8 (CH-Trp), 120.6 (C), 120.4 (CH-Trp), 119.0 (CH-Trp), 115.5 (CH-Tyr (2H)), 111.1 (CH-Trp), 109.7 (C-Trp), 70.1 (CH_2-Bz), 57.4 (CH*-Trp), 52.9 (CH*-Tyr), 37.1 (CH_2-Tyr), 29.7 (CH_2-Trp); (+)-HRMS-ESI m/z 541.2232 $(M + H)^+$, (calculated for $C_{34}H_{29}N_4O_3$, 540.2161).

(1S,4S)-4-((1H-indol-3-yl)methyl)-1-(4-(benzyloxy)benzyl)-1,2-dihydro-6H-pyrazino[2,1-b]quinazoline-3,6(4H)-diones (**9**). Yield: 119.9 mg, 21.7%; *er* = 29:71; m.p.: 165.9–166.6 °C; $[\alpha]_D^{30}$ = +205.8 (c 0.076; $CHCl_3$); ν_{max} (KBr) 3489, 3364, 2923, 1674, 1612, 1512, 1467, 1249, 774, and 695 cm^{-1}; ^{1}H NMR (300 MHz, $CDCl_3$): δ 8. 42 (dd, 1H, J = 8.0 and 1.1 Hz, CH), 8.08 (br, 1H, NH-Trp), 7.83 (ddd, 1H, J 8.5, 7.1, and 1.5 Hz, CH), 7.66 (d, J 7.7 Hz, CH), 7.62-7.52 (m, 2H, CH), 7.39 (t, 4H, J = 2.6 Hz, CH-Bz), 7.35 (ddd, J = 6.2, 3.4, and 1.5 Hz, 1H, CH-Bz), 7.30 (d, 1H, J = 8.1 Hz, CH-Trp), 7.20 (td, 1H, J = 7.6 and 1.1 Hz, CH-Trp), 6.10 (td, 1H, J = 7.5 and 1.0 Hz, CH-Trp), 6.74 (d, 2H, J = 8.7 Hz, CH-Tyr), 6.60 (d, 1H, J = 2.3 Hz, CH-Trp), 6.23 (d, 2H, J = 8.6 Hz, CH-Tyr), 5.56 (t, 1H, J = 4.2 Hz, CH*-Trp), 5.55 (s, 1H, NH-amide), 4.99 (s, 2H, CH_2-Bz), 4.33 (dt, 1H, J = 11.7 and 2.8 Hz, CH*-Tyr), 3.86 (dd, 1H, J =14.9 and 3.0 Hz, CH_2-Trp), 3.80 (dd, 1H, J = 14.9 and 4.4 Hz, CH_2-Trp), 2.95 (dd, 1H, J = 13.3 and 3.1 Hz, CH_2-Tyr), 0.53 (dd, J = 13.1 and 11.9 Hz, CH_2-Tyr); ^{13}C NMR (75 MHz, $CDCl_3$): δ 166.5 (C=O), 160.9 (C=O), 158.0 (C-Tyr), 150.2 (C=N), 147.2 (C), 136.9 (C-Trp), 135.8 (C-Bn), 134.9 (CH), 130.28 (CH-Tyr (2)), 128.6 (CH-Bz (2)), 128.1 (C-Tyr), 128.0 (CH-Bz), 127.7 (C-Trp), 127.4 (CH), 127.0 (CH-Bz (2)), 126.9 (CH), 123.5 (CH-Trp), 122.8 (CH-Trp), 120.5 (C), 120.2 (C-Trp), 119.5 (CH-Trp), 115.2 (CH-Tyr (2)), 111.4 (CH-Trp), 109.7 (C-Trp), 70.0 (CH_2-Bz), 57.9 (CH*-Tyr), 56.8 (CH*-Trp), 42.0 (CH_2-Tyr), 26.6 (CH_2-Trp(+)-HRMS-ESI m/z 541.2221 $(M + H)^+$, (calculated for $C_{34}H_{29}N_4O_3$, 540.2161).

4.5. General Conditions for the Synthesis of (1S)-4-((1H-indol-3-yl)methyl)-1-(4-hydroxybenzyl)-1,2-dihydro -6H-pyrazino[2,1-b]quinazoline-3,6(4H)-dione (10 and 11)

In an oven-dried round-bottomed flash equipped with Teflon-coated magnetic stir bar, a rubber septum, a glass stopper, and nitrogen gas inlet compound **8** or **9** (50 mg, 0.092 mmol) dissolved with anhydrous CH_2Cl_2 (5 mL) was added. After cooled the mixture to −78 °C, 1M boron trichloride in CH_2Cl_2 (190 µL, 0.19 mmol, 2eq) was added dropwise over 5 min at −78 °C. After stirring for 45 min at -78 °C, the mixture was quenched by syringe addition of $CHCl_3$/MeOH (10/1, 10 mL) at -78 °C and was warmed to ambient temperature. The solvent was evaporated, and the content was purified by flash column chromatography using hexane: EtOAc (6:4) as mobile phase. Compounds **10** or **11** were obtained as pale yellow solids. Before analysis, compounds were recrystallized from methanol.

(1R,4S)-4-((1H-indol-3-yl)methyl)-1-(4-hydroxybenzyl)-1,2-dihydro-6H-pyrazino[2,1-b]quinazoline-3,6(4H)-dione (**10**). Yield: 12.5 mg, 30%; *er* = 36:65; m.p.: 134–136 °C; $[\alpha]_D^{30}$ = +39.5 (c 0.034; CH_3OH); v_{max} (KBr) 3427, 1677, 1603, 1515, 1468, 1159, 1025, 998, and 765 cm^{-1}; ^{1}H NMR (300 MHz, DMSO-d_6): δ 10.48 (s, 1H, NH-Trp), 8.89 (d, 1H, J = 4.9 Hz, OH-Tyr), 8.32 (dd, 1H, J = 8.0 and 1.2 Hz, CH), 7.81 (dd, 1H, J = 7.7 and 1.6 Hz, CH), 7.61 (d, J = 7.9 Hz, CH), 7.56 (dd, 1H, J = 12.7 and 7.2 Hz, CH), 7.39 (dd, 1H, J = 8.0 and 5.4 Hz, CH-Trp), 7.34 (dt, 1H, J = 8.1 Hz, CH), 7.13 (dd, 1H, J = 13.4 and 6.7 Hz, CH-Trp), 6.84 (dd, 1H, J = 13.4 and 6.7 Hz, CH-Trp), 6.60 (d, 1H, J = 3.4 Hz, CH-Trp), 6.58(dd, 2H, J = 8.5 and 6.0 Hz, CH-Tyr), 6.53 (d, 1H, J = 4.4 Hz, NH-amide), 6.46 (dd, 2H, J = 7.9 and 5.7 Hz, CH-Tyr), 5.44 (dd, 1H, J = 5.0 and 2.8 Hz, CH*-Trp), 3.64 (dd, 1H, J = 14.8 and 2.7 Hz, CH_2-Trp), 3.56 (dd, 1H, J = 14.9 and 5.3 Hz), 3.26 (dt, J = 13.4 and 3.6 Hz, CH_2-Tyr), 3.03 (dt, 1H, J = 8.9 and 4.7 Hz, CH*-Tyr), 2.67 (dt, 1H, J = 13.3 and 10.5 Hz); ^{13}C NMR (75 MHz, $CDCl_3$): δ 168.1 (C=O), 160.0 (C=O), 150.5 (C-Tyr), 146.3 (C), 135.8 (C-Trp), 134.1 (CH), 129.4 (CH-Tyr (2)), 126.7 (C-Trp), 126.6 (CH), 126.4 (CH), 126.1 (CH),

124.9 (C-Tyr), 123.7 (CH-Trp), 121.2 (CH-Trp), 119.6 (CH-Trp), 118.9 (CH-Trp), 115.0 (C-Tyr), 111.2 (CH-Trp), 108.0 (C-Trp), 56.6 (CH*-Trp), 52.8 (CH*-Tyr), 36.3 (CH_2-Tyr), 26.5 (CH_2-Trp); (+)-HRMS-ESI *m/z* 451.1766 $(M + H)^+$, 473.1576 $(M + Na)^+$ (calculated for $C_{27}H_{23}N_4O_3$, 450.1692).

(1S,4S)-4-((1H-indol-3-yl)methyl)-1-(4-hydroxybenzyl)-1,2-dihydro-6H-pyrazino[2,1-b]quinazoline-3,6(4H)-dione (**11**). Yield: 25.7 mg, 68.7%; *er* = 61:39; m.p.: 102–103 °C; $[\alpha]_D^{30}$ = +75.9 (*c* 0.079; CH_3OH); v_{max} (KBr) 3428, 2927, 1667, 1610, 1592, 1474, 1337, 1232, 772, and 699 cm^{-1}; ^1H NMR (300 MHz, DMSO-d_6): δ 10.54 (s, 1H, NH-Trp), 8.86 (s, 1H, OH-Tyr), 8.34 (dd, 1H, *J* = 8.0 and 1.2 Hz, CH), 7.83 (dd, 1H, *J* = 7.7, and 1.6Hz, CH), 7.62 (d, *J* = 7.9 Hz, CH), 7.56 (dt, 1H, *J* = 7.6, 7.6, and 0.6 Hz, CH), 7.49 (d, 1H, *J* = 7.9 Hz, CH-Trp), 7.30 (d, 1H, *J* = 8.1 Hz, CH), 7.20 (d, 1H, *J* = 3.3 Hz, NH-amide), 7.09 (dt, 1H, *J* = 7.6 and 0.6 Hz, CH-Trp), 6.94 (dt, 1H, *J* = 7.5 and 0.5 Hz, CH-Trp), 6.66 (d, 1H, *J* = 2.3 Hz, CH-Trp) 6.57(d, 2H, *J* = 5.6 Hz, CH-Tyr), 6.45 (d, 2H, *J* = 8.4 Hz, CH-Tyr), 5.37 (dd, 1H, *J* = 5.2 and 3.3 Hz, CH*-Trp), 4.36 (dt, 1H, *J* = 10.5 and 3.4 Hz, CH*-Tyr), 3.63 (dd, 1H, *J* = 14.8 and 3.1 Hz, CH_2-Trp), 3.55 (dd, 1H, *J* = 14.9 and 5.4 Hz, CH_2-Trp), 2.69 (dd, *J* = 13.4 and 3.6 Hz CH_2-Tyr), 0.86 (dd, 1H, *J* = 13.3 and 10.5 Hz, CH_2-Tyr); ^{13}C NMR (75 MHz, $CDCl_3$): δ 165.8 (C=O), 160.2 (C=O), 155.8 (C-Tyr), 150.4 (C=N), 146.7 (C), 135.6 (C-Trp), 134.2 (CH), 129.9 (CH-Tyr (2)), 127.4 (CH-Trp), 126.3 (CH), 126.2 (CH), 126.1 (CH-Trp), 125.9 (C-Tyr), 123.7 (CH-Trp), 121.1 (CH-Trp), 119.5 (CH-Trp), 118.7 (CH-Trp), 114.9 (CH-Tyr (2)), 111.2 (CH-Trp), 108.0 (C-Trp), 57.2 (CH*-Trp), 56.3 (CH*-Tyr), 41.8 (CH_2-Tyr), 26.1 (CH_2-Trp); (+)-HRMS-ESI *m/z* 451.1771 $(M + H)^+$, 473.1564 $(M + Na)^+$ (calculated for $C_{27}H_{23}N_4O_3$, 450.1692).

4.6. General Conditions for the Synthesis of (1S,4R)-4-((1H-indol-3-yl)methyl)-1-(4-(benzyloxy)benzyl)-8,10-dichloro-1,2-dihydro-6H-pyrazino[2,1-b]quinazoline-3,6(4H)-dione (12)

In a closed vial 3,5-dichloro anthranilic acid (**xiii**, 41 mg, 200 µmol), *N*-Boc-*o*-Bn-Tyrosine (**xii**, 74 mg, 200 µmol), and triphenylphosphite (63 µL, 220 µmol) were added along with 1 mL of dried pyridine. The vial was heated in heating block with stirring at 55 °C for 16–24 h. After cooling the mixture to room temperature, D-tryptophan methyl ester hydrochloride (**ii**, 51 mg, 200 µmol) was added, and the mixture was irradiated in the microwave at the constant temperature at 220 °C for 1.5 min. Reaction mixtures were prepared in the same conditions and treated in parallel. After removing the solvent with toluene, the crude product was purified by flash column chromatography using hexane: EtOAc (60:40) as a mobile phase. The preparative TLC was performed using CH_2Cl_2:Me_2CO (95:5) as mobile phase. The major compound appeared as a black spot with no fluorescence under the UV light (366 nm). Compound **12** was collected as orange solids. Before analysis, compound was recrystallized from methanol. Yield: 26.2 mg, 2.2%; *er* 67:33; m.p.: 233–235 °C; $[\alpha]_D^{30}$ = +244.44 (*c* 0.045; CH_3OH); ν_{max} (KBr) 3424, 3334, 2921, 1681, 1593, 1511, 1455, 1247,1012, 695, and 420 cm^{-1}; ^1H NMR (300 MHz, $CDCl_3$): δ 8.26 (d, 1H, *J* = 2.4 Hz, CH), 8.09 (br, 1H, NH-Trp), 7.85 (d, 1H, *J* = 2.4 Hz, CH), 7.44 (dd, 5H, *J* = 6.4 and 2.0 Hz, CH-Bz), 7.39 (d, 2H, *J* = 8.2 Hz, CH-Trp), 7.19 (t, 1H, *J* = 7.7 Hz, CH-Trp), 6.96 (t, 1H, *J* = 7.5 Hz, CH-Trp), 6.75 (d, 2H, *J* = 8.7 Hz, CH-Tyr), 6.62 (d, 1H, *J* = 2.3 Hz, CH-Trp), 6.40 (d, 2H, *J* = 8.5 Hz, CH-Tyr), 5.56 (dd, 1H, *J* = 5.2 and 2.6, CH*-Trp), 5.41 (s, 1H, NH-amide), 5.06 (s, 2H, CH_2-Bz), 3.77 (dd, 1H, *J* = 15.0 and 2.6 Hz, CH_2-Trp), 3.62 (dd, 1H, *J* = 15.6 and 4.1 Hz, CH_2-Trp), 3.50 (dd, 1H, *J* = 17.5 and 4.4 Hz, CH_2-Tyr), 2.95 (dd, *J* = 10.8 and 3.6 Hz, CH*-Tyr), 2.51 (dd, 1H, *J* = 14.8 and 10.9 Hz, CH_2-Tyr); ^{13}C NMR (75 MHz, $CDCl_3$): δ 168.7 (C=O), 159.2 (C=O), 158.0 (C-Tyr), 152.0 (C=N), 142.4 (C), 136.9 (C-Bz), 136.1(C-Trp), 135.1 (CH), 133.2 (C-Cl), 132.6 (C-Cl), 129.6 (CH-Tyr (2)), 128.7 (CH-Bz (2)), 128.1 (CH-Bz), 127.4 (CH-Bz(2)), 126.9 (C-Trp), 126.6 (C-Tyr), 125.2 (CH), 123.7 (CH-Trp), 122.9 (CH-Trp), 122.4 (C), 118.9 (CH-Trp), 115.5 (CH-Tyr (2H)), 111.2 (CH-Trp), 109.4 (C-Trp), 70.1 (CH_2-Bz), 57.6 (CH*-Trp), 53.1 (CH*-Tyr), 36.9 (CH_2-Tyr), 27.1 (CH_2-Trp). (+)-HRMS-ESI *m/z* 609.1427 $(M + H)^+$, (calculated for $C_{34}H_{27}N_4O_3Cl_2$, 608.1382).

4.7. Neuroprotection Assay

Compounds **1**, **2**, **4**, **5**, **6**, **7**, **8**, **9**, **10**, **11**, and **12c** together with fiscalin B (**3**) were assayed in co-treatment with rotenone at eight concentrations per triplicate. The SH-SY5y cells (ATCC Ref.: CRL-2266) were seeded at a density of 40,000/well in a 96-well plate and were incubated in a humidified atmosphere at 37 °C with 5% CO_2 for overnight. The compounds were dissolved in DMSO at the concentration of 10 mM and the higher concentration assayed was 100 µM. The negative control was 1% DMSO. After 24 h of co-treatment, plates were treated with MTT (3-(4,5-dimethylthiazol-2-yl)-2,5-diphenyltetrazolium bromide) at 5 µg/mL in Minimum Essential Medium Eagle (MEM) for 3 h at the standard culture condition. Then, DMSO was added to the plates to solubilizing the formazan crystal formed in viable cells and plates were put in the stirring for 5 min to homogenize the solution. Absorbance at 570 nm was measured by VICTOR Multilabel Plate Reader (PerkinElmer).

4.8. Screening Test for Antitumor Activity

Compounds **1**, **2**, **4**, **5**, **6**, **7**, **8**, **10**, and **11** were reconstituted in sterile DMSO to the final concentration of 60 mM, and several aliquots were made and stored at −20 °C to avoid repeated freeze-thaw cycles. For experiments, the compounds were freshly diluted in medium to the desired concentration. Screening for tumor cell growth inhibitory activity was carried out in three human tumor cell lines (NCI-H460, BxPC3 and PANC1), with the sulforhodamine B (SRB) assay, as previously described [32]. Briefly, tumor cells were plated in 96-well plates, incubated at 37 °C for 24 h, and then treated for 48 h with 5 serial dilutions (1:2) of each compound (ranging from 25 µM to 1.5625 µM, 150 µM to 9.375 µM or 200 µM to 12.5 µM, depending on the compound and for reasons related with solubility). The effect of the vehicle solvent (DMSO) was also analyzed as a control. Cells were fixed with 10% ice-cold trichloroacetic acid, washed with water, and stained with SRB. Finally, the plates were washed with 1% acetic acid and the bound SRB was solubilized with 10 mM Tris Base. Absorbance was measured in a microplate reader (Synergy Mx, Biotek Instruments Inc., Winooski, VT, USA) at 510 nm. For each compound, the corresponding GI_{50} (concentration which inhibited 50% of net cell growth) was determined, as previously described [33].

4.9. Testing Effect of Compounds on Non-malignant Breast Cells

Compounds **5**, **7**, **10** and **11** (which presented simultaneously the best neuroprotection and antitumor effects) were tested against the non-malignant MCF-12A human breast epithelial cells. For that, cells were incubated with specific concentrations of each compound (corresponding to approximately the highest GI_{50} concentration obtained in the antitumor activity screening) for 48 h, followed by removal of the compound, addition of new medium to the cells and then 5 more days in culture. At the end of the 7 days in total, the sulforhodamine B (SRB) assay was performed, as previously described [32].

5. Conclusions

New quinazolinone alkaloid derivatives with *anti* and *syn* stereochemistry were synthetized by combining both a one-pot microwave-assisted reaction and a multi-step approach. Interestingly, fumiquinazoline G (**1**) presented a better antitumor activity in all the tumor cell lines tested, with GI_{50} values lower than 20 µM. The antitumor activity of the remaining compounds was not relevant, with GI_{50} values higher than 20 µM in the tested cell lines. The effect of the synthesized compounds in the growth of the tested non-malignant cells was smaller than the effect on the studied tumor cells. It is worth noting that among the compounds tested, only **1**, **3**, **5**, and **7** showed potential for neuroprotection in a PD in vitro model. This finding highlights new insights into marine natural products belonging to the proteomimetic quinazolinone alkaloids.

Supplementary Materials: The following are available online Figures S1–S57.

Author Contributions: E.S. conceived the study design. S.L. synthesized the compounds and elucidated their structure and, A.S., E.S., A.K. and M.M.M.P. analyzed the data. D.I.S.P.R. performed the LC analysis. R.F. and C.X. performed the cytotoxic studies in tumor cells and non-malignant breast cells. C.X. and M.H.V. analyzed data from the cytotoxic studies, discussed and wrote those results. S.L. and E.S. wrote the manuscript, while all authors gave significant contributions in discussion and revision. All authors agreed with the final version of the manuscript.

Funding: This research was partially supported by the Strategic Funding UID/Multi/04423/2013 through national funds provided by FCT—Foundation for Science and Technology and European Regional Development Fund (ERDF), in the framework of the program PT2020. The authors thank to national funds provided by FCT—Foundation for Science and Technology and European Regional Development Fund (ERDF) and COMPETE under the Strategic Funding UID/Multi/04423/2013, the projects POCI-01-0145-FEDER-028736 and PTDC/MAR-BIO/4694/2014 (POCI-01-0145-FEDER-016790; 3599-PPCDT).

Acknowledgments: S.L. thanks Erasmus Mundus Action 2 (LOTUS+, LP15DF0205) for full PhD scholarship. D.I.S.P.R. thanks for her postdoctoral scholarship (NOVELMAR/BPD_2/2016-019). To Sara Cravo for technical support.

Conflicts of Interest: The authors declare no conflict of interest.

References

1. Plun-Favreau, H.; Lewis, P.A.; Hardy, J.; Martins, L.M.; Wood, N.W. Cancer and neurodegeneration: Between the devil and the deep blue sea. *PLoS Genet.* **2010**, *6*, e1001257. [CrossRef] [PubMed]
2. Klus, P.; Cirillo, D.; Botta Orfila, T.; Tartaglia, G.G. Neurodegeneration and cancer: Where the disorder prevails. *Sci. Rep.* **2015**, *5*, 15390. [CrossRef] [PubMed]
3. Resende, D.I.S.P.; Boonpothong, P.; Sousa, E.; Kijjoa, A.; Pinto, M.M.M. Chemistry of the fumiquinazolines and structurally related alkaloids. *Nat. Prod. Rep.* **2019**, *36*, 7–34. [CrossRef] [PubMed]
4. Buttachon, S.; Chandrapatya, A.; Manoch, L.; Silva, A.; Gales, L.; Bruyère, C.; Kiss, R.; Kijjoa, A. Sartorymensin, a new indole alkaloid, and new analogues of tryptoquivaline and fiscalins produced by neosartorya siamensis (kufc 6349). *Tetrahedron* **2012**, *68*, 3253–3262. [CrossRef]
5. Tamiya, H.; Ochiai, E.; Kikuchi, K.; Yahiro, M.; Toyotome, T.; Watanabe, A.; Yaguchi, T.; Kamei, K. Secondary metabolite profiles and antifungal drug susceptibility of aspergillus fumigatus and closely related species, aspergillus lentulus, aspergillus udagawae, and aspergillus viridinutans. *J. Infect. Chemother.* **2015**, *21*, 385–391. [CrossRef] [PubMed]
6. He, F.; Sun, Y.-L.; Liu, K.-S.; Zhang, X.-Y.; Qian, P.-Y.; Wang, Y.-F.; Qi, S.-H. Indole alkaloids from marine-derived fungus aspergillus sydowii scsio 00305. *J. Antibiot.* **2012**, *65*, 109–111. [CrossRef] [PubMed]
7. Takahashi, C.; Matsushita, T.; Doi, M.; Minoura, K.; Shingu, T.; Kumeda, Y.; Numata, A. Fumiquinazolines a–g, novel metabolites of a fungus separated from a pseudolabrus marine fish. *J. Chem. Soc. Perkin Trans. 1* **1995**, *18*, 2345–2353. [CrossRef]
8. Numata, A.; Takahashi, C.; Matsushita, T.; Miyamoto, T.; Kawai, K.; Usami, Y.; Matsumura, E.; Inoue, M.; Ohishi, H.; Shingu, T. Fumiquinazolines, novel metabolites of a fungus isolated from a saltfish. *Tetrahedron Lett.* **1992**, *33*, 1621–1624. [CrossRef]
9. Han, X.-x.; Xu, X.-y.; Cui, C.-b.; Gu, Q.-q. Alkaloidal compounds produced by a marine-derived fungus, aspergillus fumigatus h1-04, and their antitumor activities. *Chinese J. Org. Chem.* **2007**, *17*, 232–237.
10. Cheng, Z.; Lou, L.; Liu, D.; Li, X.; Proksch, P.; Yin, S.; Lin, W. Versiquinazolines a–k, fumiquinazoline-type alkaloids from the gorgonian-derived fungus aspergillus versicolor lzd-14-1. *J. Nat. Prod.* **2016**, *79*, 2941–2952. [CrossRef]
11. Wong, S.M.; Musza, L.L.; Kydd, G.C.; Kullnig, R.; Gillum, A.M.; Cooper, R. Fiscalins: New sibstamce p inhibitors produced by the fungus neosartorya fischeri. Taxonomy, fermentation, structures, and biological properties. *J. Antibiot.* **1993**, *46*, 545–553. [CrossRef] [PubMed]
12. Barrow, C.J.; Sun, H.H. Spiroquinazoline, a novel substance p inhibitor with a new carbon skeleton, isolated from aspergillus flavipes. *J. Nat. Prod.* **1994**, *57*, 471–476. [CrossRef] [PubMed]
13. Thornton, E.; Vink, R. Treatment with a substance p receptor antagonist is neuroprotective in the intrastriatal 6-hydroxydopamine model of early parkinson's disease. *PLoS ONE* **2012**, *7*. [CrossRef] [PubMed]
14. Dunlop, R.A.; Cox, P.A.; Banack, S.A.; Rodgers, K.J. The non-protein amino acid bmaa is misincorporated into human proteins in place of l-serine causing protein misfolding and aggregation. *PLoS ONE* **2013**, *8*, e75376. [CrossRef] [PubMed]
15. Rodgers, K.J. Non-protein amino acids and neurodegeneration: The enemy within. *Exp. Neurol.* **2014**, *253*, 192–196. [CrossRef]
16. Long, S.; Resende, D.I.S.P.; Kijjoa, A.; Silva, A.M.S.; Pina, A.; Fernández-Marcelo, T.; Vasconcelos, M.H.; Sousa, E.; Pinto, M.M.M. Antitumor activity of quinazolinone alkaloids inspired by marine natural products. *Mar. Drugs* **2018**, *16*, 261. [CrossRef]

17. Wang, H.; Ganesan, A. Total synthesis of the fumiquinazoline alkaloids: Solution-phase studies. *J. Org. Chem.* **2000**, *65*, 1022–1030. [CrossRef]
18. Liu, J.F.; Ye, P.; Zhang, B.; Bi, G.; Sargent, K.; Yu, L.; Yohannes, D.; Baldino, C.M. Three-component one-pot total syntheses of glyantrypine, fumiquinazoline f, and fiscalin b promoted by microwave irradiation. *J. Org. Chem.* **2005**, *70*, 6339–6345. [CrossRef] [PubMed]
19. Okaya, S.; Okuyama, K.; Okano, K.; Tokuyama, H. Trichloroboron-promoted deprotection of phenolic benzyl ether using pentamethylbenzene as a non lewis-basic cation scavenger. In *Organic Syntheses*; Wiley: Hoboken, NJ, USA, 2016; Volume 93, pp. 63–74.
20. Hernández, F.; Buenadicha, F.L.; Avendao, C.; Söllhuber, M. 1-alkyl-2,4-dihydro-1h-pyrazino[2,1-b]quinazoline-3,6-diones as glycine templates. Synthesis of fiscalin b. *Tetrahedron Asymmetr.* **2002**, *12*, 3387–3398. [CrossRef]
21. Campos, A.C.; Fogaça, M.V.; Sonego, A.B.; Guimarães, F.S. Cannabidiol, neuroprotection and neuropsychiatric disorders. *Pharmacol. Res.* **2016**, *112*, 119–127. [CrossRef] [PubMed]
22. Bagli, E.; Goussia, A.; Moschos, M.M.; Agnantis, N.; Kitsos, G. Natural compounds and neuroprotection: Mechanisms of action and novel delivery systems. *In Vivo* **2016**, *30*, 535–547. [PubMed]
23. Cannon, J.R.; Tapias, V.; Na, H.M.; Honick, A.S.; Drolet, R.E.; Greenamyre, J.T. A highly reproducible rotenone model of parkinson's disease. *Neurobiol. Dis.* **2009**, *34*, 279–290. [CrossRef] [PubMed]
24. Condello, S.; Currò, M.; Ferlazzo, N.; Caccamo, D.; Satriano, J.; Ientile, R. Agmatine effects on mitochondrial membrane potential andnf-κb activation protect against rotenone-induced cell damage in human neuronal-like sh-sy5y cells. *J. Neurochem.* **2011**, *116*, 67–75. [CrossRef] [PubMed]
25. Vichai, V.; Kirtikara, K. Sulforhodamine b colorimetric assay for cytotoxicity screening. *Nat. protoc.* **2006**, *1*, 1112–1116. [CrossRef]
26. Zhou, Y.; Debbab, A.; Mándi, A.; Wray, V.; Schulz, B.; Müller, W.E.G.; Kassack, M.; Lin, W.; Kurtán, T.; Proksch, P.; et al. Alkaloids from the sponge-associated fungus aspergillus sp. *Eur. J. Org. Chem.* **2013**, *5*, 894–906. [CrossRef]
27. Rodrigues, B.S.F.; Sahm, B.D.B.; Jimenez, P.C.; Pinto, F.C.L.; Mafezoli, J.; Mattos, M.C.; Rodrigues-Filho, E.; Pfenning, L.H.; Abreu, L.M.; Costa-Lotufo, L.V.; et al. Bioprospection of cytotoxic compounds in fungal strains recovered from sediments of the brazilian coast. *Chem. Biodivers.* **2015**, *12*, 432–442. [CrossRef]
28. Fujimoto, H.; Negishi, E.; Yamaguchi, K.; Nishi, N.; Yamazaki, M. Isolation of new tremorgenic metabolites from an ascomycete, corynascus setosus. *Chem. Pharm. Bull.* **1996**, *40*, 1843–1848. [CrossRef]
29. Wu, B.; Chen, G.; Liu, Z.-g.; Pei, Y. Two new alkaloids from a marine-derived fungus neosartorya fischeri. *Rec. Nat. Prod.* **2015**, *9*, 271–275.
30. Kantharaju, B.S.P.; Suresh Babu, V.V. Synthesis of fmoc-amino acid chlorides assisted by ultrasonication, a rapid approach. *Int. J. Pept. Res. Ther.* **2002**, *9*, 227–229. [CrossRef]
31. Wang, H.; Ganesan, A. Total synthesis of the quinazoline alkaloids (−)-fumiquinazoline g and (−)-fiscalin b. *J. Org. Chem.* **1998**, *63*, 2432–2433. [CrossRef]
32. Lopes-Rodrigues, V.; Oliveira, A.; Correia-da-Silva, M.; Pinto, M.; Lima, R.T.; Sousa, E.; Vasconcelos, M.H. A novel curcumin derivative which inhibits p-glycoprotein, arrests cell cycle and induces apoptosis in multidrug resistance cells. *Bioorg. Med. Chem.* **2017**, *25*, 581–596. [CrossRef] [PubMed]
33. Kijjoa, A.; Santos, S.; Dethoup, T.; Manoch, L.; Almeida, A.P.; Vasconcelos, M.H.; Silva, A.; Gales, L.; Herz, W. Sartoryglabrins, analogs of ardeemins, from neosartorya glabra. *Nat. Prod. Commun.* **2011**, *6*, 807–812. [PubMed]

Sample Availability: Samples of the compounds **1–12** are available from the authors.

molecules

MDPI

Review

Enantiomeric Tartaric Acid Production Using *cis*-Epoxysuccinate Hydrolase: History and Perspectives

Jinsong Xuan [1,*] and Yingang Feng [2,*]

1. Department of Biological Science and Engineering, School of Chemical and Biological Engineering, University of Science and Technology Beijing, 30 Xueyuan Road, Beijing 100083, China
2. Shandong Provincial Key Laboratory of Synthetic Biology and CAS Key Laboratory of Biofuels, Qingdao Institute of Bioenergy and Bioprocess Technology, Chinese Academy of Sciences, Songling Road 189, Qingdao, Shandong 266101, China

* Correspondence: jsxuan@sas.ustb.edu.cn (J.X.); fengyg@qibebt.ac.cn (Y.F.)

Received: 31 January 2019; Accepted: 1 March 2019; Published: 5 March 2019

Abstract: Tartaric acid is an important chiral chemical building block with broad industrial and scientific applications. The enantioselective synthesis of L(+)- and D(−)-tartaric acids has been successfully achieved using bacteria presenting *cis*-epoxysuccinate hydrolase (CESH) activity, while the catalytic mechanisms of CESHs were not elucidated clearly until very recently. As biocatalysts, CESHs are unique epoxide hydrolases because their substrate is a small, mirror-symmetric, highly hydrophilic molecule, and their products show very high enantiomeric purity with nearly 100% enantiomeric excess. In this paper, we review over forty years of the history, process and mechanism studies of CESHs as well as our perspective on the future research and applications of CESH in enantiomeric tartaric acid production.

Keywords: *cis*-epoxysuccinate hydrolase; tartaric acid; enantioselectivity; stereoselectivity; regioselectivity; epoxide hydrolase; immobilization; whole cell catalyst; enzyme stability; biocatalyst

1. Introduction

Tartaric acid (TA) is a well-known organic acid that naturally occurs in many kinds of fruit, most notably in grapes. The chemical chirality of TA was first discovered by Jean-Baptiste Biot in 1832 [1]. The naturally-occurring form of the acid is L(+)-TA, while D(−)-TA rarely exists in natural sources [2,3]. L(+)-TA is widely used in the food, wine, pharmaceutical, chemical, and polyester industries. D(−)-TA is also important in pharmaceutical manufacturing [4–6]. Both are well-known chiral chemical building blocks with broad industrial and scientific applications [7,8]. In enantioselective chemical synthesis, TA serves not only as a resolving agent or chiral auxiliary in the synthesis of bioactive molecules, but also a source of new asymmetric organocatalysts [7–9]. Traditionally, L(+)-TA is obtained as a solid by-product during wine fermentation, and this kind of production method is strongly influenced by the growth of grapes and the climatic conditions. Chemical synthesis of L(+)-TA with maleic acid is also possible but this gives a much less soluble racemic product (DL-form) which is not suitable for inclusion in foods because D(−)-TA in the product is considered to be harmful to human health. Commercial application of the chemical method is limited by both the product form and the high production cost [10]. Currently, microbial methods are considered to be much simpler and more economical for the production of L(+)-TA and D(−)-TA.

Epoxide hydrolases (EHs, EC 3.3.2.3) are biocatalysts that are ubiquitous in Nature. They can hydrolyze racemic epoxides to their corresponding optically active epoxides and pure vicinal diols, which are versatile intermediates for chiral pharmaceutical synthesis. In general, this enzymatic process occurs under mild conditions without the need for any cofactors, prosthetic groups, or metal ions [11]. The high transformation rate and enantioselectivity of epoxide hydrolases have gained them increasing attention in recent years, and they have found more and more applications in the organic chemical industry [11]. Epoxide hydrolases are found in a variety of sources, such as plants, insects, mammals, and microbes [12–14]. Mammalian epoxide hydrolases have been the subject of many studies because of their key role in xenobiotic detoxification in the liver, but their use as biocatalysts has been hindered by their limited availability [15]. Lately, bacterial epoxide hydrolases have been increasingly recognized as highly versatile biocatalysts owing to their abundance, high efficiency, and environmental friendliness [16].

cis-Epoxysuccinic acid hydrolases (CESHs) are epoxide hydrolase members that catalyze the asymmetric hydrolysis of *cis*-epoxysuccinate (CES) to form an enantiomeric tartrate [17–19]. Bacteria presenting CESH activity were discovered in the 1970s, and the synthesis of L(+)-TA was the first application of an epoxide hydrolase [20]. Since then, a large number of bacteria with CESH activity have been discovered and successfully applied for industrial TA production. The sequences and mechanism of CESHs have been partly elucidated since 2000, when it was revealed that CESH[L] and CESH[D], which produce L(+)-TA and D(−)-TA, respectively, are completely different proteins in terms of both sequence and structure. Therefore, CESHs are interesting EHs not only for TA production, but also for enantiomer biosynthesis in general. In this review, we summarize the body of literature on CESHs including both process optimization for industrial application and mechanism studies to understand how their regio- and stereoselectivity makes them efficient biocatalysts. We also provide our perspective on the use of CESHs in future research and applications.

2. The History of CESH Studies

Several patents involving bacteria with CESH activity were filed by Japanese companies in the 1970s [21–26]. However, these bacteria and their associated CEHSs did not receive much attention in the following ten years. In fact, unlike mammalian EHs, which have been subject to extensive enzymatic and biochemical study, microbial EHs were not well studied until the 1990s. Several studies reported in the 1990s showed that bacterial CESHs are promising biocatalysts for industrial synthesis, and EHs were proposed as new tools for the synthesis of fine organic chemicals [16,20]. In the 1990s, several groups from China, Belgium, Japan, and Slovakia continued to report on new bacteria containing CESH activity, and some process optimization methods, such as immobilization, were performed for industrial application [3,10,27–31]. The microbial production of L(+)-tartaric acid was successfully commercialized in the late 1990s [32]. Subsequently, new bacterial species were isolated for TA production in the 21st century [18,33–39], which demonstrates that CESHs exist broadly in bacteria.

In 2000, the first CESH encoding gene was reported from *Alcaligenes* sp. MCI3611 for the production of D(−)-TA [40]. Genes from several other CESHs, including CESH[D] from *Bordetella* sp. BK-52 and CESH[L]s from *Rhodococcus opacus*, *Nocardia tartaricans*, and *Klebsiella* sp. BK-58, have also been sequenced and cloned [17,19,41–43]. The genes and the derived amino acid sequences of CESHs provided the basis for later studies of the recombinant expression, structure, and mechanisms as well as the protein engineering of CESHs. Analyses of the amino acid sequences of CESHs indicated that CESH[D]s and CESH[L]s are completely different proteins [19]. Subsequent structural analysis and mechanism studies revealed that they have different structures and catalysis mechanisms [41,43–45]. In a recent study [46], we determined a high-resolution structure and elucidated detailed catalytic mechanisms for CESH[D], but these characteristics have not yet been reported for CESH[L].

With the knowledge of CESH sequences, structures, and mechanisms, scientists began to conduct extensive protein engineering research to improve enzyme stability and activity. As a result, many mutants with good properties have been obtained for potential industrial use [42,47,48]. However, these studies were accomplished without knowledge of the high-resolution structures of CESHs; therefore, there is potential to further improve the use of CESHs as biocatalysts in industrial applications.

3. Bacteria that Produce CESHs

CESHs catalyze the enzymatic hydrolysis of *cis*-epoxysuccinate to form L(+)-TA or D(−)-TA with high product enantioselectivity. TA products obtained by the hydrolysis of CES using purified CESHs generally have enantiomeric excess (EE) values of near 100% [19,36,38,39,44,45]. Therefore, bacteria that produce CESH[D] or CESH[L] have been separately reported. More than twenty species have been isolated by researchers and are distributed among more than ten genera (Table 1). Isolated species with CESH[L] activity include both Gram-positive and Gram-negative bacteria, while to date, all species with CESH[D] activity are Gram-negative. Only the genus *Pseudomonas* has both type of species.

Table 1. Strains producing *cis*-epoxysuccinic acid hydrolases (CESHs).

CESH	Genus	Species	Gene	Reference
CESH[L]	*Acetobacter*	*Acetobacter curtus*		[26]
	Acinetobacter	*Acinetobacter tartarogenes*		[23]
	Agrobacterium	*Agrobacterium aureum*		[22]
		Agrobacterium viscosum		[22]
	Corynebacterium	*Corynebacterium* S-13		[26]
		Corynebacterium sp. JZ-1		[28]
	Klebsiella	*Klebsiella* sp. BK-58	KF977193	[38,43]
	Labrys	*Labrys sp.* BK-8		[39]
	Nocardia	*Nocardia tartaricans*		[24]
		Nocardia tartaricans SW13-57		[29]
		Nocardia tartaricans CAS-52	JQ267565	[37]
	Pseudomonas	*Pseudomonas* sp. KB-86		[22]
	Rhizobium	*Rhizobium validum*		[22,27]
	Rhodococcus	*Rhodococcus opacus*	DQ471957	[17]
		Rhodococcus ruber M1		[34]
		Rhodococcus rhodochrous		[3]
CESH[D]	*Achromobacter*	*Achromobacter tartarogenes*		[21]
		Achromobacter epoxylyticus		[21]
		Achromobacter acinus		[21]
		Achromobacter sericatus		[21]
	Alcaligenes	*Alcaligenes epoxylyticus*		[21]
		Alcaligenes margaritae		[21]
		Alcaligenes sp. MCI3611	[1]	[33,40]
	Bordetella	*Bordetella* sp. strain 1–3		[35]
		Bordetella sp. BK-52	EU053208	[19,36]
	Pseudomonas	*Pseudomonas putida*		[31]

[1] The gene and protein sequence were reported in a patent [40].

More than ten species with CESH[L] activity have been reported since the 1970s. Miura et al. discovered that microorganisms of the genus *Nocardia* can produce CESH[L]; specifically *Nocardia tartaricans nov. sp.* was identified as a preferred natural species [24]. Kamatani et al. also isolated microorganisms capable of hydrolyzing *cis*-epoxysuccinate to L(+)-TA belonging to the genera *Pseudomonas*, *Agrobacterium*, and *Rhizobium* [22]. Two strains of *Rhizopus validum* and *Corynebacterium* JZ-1 that were discovered in soil can produce L(+)-TA, and the latter has a molar conversion rate of *cis*-epoxysuccinate as high as 96% [27,28]. Screening of 65 *Nocardia* strains identified *Nocardia* sp. SW 13-57 as a high-yield strain with the ability to produce CESH[L]. Its molar conversion rate is over 90% and the CESH[L] formation is effectively induced by sodium *cis*-epoxysuccinate during fermentation [29]. In addition, *Rhodococcus ruber* M1 isolated from soil was the first strain in *Rhodococcus* reported to produce CESH[L] [34]. Further, a strain of *Klebsiella* sp. BK-58 can produce a novel CESH[L] with good thermal and pH stability, which is a promising biocatalyst for the industrial production of L(+)-TA [38].

Ten species in four genera have been reported to have CESH CESH[D] activity. Sato et al. first isolated four novel species of *Achromobacter* and two novel species of *Alcaligenes* with CESH[D] activity [21]. A strain belonging to the genus *Pseudomonas* and the microbial cells of *Alcaligenes* sp. MCI3611 also has the capability to hydrolyze *cis*-epoxysuccinate to D(−)-TA [31,33]. The DNA fragment encoding the enzyme to produce D(−)-TA was successfully obtained from the chromosomal DNA library of *Alcaligenes* sp. MCI3611 [40]. Two strains from the genus *Bordetella* (*Bordetella* sp. strain 1-3 and *Bordetella* sp. BK-52), isolated from vegetable fields in Hangzhou, were able to transform *cis*-epoxysuccinate to D(−)-TA [35,36]. Unlike traditional *Bordetella* species that are exclusively associated with humans and warm-blooded animals, both of these strains are from the natural environment. Unlike previously reported CESH[D] producing bacteria, *Bordetella* sp. strain 1-3 has also been reported to have the ability to degrade D(−)-TA as its carbon source, so some measures should be adopted to stop this degradation process to accumulate D(−)-TA. Furthermore, the molecular weight of CESH[D] from *Bordetella* sp. strain 1-3 is the same as the beta subunit of the previously reported CESH[D] from *Alcaligenes* sp. The eight amino acid sequence of the N-terminal region of CESH[D] from *Bordetella* sp. strain 1-3 has also been shown to have the same sequence as the beta subunit from *Alcaligenes* sp. [19].

The gene sequences of several CESH-producing bacteria have been reported, including CESH[L] genes from *Rhodococcus opacus*, *Nocardia tartaricans* CAS-52, and *Klebsiella* sp. BK-58, and CESH[D] genes from *Alcaligenes sp.* MCI3611 and *Bordetella* sp. BK-52 [17,19,37,40,43]. Some of these genes have been successfully expressed in *Escherichia coli* [17,19,37,43]. The amino acid sequences of CESH[L] from *Rhodococcus opacus* and *Nocardia tartaricans* CAS-52 are identical, but they only share 36% sequence identity with the CESH[L] from *Klebsiella* sp. BK-58. The two CESH[D]s from *Alcaligenes* sp. MCI3611 and *Bordetella* sp. BK-52 have identical amino acid sequences. The amino acid sequences of CESHs from most bacteria in Table 1 are still unknown.

4. Stability of CESHs

Although CESHs have excellent product enantioselectivity and high activity, pure CESHs are unstable and heat-sensitive and are thus unsuitable for industrial applications [3,17,49]. A continuous bioconversion study using *Rhodococcus rhodochrous* showed that the effect of the large increase in stability at a lower temperature was much more important than the decrease in activity [3]. To improve the stability of CESHs at the optimal pH and temperature, whole-cell immobilization was adopted for the industrial bioconversion process. Carriers including gelatin beads, pectate gel beads, and κ-carrageenan were screened, and the process of immobilization was optimized for different species [30,49–54]. Whole-cell immobilization was shown to not only increase the stability of the biocatalysts, but also improve the activity and conversion ratio.

CESH[L] and CESH[D] show different stabilities. CESH[D] from *Bordetella* sp. BK-52 has high stability and activity in a broad range of temperatures (37–45 °C) and pH values (4.6–9.0) with optimal conditions being 40 °C and pH 6.5 [19]. A comparison study indicated that CESH[D] has greater thermal and pH stability than CESH[L] [41]. However, the recently discovered novel CESH[L]s from *Klebsiella* sp. and *Labrys* sp. BK-8 have good thermal and pH stability. The former is stable up to 50 °C and at pH 5 to 11, while the latter is stable over a broad range of temperatures and pH values with the greatest activity occurring at 50 °C and 8.5 [38,39]. Therefore, they could be used as alternative biocatalysts for the production of L(+)-TA.

In addition, the stability and activity of CESHs can also be improved by protein engineering methods such as fusing a binding module to CESH or changing the protein primary structure. For example, the wild-type CESH[L] gene from *Rhodococcus opacus* has been fused with a carbohydrate binding module (CBM30), and the resulting fusion enzyme (CBM30-CESH) exhibited improved temperature and pH adaptability than free native CESH[L] [55]. The CESH[L] mutant 5X-1 from *Rhodococcus opacus* was successfully constructed by combining directed evolution with various semi-rational redesign methods. The optimal reaction temperature using mutant 5X-1 occurred at 55 °C, which is much higher than the optimal temperature at 35 °C using the wild-type enzyme. The pH range for the effective working of mutant 5X-1 extended from 8.0–9.0 to 5.0–10.0 [47]. Random mutation by error-prone PCR and high throughput screening revealed that single point mutations on the Phe10 residue of CESH[L] from *Klebsiella* sp. BK-58 resulted in different levels of enzyme thermostability and catalytic activity. The mutant F10Q had 230% higher activity but lower stability than the wild-type enzyme [48].

5. Process Optimization for TA Production Using CESHs

Since the discovery of CESH activity in the 1970s, CESH-producing bacteria and CESHs have been utilized to produce TA with high enantiopurity. The cell lysate or crude enzyme solution is not suitable for TA production because the cellular protease degrades CESH rapidly. Therefore, significant effort has been made to optimize TA production using CESHs or CESH-producing bacteria. TA production was established in the 1970s including the surfactant addition and recovery of TA from the media [23,25]. Subsequent studies have found that these enzymes have low stability [3], so studies on process optimization have mainly focused on the methods of immobilization and recombination.

Bacterial cells can be immobilized by different carriers and show different levels of efficiency and stability. Different microorganisms have different optimal immobilization methods, for example, the best cell immobilization carriers for *Nocardia tartaricans*, *Corynebacterium*, and *Rhizobium* are gelatin, κ-carrageenan, and sodium alginate, respectively [30,52,56]. The immobilization of *Labrys* and recombinant *E. coli* cells with carrageenan is also an excellent process for TA production with high efficiency and stability [39,54]. Aside from the carriers, the subsequent processes of immobilization also have important effects on both the activity and stability. Rosenberg et al. found that although κ-carrageenan is an excellent carrier for the immobilization of *Nocardia tartaricans* cells, the use of cross-linked calcium pectate gel (CPG) is advantageous for the preparation of spherical particles with high activity and stability [50,51]. Sodium alginate–cellulose sulfate-poly(methylene-*co*-guanidine) (SA-CS/PMCG) capsules have been shown immobilize *Nocardia tartaricans* with a better performance than CPG [49]. Additionally, various surfactants can greatly enhance the activity of the immobilized cells, mainly through a change in the permeability of the cell membrane [50,52,57]. Dong et al. reported that ultrasound treatment could be used to change the cell permeability and improve the bioconversion efficiency of immobilized *E. coli* cells containing expressed recombinant CESH[D] [58].

The genes of CESH[L] and CESH[D] have been cloned and expressed in *E. coli* successfully [17,19,40,41,43]. Therefore, recombinant CESHs also have good potential to be used in industrial TA production. Some studies have reported that the stability of enzymes can be improved by immobilization. For example, we significantly improved the stability of CESH[L] by fusing it with CBM [55], and then purified and immobilized it on cellulose in one step. Wang et al. immobilized CESH[L] on agarose Ni-IDA to enhance its stability [59]. The preparation of recombinant CESH[L] was also improved by the utilization of the heat-induced promoter to avoid the chemical induction of protein expression [60]. Still, more effort is needed to optimize the stability of recombinant CESH and the process of TA production using the recombinant enzyme.

6. Structure and Catalytic Mechanism of CESH[L]

Until now, there have been no reports on the structure of CESH[L]. Only two CESH[L] sequences have been reported and they share a 36% sequence identity [17,37,43]. Both have about a 30% sequence identity with L-2-haloacid dehalogenase, of which the structure is known. Therefore, homology modeling has been performed to elucidate the catalytic mechanism of CESH[L] [41,42,44]. L-2-haloacid dehalogenase has an α/β hydrolase fold that is adopted by most EHs [61]. Therefore, it is likely that CESH[L] also adopts the α/β hydrolase fold, and the catalytic mechanism of CESH[L] has been proposed to be similar to most EHs, i.e., a two-step mechanism including an ester intermediate [61]. The two-step mechanism was confirmed by ^{18}O experiments for both CESH[L]s [43,44].

As CESH[L] and L-2-haloacid dehalogenase only have about a 30% sequence identity and different substrates, the catalytic residues cannot be deduced from homology modeling. Two mutagenesis analyses revealed that D18, H190, and D193 are essential for the activity; therefore, they were proposed to be a catalytic triad, with D18 activating the attacking water molecule [42,44]. However, as the CES substrate is a polar hydrophilic molecule, the active site of CESH[L] contains many charged and hydrophilic residues, making it difficult to elucidate the role of each residue without a high-resolution crystal structure of the CESH[L]-substrate complex. The proton donor that may facilitate the ring opening is still not clear, and how the CES is fixed in the active site to ensure the stereoselectivity also remains unknown.

7. Structure and Catalytic Mechanism of CESH[D]

Although CESH[D] and its gene sequence were reported earlier than CESH[L], knowledge of CESH[D] catalysis was not obtained until very recently. The protein sequences of CESH[D]s from *Bordetella* sp. BK-52 and *Alcaligenes* sp. MCI3611 have about 30% sequence identity to the Kce enzyme, whose function is totally different [19,40,41]. Homology modeling using Kce as a template indicated that CESH[D] has a TIM barrel fold with a metal ion which is crucial for its activity [41,45]. The metal ion was identified to be a divalent ion, either zinc, calcium, or magnesium [41,45], which is coordinated by three residues in Kce; however, only two of these are conserved in CESH[D] [41,45]. The third coordinative residue could not be identified before the CESH[D] structure was determined. An ^{18}O labeling experiment indicated that CESH[D] hydrolyzes CES through a one-step mechanism [45] instead of the two-step mechanism of CESH[L]. As CESH[D] and Kce have different substrates, their active sites are very different and their catalytic residues cannot be deduced by homology modeling. Mutagenesis studies have identified a large number of essential residues, so it is difficult to identify the key catalytic residue and stereoselective catalytic mechanism from these studies without a high-resolution CESH[D] structure.

The catalytic mechanism of CESH[D] was not elucidated in detail until our recent report on high-resolution CESH[D] structures [46]. The structure of substrate-free CESH[D] revealed not only the third metal coordinative residue (Glu14), but also three coordinative water molecules that formed an equilateral triangle. Trials using an inactive mutant and CES co-crystallization obtained an unexpected CESH[D] structure in complex with its reaction product, D(−)-TA. In the complex structure, three oxygen atoms of TA occupy the positions of the three coordinative water molecules in the substrate-free CESH[D] structure. The identification of the structure of the product–enzyme complex provided the details of substrate binding and positioning, from which the key catalytic residues and substrate recognition residues were elucidated. Instead of the previously proposed catalytic residue D251 [45], the catalytic residues were identified as D115 and E190, while R11 provided the proton and facilitated the ring opening. D251 played a crucial role in fixing the position of R11, which supports the importance of this residue, as identified in the previous mutagenesis analysis. This structure and catalytic mechanism explained the stereoselectivity, regioselectivity, and substrate selectivity of CESH[D] [46].

The structure of CESH[D] has some distinct features in comparison with known EHs. In contrast with the α/β hydrolase fold or LEH fold adopted by most EHs, except for LA4H, ChEH, and FosX, CESH[D] adopts a TIM-barrel fold [61]. Also, CESH[D] has a one-step mechanism, while α/β hydrolase fold EHs have a two-step mechanism. LEH adopts a one-step mechanism, but the substrate specificity is determined by the hydrophobic interaction (molecular shapes), and no metal ion is needed for LEH [62]. The EH with the most similar mechanism is FosX, which also contains a metal ion for substrate binding and adopts a one-step mechanism (Figure 1) [63]. However, FosX has a dimeric VOC family fold with paired $\beta\alpha\beta\beta\beta$ where the active sites are located at the dimer interface, and the substrate of FosX only occupies two coordination sites of the metal ion with square pyramidal coordination geometry [63]. Although CESH[D] is also a dimeric protein, its active sites are located at the center of each dimer's subunit, and its metal elements have octahedral coordination geometry where three of the coordination sites are occupied by the substrates [46]. Therefore, CESH[D] is a unique EH in terms of both its protein fold and catalytic mechanism.

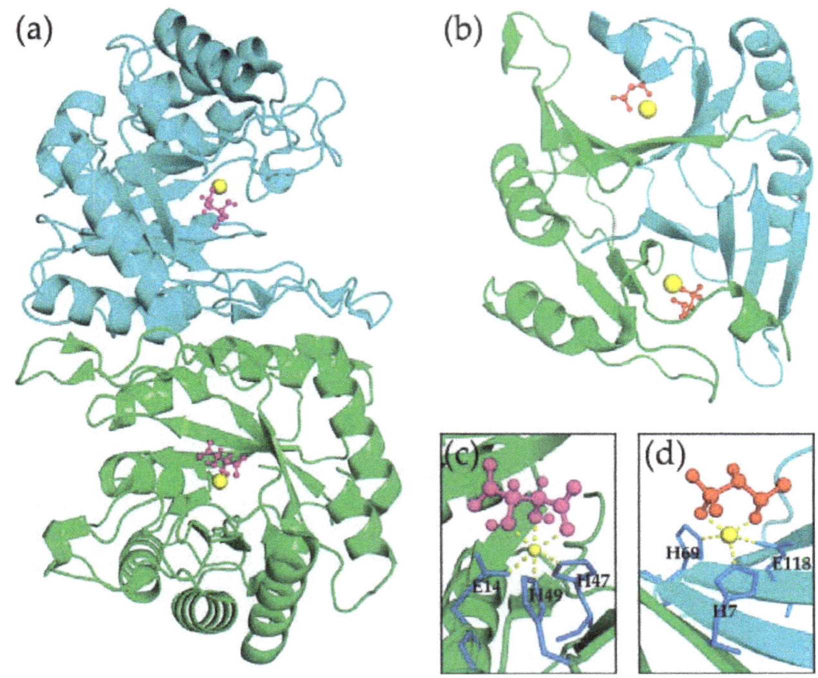

Figure 1. Comparison of CESH[D] and FosX: (**a**) structure of CESH[D]; (**b**) structure of FosX; (**c**) the active site of CESH[D]; (**d**) the active site of FosX. Both proteins are dimeric and colored in green and cyan for each monomer. The metal ions are shown as yellow balls. The products are shown as magenta and red ball and sticks for CESH[D] and FosX, respectively. In (**c**,**d**), the three coordinative residues of the metal ion are shown as blue sticks.

8. Perspective for CESH Research and Application

CESHs are unique among the known EHs because the CES substrate is highly hydrophilic and mirror-symmetric. The structural features of CES suggest that CESHs have specific substrates; in other words, the CES molecule is fixed in CESH with an exact position, which leads to the high stereoselectivity and regioselectivity of CESHs. This feature is of great interest for enantioselective synthesis. Although CESHs and their host bacteria have been successfully applied in industry for TA production, there are still many questions to be addressed, and there is a lot of room for growth in the production of TA using CESHs.

Currently, the structure and the catalytic mechanism of CESH[L] are still not fully understood. Determination of the high-resolution structure of CESH[L], particularly of complexes with a substrate or product, is necessary to elucidate its catalytic mechanism. With this structure, rational engineering to enhance the stability will be possible. Furthermore, CESHs could also potentially be engineered as biocatalysts to catalyze different substrates, but this potential has not been explored in past studies.

Although many microorganisms have been reported to have CESH activity, only very few of them have been sequenced. Therefore, much work is still needed to isolate the CESHs and analyze their protein/gene sequences. The new sequences of CESHs may provide other new features that will help us to understand these enzymes and promote their applications.

Currently, the production of TA using whole cell catalysts is performed by wild strains. No metabolic engineering of these microorganisms has previously been reported. Therefore, understanding the features of these microorganisms and the development of genetic engineering are important topics for future studies. CESHs are intracellular enzymes that are present in certain microorganisms, which cause the activity of whole cell catalysts to depend on the cell permeability. If the CESHs could be engineered as a secretive protein, or immobilized on the cell surface, their activity would be greatly enhanced. This engineering could be done in either the original species or the recombinant *E. coli* cells, and it will improve the production of enantiomerically pure TA using CESHs.

Funding: This research was funded by the National Natural Science Foundation of China (grant numbers 31670735 and 31661143023 to YF); and the Undergraduate Education and Teaching Reform Research Project, USTB (grant number JG2018M39 to JX).

Conflicts of Interest: The authors declare no conflict of interest. The funders had no role in the design of the study; in the collection, analyses, or interpretation of data; in the writing of the manuscript, or in the decision to publish the results.

References

1. Derewenda, Z.S. On wine, chirality and crystallography. *Acta Crystallogr. Sect. A Found. Crystallogr.* **2008**, *64*, 246–258. [CrossRef] [PubMed]
2. Kodama, S.; Yamamoto, A.; Matsunaga, A.; Hayakawa, K. Direct chiral resolution of tartaric acid in food products by ligand exchange capillary electrophoresis using copper(II)-D-quinic acid as a chiral selector. *J. Chromatogr.* **2001**, *932*, 139–143. [CrossRef]
3. Willaert, R.; De Vuyst, L. Continuous production of L(+)-tartaric acid from *cis*-epoxysuccinate using a membrane recycle reactor. *Appl. Microbiol. Biotechnol.* **2006**, *71*, 155–163. [CrossRef] [PubMed]
4. Skarzewski, J.; Gupta, A. Synthesis of C_2 symmetric primary vicinal diamines. Double stereospecific Mitsunobu reaction on the heterocyclic diols derived from tartaric acid. *Tetrahedron Asymmetry* **1997**, *8*, 1861–1867. [CrossRef]
5. Su, Y.L.; Yang, C.S.; Teng, S.J.; Zhao, G.; Ding, Y. Total synthesis of four diastereoisomers of Goniofufurone from D-(−)- or L-(+)-tartaric acid. *Tetrahedron* **2001**, *57*, 2147–2153. [CrossRef]
6. Pabba, J.; Vasella, A. Synthesis of D-gluco-, L-ido-, D-galacto-, and L-altro-configured glycaro-1,5-lactams from tartaric acid. *Tetrahedron Lett.* **2005**, *46*, 3619–3622. [CrossRef]
7. Gawronski, J.; Gawronska, K. *Tartaric and Malic Acids in Synthesis: A Source Book of Building Blocks, Ligands, Auxiliaries, and Resolving Agents*, 1st ed.; Wiley-Interscience: New York, NY, USA, 1999.
8. Ghosh, A.K.; Koltun, E.S.; Bilcer, G. Tartaric acid and tartrates in the synthesis of bioactive molecules. *Synthesis-Stuttgart* **2001**, *2001*, 1281–1301. [CrossRef] [PubMed]
9. Gratzer, K.; Gururaja, G.N.; Waser, M. Towards tartaric-acid-derived asymmetric organocatalysts. *Eur. J. Org. Chem.* **2013**, *2013*, 4471–4482. [CrossRef] [PubMed]
10. Mikova, H.; Rosenberg, M.; Kristofikova, L.; Liptaj, T. Influence of molybdate and tungstate ions on activity of *cis*-epoxysuccinate hydrolase in *Nocardia tartaricans*. *Biotechnol. Lett.* **1998**, *20*, 833–835. [CrossRef]
11. Steinreiber, A.; Faber, K. Microbial epoxide hydrolases for preparative biotransformations. *Curr. Opin. Biotechnol.* **2001**, *12*, 552–558. [CrossRef]
12. Blee, E.; Schuber, F. Stereocontrolled hydrolysis of the linoleic-acid monoepoxide regioisomers catalyzed by soybean epoxide hydrolase. *Eur. J. Biochem.* **1995**, *230*, 229–234. [CrossRef] [PubMed]
13. Linderman, R.J.; Walker, E.A.; Haney, C.; Roe, R.M. Determination of the regiochemistry of insect epoxide hydrolase catalyzed epoxide hydration of juvenile-hormone by ^{18}O-labeling studies. *Tetrahedron* **1995**, *51*, 10845–10856. [CrossRef]
14. Weijers, C.A.G.M.; de Bont, J.A.M. Epoxide hydrolases from yeasts and other sources: Versatile tools in biocatalysis. *J. Mol. Catal. B Enzym.* **1999**, *6*, 199–214. [CrossRef]
15. Arand, M.; Cronin, A.; Adamska, M.; Oesch, F. Epoxide hydrolases: Structure, function, mechanism, and assay. *Methods Enzymol.* **2005**, *400*, 569–588. [PubMed]
16. Faber, K.; Mischitz, M.; Kroutil, W. Microbial epoxide hydrolases. *Acta Chem. Scand.* **1996**, *50*, 249–258. [CrossRef]
17. Liu, Z.; Li, Y.; Xu, Y.; Ping, L.; Zheng, Y. Cloning, sequencing, and expression of a novel epoxide hydrolase gene from *Rhodococcus opacus* in *Escherichia coli* and characterization of enzyme. *Appl. Microbiol. Biotechnol.* **2007**, *74*, 99–106. [CrossRef] [PubMed]
18. Li, X.; Xu, T.; Lu, H.; Ma, X.; Kai, L.; Guo, K.; Zhao, Y. Purification and characterization of a *cis*-epoxysuccinic acid hydrolase from *Bordetella* sp. strain 1-3. *Protein Expression Purif.* **2010**, *69*, 16–20. [CrossRef] [PubMed]

19. Pan, H.; Bao, W.; Xie, Z.; Zhang, J.; Li, Y. Molecular cloning and characterization of a *cis*-epoxysuccinate hydrolase from *Bordetella* sp. BK-52. *J. Microbiol. Biotechnol.* **2010**, *20*, 659–665. [CrossRef] [PubMed]
20. Archelas, A.; Furstoss, R. Epoxide hydrolases: New tools for the synthesis of fine organic chemicals. *Trends Biotechnol.* **1998**, *16*, 108–116. [CrossRef]
21. Sato, E.; Yanai, A. Method for Preparing D-tartaric Acid. U.S. Patent 3,957,579, 18 May 1976. CAN 84:89630.
22. Kamatani, Y.; Okazaki, H.; Imai, K.; Fujita, N.; Yamazaki, Y.; Ogino, K. Production of L(+)-tartaric Acid. U.S. Patent 4,011,135, 8 March 1977. CAN 85:175611.
23. Kamatani, Y.; Okazaki, H.; Imai, K.; Fujita, N.; Yamazaki, Y.; Ogino, K. Method for producing L(+)-tartaric acid. U.S. Patent 4,028,185, 7 June 1977. CAN 85:175611.
24. Miura, Y.; Yutani, K.; Izumi, Y. Process for preparing L-tartaric acid. U.S. Patent 4,010,072, 1 March 1977. CAN 86:15234.
25. Miura, Y.; Yutani, K.; Takesue, H.; Fujii, K. Microbiological process for preparing L-tartaric acid in presence of surfactants. U.S. Patent 4,017,362, 12 April 1977.
26. Tsurumi, Y.; Fujioka, T. Process for manufacture of L(+)-tartaric acid or salts thereof. U.S. Patent 4,092,220, 30 May 1978.
27. Huang, T.; Qian, X. Production of L(+)-tartaric acid. *Ind. Microbiol.* **1990**, *20*, 14–17.
28. Zhang, J.; Huang, T. Productivity of L(+)-tartaric acid using microbial conversion method. *Ind. Microbiol.* **1990**, *20*, 7–12, 24.
29. Zheng, P.; Sun, Z. Enzymatic conversion of *cis*-epoxysuccinate into L(+)-tartarate by *Nocardia* sp. SW13-57. *Ind. Microbiol.* **1994**, *24*, 12–17.
30. Sun, Z.; Zheng, P.; Dai, X.; Li, H.; Jin, M. Production of L(+)-tartaric acid by immobilized *Nocardia tartaricans* SW13-57. *Chin. J. Biotechnol./Shengwu Gongcheng Xuebao* **1995**, *11*, 372–376.
31. Yamagishi, K.; Cho, H.; Takai, Y.; Kawaguchi, T. Production of D(−)-tartaric acid. Jp. Patent 08-245497, 24 September 1996. CAN 126:7698.
32. Lou, J.-F.; Zhang, J.-G. Research on microbial productions of L(+)-tartaric acid. *Food Sci. Technol.* **2006**, *2006*, 162–164.
33. Ikuta, M.; Sakamoto, T.; Ueda, M.; Sashita, R. Production of D(−)-tartaric acid. Jp. Patent 2000-014391, 18 January 2000. CAN 132:92374.
34. Pan, K.-X.; Min, H.; Xia, Y.; Xu, X.-Y.; Ruan, A.-D. Isolation, identification and phylogenetic analysis of *Rhodococcus* sp. strain M1 producing *cis*-epoxysuccinate hydrolase and optimization of production conditions. *Acta Microbiol. Sin.* **2004**, *44*, 276–280.
35. Li, X.; Ma, X.; Zhao, Y.; Jia, X.; Kai, L.; Guo, K.; Zhao, W. Isolation and characterization of a new bacterium capable of biotransforming *cis*-epoxysuccinic acid to D(−) -tartaric acid. *FEMS Microbiol. Lett.* **2007**, *267*, 214–220. [CrossRef] [PubMed]
36. Pan, H.; Xie, Z.; Bao, W.; Zhang, J. Isolation and identification of a novel cis-epoxysuccinate hydrolase producing *Bordetella* sp. BK-52 and optimization of enzyme production. *Acta Microbiol. Sin.* **2008**, *48*, 1075–1081.
37. Wang, Z.; Wang, Y.; Su, Z. Purification and characterization of a *cis*-epoxysuccinic acid hydrolase from *Nocardia tartaricans* CAS-52, and expression in *Escherichia coli*. *Appl. Microbiol. Biotechnol.* **2013**, *97*, 2433–2441. [CrossRef] [PubMed]
38. Cheng, Y.; Wang, L.; Pan, H.; Bao, W.; Sun, W.; Xie, Z.; Zhang, J.; Zhao, Y. Purification and characterization of a novel *cis*-epoxysuccinate hydrolase from *Klebsiella* sp. that produces L(+)-tartaric acid. *Biotechnol. Lett.* **2014**, *36*, 2325–2330. [CrossRef] [PubMed]
39. Bao, W.; Pan, H.; Zhang, Z.; Cheng, Y.; Xie, Z.; Zhang, J. Isolation of the stable strain *Labrys* sp. BK-8 for L(+)-tartaric acid production. *J. Biosci. Bioeng.* **2015**, *119*, 538–542. [CrossRef] [PubMed]
40. Asai, Y.; Kobayashi, M.; Uchida, K.; Terasawa, M. DNA encoding cis-epoxysuccinic acid hydrolase and its utilization. Jp. Patent 2000-295992, 24 October 2000. CAN 133:307122.
41. Cui, G.Z.; Wang, S.; Li, Y.; Tian, Y.J.; Feng, Y.; Cui, Q. High yield recombinant expression, characterization and homology modeling of two types of *cis*-epoxysuccinic acid hydrolases. *Protein J.* **2012**, *31*, 432–438. [CrossRef] [PubMed]
42. Vasu, V.; Kumaresan, J.; Babu, M.G.; Meenakshisundaram, S. Active site analysis of *cis*-epoxysuccinate hydrolase from *Nocardia tartaricans* using homology modeling and site-directed mutagenesis. *Appl. Microbiol. Biotechnol.* **2012**, *93*, 2377–2386. [CrossRef] [PubMed]

43. Cheng, Y.; Pan, H.; Bao, W.; Sun, W.; Xie, Z.; Zhang, J.; Zhao, Y. Cloning, homology modeling, and reaction mechanism analysis of a novel *cis*-epoxysuccinate hydrolase from *Klebsiella* sp. *Biotechnol. Lett.* **2014**, *36*, 2537–2544. [CrossRef] [PubMed]
44. Pan, H.; Xie, Z.; Bao, W.; Cheng, Y.; Zhang, J.; Li, Y. Site-directed mutagenesis of epoxide hydrolase to probe catalytic amino acid residues and reaction mechanism. *FEBS Lett.* **2011**, *585*, 2545–2550. [CrossRef] [PubMed]
45. Bao, W.; Pan, H.; Zhang, Z.; Cheng, Y.; Xie, Z.; Zhang, J.; Li, Y. Analysis of essential amino acid residues for catalytic activity of *cis*-epoxysuccinate hydrolase from *Bordetella* sp. BK-52. *Appl. Microbiol. Biotechnol.* **2014**, *98*, 1641–1649. [CrossRef] [PubMed]
46. Dong, S.; Liu, X.; Cui, G.Z.; Cui, Q.; Wang, X.; Feng, Y. Structural insight into the catalytic mechanism of a *cis*-epoxysuccinate hydrolase producing enantiomerically pure D(−)-tartaric acid. *Chem. Commun.* **2018**, *54*, 8482–8485. [CrossRef] [PubMed]
47. Qiao, P.; Wu, M.; Zhu, L.; Zhang, Y.; Yang, L.; Fei, H.; Lin, J. Enhancing the thermal tolerance of a *cis*-epoxysuccinate hydrolase *via* combining directed evolution with various semi-rational redesign methods. *J. Mol. Catal. B Enzym.* **2015**, *121*, 96–103. [CrossRef]
48. Zhang, C.; Pan, H.; Yao, L.; Bao, W.; Wang, J.; Xie, Z.; Zhang, J. Single point mutations enhance activity of *cis*-epoxysuccinate hydrolase. *Biotechnol. Lett.* **2016**, *38*, 1301–1306. [CrossRef] [PubMed]
49. Bučko, M.; Vikartovská, A.; Lacík, I.; Kolláriková, G.; Gemeiner, P.; Pätoprstý, V.; Brygin, M. Immobilization of a whole-cell epoxide-hydrolyzing biocatalyst in sodium alginate−cellulose sulfate−poly(methylene-co-guanidine) capsules using a controlled encapsulation process. *Enzyme Microb. Technol.* **2005**, *36*, 118–126. [CrossRef]
50. Rosenberg, M.; Miková, H.; Krištofíková, L. Production of L-tartaric acid by immobilized bacterial cells *Nocardia tartaricans*. *Biotechnol. Lett.* **1999**, *21*, 491–495. [CrossRef]
51. Kurillová, L.; Gemeiner, P.; Vikartovská, A.; Miková, H.; Rosenberg, M.; Ilavský, M. Calcium pectate gel beads for cell entrapment. 6. Morphology of stabilized and hardened calcium pectate gel beads with cells for immobilized biotechnology. *J. Microencaps.* **2000**, *17*, 279–296.
52. Zhang, J.-G.; Qian, Y.-J. Production of L(+)-tartaric acid by immobilized *Corynebacterium* sp. JZ-1. *Chin. J. Biotechnol./Shengwu Gongcheng Xuebao* **2000**, *16*, 188–192.
53. Vikartovská, A.; Bučko, M.; Gemeiner, P.; Nahálka, J.; Pätoprstý, V.; Hrabárová, E. Flow calorimetry—A useful tool for determination of immobilized *cis*-epoxysuccinate hydrolase activity from *Nocardia tartaricans*. *Artif. Cells Blood Substit. Biotechnol.* **2004**, *32*, 77–89. [CrossRef]
54. Wang, Z.; Wang, Y.; Shi, H.; Su, Z. Improvement of the production efficiency of L-(+)-tartaric acid by heterogeneous whole-cell bioconversion. *Appl. Biochem. Biotechnol.* **2014**, *172*, 3989–4001. [CrossRef] [PubMed]
55. Wang, S.; Cui, G.Z.; Song, X.F.; Feng, Y.; Cui, Q. Efficiency and stability enhancement of *cis*-epoxysuccinic acid hydrolase by fusion with a carbohydrate binding module and immobilization onto cellulose. *Appl. Biochem. Biotechnol.* **2012**, *168*, 708–717. [CrossRef] [PubMed]
56. Lan, X.; Bao, W.; Pan, H.; Xie, Z.; Zhang, J. Production of L(+)-tartaric acid by immobilized *Rhizobium* strain BK-20. *Chin. J. Biotechnol./Shengwu Gongcheng Xuebao* **2014**, *32*, 315–319.
57. Jin, T.-L.; Zhang, P.-D. Studies on increasing turnover rate of endoenzyme of immobilized cells by permealilizing and crosslinking method. *Ind. Microbiol.* **2003**, *33*, 14–19, 22.
58. Dong, W.; Zhao, F.; Xin, F.; He, A.; Zhang, Y.; Wu, H.; Fang, Y.; Zhang, W.; Ma, J.; Jiang, M. Ultrasound-assisted D-tartaric acid whole-cell bioconversion by recombinant *Escherichia coli*. *Ultrason. Sonochem.* **2018**, *42*, 11–17. [CrossRef] [PubMed]
59. Wang, Z.; Su, M.; Li, Y.; Wang, Y.; Su, Z. Production of tartaric acid using immobilized recominant *cis*-epoxysuccinate hydrolase. *Biotechnol. Lett.* **2017**, *39*, 1859–1863. [CrossRef] [PubMed]
60. Wang, Z.; Wang, Y.; Shi, H.; Su, Z. Expression and production of recombinant *cis*-epoxysuccinate hydrolase in *Escherichia coli* under the control of temperature-dependent promoter. *J. Biotechnol.* **2012**, *162*, 232–236. [CrossRef] [PubMed]
61. Widersten, M.; Gurell, A.; Lindberg, D. Structure-function relationships of epoxide hydrolases and their potential use in biocatalysis. *Biochim. Biophys. Acta* **2010**, *1800*, 316–326. [CrossRef] [PubMed]
62. Arand, M.; Hallberg, B.M.; Zou, J.; Bergfors, T.; Oesch, F.; van der Werf, M.J.; de Bont, J.A.M.; Jones, T.A.; Mowbray, S.L. Structure of *Rhodococcus erythropolis* limonene-1,2-epoxide hydrolase reveals a novel active site. *EMBO J.* **2003**, *22*, 2583–2592. [CrossRef] [PubMed]

63. Fillgrove, K.L.; Pakhomova, S.; Schaab, M.R.; Newcomer, M.E.; Armstrong, R.N. Structure and mechanism of the genomically encoded fosfomycin resistance protein, FosX, from *Listeria monocytogenes*. *Biochemistry* **2007**, *46*, 8110–8120. [CrossRef] [PubMed]

molecules

MDPI

Review

Synthetic Chiral Derivatives of Xanthones: Biological Activities and Enantioselectivity Studies

Carla Fernandes [1,2,*], Maria Letícia Carraro [1,†], João Ribeiro [1,†], Joana Araújo [1,†], Maria Elizabeth Tiritan [1,2,3,*] and Madalena M. M. Pinto [1,2]

1 Laboratório de Química Orgânica e Farmacêutica, Departamento de Ciências Químicas, Faculdade de Farmácia, Universidade do Porto, Rua de Jorge Viterbo Ferreira, 228, 4050-313 Porto, Portugal; leticiacarraro@hotmail.com (M.L.C.); joaobigi@gmail.com (J.R.); up201703236@ff.up.pt (J.A.); madalena@ff.up.pt (M.M.M.P.)

2 Interdisciplinary Centre of Marine and Environmental Research (CIIMAR), Edifício do Terminal de Cruzeiros do Porto de Leixões, Av. General Norton de Matos s/n, 4050-208 Matosinhos, Portugal

3 Cooperativa de Ensino Superior, Politécnico e Universitário (CESPU), Instituto de Investigação e Formação Avançada em Ciências e Tecnologias da Saúde (IINFACTS), Rua Central de Gandra, 1317, 4585-116 Gandra PRD, Portugal

* Correspondence: elizabeth.tiritan@iucs.cespu.pt (M.E.T.); cfernandes@ff.up.pt (C.F.); Tel.: +351-22-415-7224 (M.E.T.); +351-22-042-8688 (C.F.)

† These authors contributed equally to this work.

Academic Editor: Maurizio Benaglia

Received: 1 February 2019; Accepted: 19 February 2019; Published: 22 February 2019

Abstract: Many naturally occurring xanthones are chiral and present a wide range of biological and pharmacological activities. Some of them have been exhaustively studied and subsequently, obtained by synthesis. In order to obtain libraries of compounds for structure activity relationship (SAR) studies as well as to improve the biological activity, new bioactive analogues and derivatives inspired in natural prototypes were synthetized. Bioactive natural xanthones compromise a large structural multiplicity of compounds, including a diversity of chiral derivatives. Thus, recently an exponential interest in synthetic chiral derivatives of xanthones (CDXs) has been witnessed. The synthetic methodologies can afford structures that otherwise could not be reached within the natural products for biological activity and SAR studies. Another reason that justifies this trend is that both enantiomers can be obtained by using appropriate synthetic pathways, allowing the possibility to perform enantioselectivity studies. In this work, a literature review of synthetic CDXs is presented. The structures, the approaches used for their synthesis and the biological activities are described, emphasizing the enantioselectivity studies.

Keywords: synthetic xanthones; chiral derivatives of xanthones; bioactivities; enantioselectivity; enantiomeric purity

1. Introduction

In the last few years, the relationship between chirality and biological activity has been of increasing importance in Medicinal Chemistry [1]. Chirality can now be considered as one of the majors' topics in the design, discovery, development and marketing of new drugs [2–4]. Enantiomers of drugs often present different behaviours within pharmacodynamics and/or pharmacokinetics events [5–7] as well as different levels of toxicity [8–10]. These differences makes in the majority of cases therapeutics with single enantiomers of unquestionable advantages [11].

The importance of enantioselective studies and the increase of chiral drugs in the pharmaceutical market upsurges each year due to the advantages in potency, efficacy, selectivity and safety associated with the use of single enantiomers. The advances in enantioselective synthesis [12–15] as well as enantioresolution methodologies [16–19] aligned to the stricter requirements from regulatory authorities to patent new chiral drug [20–22] boosted the research in this field.

The development of methods to obtain and analyse both pure enantiomers has acquired crucial importance in the early stage of drug development, as biological and pharmacological activity evaluation of both enantiomers are required. Thus, despite the diversity of structures with different groups and positions at the base scaffolds, the library of compounds for structure-activity relationship (SAR) studies should also include stereoisomers for enantioselectivity evaluation. To perform that type of studies, it is necessary to obtain both enantiomers with very high enantiomeric purity. The improvement of chromatographic instrumentation and the development of efficient chiral stationary phases (CSPs) [23–26] have made liquid chromatography (LC) the first choice for determination of enantiomeric purity.

Chemically, xanthones (9*H*-xanthen-9-ones) are compounds with an oxygen-containing dibenzo-γ-pyrone heterocyclic scaffold [27], being considered as a privileged structure [28,29]. Within this class of compounds a broad range of biological and pharmacological activities has been reported [30–34]. Additionally, other applications have been described for xanthone derivatives, such as preparation of fluorescent probes [35,36] or stationary phases for LC [23,24,37].

Many naturally occurring xanthones, isolated from terrestrial as well as marine sources, are chiral, and present interesting biological activities [38–40]. The biosynthetic pathway of xanthones only allows the presence of specific groups in particular positions of the xanthone scaffold, which is a limiting factor for structural diversity. For this reason, in order to enlarge chemical space in this field, total synthesis needs to be considered [41,42] allowing the access to structures that otherwise could not be reached only with natural product as a basis for molecular modification. Moreover, higher number of compounds can be obtained for SAR studies.

2. Synthetic Chiral Derivatives of Xanthones

Recently, there has been an increase interest in new bioactive xanthones obtained by synthesis, particularly in chiral derivatives of xanthones (CDXs). Some reasons that can justify this trend were the importance of this class of compounds in Medicinal Chemistry align with the fact that nature usually gives only one enantiomer and the synthetic procedures allow the preparation of both enantiomers to explore the enantioselectivity in biological screening assays. The promising biological and pharmacological activities of some chiral members of this family, the clinical advantages of a single enantiomer than a racemate, the scarce examples of synthetic CDXs described, and the possibility to perform enantioselectivity studies strengthens the obtaining of new synthetic CDXs. This review aims to gather the research findings on synthetic CDXs, reporting their biological and pharmacological activities. The structures and the bioactivity differences associated to the stereochemistry of the CDXs (enantioselectivity) as well as their enantiomeric purity are highlighted.

2.1. *Synthetic CDXs Inspired in Naturally Occurring Xanthones*

Natural compounds have always been a source of inspiration for the discovery of new therapeutic agents [43]. Historically, the first proposed synthesis of a xanthone was achieved by Michael, in 1883, and later by Kostanecki and Nessler, in 1891, through the distillation of *O*-hydroxy-benzoic acid, acetic anhydride and a phenol [44,45], while the first total synthesis of a naturally occurring xanthone was a euxanthone, described by Ullmann and Panchaud, in 1906 [46]. Several synthetic CDXs analogues of natural xanthones are described. The biological and pharmacological activities evaluated are summarized in Table 1.

2.1.1. Synthetic Xanthonolignoids

Xanthonolignoids are a natural class of compounds isolated from plants of the *Clusiaceae* family (*Guttiferae*) [47]. They possess a phenylpropane core linked to a xanthone scaffold by a dioxane ring, formed by radical oxidative coupling [48,49]. Natural xanthonolignoids include kielcorins, cadensins, subalatin, calophyllumins and gemixanthone [49]. The first xanthonolignoid described was based in 2,3,4-trioxygenated xanthone being isolated, in 1969, from *Kielmeyera species* [50].

Considering that xanthonolignoids are bioactive molecules and very interesting templates for molecular modifications, several xanthonolignoids were isolated and synthesized [49]. Initially the main goal of their synthesis was to help in the structure elucidation of this class of compounds but subsequently, also to improve their biological and physicochemical properties. Both classic synthesis and biomimetic approaches have been used to obtain xanthonolignoids, mainly kielcorin derivatives [49]. The total synthesis of kielcorin derivatives requires several steps and drastic reaction conditions while the biomimetic way is based on natural building blocks and is achieved by an oxidative coupling of a suitable dihydroxyxanthone and a cinnamyl alcohol derivative, in the presence of an oxidizing agent at room temperature [49].

Pinto et al. [51], in 1987, reported the first biomimetic synthesis of xanthonolignoids of the kielcorin group, specifically kielcorin (**1**) and its stereoisomer, *cis*-kielcorin (**2**) (Figure 1). Both kielcorins **1** and **2** were also obtained by a classical method [52,53]. In 1999, by using a biomimetic approach, the synthesis of *trans*-kielcorin B (**3**) and *trans*-isokielcorin B (**4**) (Figure 1) was described [54].

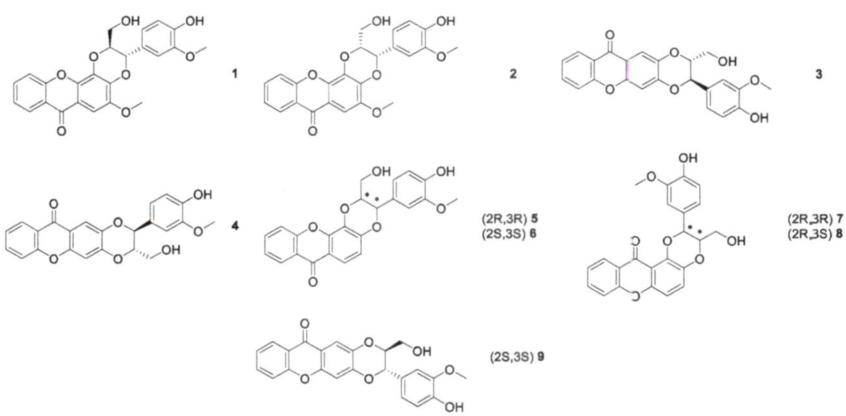

1 2 3 4 (2R,3R) 5 (2S,3S) 6 (2R,3R) 7 (2R,3S) 8 (2S,3S) 9

Figure 1. Structures of kielcorin (**1**) and synthetic derivatives **2–9**.

To obtain related bioactive compounds with a kielcorin framework, other constitutional isomers were synthesized by our group, namely *trans*-kielcorin C (**5**), *trans*-kielcorin D (**6**), *trans*-isokielcorin D (**7**), *cis*-kielcorin C (**8**), and *trans*-kielcorin E (**9**) [55] (Figure 1). Once again, the synthetic approach, involving an oxidative coupling of coniferyl alcohol with an appropriate xanthone, was modelled on a biomimetic pathway. Different oxidizing agents were used (e.g., Ag_2O, Ag_2CO_3, and $K_3[Fe(CN)_6]$) to investigate the oxidative coupling reactions. The synthesis of *trans*-kielcorin C (**5**) (also designated as demethoxykielcorin) using a classic approach was previously reported by Vishwakarma et al. [53].

Kielcorins **5–9** were evaluated for their *in vitro* effect on the growth of three human tumor cell lines, MCF-7 (breast), TK-10 (renal), UACC-62 (melanoma), and on the proliferation of human lymphocytes [56]. The growth inhibitory effect was moderate but dose-dependent and influenced by the isomerism of the tested compounds. The *trans*-kielcorins C (**5**) and D (**6**) were the most active. The inhibition of human lymphocyte proliferation induced by phytohemagglutinin (PHA) was detected [56]. The high potency and selectivity observed for these compounds suggested that kielcorins may be an important model for developing potent and isoform-selective protein kinase C (PKC) inhibitors [31]. Accordingly, kielcorins **5-9** revealed an effect compatible with PKC inhibition similar to that exhibited by the well-established PKC inhibitor chelerythrine [57]. The *trans*-kielcorins C (**5**) and

E (**9**) were evaluated and demonstrated protectives effects against *tert*-butylhydroperoxide-induced toxicity in freshly isolated rat hepatocytes [58].

In order to study if the growth inhibitory effects of the kielcorins **5–9** depended on the stereochemistry, analytical LC methods using four carbamates of polysaccharide derivative CSPs and multimodal elution conditions were developed for their enantioresolution [59]. An amylose *tris*-3,5-dimethylphenylcarbamate CSP was chosen for a preparative resolution scale-up considering not only the highest enantioselectivity obtained for these chiral compounds, but also due to its low retention factors [59]. Consequently, the enantiomers of the kielcolins **5–7** and **9** were efficiently separated by chiral LC on a multimilligram scale. A solid-phase injection system was developed and combined with a closed loop recycling system to increase the productivity and recovery of the preparative process [60]. The enantiomeric purity was also measured being higher than 99% for each enantiomer, except for compound **5** [60].

The inhibitory activity of the racemates **5–7** and **9** as well as of the corresponding enantiomers on *in vitro* growth of the human breast adenocarcinoma cell line MCF-7 was evaluated and compared. The most evident enantioselectivity was noticed between the racemate of *trans*-kielcorin D (**6**) (inactive) and the active enantiomers (+)-**6** and (−)-**6** [60].

2.1.2. Derivatives of Psorospermin

Psorospermin (**10**) was isolated from the bark and roots of the African plant *Psorospermum febrifugum*, in 1980 [61,62]. It is a natural fused tetracyclic xanthone containing two stereogenic centers with (2*R*,3′*R*)-stereochemistry and a reactive epoxide (Figure 2). The importance of the configuration and the functionality of the epoxydihydrofuran group for the *in vivo* activity have been evaluated [31,50]. The total synthesis of psorospermin (**10**) was reported for the first time in 2005, by obtaining the xanthone skeleton by the method of Grover et al. [62], including thirteen steps and with an overall yield of 1.7%. Psorospermin (**10**) revealed interesting biological activities showing antileukaemic, and antitumor activity in several human cell lines [31,62].

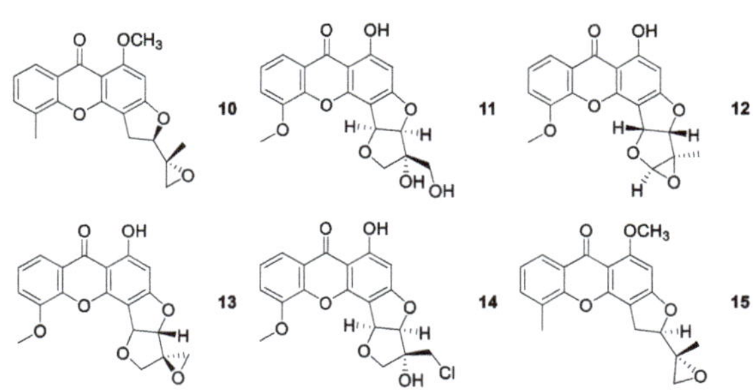

Figure 2. Structures of psorospermin (**10**) and synthetic derivatives **11–15**.

Additionally, the (*R*,*R*)-stereochemistry of psorospermin (**10**) gave optimum DNA alkylation and antitumor activity, although all four possible stereoisomers show topoisomerase II-dependent alkylation [63].

Two ring-constrained derivatives of psorospermin were also synthesized, namely, stereoisomer **12**, a ring-constrained (2*R*,3*R*)-form, and **13**, a ring-constrained (2*R*,3*S*)-compound (Figure 2) [63]. The chlorohydrin **14** retains psorospermin-like DNA alkylation characteristics despite its rigid structure and high affinity for DNA. The chlorohydrin **14** and epoxide **13** showed increased cytotoxicity against a range number of human tumor cell lines, compared to isohydroxypsorofebrin (**11**) [63].

Another study described the synthesis of two diastereisomeric pairs of *O*-5-methyl psorospermin and evaluation of *in vitro* activity against a range of solid and hematopoietic tumors. The diastereisomeric pair having the naturally occurring enantiomer (2*R*,3*R*) (**15**) (Figure 2) was the most active across all the cell lines tested. In subsequent studies using the four isomers of *O*-5-methyl psorospermin, the order of biological potency was (2*R*,3*R*) > (2*R*,3*S*) = (2*S*,3*R*) > (2*S*,3*S*) [64]. The compound (2*R*,3*R*) psorospermin (**15**) showed to be as effective as gemcitabine (chemotherapeutic drug) in slowing tumor growth *in vivo* in pancreatic cancer model [64].

2.1.3. Derivatives of Muchimangins

In many tribes and folk medicine use, plants and other organisms are commonly used to treat several conditions. For example, in Africa, the roots of *Securidaca longepedunculata* are used to treat sneezing, syphilis, gonorrhea, rheumatic pain, headache, feverish pain, malaria, sleeping sickness, among other conditions [65]. Muchimangins are a minor constituent of this specie and their biological activities have not been fully explored [66]. Dibwe et al. [67] reported the promising antiausteric activity of one natural occurring muchimangin against human pancreatic cancer PANC-1 cell line. Besides the anticancer promising activity, Kodama et al. [66] explored the antimicrobial activity of these structures and performed SAR studies. Accordingly, they synthesized several muchimangins derivatives **16–20** (Figure 3), and analyzed their antimicrobial activity.

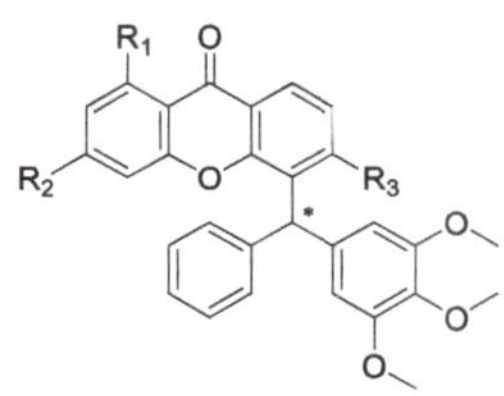

R_1	R_2	R_3		R_1	R_2	R_3		R_1	R_2	R_3	
OH	OH	OH	**16** (+)-**16** (-)-**16**	OH	H	OH	**17** (+)-**17** (-)-**17**	H	H	OH	**18** (+)-**18** (-)-**18**
OCH_3	OCH_3	OH	**19**	OCH_3	OCH_3	OCH_3	**20**				

Figure 3. Structures of muchimangin derivatives **16–20**.

To synthesize the muchimangins derivatives **16–20**, they etherified commercially available 1,2,4-trihydroxybenzene with dimethyl sulfate, producing 1,2,4-trimethoxybenzene. Then, by acylation 2,4,5-trimethoxybenzophenone was obtained. This compound was further reduced to afford 2,4,5-trimethoxydiphenylmethanol, part of the muchimangin skeleton. Afterwards, the corresponding xanthone moiety was obtained using Eaton's reagent. To finalize, both structural moieties were coupled by a Bronsted acid-catalyzed nucleophilic substitution, to produce the corresponding racemates [66]. In order to clarify the effect of chirality, Kodama et al. [66] separated the most promising derivatives using a CSP and to identify their optical rotation via polarimetry.

The preliminary SAR studies suggested that the presence of a hydroxyl group at C-6 was important for the antibacterial activity. Moreover, enantioselectivity occurred for compound **18**, with the dextro (+) enantiomer being more active against *S. aureus* than the levo (-) enantiomer and the racemate [66].

2.1.4. Derivatives of Mangiferin

Mangiferin (**21**, Figure 4) is a natural occurring chiral xanthone with a large spectrum of biological activities, which have been explored for many years [68–71]. Several authors have compiled information about the biological properties of mangiferin and derivatives [72,73].

As previously reported by Araújo et al. [74], mangiferin derivatives present a large spectrum of antimicrobial activities. Singh et al. [75,76] developed new mangiferin derivatives **22–27** (Figure 4) and screened their antipyretic and antimicrobial activities. The synthetic strategy used equivalent molar proportions of mangiferin and an appropriate base (*R*-aromatic amine) at reflux to give the corresponding derivative.

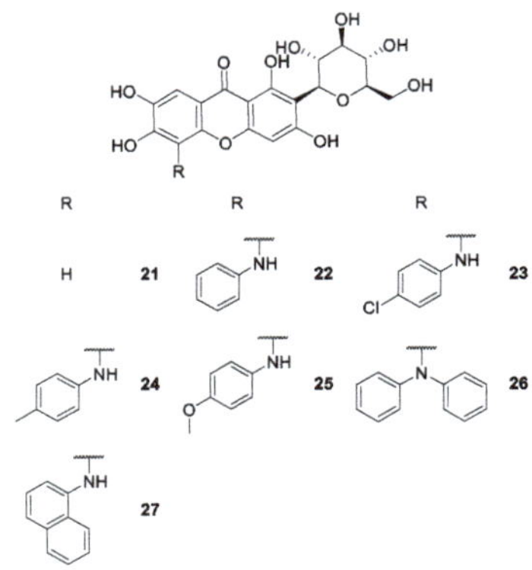

Figure 4. Structures of mangiferin (**21**) and some synthetic derivatives **22–27** with antipyretic and antimicrobial activities.

According to the results, it was suggested that the antipyretic activity may be attributed to the anti-inflammatory and antioxidant potential of mangiferin and its derivatives [76]. However, further investigations are required to understand the mechanism of action to prove its antipyretic activity. Regarding to the antimicrobial activity, the same compounds showed powerful inhibition of the growth of *S. virchow* and significant antibacterial activity against *B. pumilus* and *B. cereus*. On the other hand, all tested compounds revealed poor growth inhibition of *P. aeruginosa* and low antifungal activity [75].

In other studies, the analgesic, antioxidant and anti-inflammatory activities of other mangiferin derivatives **28–34** (Figure 5) were explored [77,78]. Dar et al. [77] analyzed the analgesic and antioxidant activities of acetylated (**28**), methylated (**29**), and cinnamoylated (**30**) mangiferin, where compound **12** was acetylated to afford **14.** Mahendran et al. [78] observed the analgesic and anti-inflammatory activities of mangiferin with benzoyl (**31**), benzyl (**32**), and acetyl (**33**) groups.

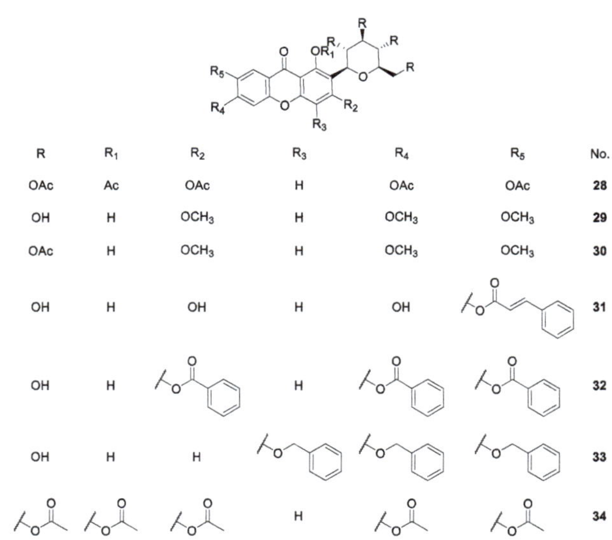

Figure 5. Structures of mangiferin derivatives **28–34** with antioxidant, anti-inflammatory and analgesic activities.

The results demonstrated that mangiferin derivatives substituted with benzoyl (compound **32**) and acetyl (compound **34**) groups displayed better antioxidant activity than mangiferin (**21**) in lipid peroxidation, p-NDA, deoxyribose and alkaline DMSO assays, while neither compound had analgesic nor anti-inflammatory activities. In all of these methods, standard drugs showed better activity than mangiferin and its derivatives **28–34** [78].

Mangiferin (**21**) is also known to possess antidiabetic activity [79,80]. This biological activity was further investigated for other mangiferin derivatives **35–42** and **46–53**, by Hu et al. [81,82] and **43–45** by Li et al. [83] (Figure 6). These works evaluated the properties of mangiferin derivatives as protein tyrosine phosphatase 1B (PTP1B) inhibitors in order to demonstrate their hypoglycaemic activity. The PTP1B has an important role in type 2 diabetes and obesity [84], which is primarily responsible for the dephosphorylation of the activated insulin receptor and thus downregulates insulin signalling [85]. For this reason, PTP1B inhibitors are a good strategy for diabetes mellitus treatment.

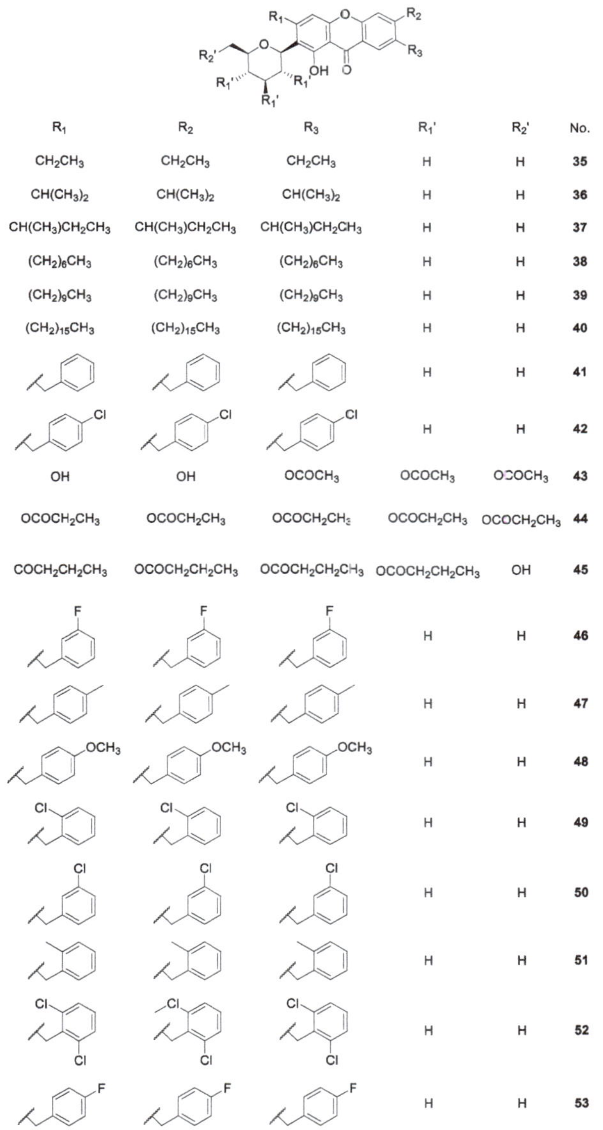

Figure 6. Structures of mangiferin derivatives **35–53** with antidiabetic activity.

According to Hu et al. [81,82], mangiferin (**21**) is a weak PTP1B inhibitor, whereas some derivatives such as **36**, **38** and **39** showed good inhibition of this protein. The SAR studies suggested that the substitution of free hydroxyl at C-3, C-6, C-7 of mangiferin (**21**) remarkably enhanced the inhibition, and the mono- or dichloro benzylated derivatives displayed better inhibitory activity than other groups [81,82]. However, further modification and biological studies are still in progress [81].

Li et al. [83] also demonstrated that the esterified-derivatives of mangiferin **43–45** (Figure 6) could repair damaged islet cells, and had higher lipid-solubility, and more potent hypoglycaemic activity than the mangiferin (**21**). The SAR studies indicated that the larger the esterification moieties or the higher lipid-solubility, the more potent hypoglycaemic activity was displayed by the derivative. Thus, esterification proved to be an effective way to improve the activity of mangiferin as a potential antidiabetic drug [83].

Correia-da-Silva et al. [86] developed new sulfated xanthones **54–57**, inspired by the mangiferin scaffold, to study their anticoagulant and antiplatelet properties (Figure 7). The synthetic approach included the sulfation of commercially available mangiferin affording mangiferin-2′,3′,4′,6′-tetrasulfate. It was found that an increase of the quantities of sulfating agent furnished the 2′,3, 3′,4′,6,6′,7-heptasulfated derivative. The precursor of the other derivatives could be a suitable xanthone scaffold, where 3,6-(*O*-β-glucopyranosyl)xanthone was obtained after deprotection of the glycosylated xanthone 3,6-(2,3,4,6-tetra-*O*-acetyl-β-D-glucopyranosyl)xanthone. The sulfation of 3,6-dihydroxyxanthone allowed preparation of persulfated 3,6-(O-β-glucopyranosyl)xanthone. Two polysulfated xanthonosides proved to be inhibitors of thrombosis, combining anticoagulant and antiplatelet effects in a single molecule [86].

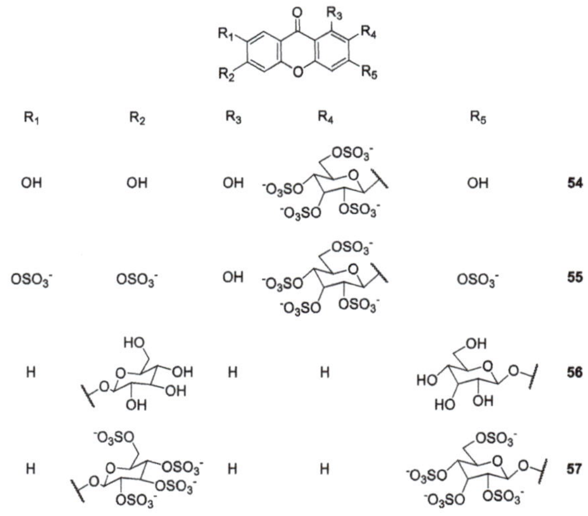

Figure 7. Structures of mangiferin derivatives **54–57** with anticoagulant and antiplatelet activities.

2.1.5. Derivatives of α-Mangostin

One of the most studied xanthones found in Nature is α-mangostin, isolated from tropical fruits of *Garcinia mangostana* which have been used for centuries in folk medicine to treat many conditions [87,88]. Several studies have reported its anticancer and antimicrobial activities, among others, which have prompted researchers all over the world to synthesize diverse derivatives [74,87–94].

One strategy concerned cationic antimicrobial peptides (CAMPs),amphipathic structures whose hydrophobic moiety penetrates the membrane core, while the cationic residues disrupt bacterial membranes [95–98]. Due to the manufacturing costs and poor stability of peptides, Koh et al. [95,99] developed small-molecules **58–70** with CAMP characteristics by combining the α-mangostin core with basic amino acids moieties (Figure 8). The purpose of the work was to confirm if lipophilic chains enhance the membrane permeability and to examine the role of the cationic moieties conjugating the xanthone scaffold with basic amino acids [99]. This strategy was also used to develop new anti-mycobacterial derivatives **65–70** [95], and by Lin et al. [97] for studies seeking new antimicrobial and hemolytic compounds.

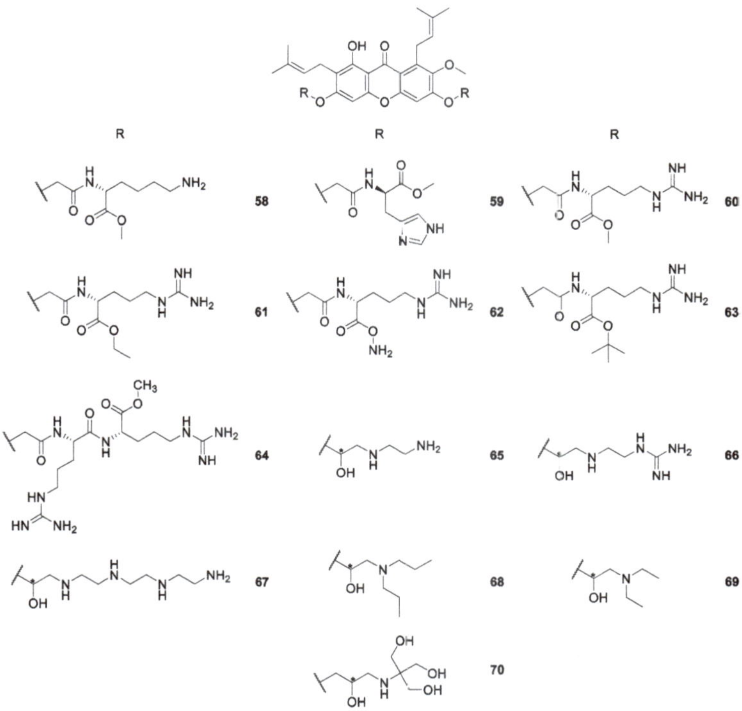

Figure 8. Structures of derivatives of α-mangostin **58–70**.

According to the results, the amphiphilic xanthone derivatives **65–70** (Figure 8) possessed promising mycobacterial activity without resistance mechanisms, which may contribute to the development of an entirely new class of therapeutics for tuberculosis [95]. Besides the interesting activity of these compounds, the cationic and hydrophobic moieties enhanced the water-solubility, and also lead to high membrane selectivity and excellent antibacterial activity against Gram-positive bacterial strains, including methicillin-resistant Staphylococcus aureaus and vancomycin-resistant Enterococcus. It is important to point out that the membrane selectivity of these compounds was higher than several membrane-active antimicrobial agents in clinical trials [97,99].

2.1.6. Derivatives of Caged Xanthones

The *Garcinia* genus contains caged xanthones which mainly occur in a few species like *G. morella*, *G. hanburyi*, *G. bracteata*, *G. gaudichaudii*, and *G. scortechinii*, widely distributed in Southeast Asia [100–103]. The caged core is responsible for the vast range of bioactivities of this class of compounds, such as anti-viral, and antibacterial effects, among others [104–107]. Many reports have described the potential antitumor activity of gambogic and morrelic acids [100,101,104,105,107,108]. According to this, many caged xanthones **71–155** have been synthesized and studied through the last years, with a diversity of purposes (Figures 9–14).

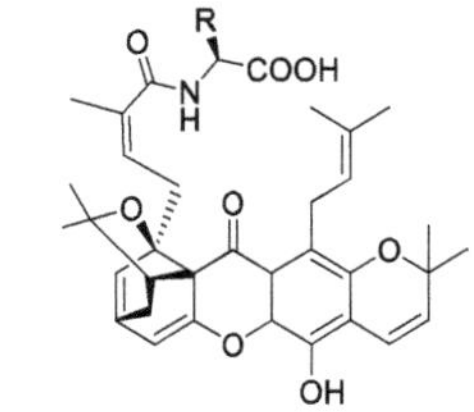

R	No.	R	No.	R	No.	R	No.
$CH(CH_3)_2$	**71**	$CH_2(C_6H_5)$	**72**	CH_2CH_2COOBz	**73**	CH_2CH_2COOH	**74**
$CH_2CH_2SCH_3$	**75**	$CH_2CH(CH_3)_2$	**76**	CH_2OH	**77**	$CH(CH_3)CH_2CH_3$	**78**
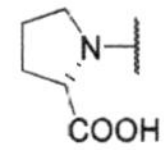	**79**	$CH_2CH_2CONTrt$	**80**	$CH_2CH_2CONH_2$	**81**	$(CH_2)_4NH_2$	**82**

Figure 9. Structures of caged xanthones **71–82** with antimicrobial activity.

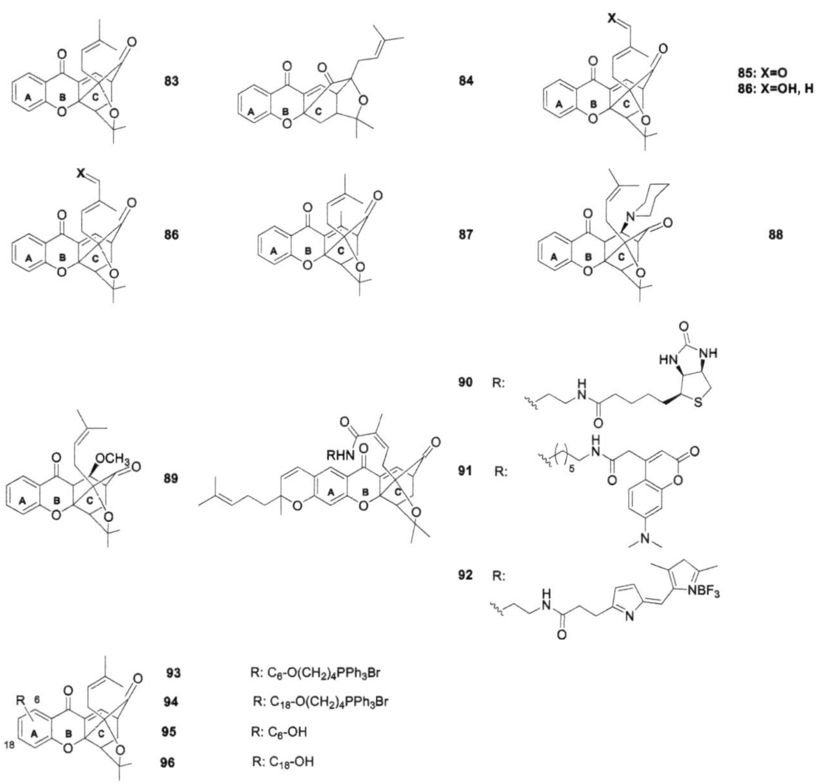

Figure 10. Structures of caged xanthone **83–96**, with antimalarial and antitumor activities.

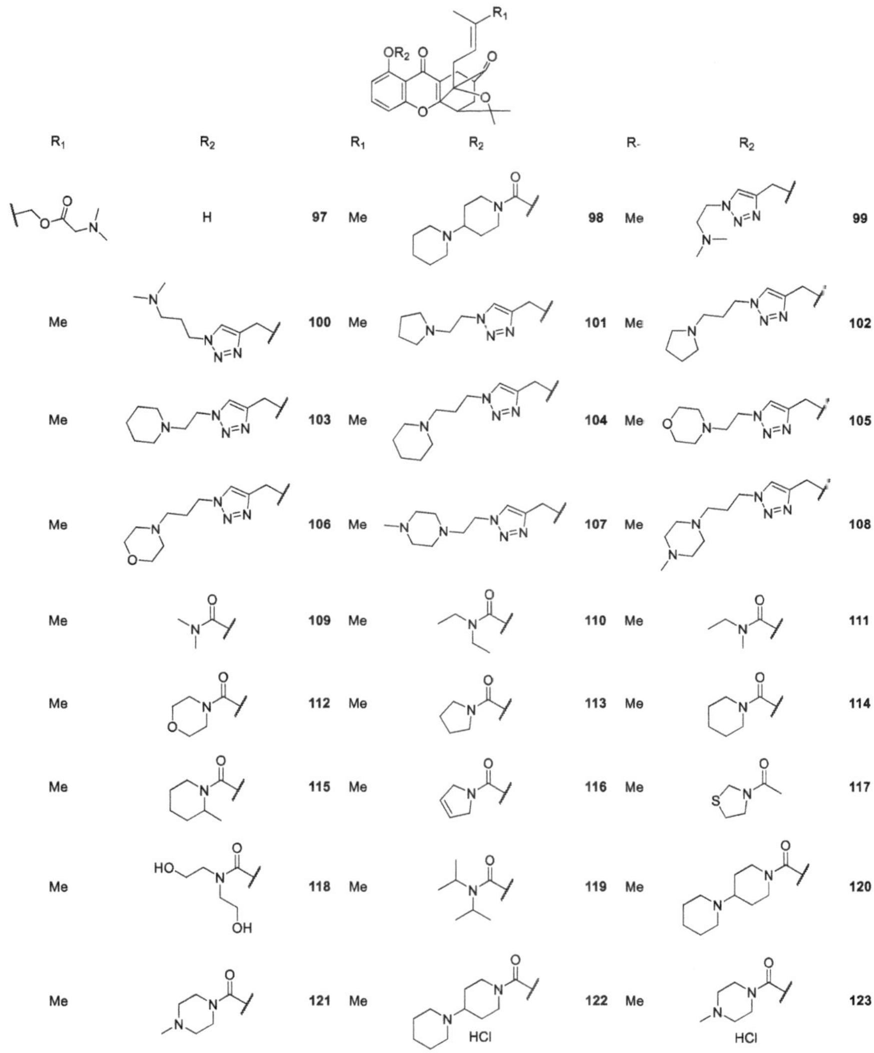

Figure 11. Structures of caged xanthones **97–123**, with antitumor activity.

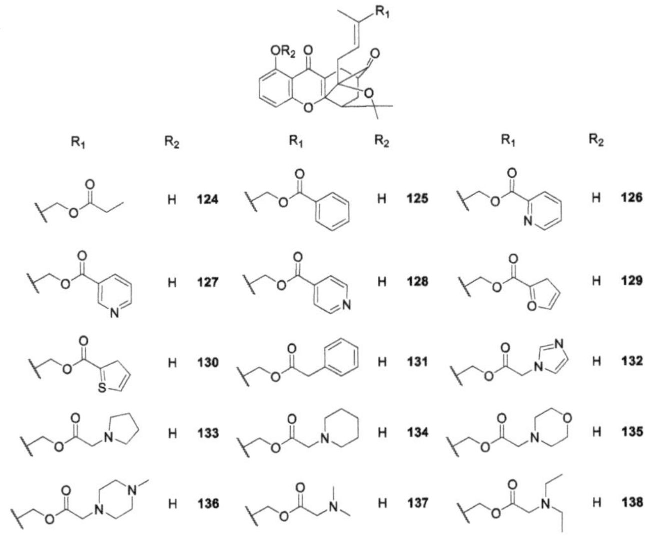

Figure 12. Structures of caged xanthones **124–138**, with antitumor activity.

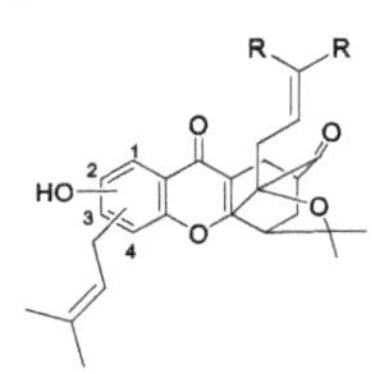

R	-OH	-Prenyl		R	-OH	-Prenyl		R	-OH	-Prenyl	
Me	C2	C1	**139**	Me	C3	C4	**140**	Me	C4	C3	**141**
Me	C1 C3	C4	**142**	Me	C1 C3	C2	**143**	H	C2	C1	**144**
H	C3	C4	**145**	H	C4	C3	**146**	H	C1 C3	C4	**147**
				H	C1 C3	C2	**148**				

Figure 13. Structures of caged xanthones **139–148**, with antitumor activity.

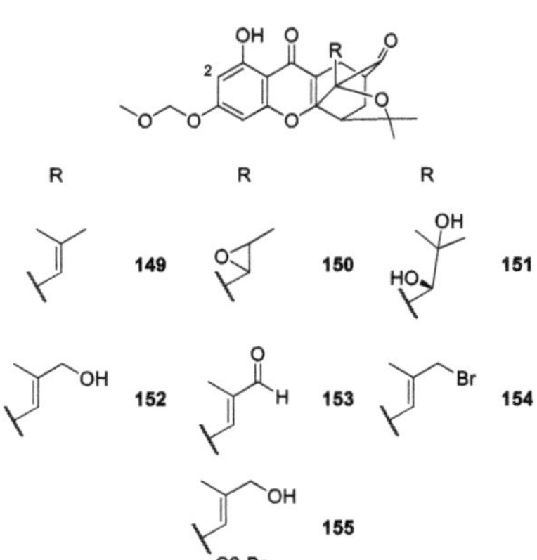

Figure 14. Structures of gambogic acid derivatives **149–155**, with antitumoral activity.

Chaiyakunvat et al. [105] inspired by the biological properties of caged xanthones, synthesized a few morrelic acid derivatives and evaluated their antimicrobial activity. They started with the synthesis of methylated morrellic acid and, afterwards, they synthesized derivatives **71–82** with amino acid moieties via solid-phase synthesis (Figure 9).

The morellic acid derivatives showing more inhibition of bacterial growth were the ones with an amino acid-containing hydrophobic side chain like **71**, **72**, **76**, **78** and **79** (Figure 9) [105]. This is in agreement with previous reports where the antimicrobial activity was higher in the structures with hydrophobic and/or aromatic amino acids [99,105].

Theodorakis et al. [106,109–111] synthesized new caged xanthones and studied their properties. They developed a Claisen/Diels–Alder reaction cascade that, in combination with a Pd(0)-catalyzed reverse prenylation, provided a rapid and efficient access to the caged xanthone pharmacophore. Afterwards, various A-ring oxygenated derivatives of cluvenone (**83**) were further synthesized and analyzed (Figure 10) [106,109–111].

The SAR studies showed that their activity could be substantially improved by attaching a triphenylphosphonium group at the A ring of the caged xanthone. Derivatives **93** and **94** (Figure 10) were found to be highly effective as antimalarials against *Plasmodium falciparum* [106]. The conjugation of these compounds with a phosphonium salt improved their efficacy, resulting in lead compounds with a promising therapeutic window [106]. It was suggested that, further modification of the caged xanthone could increase the selective cytotoxicity and lead to a promising lead candidate [106].

Cluvenone (**83**) was also reported to induce cell death via apoptosis, presenting similar cytotoxicity in multidrug-resistant and sensitive leukemia cells [109,110]. The caged xanthone derivatives proved to be active with cytotoxicity at low to sub-micromolar concentrations in solid and non-solid tumor cell lines, respectively. Additionally, they induced apoptosis in HUVE cells. Remarkably, similar IC_{50} values were obtained for the compounds tested in the HL-60 and HL-60/ADR cell lines, suggesting that these compounds were not subject to a drug resistance mechanism. Therefore, it was suggested that members of this family of compounds may have therapeutic potential in relapsed cancers typically resistant to standard chemotherapeutic agents. In addition, the cytotoxicity observed in HUVE cells suggested that these compounds may be interesting leads for the development of new angiogenesis inhibitors [111]. Elbel et al. [112] synthesized selected A-ring hydroxylated analogues and evaluated their effect on cell growth, mitochondrial fragmentation, mitochondriotoxicity and Hsp90 client protein degradation. They found out that both the C6 and C18 hydroxylated cluvenones inhibited the growth of CEM cells at low micromolar concentrations and induced cell death via the mitochondrial pathway. In addition, cluvenone (**83**) and the hydroxylated cluvenones induced Hsp90-dependent protein client degradation at low micromolar concentrations [112].

Table 1. Summary of the biological activities of synthetic CDXs inspired in natural xanthones.

CDXs	Biological Activities	Ref.
Kielcorin derivatives **2–9**	Antitumor and protein kinase C inhibition	[56,58,59,61]
Psorospermin derivatives **11–15**	Antitumor	[64]
Muchimangin derivatives **16–20**	Antimicrobial	[66]
Mangiferin derivatives **21–57**	Antipyretic, antimicrobial, analgesic, antioxidant, anti-inflammatory, antidiabetic, anticoagulant and antiplatelet	[68,72,73,75–78,81–83,86]
α-Mangostin derivatives **58–70**	Antimicrobial, hemolytic and antimycobacterial	[91,95–99,118]
Caged xanthones **71–155**	Antimalarial, antitumor, anti-proliferation and anti-angiogenesis	[106,108,109,111,113–117,119,120]

Zhang et al. [113–116] synthesized a series of caged xanthone derivatives to improve the physicochemical properties and *in vivo* cytotoxic potency. For that, they relied on MAD28 synthesis and characterization of the derivatives. The structural modifications revealed that the presence of a carbamate moiety was useful for obtaining comparable cytotoxicity and improved aqueous solubility and permeability (Figure 11).

It is important to highlight that compound **137** (named DDO-6306, Figure 12) displays growth inhibition in Heps transplanted mice, and is now undergoing further evaluation as a candidate for cancer chemotherapy [115]. In a more recent study, compound **105** (Figure 11), considered as the lead compound and called MAD28, successfully led to the discovery of a novel series of natural-product-like triazole-bearing caged xanthones with improved drug-like properties as orally-active antitumor agents *in vivo* [115].

Regarding the caged xanthone derivatives containing carbamate scaffolds **109–123** (Figure 11), the results showed a potent antiproliferative activity and good physicochemical properties, which contributed to improving their *in vivo* activities. The compound **122** (DDO-6337) showed moderate inhibitory activity toward Hsp90 ATPase and resulted in the degradation of Hsp90 client proteins, such as HIF-1, which ultimately contributed to its antitumor and anti-angiogenesis activities [116].

Compounds **140–143** (Figure 13) exhibited micromolar inhibition against several cancer cell lines. Some interesting SAR considerations have been highlighted, such as the importance of the periphery gem-dimethyl groups in maintaining the anti-tumor activity, the effect of the substituent at C-1 position of B-ring on activity, since hydroxyl group at C-1 position enhanced the potency while prenyl group reduces it, and, that the change of hydroxyl or prenyl groups in carbons C-2, C-3 and C-4 had no significant effect on the anti-tumor activity. These events indicated that referred sites can be used to improve drug-like properties [114].

In another study, Miao et al. [117] developed small molecule entities inspired by the structure of gambogic acid. They focused on modifications of the prenyl moiety of the caged xanthones which led to synthesize seven derivatives **149–155** (Figure 14), which were tested for anti-tumor activity [117].

The SAR studies suggested that compounds **151**, **153**, and **154** showed similar cytotoxicity to gambogic acid against A549 cells, whereas compounds **149–151** and **152** were less active than gambogic acid. Although these experiments were preliminary, the results suggested that promising agents with anti-tumor activities could be obtained by modification at C-2 position of the B ring and at C-21/22 or C-23 position of the prenyl group in the caged scaffold. The formation of dihydroxy and epoxy groups of the double bond at C-21/22 and the introduction of an electron-withdrawing group at C-23 evidently affected the anti-proliferation activity [117]. In Table 1 a summary of the synthetic CDXs inspired in natural xanthones and their described biological activities are presented, as well as the associated references.

2.2. Synthetic CDXs Obtained by Binding/Coupling Chiral Moieties to the Xanthone Scaffold

Another approach to acquire synthetic CDXs is by binding chiral moieties to the xanthone scaffold using different strategies. The biological and pharmacological activities evaluated of the various CDXs obtained by this strategy are included in Table 2.

2.2.1. XAA and DMXAA Analogues

DMXAA (5,6-dimethylxanthone-4-acetic acid, vadimezan, ASA404, **157**, Figure 15) is a simple carboxylated xanthone discovered by SAR studies involving a series of xanthone-4-acetic acids (XAA, **156**, Figure 15) related to the parent compound flavone acetic acid [121]. DMXAA is one of the most studied xanthones considering not only its remarkable pharmacological profile [122–131], but also its physicochemical and pharmacokinetic properties [130,132,133]. It is a tumor vascular-disrupting agent leading to a fast, vascular collapse and tumor necrosis by immunomodulation and cytokines induction. DMXAA (**157**) also demonstrated antiviral [134], antiplatelet and antithrombotic [135] activities. This synthetic xanthone is not chiral but, it is evident that structurally and from a biological activity perspective, it may be an attractive scaffold for the development of other bioactive analogues and derivatives.

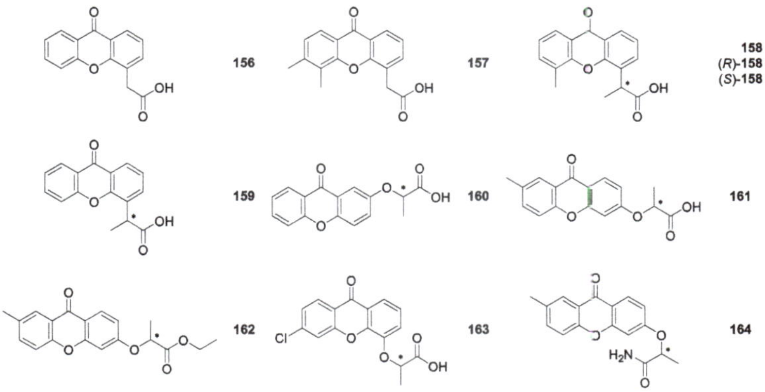

Figure 15. Structures of XAA (**156**), DMXAA (**157**) and chiral analogues **158–164**.

Rewcastle et al. [136], in 1991, synthetized the DMXAA chiral analogues 2-(5-methyl-9-oxo-9*H*-xanthen-4-yl)propanoic acid (**158**) and 2-(9-oxo-9*H*-xanthen-4-yl)propanoic acid (**159**) as racemates (Figure 15). CDX **159** was synthetized via bromination of 4-ethylxanthenone followed by conversion of the resulting 4-(1-bromoethyl)xanthenone to compound **159** via the nitrile. CDX **158** was obtained by reaction of 2′-hydroxyacetophenone with benzyl chloride under phase-transfer conditions, giving 2′-benzyloxyacetophenone, which was converted successively to an alcohol, using $NaBH_4$, and a chloride, with anhydrous $CaCl_2$ and HCl. After three reaction steps, 2-(2-hydroxyphenyl)propanoic acid was obtained and then reacted with 2-iodo-3-methylbenzoic acid via a copper/TDA-catalysed condensation. Finally, the resulting diacid was ring-closed using H_2SO_4 to give the racemic compound **158**. For this CDX both enantiomers were separated by indirect method employing (*R*)-(−)-pantolactone as chiral resolving agent. The obtained mixture of diastereomers was separated by chromatography on silica gel. Further, hydrolysis of the esters under non-enolizing conditions afforded both enantiomers, (*S*)-(+)-**158** and (*R*)-(−)-**158** (Figure 15) [136].

The racemic compounds **158** and **159**, as well as enantiomers (*S*)-(+)-**158** and (*R*)-(-)-**158** were tested in *in vitro* and *in vivo* tumor assays [136]. It was found that all the compounds were active. Moreover, enantioselectivity was observed, being the *S*-(+)-enantiomer much more dose-dependent than the *R*-(-)-enantiomer. It was suggested that the enantiomers have different intrinsic activities, rather than differing in their metabolism [136]. CDX **159** had been tested previously as anti-inflammatory agent [137].

Marona et al. reported the synthesis [138] of three new chiral analogues of XAA **160–162** (Figure 15) and the evaluation of their cytotoxicity against J7774A.1 cells [139]. Compounds **160** and **161** were obtained by condensation of 2-hydroxyxanthone and 2-methyl-6-hydroxyxanthone, respectively, with α-bromopropionic acid and compound **162** by esterification of compound **161** [138]. Regarding the biological activity tested, it was found that all CDXs showed weak cytotoxicity [139].

Recently, Zelaszczyk et al. [140] synthesized two new chiral XAA derivatives **163** and **164** (Figure 15) by the reaction of 3-hydroxyxanthone with ethyl 2-bromopropanoate, followed by ester hydrolysis [140]. These compounds were found to have anti-inflammatory and analgesic activities [34,140].

2.2.2. Synthetic Aminoalkanolic CDXs

Our group has a vast experience in synthesis and biological/pharmacological activity evaluation of xanthone derivatives [86,141–147] and, recently, reported the synthesis of a library of CDXs **165-179** in an enantiomerically pure form (Figures 16 and 17) [148,149]. Among the synthesized CDXs, the compounds **166-171** and **174–179** are aminoalkanolic, while CDXs **165, 172** and **173** comprise of simple amines with a *p*-tolyl moiety (compounds **165** and **173**) or an aminoester (compound **172**) [148,149].

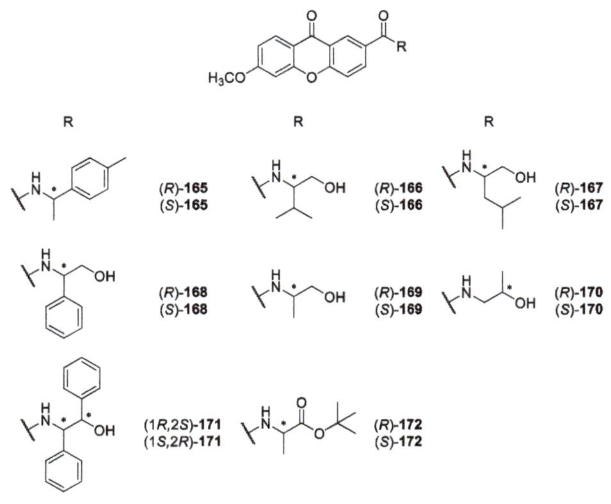

Figure 16. Structures of aminoalkanolic CDXs **166–171** and analogues **165** and **172**.

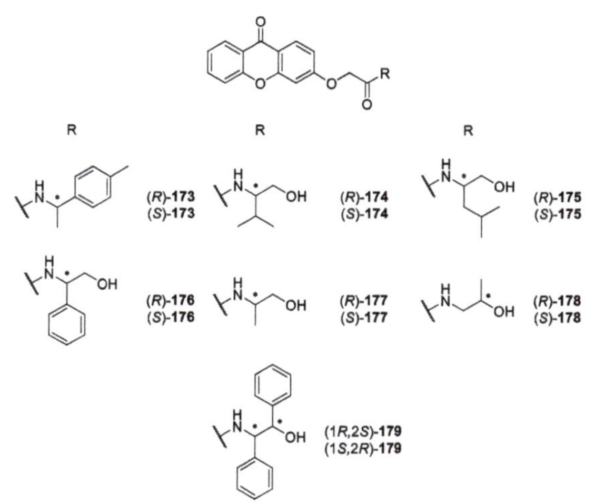

Figure 17. Structures of aminoalkanolic CDXs **174–179** and analogue **173**.

Considering that carboxyxanthone derivatives are suitable molecular entities to perform molecular modifications to obtain new bioactive derivatives [34], the synthesis of all CDXs **165-179** were achieved by using two carboxyxanthone derivatives as substrates, namely 6-methoxy-9-oxo-9*H*-xanthene-2-carboxylic acid and 2-((9-oxo-9*H*-xanthen-3-yl)oxy)acetic acid. The synthetic strategy used was the coupling of the carboxyxanthone derivatives with both enantiomers of eight commercially available chiral building blocks, using *O*-(benzotriazol-1-yl)-*N,N,N′,N′*-tetramethyluronium tetrafluoroborate (TBTU) as coupling reagent [148,149]. TBTU has been widely used as efficient reagent for the synthesis of diverse classes of compounds, including peptides [150], esters [151,152], phenylhydrazides [153], acid azides [154], among others. However, this was the first report of the use of TBTU to synthesize CDXs [149]. All used commercial chiral blocks included both enantiomers of enantiomerically pure building blocks with no tendency towards racemization or enantiomeric interconversion and having a primary amine as reactive group for amide formation. Amino alcohols, amines, and amino esters were selected. The coupling reactions were performed at room temperature, showing short reactions times and excellent yields (ranging from 94% to 99%) [148,149]. The synthetic methodology used to obtain the referred CDXs provided to be very efficient, broad-scope applicability, and operationally simplest. Moreover, it was found that the synthesis of the CDXs was easily scaled-up for both enantiomers in order to obtain available quantities for biological and pharmacological assays as well as other applications.

LC using different types of CSPs, namely polysaccharide-based [149], macrocyclic antibiotics [155,156], and Pirkle-type [157,158] was used for enantioresolution studies and determination of the enantiomeric purity of the synthesized CDXs. The enantioselective LC method using polysaccharide-based CSPs under multimodal elution conditions afforded very high resolutions with short chromatographic runs. The best performances were achieved on amylose *tris*-3,5-dimethylphenylcarbamate stationary phase under polar organic elution conditions. The resolution achieved allowed the determination of the enantiomeric purity for all CDXs, affording values higher than 99% [149].

Considering the macrocyclic antibiotic-based CSPs, four commercially available columns were used, namely Chirobiotic T™, Chirobiotic R™, Chirobiotic V™ and Chirobiotic TAG™, under multimodal elution conditions. The optimized chiral LC conditions were successfully employed for the accurate determination of the enantiomeric purity, always higher than 99%. The studies also explored the influence of different mobile phase compositions and pH on the chromatographic parameters as well as of the structural features of the CDXs on their chiral discrimination by the macrocyclic antibiotic-based CSPs [155,156].

Regarding the Pirkle-type CSPs, the (*S*,*S*)-Whelk-O1® CSP showed the best performance for the resolution of the CDXs evaluated, presenting very high enantioselectivity for CDXs with aromatic group linked to the chiral moiety. Polar organic elution mode presented the best chromatographic parameters allowing good resolutions and lower run time [157,158].

The overall results proved that, for the same enantiomeric pair of CDXs, the polysaccharide-based CSPs were the most efficient to separate the enantiomers of this group of compounds, since all the CDXs were enantioseparated with excellent enantioselectivity and resolution [159].

For each enantiomeric pair of the synthesized CDXs **165–179** the *in vitro* growth inhibitory activity in three human tumor cell lines, A375-C5 (melanoma), MCF-7 (breast adenocarcinoma), and NCI-H460 (non-small cell lung cancer), were evaluated [149]. The results obtained demonstrate that some CDXs exhibited interesting growth inhibitory effects on the tumor cell lines. The most active CDX in all human tumor cell lines was compound **(1*R*,2*S*)-179**. Furthermore, it is important to highlight that the effects on the growth of the human tumor cell lines were attributed not only to the nature and positions of substituents on the xanthonic scaffold, but also, in some cases, were associated with the stereochemistry of the CDXs concerning enantioselectivity results. Interesting examples of enantioselectivity were observed between the enantiomeric pairs of CDXs **165**, **167**, and **171** [149].

Recently, other enantioselectivity studies associated with biological activity were conducted, specifically the *in vitro* and *in silico* inhibition of cyclooxygenases (COX-1 and COX-2) for the enantiomeric pairs of CDXs **166**, **168** and **169** [160]. All CDXs exhibited COX-1 and COX-2 inhibition being, in general, the inhibitory effects similar for both COXs. The only exception was the enantiomeric pair of compound **166**, being the (*R*)-(+)-enantiomer more active at inhibiting COX-2 than COX-1. Interestingly, all pairs demonstrated enantioselectivity for COX-1. Concerning COX-2, the % of inhibition was also dependent of the stereochemistry being (*S*)-(-)-**166** and (*S*)-(+)-**169** more active [160].

Additionally, for the same enantiomeric pairs of CDXs **166**, **168** and **169**, human serum albumin (HSA) binding affinity was evaluated by spectrofluorimetry and *in silico* studies, by a docking approach [160]. All CDXs demonstrated to bind with high affinity to HSA and enantioselectivity was observed for compound **168**.

Taking into account that these CDXs have molecular moieties structurally very similar to local anaesthetics, the ability to block compound action potentials (CAP) at the isolated rat sciatic nerve was also investigated [161]. The CDXs (*S*)-**165,** (*S*)-**166** and (*S*)-**167** were chosen for biological evaluation and the results suggested that the nerve conduction blockade might result predominantly from an action on Na^+ ionic currents. It was also investigated if the CDXs could prevent hypotonic haemolysis on rat erythrocytes. However, data suggested that all tested CDXs caused no significant protection against hypotonic when applied in concentrations high enough to block the sciatic nerve conduction in the rat [161].

Besides the potential as new drugs, CDXs present structural features with interest as chiral selectors for LC [23]. In this context, some of these small molecules ((*S*)-**167**, (*R*)-**168**, (*S*)-**168**, (*R*)-**176**, (1*R*,2*S*)-**179** and (1*S*,2*R*)-**179**) were selected and bound to a chromatographic support for a new application as CSPs in LC [24]. The new xanthonic CSPs afforded promising enantioresolution results, high stability and reproducibility. Accordingly, CDXs have important applications in the field of Medicinal Chemistry, not only as candidates for potential new drugs but also as analytical tools for enantioseparation in LC [37].

Recently, our group also performed enantioselectivity studies with chiral thioxanthones, *S*-analogues of xanthones, as modulators of P-glycoprotein (P-gp) [162]. It was found that one of the enantiomers modulated P-gp expression differently from its pair.

Marona *et col.* [163–169] also described the synthesis and biological activity evaluation of a series of aminoalkanolic CDXs **180–205, 217–220, 238** (Figures 18–20). More recently, they synthesized the aminoalkanolic CDXs **206–243** (Figures 18–20), being tested for anticonvulsant, antimicrobial and cardiovascular activities [170–174]. CDXs **180–194, 198–202** were synthesized by condensation of an appropriate 2-bromomethylxanthone or 2-bromomethyl-7-chloroxanthone with the adequate aminoalkanol in toluene, in the presence of anhydrous potassium carbonate [163,164,175]. The exchange of secondary amino group of compound **189** for a tertiary amino group (compound **193**) was generated by reductive *N*-methylation [163]. Compound **210**, however, was synthesized through the chlorination of compound **186** with thionyl chloride in toluene [170].

Compounds **195–197, 221–223, 228–232** and **241–243** were synthesized by the aminolysis of 3- or 4-((oxiran-2-yl)methoxy)xanthone in *n*-propanol, or by the amination of 2-methyl-6-hydroxy-xanthone using propylene epichlorohydrin, in the presence of sodium hydroxide and water [167,172–174].

The synthesis of compounds **203–204, 206–209, 213–216, 224–227** and **240** involved a multi-step process. At first, a substituted benzoic acid reacted with 2- or 4-methylphenol in two steps involving an Ullmann condensation and electrophilic addition. The intermediate methyl derivatives of substituted xanthone were used in the reaction with *N*-bromosuccinimide giving appropriate bromide derivatives. The last step comprised an aminolysis by means of appropriate aminoalkanol carried out in toluene in the presence of anhydrous K_2CO_3 [171,176].

Most of the synthesized CDXs **180–209, 211–220, 224–227, 233–238, 240** were evaluated for anticonvulsant activity [163–167,169,171,175,177–179]. The studies involved three kind of tests: maximal electroshock-induced seizures (MES), subcutaneous pentetrazole seizure threshold (scMet), and neurological toxicity (TOX).

In one of the first MES assays in mice, 2-amino-1-propanol-, 1-amino-2-propanol and 1-amino-2-butanol derivatives of 6-methoxy- or 6-chloroxanthone were the most interesting compounds. In fact, the results indicated that compound **184** was the most active [163]. Further study compared the anticonvulsant activity of CDX **184** (racemate) with the single enantiomers ((*R*)-**184**, (*S*)-**184**). All the compounds showed excellent results, and no significant differences were observed in the anticonvulsant activity of the single enantiomers compared with the racemate [166].

Additionally, the enantiomeric purity of (*R*)-**184** and (*S*)-**184** was determined by a liquid chromatography–mass spectrometry method with an electrospray ionization interface (ESI-LC/MS). The separation of the two enantiomers ((*R*)-**184** and (*S*)-**184**) was carried out on the commercially available cellulose *tris*-(3,5-dimethylphenyl carbamate) CSP, Chiralcel® OD-RH, giving enantiomeric purity values higher than 99.9% [166].

Interesting anticonvulsant results were also observed with alkanolic chiral derivatives **189** and **193** of 7-chloroxanthone, which displayed anti-MES activity corresponding with that for phenytoin, carbamazepine and valproate [164]. Moreover, it is important to highlight that in this study some cases of enantioselectivity were observed. For example, although enantiomers (*R*)-**189** and (*S*)-**189** showed anticonvulsant activity, the (*S*)-enantiomer was more neurotoxic. Furthermore, considering the compound **193** (racemate), the (*R*)-enantiomer ((*R*)-**193**) in comparison to (*S*)-enantiomer ((S)-**193**) showed higher anticonvulsant activity [164,177]. Recently, several aminoalkanolic chiral derivatives substituted in positions 2, 3, or 4 evaluated for their anticonvulsant activities with 2-xanthonoxy derivatives **206, 207** and **209**, being the most active and exhibiting neurotoxicity after 30 min after administration at the dose of 100 mg/kg [171]. A further study including chiral aminoalkanol derivatives substituted in position 2 of the xanthonic scaffold (structures not shown) also emphasized the importance to examine biologically enantiomers other than racemates [178].

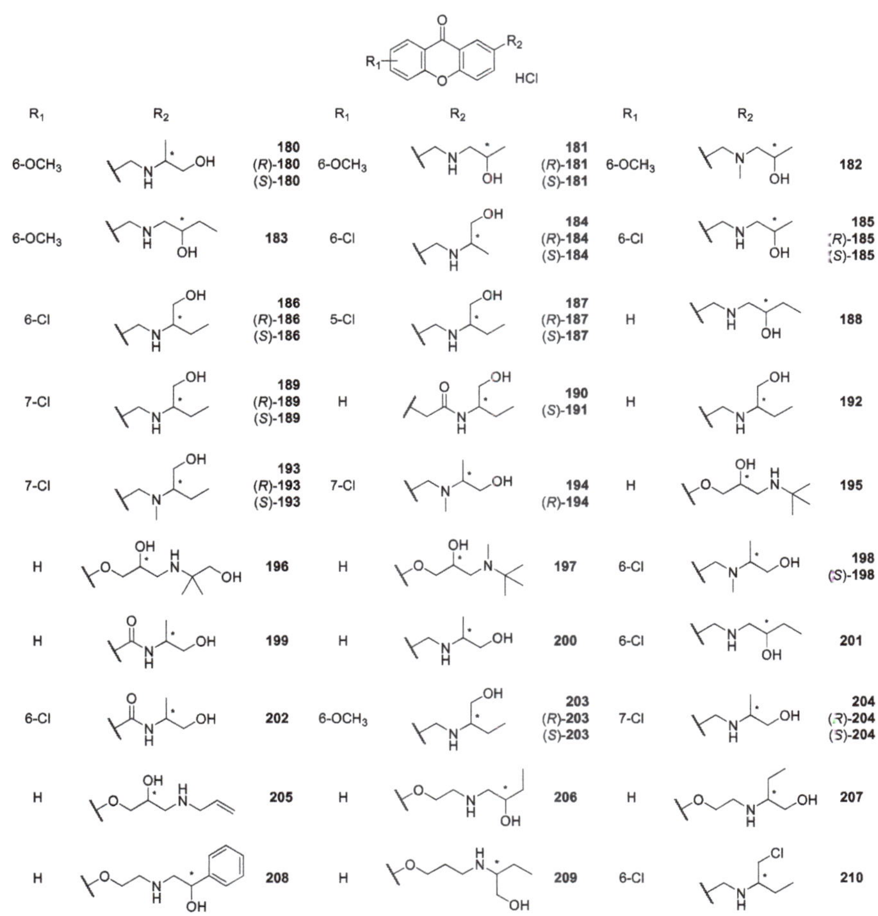

Figure 18. Structures of 2-aminoalkanolic CDXs **180–210**.

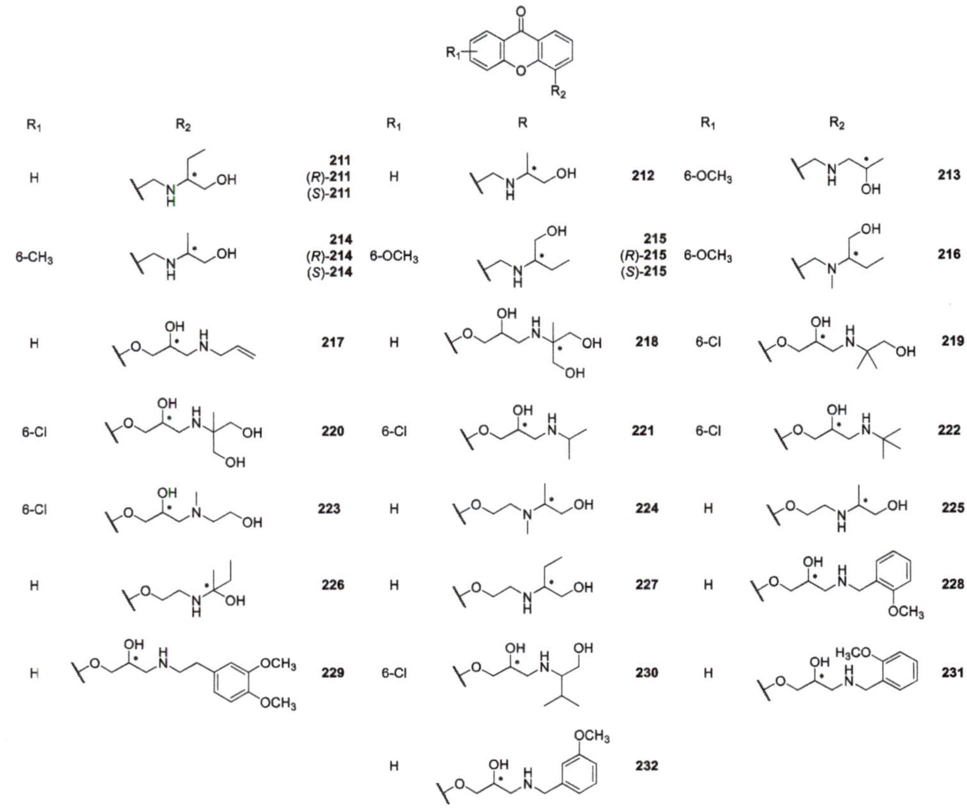

Figure 19. Structures of 4-aminoalkanolic CDXs **211–232**.

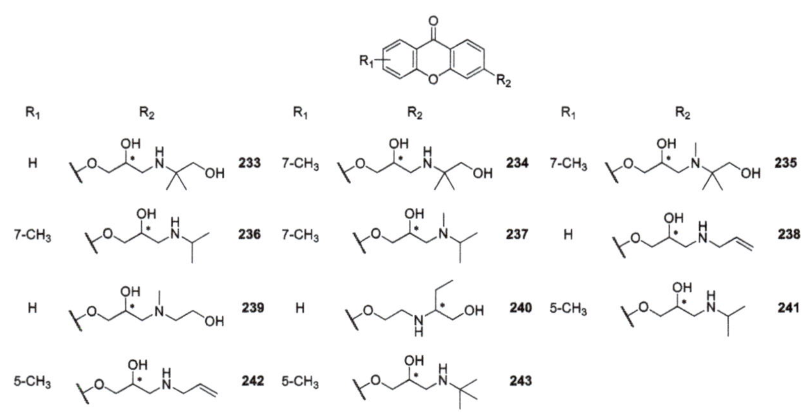

Figure 20. Structures of 3-aminoalkanolic CDXs **233–243**.

Additionally, a structure-anticonvulsant activity relationship study was described including series of aminoalkanol derivatives **204, 213–216** of 6-methoxy- or 7-chloro-2-methylxanthone as well as 6-methoxy-4-methylxanthone [176]. All the compounds showed activity in the MES screening which is recognized as one of the two most widely used seizures models for early identification of candidates as anticonvulsants. The tested compounds were evaluated in the form of racemic mixture and some additionally in the form of single enantiomers to determine stereochemistry-activity relationship. In fact, as demonstrated before [176], stereochemistry is one of the factors that can potentially influenced anticonvulsant activity of the CDXs. However, considering anti-MES activity it was not possible to establish relationship between stereochemistry and anticonvulsant properties because all sets of compounds gave different results. Racemate and enantiomers showed either similar results or diverged in duration of activity or lower effective doses. However, the anticonvulsant activity was associated with both aminoalkanol type and respective configuration as well as the location of substitution in the xanthone scaffold [176]. The overall results from several studies of Marona *et col.* [163–167,171,175–179] are quite encouraging and suggested that in the group of xanthone derivatives new potential anticonvulsants might be found.

Some of alkanolic CDXs were also evaluated for cardiovascular activity [167,173–175,179,180], including antiarrhythmic, hypotensive, α_1- and β_1-adrenergic blocking activities, effect on the normal electrocardiogram and influence on the central nervous system (CNS) [169]. Among the investigated compounds, some of them exhibited significant antiarrhythmic and/or hypotensive activity. For example, compounds **218** and **219** revealed the strongest anti-arrhythmic activity in the adrenaline-induced model of arrhythmia. Additionally, compound **219** was also the most potent concerning hypotensive activity [169]. Recently, compounds **231** and **232** were also evaluated for their cardiovascular activity, through both α_1- and β_1-adrenergic blocking. These compounds, classified as beta-blockers with vasodilating properties, exhibited also hypotensive vasorelaxant activities comparable to those of carvedilol [173].

The effects on platelet aggregation of racemic CDXs **180–182**, **195** and **196**, and the enantiomeric pure CDX (*R*)-**193** were evaluated and showed motivating results. The most active and promising compound was (*R*)-**193** which nearly completely inhibited the thrombin aggregation concentration (TAC) [181]. The results indicated that the presence of the 2-*N*-methylamino-1-butanol at position 2 and the chloride atom at the 7-position of the xanthone scaffold promoted antiplatelets activity.

Alkanolic CDXs **180**, **183**, **184** and **194** were used to assess mutagenic and antimutagenic activities in assays using the *Vibrio harveyi* test [182]. According to the obtained results, the most beneficial mutagenic and antimutagenic profiles were observed for compound **194**. This compound was shown to have strong antimutagenic activity towards the BB7 *V. harveyi* strain while failing to induce mutagenic responses in the tested strains. The modification of the chemical structure of compound **194** through chlorination of the hydroxyl group, improved considerably the antimutagenic activity maintaining the inability to induce mutagenic responses in the strains. Thus, antimutagenic potency reached a maximum with the presence of tertiary amine and one chloride atom in the side chain. Minimal activity was showed to compound **184** and no antimutagenic activity was observed for compound **180** [182].

In recent years, several aminoalkanolic CDXs **210**, **228–230**, **241–242** have been evaluated for their antibacterial and antifungal activities [170,172]. Compound **210** was evaluated against several dermatophytes, moulds and yeasts, exhibiting good activity results against ten strains of the *Aspergillus fumigatus*, *niger* and *flavus* moulds, being among the tested compounds the most active [170]. This CDX also exhibited moderate to good activity against some strains of the *Trichophyton mentagrophytes* and *rubum* dermatophytes, while being inactive against the *Candida albicans* yeast [170].

The antimicrobial activity of CDXs **228–230**, **241–242** against 12 strains of the bacteria *Helicobacter pylori* was evaluated through the Kirby-Bauer method, by measuring the diameters of inhibition zones, showing that compounds **228**, **230**, **241–242** exhibited strong activity against the strains ATCC 43504, 700684 and 43504. Actually, those CDXs were considered the best of the tested compounds, while compound **229**, demonstrated weak antibacterial activity [172].

In a recent study on the influence of reactive oxygen species (ROS) in the anticancer activity of aminoalkanolic derivatives, it was reported that in the case of CDXs **219** and **223**, ROS were of great importance to their proapoptotic activity [183]. These encouraging results suggested that aminoalkanolic CDXs might be interesting structures for potential use in anticancer therapy [183].

2.2.3. CDXs Conjuged with Amines, Amino Esters and Amino Acids

Inspired by natural xanthones properties, Rakesh et al. [184] synthesize xanthone derivatives with conjugated L-amino acids (**244–263**, Figure 21), to determine the corresponding antimicrobial and anti-inflammatory activities. The same research group recently reported the evaluation of *in vitro* anticancer activity of those compounds, against three different cancer cell lines, MCF-7, MDA-MB-435 and A549, validated by DNA binding and molecular docking approaches [185].

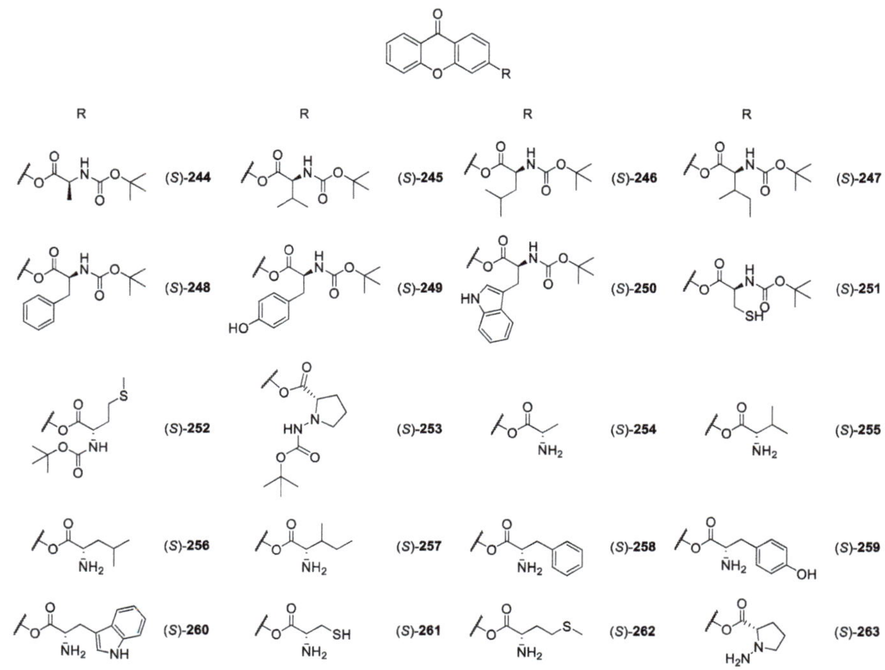

Figure 21. Structures of CDXs **244–263**, with antimicrobial and anti-inflammatory activities.

The synthetic strategy used to obtain the compounds was accomplished using 2-chlorobenzoic acid and resorcinol in anhydrous zinc chloride to give 2-chlorophenyl-(2,4-dihydroxyphenyl) methanone and cyclized with DMSO and NaOH to give 3-hydroxyxanthone. This chemical substrate was, afterwards, conjugated with different protected amino acids [184,185].

The compounds with the best antimicrobial and anti-inflammatory activities were those conjugated with L-phenylalanine, L-tyrosine and L-tryptophan, followed by compounds conjugated with L-cysteine, L-methionine and L-proline [184]. Additionally, the compounds with amino acids with high aromaticity and hydrophobicity, presented more stable amphiphilic structures.

The antimicrobial effect comes from the penetration of the amino acid hydrophobic chains in the bacterial membranes where the cationic moiety of the amino acids interacts with the membrane phospholipids disturbing bacterial membrane. This strategy proved to be effective to develop new antimicrobial agents [184], as the microorganism die without developing mutations or resulting in loss of recognition by the antibiotics [96]. Other studies accomplished the same conclusions [168,172,186]. Regarding the antitumor activity, the compounds (*S*)-**248**, (*S*)-**249** and (*S*)-**250** exhibited potent inhibition against the tested tumor cell lines as well as DNA binding. The SAR studies showed that the aromatic and hydrophobic amino acids, such as phenylalanine, tyrosine, and tryptophan, favored the DNA binding studies and antitumor activities; whereas, aliphatic amino acids showed lower activity. The derivatives with glycine, alanine, valine, leucine, and isoleucine showed less or moderate anticancer properties [185].

2.2.4. CDXs Containing Piperazine Moieties and Analogues

Several CDXs containing piperazine moieties (**267–270, 272–294**) and analogues (**264, 265, 271**) were synthesized (Figures 22–24) and their biological activity evaluated by Marona's group [165,167,168,172,173,181,187–192].

2-Hydroxyxanthone was the building block used to synthesize compounds **267–273** and **277** using epichorhydrin in the presence of pyridine [165]. Compounds **274–276, 278, 280–282** were synthesized by amination of 2-(2-bromoethoxy)-9*H*-xanthen-9-one and derivatives in *n*-propanol or toluene in the presence of K_2CO_3 [187]. The compounds **283–284, 292–293** were obtained by aminolysis of 4-[(2,3-epoxy)propoxy]xanthone with appropriate 1-piperazine derivatives in *n*-propanol, while **279** was synthesized using the same methodology through the aminolysis of 3-[(2,3-epoxy)propoxy]xanthone [172,173,188].

Chiral compounds **287–291** were obtained by amination of respective parent compounds [189] with appropriate amines in *n*-propanol. In addition, compound **287** was obtained from compound **286** by acetylation. In order to optimize synthetic methodologies, CDX **286** was obtained using an alternative method including (*R,S*)-4-(3-chloro-2-hydroxypropoxy)-9*H*-xanthen-9-one as intermediate [189].

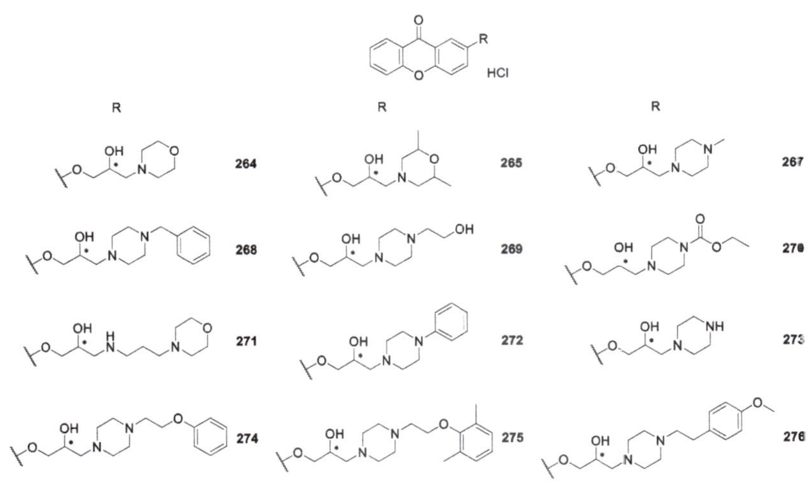

Figure 22. Structures of 2-piperazine derivatives **264–276**.

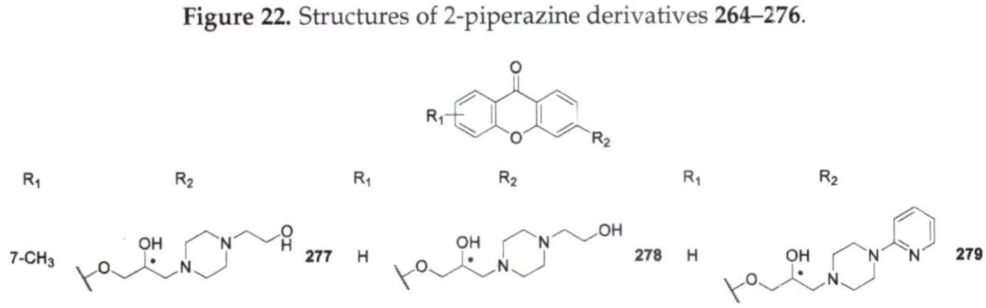

Figure 23. Structures of 3-piperazine derivatives **277–279**.

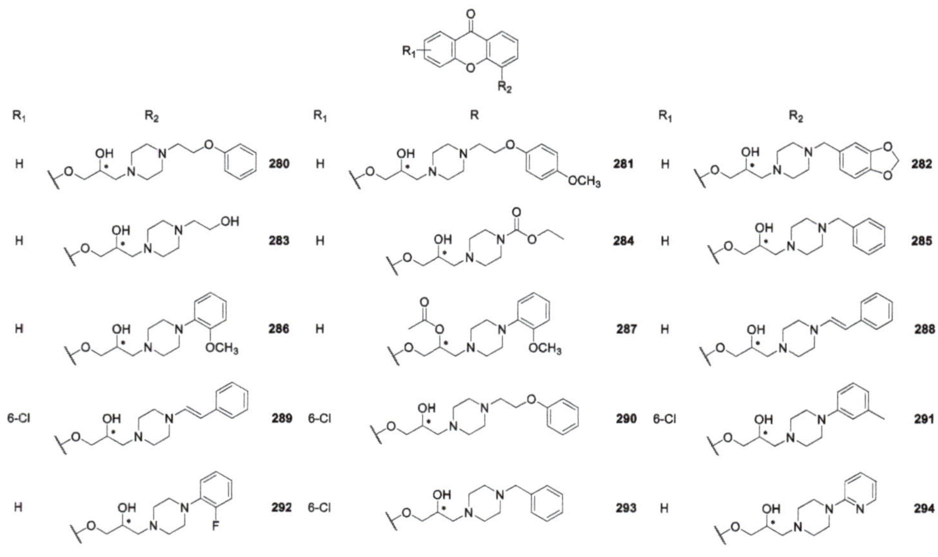

Figure 24. Structures of 4-piperazine derivatives **280–294**.

CDXs **267–273** and **277** were evaluated for anticonvulsant activity in the MES- and subcutaneous pentylenetrazole-induzed seizures in mice and rats [165]. Among them, the most promising compound was CDX **268** which was active in both the anticonvulsant tests.

Moreover, the influence on the platelet aggregation of CDXs **264, 265, 269** and **271** was evaluated by using adenosine-5′-diphosphate (ADP), sodium arachidonate (AA) or thrombin as the aggregating agents [181]. CDXs **265** and **271** were active, inducing 80-90% inhibition of thrombin-stimulated platelet aggregation.

Considering that the xanthone itself proved to possess vasorelaxing properties in thoracic aorta isolated from rats [193] and the strongest hypotensive effects were observed for compounds containing piperazine moiety [189], several compounds that combine the xanthone nucleus and piperazine rings (**274–276, 278–291**) were evaluated for anti-arrhythmic and/or antihypertensive activities. It is important to emphasize that CDXs **274** and **280** demonstrated to possess significant anti-arrhythmic activity in the adrenaline-induced model of arrhythmia [187]. The strongest hypotensive activity which persisted for 60 min was also associated to compound **89**.

Additionally, in another study related to the same biological activities, compound **96** was the most promising considering its effect on circulatory system. Moreover, this CDX diminished arterial blood pressure by about 40% during one hour [188].

A recent cardiovascular activity study of several CDXs **286–291** was described, including the following pharmacological experiments: the binding affinity for adrenoceptors, the influence on the normal electrocardiogram, the effect on the arterial blood pressure and prophylactic antiarrhythmic activity in adrenaline induced model of arrhythmia (rats, iv) [189]. The CDXs **286** and **287** revealed to act as potential antiarrhythmics in adrenaline induced model of arrhythmia in rats after intravenous injection. In another study, CDX **279** was reported as a promising hypotensive with its activity attributed to the blockage of α1-adrenoreceptors [173]. The results obtained were quite encouraging and suggested that in the group of xanthone derivatives new potential antiarrhythmics and hypotensives might be found.

Table 2. Summary of the biological activities of synthetic CDXs obtained by binding/coupling chiral moieties to the xanthone scaffold.

CDXs	Biological Activities	Ref.
XAA, DMXAA and analogues **156–164**	Antitumor, antiviral, antiplatelet, antithrombotic, anti-inflammatory and analgesic	[122–133,136] [134] [135] [137] [140]
Aminoalkanolic CDXs **165–243**	Cyclooxygenases inhibition, antitumor, anticonvulsant, cardiovascular, antiplatelet aggregation, antimutagenic, antifungal, antibacterial and anticancer	[160] [148] [139–163,167,171,175–179] [167,169,173–175,179,180] [181] [182] [170] [172] [183]
CDXs conjugated with amines, amino esters and amino acids **244–263**	Anti-inflammatory	[168,172,186]
CDXs containing piperazine moieties and analogues **264–294**	Anticonvulsant, antiplatelet aggregation, cardiovascular, antifungal and antibacterial	[165] [181] [173,187–190] [188] [172]
CDXs containing other moieties **295–310**	No activities reported	[194]

A series of some chiral derivatives of 2-xanthones **267–273, 277** with a piperazine moiety was evaluated for their activity against *M. tuberculosis*. The highest level of activity against *M. tuberculosis* was observed for compound **270**, which exhibited 94% growth inhibition. This compound was also examined for its anti-*M. avium* activity as well as cytotoxicity, showing insignificant anti *M. avium* activity and cytotoxic effects [188].

Recently, CDXs **284–285, 290, 292, 293** were evaluated for their antibacterial activity against 12 strains of the bacteria *Helicobacter pylori* [172]. CDX **285** showed strong activity against *H. pylori* strain ATCC 43504 and 700684, being the only compound to show higher activity against clarithromycin-resistant *H. Pylori* strains, than to the one resistant to metronidazole [172].

2.2.5. CDXs Containing Other Moieties

Recently, Cherkadu et al. [194] reported the synthesis of some CDXs **295–310** (Figure 25) containing moieties other than piperazine and aminoalcohols [194]. The synthesis of these CDXs, a series of 2-(aminobenzothiazol)methylxanthones, was performed through the reaction of 3-hydroxy-xanthone, aromatic aldehydes and 2-aminobenzothiazoles in DMF at 120°C with $FeCl_3$ as catalyst [194]. To the best of our knowledge, no biological activities were reported for those CDXs.

A summary of the synthetic CDXs obtained by binding/coupling chiral moieties to the xanthone scaffold, their biological activities as well as the associated references are presented in Table 2.

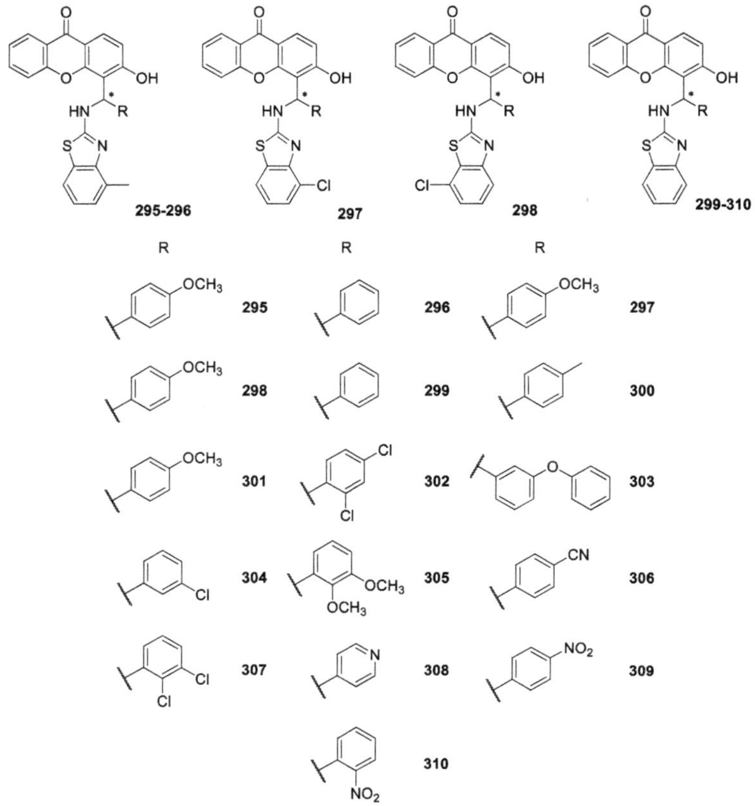

Figure 25. Structures of CDXs **295–310**.

3. Conclusions

The synthetic CDXs, inspired in natural sources, and obtained by coupling chiral moieties to the xanthone scaffold, demonstrated potential to perform a large variety of biological activities, including antitumor, antimicrobial, anticonvulsant, antimalarial, anti-inflammatory, antiplatelet, anti- thrombotic, antipyretic, analgesic, antioxidant, antidiabetic, anticoagulant, among others. Nevertheless, for this family of compounds the main biological activities reported were antitumor and antimicrobial.

The more studied chiral moieties were amines, amino alcohols and amino acids. The cationic moieties of the amino acids have been indicated as a good approach to develop new antimicrobial agents both for CDXs inspired in natural xanthones and obtained by coupling chiral moieties to the xanthone scaffold. Regarding enantioselectivity, some studies reported the importance in SAR studies, but the majority neglected the influence of stereochemistry in the biological activity.

Author Contributions: C.F, M.L.C, J.R. and J.A collected the primary data and contributed in writing of the manuscript. M.E.T., C.F. and M.M.M.P. supervised the development of the manuscript, and assisted in data interpretation, manuscript evaluation, and editing.

Funding: This work was supported by the Strategic Funding UID/Multi/04423/2013 through national funds provided by FCT – Foundation for Science and Technology and European Regional Development Fund (ERDF), in the framework of the programme PT2020, the project PTDC/MAR-BIO/4694/2014 (reference POCI-01-0145-FEDER-016790; Project 3599 – Promover a Produção Científica e Desenvolvimento Tecnológico e a Constituição de Redes Temáticas (3599-PPCDT)) as well as by Project No. POCI-01-0145-FEDER-028736, co-financed by COMPETE 2020, under the PORTUGAL 2020 Partnership Agreement, through the European Regional Development Fund (ERDF), and CHIRALXANT-CESPU-2018.

Conflicts of Interest: The authors declare no conflict of interest.

References

1. Tiritan, M.E.; Ribeiro, A.R.; Fernandes, C.; Pinto, M. Chiral pharmaceuticals. In *Kirk-Othmer Encyclopedia of Chemicl Technology*; John Wiley & Sons, Inc.: Hoboken, NJ, USA, 2016.
2. Calcaterra, A.; D'Acquarica, I. The market of chiral drugs: Chiral switches versus de novo enantiomerically pure compounds. *J. Pharm. Biomed. Anal.* **2018**, *147*, 323–340. [CrossRef] [PubMed]
3. Caner, H.; Groner, E.; Levy, L.; Agranat, I. Trends in the development of chiral drugs. *Drug Discov. Today* **2004**, *9*, 105–110. [CrossRef]
4. Blaser, H.U. Chirality and its implications for the pharmaceutical industry. *Rend. Lincei.* **2013**, *24*, 213–216. [CrossRef]
5. Mannschreck, A.; Kiesswetter, R.; von Angerer, E. Unequal activities of enantiomers via biological receptors: Examples of chiral drug, pesticide, and fragrance molecules. *J. Chem. Ed.* **2007**, *84*, 2012–2017. [CrossRef]
6. Triggle, D.J. Stereoselectivity of drug action. *Drug Discov. Today* **1997**, *2*, 138–147. [CrossRef]
7. Cordato, D.J.; Mather, L.E.; Herkes, G.K. Stereochemistry in clinical medicine: A neurological perspective. *J. Clin. Neurosci.* **2003**, *10*, 649–654. [CrossRef]
8. Smith, S.W. Chiral toxicology: It's the same thing only different. *Toxicol. Sci.* **2009**, *110*, 4–30. [CrossRef] [PubMed]
9. Silva, B.; Fernandes, C.; Tiritan, M.E.; Pinto, M.M.M.; Valente, M.J.; Carvalho, M.; de Pinho, P.G.; Remião, F. Chiral enantioresolution of cathinone derivatives present in "legal highs", and enantioselectivity evaluation on cytotoxicity of 3,4-methylenedioxypyrovalerone (mdpv). *Forensic. Toxicol.* **2016**, *34*, 372–385. [CrossRef] [PubMed]
10. Silva, B.; Fernandes, C.; de Pinho, P.G.; Remião, F. Chiral resolution and enantioselectivity of synthetic cathinones: A brief review. *J. Anal. Toxicol.* **2018**, *42*, 17–24. [CrossRef] [PubMed]
11. Caldwell, J. Importance of stereospecific bioanalytical monitoring in drug development. *J. Chromatogr. A* **1996**, *719*, 3–13. [CrossRef]
12. Andrushko, V.; Andrushko, N. *Stereoselective Synthesis of Drugs and Natural Products*; Jonh Wiley & Sons: Hoboken, NJ, USA, 2013; Volume 2, p. 1836.
13. Bhadra, S.; Yamamoto, H. Substrate directed asymmetric reactions. *Chem. Rev.* **2018**, *118*, 3391–3446. [CrossRef] [PubMed]
14. Karimi, B.; Tavakolian, M.; Akbari, M.; Mansouri, F. Ionic liquids in asymmetric synthesis: An overall view from reaction media to supported ionic liquid catalysis. *ChemCatChem* **2018**, *10*, 3173–3205. [CrossRef]
15. Xue, Y.P.; Cao, C.H.; Zheng, Y.G. Enzymatic asymmetric synthesis of chiral amino acids. *Chem. Soc. Rev.* **2018**, *47*, 1516–1561. [CrossRef] [PubMed]
16. Fernandes, C.; Tiritan, M.E.; Pinto, M.M.M. Chiral separation in preparative scale: A brief overview of membranes as tools for enantiomeric separation. *Symmetry* **2017**, *9*, 206. [CrossRef]
17. Ward, T.J.; Ward, K.D. Chiral separations: A review of current topics and trends. *Anal. Chem* **2012**, *84*, 626–635. [CrossRef] [PubMed]
18. Francotte, E.R. Enantioselective chromatography as a powerful alternative for the preparation of drug enantiomers. *J. Chromatogr. A* **2001**, *906*, 379–397. [CrossRef]
19. Haginaka, J. Pharmaceutical and biomedical applications of enantioseparations using liquid chromatographic techniques. *J. Pharm. Biomed. Anal.* **2002**, *27*, 357–372. [CrossRef]
20. Council of the European Communities. *Investigation of Chiral Active Substances. Directive 75/318/EEC*; Council of the European Communities: Brussels, Belgium, 1993.
21. FDA. Fda's policy statement for the development of new stereoisomeric drugs. *Fed. Reg.* **1992**, *57*, 249.
22. Shimazawa, R.; Nagai, N.; Toyoshima, S.; Okuda, H. Present state of new chiral drug development and review in Japan. *J. Health Sci.* **2008**, *54*, 23–29. [CrossRef]
23. Fernandes, C.; Phyo, Y.Z.; Silva, A.S.; Tiritan, M.E.; Kijjoa, A.; Pinto, M.M.M. Chiral stationary phases based on small molecules: An update of the last 17 years. *Sep. Purif. Rev.* **2018**, *47*, 89–123. [CrossRef]
24. Fernandes, C.; Tiritan, M.E.; Cravo, S.; Phyo, Y.Z.; Kijjoa, A.; Silva, A.M.S.; Cass, Q.B.; Pinto, M.M.M. New chiral stationary phases based on xanthone derivatives for liquid chromatography. *Chirality* **2017**, *29*, 430–442. [CrossRef] [PubMed]

25. Ribeiro, J.; Tiritan, M.E.; Pinto, M.M.M.; Fernandes, C. Chiral stationary phases for liquid chromatography based on chitin- and chitosan-derived marine polysaccharides. *Symmetry* **2017**, *9*, 190. [CrossRef]
26. Lämmerhofer, M. Chiral recognition by enantioselective liquid chromatography: Mechanisms and modern chiral stationary phases. *J. Chromatogr. A* **2010**, *1217*, 814–856. [CrossRef] [PubMed]
27. Gales, L.; Damas, A.M. Xanthones–a structural perspective. *Curr. Med. Chem.* **2005**, *12*, 2499–2515. [CrossRef] [PubMed]
28. Muthukrishnan, M.; Basavanag, U.M.V.; Puranik, V.G. The first ionic liquid-promoted kabbe condensation reaction for an expeditious synthesis of privileged bis-spirochromanone scaffolds. *Tetrahedron. Lett.* **2009**, *50*, 2643–2648. [CrossRef]
29. Horton, D.A.; Bourne, G.T.; Smythe, M.L. The combinatorial synthesis of bicyclic privileged structures or privileged substructures. *Chem. Rev.* **2003**, *103*, 893–930. [CrossRef] [PubMed]
30. Waseem, S.; Ahmad, I. Recent insight into the biological activities of synthetic xanthone derivatives. *Eur. J. Med. Chem.* **2016**, *116*, 267–280.
31. Pinto, M.; Sousa, M.; Nascimento, M. Xanthone derivatives: New insights in biological activities. *Curr. Med. Chem.* **2005**, *12*, 2517–2538. [CrossRef]
32. Wezeman, T.; Brase, S.; Masters, K.S. Xanthone dimers: A compound family which is both common and privileged. *Nat. Prod. Rep.* **2015**, *32*, 6–28. [CrossRef] [PubMed]
33. Na, Y. Recent cancer drug development with xanthone structures. *J. Pharm. Pharmacol.* **2009**, *61*, 707–712. [CrossRef] [PubMed]
34. Ribeiro, J.; Veloso, C.; Fernandes, C.; Tiritan, M.E.; Pinto, M.M.M. Carboxyxanthones: Bioactive agents and molecular scaffold for synthesis of analogues and derivatives. *Molecules* **2019**, *24*, 180. [CrossRef] [PubMed]
35. Sathyadevi, P.; Chen, Y.J.; Wu, S.C.; Chen, Y.H.; Wang, Y.M. Reaction-based epoxide fluorescent probe for in vivo visualization of hydrogen sulfide. *Biosens. Bioelectron.* **2015**, *68*, 681–687. [CrossRef] [PubMed]
36. Takashima, I.; Kawagoe, R.; Hamachi, I.; Ojida, A. Development of an and logic-gate-type fluorescent probe for ratiometric imaging of autolysosome in cell autophagy. *Chem. Eur. J.* **2015**, *21*, 2038–2044. [CrossRef] [PubMed]
37. Fernandes, C.; Tiritan, M.E.; Pinto, M.M.M. Chiral derivatives of xanthones: Applications in medicinal chemistry and a new approach in liquid chromatography. *Sci. Chromatogr.* **2015**, *7*, 1–14.
38. Masters, K.S.; Brase, S. Xanthones from fungi, lichens, and bacteria: The natural products and their synthesis. *Chem. Rev.* **2012**, *112*, 3717–3776. [CrossRef] [PubMed]
39. Vieira, L.M.; Kijjoa, A. Naturally-occurring xanthones: Recent developments. *Curr. Med. Chem.* **2005**, *12*, 2413–2446. [CrossRef] [PubMed]
40. Pinto, M.M.M.; Castanheiro, R.A.P.; Kijjoa, A. Xanthones from marine-derived microorganisms: Isolation, structure elucidation, and biological activities. In *Encyclopedia of Analytical Chemistry*, 1st ed.; John Wiley & Sons: Hoboken, NJ, USA, 2014; Volume 27, pp. 1–21.
41. Sousa, M.; Pinto, M. Synthesis of xanthones: An overview. *Curr. Med. Chem.* **2005**, *12*, 2447–2479. [CrossRef] [PubMed]
42. Azevedo, C.; Afonso, C.; Pinto, M. Routes to xanthones: An update on the synthetic approaches. *Curr. Org. Chem.* **2012**, *16*, 2818–2867. [CrossRef]
43. Lanzotti, V. Drugs based on natural compounds: Recent achievements and future perspectives. *Phytochem. Rev.* **2014**, *13*, 725–726. [CrossRef]
44. Michael, A. On the action of aromatic oxy-acids on phenols. *Am. Chem. J.* **1883**, *5*, 81–97.
45. V Kostanecki, S. Über das gentisin. *Monatsh. Chem.* **1891**, *12*, 205–210. [CrossRef]
46. Goldberg, I. Ueber Phenylirungen bei Gegenwart von Kupfer als Katalysator. *Chem. Ber.* **1906**, *39*, 1691–1692. [CrossRef]
47. Chen, M.T.; Kuoh, Y.P.; Wang, C.H.; Chen, C.M.; Kuoh, C.S. Additional constituents of hypericum subalatum. *J. Chin. Chem. Soc.* **1988**, *36*, 165–168. [CrossRef]
48. Wu, Q.-L.; Wang, S.-P.; Du, L.-J.; Yang, J.-S.; Xiao, P.-G. Xanthones from hypericum japonicum and h. Henryi. *Phytochemistry* **1998**, *49*, 1395–1402. [CrossRef]
49. Pinto, M.; Sousa, E. Natural and synthetic xanthonolignoids: Chemistry and biological activities. *Curr. Med. Chem.* **2003**, *10*, 1–12. [CrossRef] [PubMed]
50. Gottlieb, O.R.; Mesquita, A.A.L.; Silva, E.M.; Melo, T. Xanthones of *kielmeyera ferruginea*. *Phytochemistry* **1969**, *8*, 665–666. [CrossRef]

51. Pinto, M.M.D.M.; Mesquita, A.A.L.; Gottlieb, O.R. Xanthonolignoids from *kielmeyera coriacea*. *Phytochemistry* **1987**, *26*, 2045–2048. [CrossRef]
52. Tanaka, H.; Ishihara, M.; Ichino, K.; Ohiwa, N.; Ito, K. Total synthesis of a xanthonolignoid, kielcorin. *Chem. Pharm. Bull.* **1989**, *37*, 1916–1918. [CrossRef]
53. Vishwakarma, R.; Kapil, R.S.; Popli, S.P. Total synthesis of Keilcorin. *Indian J. Chem. Sect. B* **1986**, *25*, 1021.
54. Fernandes, E.G.R.; Pinto, M.M.M.; Silva, A.M.S.; Cavaleiro, J.A.S.; Gottlieb, O.R. Synthesis and structural elucidation of xanthonolignoids: Trans-(±)- kielcorin b and trans-(±)-isokielcorin b. *Heterocycles* **1999**, *51*, 821–828.
55. Saraiva, L.; Fresco, P.; Pinto, E.; Sousa, E.; Pinto, M.; Gonçalves, J. Synthesis and in vivo modulatory activity of protein kinase c of xanthone derivatives. *Bioorg. Med. Chem.* **2002**, *10*, 3219–3227. [CrossRef]
56. Sousa, E.P.; Silva, A.; Pinto, M.M.; Pedro, M.M.; Cerqueira, F.A.; Nascimento, M.S. Isomeric kielcorins and dihydroxyxanthones: Synthesis, structure elucidation, and inhibitory activities of growth of human cancer cell lines and on the proliferation of human lymphocytes in vitro. *Helv. Chim. Acta* **2002**, *85*, 2862–2876. [CrossRef]
57. Saraiva, L.; Fresco, P.; Pinto, E.; Sousa, E.; Pinto, M.; Gonçalves, J. Inhibition of α, βi, δ, η and ζ protein kinase c isoforms by xanthonolignoids. *J. Enzyme. Inhib. Med. Chem.* **2003**, *18*, 357–370. [CrossRef] [PubMed]
58. Fernandas, E.R.; Carvalho, F.D.; Remião, F.G.; Bastos, M.L.; Pinto, M.M.; Gottlieb, O.R. Hepatoprotective activity of xanthones and xanthonolignoids against tert-butylhydroperoxide-induced toxicity in isolated rat hepatocytes—comparison with silybin. *Pharm. Res.* **1995**, *12*, 1756–1760. [CrossRef]
59. Sousa, E.P.; Tiritan, M.; Oliveira, R.V.; Afonso, C.; Cass, Q.; Pinto, M.M. Enantiomeric resolution of kielcorin derivatives by hplc on polysaccharide stationary phases using multimodal elution. *Chirality* **2004**, *16*, 279–285. [CrossRef]
60. Sousa, M.E.; Tiritan, M.E.; Belaz, K.R.A.; Pedro, M.; Nascimento, M.S.J.; Cass, Q.B.; Pinto, M.M.M. Multimilligram enantioresolution of low-solubility xanthonolignoids on polysaccharide chiral stationary phases using a solid-phase injection system. *J. Chromatogr. A* **2006**, *1120*, 75–81. [CrossRef]
61. Kupchan, S.M.; Streelman, D.R.; Sneden, A.T. Psorospermin, a new antileukemic xanthone from psorospermum febrifugum. *J. Nat. Prod.* **1980**, *43*, 296–301. [CrossRef]
62. Schwaebe, M.K.; Moran, T.J.; Whitten, J.P. Total synthesis of psorospermin. *Tetrahedron. Lett.* **2005**, *46*, 827–829. [CrossRef]
63. Heald, R.A.; Dexheimer, T.S.; Vankayalapati, H.; Siddiqui-Jain, A.; Szabo, L.Z.; Gleason-Guzman, M.C.; Hurley, L.H. Conformationally restricted analogues of psorospermin: Design, synthesis, and bioactivity of natural-product-related bisfuranoxanthones. *J. Med. Chem.* **2005**, *48*, 2993–3004. [CrossRef] [PubMed]
64. Fellows, I.M.; Schwaebe, M.; Dexheimer, T.S.; Vankayalapati, H.; Gleason-Guzman, M.; Whitten, J.P.; Hurley, L.H. Determination of the importance of the stereochemistry of psorospermin in topoisomerase ii–induced alkylation of DNA and in vitro and in vivo biological activity. *Mol. Cancer Ther.* **2005**, *4*, 1729–1739. [CrossRef] [PubMed]
65. Ayensu, E.S. *Medicinal Plants of West Africa*; Reference Publications Inc.: Algonac, MI, USA, 1978.
66. Kodama, T.; Ito, T.; Dibwe, D.F.; Woo, S.Y.; Morita, H. Syntheses of benzophenone-xanthone hybrid polyketides and their antibacterial activities. *Bioorg. Med. Chem. Lett.* **2017**, *27*, 2397–2400. [CrossRef]
67. Dibwe, D.; Awale, S.; Kadota, S.; Tezuka, Y. Muchimangins a–d: Novel diphenylmethyl-substituted xanthones from securidaca longepedunculata. *Tetrahedron. Lett.* **2012**, *53*, 6186–6190. [CrossRef]
68. Gold-Smith, F.; Fernandez, A.; Bishop, K. Mangiferin and cancer: Mechanisms of action. *Nutrients* **2016**, *8*, 396. [CrossRef] [PubMed]
69. Zheng, M.; Lu, Z. Antiviral effect of mangiferin and isomangiferin on herpes simplex virus. *Chin. Med. J.* **1990**, *103*, 160–165. [PubMed]
70. Zhu, X.; Song, J.; Huang, Z.; Wu, Y.; Yu, M. Antiviral activity of mangiferin against herpes simplex virus type 2 in vitro. *Acta Pol. Pharm. Sinica.* **1993**, *14*, 452–454.
71. Yoosook, C.; Bunyapraphatsara, N.; Boonyakiat, Y.; Kantasuk, C. Anti-herpes simplex virus activities of crude water extracts of thai medicinal plants. *Phytomedicine* **2000**, *6*, 411–419. [CrossRef]
72. Benard, O.; Chi, Y. Medicinal properties of mangiferin, structural features, derivative synthesis, pharmacokinetics and biological activities. *Mini-Rev. Med. Chem.* **2015**, *15*, 582–594. [CrossRef]
73. Imran, M.; Arshad, M.S.; Butt, M.S.; Kwon, J.H.; Arshad, M.U.; Sultan, M.T. Mangiferin: A natural miracle bioactive compound against lifestyle related disorders. *Lip. Health Dis.* **2017**, *16*, 84. [CrossRef]

74. Araújo, J.; Fernandes, C.; Pinto, M.; Tiritan, M. Chiral derivatives of xanthones with antimicrobial activity. *Molecules* **2019**, *24*, 314. [CrossRef]
75. Singh, S.; Tiwari, R.; Sinha, S.; Danta, C.; Prasad, S. Antimicrobial evaluation of mangiferin and its synthesized analogues. *Asian Pac. J. Trop. Biomed.* **2012**, *2*, S884–S887. [CrossRef]
76. Singh, S.; Sinha, S.; Prasad, S.; Kumar, R.; Bithu, B.; Kumar, S.; Singh, P. Synthesis and evaluation of novel analogues of mangiferin as potent antipyretic. *Asian Pac. J. Trop. Biomed.* **2011**, *4*, 866–869. [CrossRef]
77. Dar, A.; Faizi, S.; Naqvi, S.; Roome, T.; Zikr-ur-Rehman, S.; Ali, M.; Firdous, S.; Moin, S. Analgesic and antioxidant activity of mangiferin and its derivatives: The structure activity relationship. *Biol. Pharm. Bull.* **2005**, *28*, 596–600. [CrossRef] [PubMed]
78. Mahendran, S.; Badami, S.; Ravi, S.; Thippeswamy, B.; Veerapur, V. Synthesis and evaluation of analgesic and anti-inflammatory activities of most active free radical scavenging derivatives of mangiferin. *Bri. J. Appl. Sci. Technol.* **2014**, *4*, 4959–4973. [CrossRef]
79. Miura, T.; Ichiki, H.; Iwamoto, N.; Kato, M.; Kubo, M.; Sasaki, H.; Okada, M.; Ishida, T.; Seino, Y.; Tanigawa, K. Antidiabetic activity of the rhizoma of anemarrhena asphodeloides and active components, mangiferin and its glucoside. *Biol. Pharm. Bull.* **2001**, *24*, 1009–1011. [CrossRef] [PubMed]
80. Miura, T.; Ichiki, H.; Hashimoto, I.; Iwamoto, N.; Kao, M.; Kubo, M.; Ishihara, E.; Komatsu, Y.; Okada, M.; Ishida, T.; et al. Antidiabetic activity of a xanthone compound, mangiferin. *Phytomedicine* **2001**, *8*, 85–87. [CrossRef] [PubMed]
81. Hu, H.G.; Wang, M.J.; Zhao, Q.J.; Yu, S.C.; Liu, C.M.; Wu, Q.Y. Synthesis of mangiferin derivates and study their potent ptp1b inhibitory activity. *Chin. Chem. Lett.* **2007**, *18*, 1323–1326. [CrossRef]
82. Hu, H.; Wang, M.; Zhao, Q.; Liao, H.; Cai, L.; Song, Y.; Zhang, J.; Yu, S.; Chen, W.; Liu, C.; et al. Synthesis of mangiferin derivatives as protein tyrosine phosphatase 1b inhibitors. *Chem. Nat. Compd.* **2007**, *43*. [CrossRef]
83. Li, X.; Du, Z.; Huang, Y.; Liu, B.; Hu, W.; Lu, W.; Deng, J. Synthesis and hypoglycemic activity of esterified-derivatives of mangiferin. *Chin. J. Nat. Med.* **2013**, *11*, 296–301. [CrossRef]
84. Klaman, L.; Boss, O.; Peroni, O.; Kim, J.; Martino, J.; Zabolotny, J.; Moghal, N.; Lubkin, M.; Kim, Y.; Sharpe, A.; et al. Increased energy expenditure, decreased adiposity, and tissue-specific insulin sensitivity in protein-tyrosine phosphatase 1b-deficient mice. *Mol. Cell. Biol.* **2000**, *20*, 5479–5489. [CrossRef]
85. Elchebly, M.; Payette, P.; Michaliszyn, E.; Cromlish, W.; Collins, S.; Loy, A.; Normandin, D.; Cheng, A.; Himms-Hagen, J.; Chan, C.; et al. Increased insulin sensitivity and obesity resistance in mice lacking the protein tyrosine phosphatase-1b gene. *Science* **1999**, *283*, 1544–1548. [CrossRef]
86. Correia-da-Silva, M.; Sousa, E.; Duarte, B.; Marques, F.; Carvalho, F.; Cunha-Ribeiro, L.; Pinto, M. Polysulfated xanthones: Multipathway development of a new generation of dual anticoagulant/antiplatelet agents. *J. Med. Chem.* **2011**, *54*, 5373–5384. [CrossRef] [PubMed]
87. Kaomongkolgit, R.; Jamdez, K.; Chaisomboon, N. Antifungal activity of alpha-mangostin against candida albicans. *J. Oral. Sci.* **2009**, *51*, 401–406. [CrossRef] [PubMed]
88. Jung, H.; Su, B.; Keller, W.; Mehta, R.; Kinghorn, A. Antioxidant xanthones from the pericarp of garcinia mangostana (mangosteen). *J. Agric. Food. Chem.* **2006**, *54*. [CrossRef] [PubMed]
89. Guzman-Beltran, S.; Rubio-Badillo, M.A.; Juarez, E.; Hernandez-Sanchez, F.; Torres, M. Nordihydroguaiaretic acid (ndga) and alpha-mangostin inhibit the growth of mycobacterium tuberculosis by inducing autophagy. *Int. Immunopharmacol.* **2016**, *31*, 149–157. [CrossRef] [PubMed]
90. Xu, Z.; Huang, L.; Chen, X.H.; Zhu, X.F.; Qian, X.J.; Feng, G.K.; Lan, W.J.; Li, H.J. Cytotoxic prenylated xanthones from the pericarps of garcinia mangostana. *Molecules* **2014**, *19*, 1820–1827. [CrossRef]
91. Sudta, P.; Jiarawapi, P.; Suksamrarn, A.; Hongmanee, P.; Suksamrarn, S. Potent activity against multidrug-resistant mycobacterium tuberculosis of α-mangostin analogs. *Chem. Pharm. Bull.* **2013**, *61*, 194–203. [CrossRef] [PubMed]
92. Chen, L.; Yang, L.; Wang, C. Anti-inflammatory activity of mangostins from garcinia mangostana. *Food Chem. Toxicol.* **2008**, *46*, 688–693. [CrossRef] [PubMed]
93. Iikubo, K.; Ishikawa, Y.; Ando, N.; Umezawab, K.; Nishiyama, S. The first direct synthesis of alfa-mangostin, a potent inhibitor of the acidic sphingomyelinase. *Tetrahedron. Lett.* **2002**, *43*. [CrossRef]
94. Suksamrarn, S.; Suwannapoch, N.; Phakhodee, W.; Thanuhiranlert, J.; Ratananukul, P.; Chimnoi, N.; Suksamrarn, A. Antimycobacterial activity of prenylated xanthones from the fruits of garcinia mangostana. *Chem. Pharm. Bull.* **2003**, *51*, 857–859. [CrossRef] [PubMed]

95. Koh, J.J.; Zou, H.; Mukherjee, D.; Lin, S.; Lim, F.; Tan, J.K.; Tan, D.; Stocker, B.L.; Timmer, M.; Corkran, H.M.; et al. Amphiphilic xanthones as a potent chemical entity of anti-mycobacterial agents with membrane-targeting properties. *Eur. J. Med. Chem.* **2016**, *123*, 684–703. [CrossRef] [PubMed]
96. Li, J.; Liu, S.; Koh, J.J.; Zou, H.; Lakshminarayanan, R.; Bai, Y.; Pervushin, K.; Zhou, L.; Verma, C.; Beuerman, R.W. A novel fragment based strategy for membrane active antimicrobials against mrsa. *BBA-Biomembranes* **2015**, *1848*, 1023–1031. [CrossRef] [PubMed]
97. Lin, S.; Koh, J.J.; Aung, T.T.; Lim, F.; Li, J.; Zou, H.; Wang, L.; Lakshminarayanan, R.; Verma, C.; Wang, Y.; et al. Symmetrically substituted xanthone amphiphiles combat gram-positive bacterial resistance with enhanced membrane selectivity. *J. Med. Chem.* **2017**, *60*, 1362–1378. [CrossRef] [PubMed]
98. Zou, H.; Koh, J.J.; Li, J.; Qiu, S.; Aung, T.T.; Lin, H.; Lakshminarayanan, R.; Dai, X.; Tang, C.; Lim, F.H.; et al. Design and synthesis of amphiphilic xanthone-based, membrane-targeting antimicrobials with improved membrane selectivity. *J. Med. Chem.* **2013**, *56*, 2359–2373. [CrossRef] [PubMed]
99. Koh, J.J.; Lin, S.; Aung, T.T.; Lim, F.; Zou, H.; Bai, Y.; Li, J.; Lin, H.; Pang, L.M.; Koh, W.L.; et al. Amino acid modified xanthone derivatives: Novel, highly promising membrane-active antimicrobials for multidrug-resistant gram-positive bacterial infections. *J. Med. Chem.* **2015**, *58*, 739–752. [CrossRef] [PubMed]
100. Ren, Y.; Lantvit, D.D.; Carcache de Blanco, E.J.; Kardono, L.B.; Riswan, S.; Chai, H.; Cottrell, C.E.; Farnsworth, N.R.; Swanson, S.M.; Ding, Y.; et al. Proteasome-inhibitory and cytotoxic constituents of garcinia lateriflora: Absolute configuration of caged xanthones. *Tetrahedron* **2010**, *66*, 5311–5320. [CrossRef] [PubMed]
101. Ren, Y.; Yuan, C.; Chai, H.B.; Ding, Y.; Li, X.C.; Ferreira, D.; Kinghorn, A.D. Absolute configuration of (−)-gambogic acid, an antitumor agent. *J. Nat. Prod.* **2011**, *74*, 460–463. [CrossRef] [PubMed]
102. Rukachaisirikul, V.; Kaewnok, W.; Koysomboon, S.; Phongpaichit, S.; Taylor, W.C. Caged-tetraprenylated xanthones from garcinia scortechinii. *Tetrahedron* **2000**, *56*, 8539–8543. [CrossRef]
103. Sukpondma, Y.; Rukachaisirikul, V.; Phongpaichit, S. Antibacterial caged-tetraprenylated xanthones from the fruits of garcinia hanburyi. *Chem. Pharm. Bull.* **2005**, *53*, 850–852. [CrossRef] [PubMed]
104. Banik, K.; Harsha, C.; Bordoloi, D.; Lalduhsaki Sailo, B.; Sethi, G.; Leong, H.C.; Arfuso, F.; Mishra, S.; Wang, L.; Kumar, A.P.; et al. Therapeutic potential of gambogic acid, a caged xanthone, to target cancer. *Cancer Lett.* **2018**, *416*, 75–86. [CrossRef] [PubMed]
105. Chaiyakunvat, P.; Anantachoke, N.; Reutrakul, V.; Jiarpinitnun, C. Caged xanthones: Potent inhibitors of global predominant mrsa usa300. *Bioorg. Med. Chem. Lett.* **2016**, *26*, 2980–2983. [CrossRef]
106. Ke, H.; Morrisey, J.M.; Qu, S.; Chantarasriwong, O.; Mather, M.W.; Theodorakis, E.A.; Vaidya, A.B. Caged garcinia xanthones, a novel chemical scaffold with potent antimalarial activity. *Antimicrob. Agents Chemother.* **2017**, *61*. [CrossRef] [PubMed]
107. Reutrakul, V.; Anantachoke, N.; Pohmakotr, M.; Jaipetch, T.; Sophasan, S.; Yoosook, C.; Kasisit, J.; Napaswat, C.; Santisuk, T.; Tuchinda, P. Cytotoxic and anti-hiv-1 caged xanthones from the resin and fruits of garcinia hanburyi. *Planta Med.* **2007**, *73*, 33–40. [CrossRef] [PubMed]
108. Alam, S.; Khan, F. Virtual screening, docking, admet and system pharmacology studies on garcinia caged xanthone derivatives for anticancer activity. *Sci. Rep.* **2018**, *8*, 5524. [CrossRef] [PubMed]
109. Batova, A.; Altomare, D.; Chantarasriwong, O.; Ohlsen, K.L.; Creek, K.E.; Lin, Y.C.; Messersmith, A.; Yu, A.L.; Yu, J.; Theodorakis, E.A. The synthetic caged garcinia xanthone cluvenone induces cell stress and apoptosis and has immune modulatory activity. *Mol. Cancer Ther.* **2010**, *9*, 2869–2878. [CrossRef] [PubMed]
110. Chantarasriwong, O.; Cho, W.C.; Batova, A.; Chavasiri, W.; Moore, C.; Rheingold, A.L.; Theodorakis, E.A. Evaluation of the pharmacophoric motif of the caged garcinia xanthones. *Org. Biomol. Chem.* **2009**, *7*, 4886–4894. [CrossRef]
111. Batova, A.; Lam, T.; Wascholowski, V.; Yu, A.L.; Giannis, A.; Theodorakis, E.A. Synthesis and evaluation of caged garcinia xanthones. *Org. Biomol. Chem.* **2007**, *5*, 494–500. [CrossRef] [PubMed]
112. Elbel, K.M.; Guizzunti, G.; Theodoraki, M.A.; Xu, J.; Batova, A.; Dakanali, M.; Theodorakis, E.A. A-ring oxygenation modulates the chemistry and bioactivity of caged garcinia xanthones. *Org. Biomol. Chem.* **2013**, *11*, 3341–3348. [CrossRef]
113. Li, X.; Wu, Y.; Wang, Y.; You, Q.; Zhang, X. 'Click chemistry' synthesis of novel natural product-like caged xanthones bearing a 1,2,3-triazole moiety with improved druglike properties as orally active antitumor agents. *Molecules* **2017**, *22*, 1834. [CrossRef]

114. Li, X.; Zhang, X.; Wang, X.; Li, N.; Lin, C.; Gao, Y.; Yu, Z.; Guo, Q.; You, Q. Synthesis and anti-tumor evaluation of b-ring modified caged xanthone analogues of gambogic acid. *Chin. J. Chem.* **2012**, *30*, 35–42. [CrossRef]
115. Wu, Y.; Hu, M.; Yang, L.; Li, X.; Bian, J.; Jiang, F.; Sun, H.; You, Q.; Zhang, X. Novel natural-product-like caged xanthones with improved druglike properties and in vivo antitumor potency. *Bioorg. Med. Chem. Lett.* **2015**, *25*, 2584–2588. [CrossRef]
116. Xu, X.; Wu, Y.; Hu, M.; Li, X.; Bao, Q.; Bian, J.; You, Q.; Zhang, X. Novel natural product-like caged xanthones bearing a carbamate moiety exhibit antitumor potency and anti-angiogenesis activity in vivo. *Sci. Rep.* **2016**, *6*, 35771. [CrossRef] [PubMed]
117. Miao, G.; Ma, J.; Yang, K.; Huang, Z.; Gu, Q.; Wang, Y.; Guo, Q.; You, Q.; Wang, J. Synthesis and bioevaluation of novel oxa-caged garcinia xanthones as anti-tumour agents. *Aust. J. Chem.* **2015**, *68*, 872. [CrossRef]
118. Koh, J.J.; Lin, S.; Bai, Y.; Sin, W.; Aung, T.T.; Li, J.; Chandra, V.; Pervushin, K.; Beuerman, R.; Liu, S. Antimicrobial activity profiles of amphiphilic xanthone derivatives are a function of their molecular oligomerization. *BBA-Biomembranes* **2018**, *860*, 2281–2298. [CrossRef] [PubMed]
119. Cho, W.C. *Synthesis of Caged Garcinia Xanthone Analogues*; University of California: San Diego, CA, USA, 2009.
120. Zhang, X.; Li, X.; Sun, H.; Jiang, Z.; Tao, L.; Gao, Y.; Guo, Q.; You, Q. Synthesis and evaluation of novel aza-caged garcinia xanthones. *Org. Biomol. Chem.* **2012**, *10*, 3288–3299. [CrossRef] [PubMed]
121. Rewcastle, G.W.; Atwell, G.J.; Zhuang, L.; Baguley, B.C.; Denny, W.A. Potential antitumor agents. 61. Structure-activity relationships for in vivo colon 38 activity among disubstituted 9-oxo-9h-xanthene-4-acetic acids. *J. Med. Chem.* **1991**, *34*, 217–222. [CrossRef] [PubMed]
122. Baguley, B.C.; McKeage, M.J. Asa404: A tumor vascular-disrupting agent with broad potential for cancer therapy. *Future Oncol.* **2010**, *6*, 1537–1543. [CrossRef] [PubMed]
123. Head, M.; Jameson, M.B. The development of the tumor vascular-disrupting agent asa404 (vadimezan, dmxaa): Current status and future opportunities. *Expert Opin. Investig. Drugs* **2010**, *19*, 295–304. [CrossRef]
124. Rehman, F.; Rustin, G. Asa404: Update on drug development. *Expert Opin. Investig. Drugs* **2008**, *17*, 1547–1551. [CrossRef]
125. Ching, L.M. Asa404. Vascular-disrupting agent, oncolytic. *Drugs Future* **2008**, *33*, 561–569. [CrossRef]
126. McKeage, M.J.; Kelland, L.R. 5,6-dimethylxanthenone-4-acetic acid (dmxaa): Clinical potential in combination with taxane-based chemotherapy. *Am. J. Cancer* **2006**, *5*, 155–162. [CrossRef]
127. Baguley, B.C.; Wilson, W.R. Potential of dmxaa combination therapy for solid tumors. *Expert Rev. Anticanc.* **2002**, *2*, 593–603. [CrossRef] [PubMed]
128. Baguley, B.C. Antivascular therapy of cancer: *Dmxaa. Lancet. Oncol.* **2003**, *4*, 141–148. [CrossRef]
129. McKeage, M. Clinical trials of vascular disrupting agents in advanced non-small-cell lung cancer. *Clin. Lung Cancer* **2011**, *12*, 143–147. [CrossRef] [PubMed]
130. Zhou, S.; Kestell, P.; Baguley, B.C.; Paxton, J.W. 5,6-dimethylxanthenone-4-acetic acid (dmxaa): A new biological response modifier for cancer therapy. *Invest. New Drugs* **2002**, *20*, 281–295. [CrossRef] [PubMed]
131. Daei Farshchi Adli, A.; Jahanban-Esfahlan, R.; Seidi, K.; Samandari-Rad, S.; Zarghami, N. An overview on vadimezan (dmxaa): The vascular disrupting agent. *Chem. Biol. Drug. Des.* **2018**, *91*, 996–1006. [CrossRef] [PubMed]
132. Gomes, A.S.; Brandao, P.; Fernandes, C.S.G.; da Silva, M.; de Sousa, M.; Pinto, M.M.M. Drug-like properties and adme of xanthone derivatives: The antechamber of clinical trials. *Curr. Med. Chem.* **2016**, *23*, 3654–3686. [CrossRef] [PubMed]
133. Jameson, M.B.; Head, M. Pharmacokinetic evaluation of vadimezan (asa404, 5,6-dimethylxanthenone-4-acetic acid, dmxaa). *Expert Opin. Drug Metab. Toxicol.* **2011**, *7*, 1315–1326. [CrossRef] [PubMed]
134. Shirey, K.A.; Nhu, Q.M.; Yim, K.C.; Roberts, Z.J.; Teijaro, J.R.; Farber, D.L.; Blanco, J.C.; Vogel, S.N. The anti-tumor agent, 5,6-dimethylxanthenone-4-acetic acid (dmxaa), induces ifn-β-mediated antiviral activity in vitro and in vivo. *J. Leukoc. Biol.* **2011**, *89*, 351–357. [CrossRef] [PubMed]
135. Zhang, S.H.; Zhang, Y.; Shen, J.; Zhang, S.; Chen, L.; Gu, J.; Mruk, J.S.; Cheng, G.; Zhu, L.; Kunapuli, S.P.; et al. Tumor vascular disrupting agent 5,6-dimethylxanthenone-4-acetic acid inhibits platelet activation and thrombosis via inhibition of thromboxane a2 signaling and phosphodiesterase. *J. Thromb. Haemo.* **2013**, *11*, 1855–1866.

136. Rewcastle, G.W.; Atwell, G.J.; Baguley, B.C.; Boyd, M.; Thomsen, L.L.; Zhuang, L.; Denny, W.A. Potential antitumor agents. 63. Structure-activity relationships for side-chain analogs of the colon 38 active agent 9-oxo-9h-xanthene-4-acetic acid. *J. Med. Chem.* **1991**, *34*, 2864–2870. [CrossRef]
137. Nakanishi, M.; Oe, T.; Tsuruda, M.; Matsuo, H.; Sakuragi, S. Studies on anti inflammatory agents. Xxxi. Studies on the synthesis and anti inflammatory activity of xanthenyl and benzo pyranopyridinylacetic acid derivatives (japanese). *yakugaku zasshi* **1976**, *96*, 99–109. [CrossRef] [PubMed]
138. Marona, H. [synthesis of some 2-xanthonylmethylthioalkanoic acids.]. *Acta Pol. Pharm.* **1988**, *45*, 13–34.
139. Marona, H.; Pękala, E.; Gunia, A.; Czuba, Z.; Szneler, E.; Sadowski, T.; Król, W. The influence of some xanthone derivatives on the activity of j-774a.1 cells. *Sci. Pharm.* **2009**, *77*, 743–754. [CrossRef]
140. Żelaszczyk, D.; Lipkowska, A.; Szkaradek, N.; Słoczyńska, K.; Gunia-Krzyżak, A.; Librowski, T.; Marona, H. Synthesis and preliminary anti-inflammatory evaluation of xanthone derivatives. *Heterocycl. Commun.* **2018**, *24*, 231–236. [CrossRef]
141. Sousa, E.; Paiva, A.; Nazareth, N.; Gales, L.; Damas, A.M.; Nascimento, M.S.; Pinto, M. Bromoalkoxyxanthones as promising antitumor agents: Synthesis, crystal structure and effect on human tumor cell lines. *Eur. J. Med. Chem.* **2009**, *44*, 3830–3835. [CrossRef] [PubMed]
142. Sousa, E.; Palmeira, A.; Cordeiro, A.S.; Sarmento, B.; Ferreira, D.; Lima, R.T.; Vasconcelos, M.H.; Pinto, M. Bioactive xanthones with effect on p-glycoprotein and prediction of intestinal absorption. *Med. Chem. Res.* **2013**, *22*, 2115–2123. [CrossRef]
143. Cruz, I.; Puthongking, P.; Cravo, S.; Palmeira, A.; Cidade, H.; Pinto, M.; Sousa, E. Xanthone and flavone derivatives as dual agents with acetylcholinesterase inhibition and antioxidant activity as potential anti-alzheimer agents. *J. Chem-Ny* **2017**, *2017*. [CrossRef]
144. Neves, M.P.; Cidade, H.; Pinto, M.; Silva, A.M.; Gales, L.; Damas, A.M.; Lima, R.T.; Vasconcelos, M.H.; de Sao Jose Nascimento, M. Prenylated derivatives of baicalein and 3,7-dihydroxyflavone: Synthesis and study of their effects on tumor cell lines growth, cell cycle and apoptosis. *Eur. J. Med. Chem.* **2011**, *46*, 2562–2574. [CrossRef]
145. Paiva, A.M.; Sousa, M.E.; Camoes, A.; Nascimento, M.S.J.; Pinto, M.M.M. Prenylated xanthones: Antiproliferative effects and enhancement of the growth inhibitory action of 4-hydroxytamoxifen in estrogen receptor-positive breast cancer cell line. *Med. Chem. Res.* **2012**, *21*, 552–558. [CrossRef]
146. Azevedo, C.M.; Afonso, C.M.; Soares, J.X.; Reis, S.; Sousa, D.; Lima, R.T.; Vasconcelos, M.H.; Pedro, M.; Barbosa, J.; Gales, L.; et al. Pyranoxanthones: Synthesis, growth inhibitory activity on human tumor cell lines and determination of their lipophilicity in two membrane models. *Eur. J. Med. Chem.* **2013**, *69*, 798–816. [CrossRef]
147. Urbatzka, R.; Freitas, S.; Palmeira, A.; Almeida, T.; Moreira, J.; Azevedo, C.; Afonso, C.; Correia-da-Silva, M.; Sousa, E.; Pinto, M.; et al. Lipid reducing activity and toxicity profiles of a library of polyphenol derivatives. *Eur. J. Med. Chem.* **2018**, *151*, 272–284. [CrossRef] [PubMed]
148. Fernandes, C.; Masawang, K.; Tiritan, M.E.; Sousa, E.; De Lima, V.; Afonso, C.; Bousbaa, H.; Sudprasert, W.; Pedro, M.; Pinto, M.M. New chiral derivatives of xanthones: Synthesis and investigation of enantioselectivity as inhibitors of growth of human tumor cell lines. *Bioorg. Med. Chem.* **2014**, *22*, 1049–1062. [CrossRef] [PubMed]
149. Fernandes, C.; Brandão, P.; Santos, A.; Tiritan, M.E.; Afonso, C.; Cass, Q.B.; Pinto, M.M. Resolution and determination of enantiomeric purity of new chiral derivatives of xanthones using polysaccharide-based stationary phases. *J. Chromatogr. A* **2012**, *1269*, 143–153. [CrossRef] [PubMed]
150. Arabanian, A.; Mohammadnejad, M.; Balalaie, S. A novel and efficient approach for the amidation of c-terminal peptides. *J. Iran Chem. Soc.* **2010**, *7*, 840–845. [CrossRef]
151. Twibanire, J.D.K.; Grindley, T.B. Efficient and controllably selective preparation of esters using uronium-based coupling agents. *Org. Lett.* **2011**, *13*, 2988–2991. [CrossRef] [PubMed]
152. Balalaie, S.; Mahdidoust, M.; Eshaghi-Najafabadi, R. 2-(1h-benzotriazole-1-yl)-1,1,3,3-tetramethyluronium tetrafluoro borate (tbtu) as an efficient coupling reagent for the esterification of carboxylic acids with alcohols and phenols at room temperature. *Chin. J. Chem.* **2008**, *26*, 1141–1144. [CrossRef]
153. Balalaie, S.; Mahdidoust, M.; Eshaghi-Najafabadi, R. 2-(1h-benzotriazole-1-yl)-1, 1, 3, 3-tetramethyluronium tetrafluoroborate as an efficient coupling reagent for the amidation and phenylhydrazation of carboxylic acids at room temperature. *J. Iran Chem. Soc.* **2007**, *4*, 364–369. [CrossRef]

154. Naik, S.A.; Lalithamba, H.S.; Sureshbabu, V.V. Application of 2-(1h-benzotriazol-1-yl)-1,1,3,3-tetramethyl uronium tetrafluoroborate (tbtu) for the synthesis of acid azides. *Indian J. Chem. Sect.* **2011**, *50*, 103–109. [CrossRef]
155. Fernandes, C.; Tiritan, M.E.; Cass, Q.; Kairys, V.; Fernandes, M.X.; Pinto, M. Enantioseparation and chiral recognition mechanism of new chiral derivatives of xanthones on macrocyclic antibiotic stationary phases. *J. Chromatogr. A* **2012**, *1241*, 60–68. [CrossRef]
156. Phyo, Y.Z.; Cravo, S.; Palmeira, A.; Tiritan, M.E.; Kijjoa, A.; Pinto, M.M.M.; Fernandes, C. Enantiomeric resolution and docking studies of chiral xanthonic derivatives on chirobiotic columns. *Molecules* **2018**, *23*, 142. [CrossRef]
157. Fernandes, C.; Palmeira, A.; Santos, A.; Tiritan, M.E.; Afonso, C.; Pinto, M.M. Enantioresolution of chiral derivatives of xanthones on (s,s)-whelk-o1 and l -phenylglycine stationary phases and chiral recognition mechanism by docking approach for (s,s)-whelk-o1. *Chirality* **2013**, *25*, 89–100. [CrossRef] [PubMed]
158. Carraro, M.L.; Palmeira, A.; Tiritan, M.E.; Fernandes, C.; Pinto, M.M.M. Resolution, determination of enantiomeric purity and chiral recognition mechanism of new xanthone derivatives on (s,s)-whelk-o1 stationary phase. *Chirality* **2017**, 1–10. [CrossRef] [PubMed]
159. Fernandes, C.; Pinto, M.M.M.; Tiritan, M.E. Enantioresolution of chiral derivatives of xanthones on different types of liquid chromatography stationary phases: a comparative study. *Curr. Chromatogr.* **2014**, *1*, 139–150. [CrossRef]
160. Fernandes, C.; Palmeira, A.; Ramos, I.I.; Carneiro, C.; Afonso, C.; Tiritan, M.E.; Cidade, H.; Pinto, P.C.A.G.; Saraiva, M.L.M.F.S.; Reis, S.; et al. Chiral derivatives of xanthones: Investigation of the effect of enantioselectivity on inhibition of cyclooxygenases (cox-1 and cox-2) and binding interaction with human serum albumin. *Pharmaceuticals* **2017**, *10*, 50. [CrossRef] [PubMed]
161. Fernandes, C.; Oliveira, L.; Tiritan, M.E.; Leitao, L.; Pozzi, A.; Noronha-Matos, J.B.; Correia-de-Sá, P.; Pinto, M.M. Synthesis of new chiral xanthone derivatives acting as nerve conduction blockers in the rat sciatic nerve. *Eur. J. Med. Chem.* **2012**, *55*, 1–11. [CrossRef]
162. Lopes, A.; Martins, E.; Silva, R.; Pinto, M.M.M.; Remião, F.; Sousa, E.; Fernandes, C. Chiral thioxanthones as modulators of p-glycoprotein: Synthesis and enantioselectivity studies. *Molecules* **2018**, *23*, 626. [CrossRef]
163. Marona, H. Synthesis and anticonvulsant effects of some aminoalkanolic derivatives of xanthone. *Pharmazie* **1998**, *53*, 672–676.
164. Marona, H. Evaluation of some 2-substituted derivatives of xanthone for anticonvulsant properties. *Pharmazie* **1998**, *53*, 405–409.
165. Marona, H.; Gorka, Z.; Szneler, E. Aminoalkanolic derivatives of xanthone with potential antiepileptic activity. *Pharmazie* **1998**, *53*, 219–223.
166. Marona, H.; Pękala, E.; Antkiewicz-Michaluk, L.; Walczak, M.; Szneler, E. Anticonvulsant activity of some xanthone derivatives. *Bioorg. Med. Chem.* **2008**, *16*, 7234–7244. [CrossRef]
167. Marona, H.; Pekala, E.; Filipek, B.; Maciag, D.; Szneler, E. Pharmacological properties of some aminoalkanolic derivatives of xanthone. *Pharmazie* **2001**, *56*, 567. [PubMed]
168. Marona, H.; Szkaradek, N.; Karczewska, E.; Trojanowska, D.; Budak, A.; Bober, P.; Przepiórka, W.; Cegla, M.; Szneler, E. Antifungal and antibacterial activity of the newly synthesized 2-xanthone derivatives. *Archiv. Der. Pharmazie* **2009**, *342*, 9–18. [CrossRef] [PubMed]
169. Marona, H.; Szkaradek, N.; Rapacz, A.; Filipek, B.; Dybała, M.; Siwek, A.; Cegła, M.; Szneler, E. Preliminary evaluation of pharmacological properties of some xanthone derivatives. *Bioorg. Med. Chem.* **2009**, *17*, 1345–1352. [CrossRef] [PubMed]
170. Klesiewicz, K.; Żelaszczyk, D.; Trojanowska, D.; Bogusz, B.; Małek, M.; Waszkielewicz, A.; Szkaradek, N.; Karczewska, E.; Marona, H.; Budak, A. Preliminary antifungal activity assay of selected chlorine-containing derivatives of xanthone and phenoxyethyl amines. *Chem. Biol. Drug. Des.* **2018**, *92*, 1867–1875. [CrossRef] [PubMed]
171. Waszkielewicz, A.M.; Słoczyńska, K.; Pękala, E.; Żmudzki, P.; Siwek, A.; Gryboś, A.; Marona, H. Design, synthesis, and anticonvulsant activity of some derivatives of xanthone with aminoalkanol moieties. *Chem. Biol. Drug. Des.* **2017**, *89*, 339–352. [CrossRef] [PubMed]
172. Klesiewicz, K.; Karczewska, E.; Budak, A.; Marona, H.; Szkaradek, N. Anti-helicobacter pylori activity of some newly synthesized derivatives of xanthone. *J. Antibiot.* **2016**, *69*, 825–834. [CrossRef] [PubMed]

173. Kubacka, M.; Szkaradek, N.; Mogilski, S.; Pańczyk, K.; Siwek, A.; Gryboś, A.; Filipek, B.; Żmudzki, P.; Marona, H.; Waszkielewicz, A.M. Design, synthesis and cardiovascular evaluation of some aminoisopropanoloxy derivatives of xanthone. *Bioorg. Med. Chem.* **2018**, *26*, 3773–3784. [CrossRef]
174. Szkaradek, N.; Rapacz, A.; Pytka, K.; Filipek, B.; Zelaszczyk, D.; Szafrański, P.; Słoczyńska, K.; Marona, H. Cardiovascular activity of the chiral xanthone derivatives. *Bioorg. Med. Chem.* **2015**, *23*, 6714–6724. [CrossRef]
175. Librowski, T.; Czarnecki, R.; Jastrzebska-Wiesek, M.; Opoka, W.; Marona, H. The influence of some aminoalkanolic xanthone derivatives on central nervous and cardiovascular systems in rodents. *Boll. Chim. Farm* **2004**, *143*, 267–274.
176. Szkaradek, N.; Gunia, A.; Waszkielewicz, A.M.; Antkiewicz-Michaluk, L.; Cegła, M.; Szneler, E.; Marona, H. Anticonvulsant evaluation of aminoalkanol derivatives of 2-and 4-methylxanthone. *Bioorg. Med. Chem.* **2013**, *21*, 1190–1198. [CrossRef]
177. Rajtar, G.; Zolkowska, D.; Kleinrok, Z.; Marona, H. Pharmacological properties of some xanthone derivatives. *Acta Pol. Pharm.* **1999**, *56*, 311–318. [PubMed]
178. Jastrzêbska-Wiêsek, M.; Librowski, T.; Czarnecki, R.; Marona, H.; Nowak, G. Central activity of new xanthone derivatives with chiral center in some pharmacological tests in mice. *Pol. J. Pharmacol.* **2003**, *55*, 461–466. [PubMed]
179. Jastrzebska-Wiesek, M.; Czarnecki, R.; Marona, H. The anticonvulsant, local anesthetic and hemodynamic properties of some chiral aminobutanol derivatives of xanthone. *Acta Pol. Pharm.* **2008**, *65*, 591–600. [PubMed]
180. Librowski, T.; Czamecki, R.; Jastrzebska, M. Chiral 2-amino-1-butanol xanthone derivatives as potential antiarrhythmic and hypotensive agents. *Acta Pol. Pharm.* **1999**, *56*, 87–90. [PubMed]
181. Rajtar, G.; Zolkowska, D.; Kleinrok, Z.; Marona, H. Antiplatelets activity of some xanthone derivatives. *Acta Pol. Pharm.* **1999**, *56*, 319–324. [PubMed]
182. Słoczyńska, K.; Pękala, E.; Wajda, A.; Węgrzyn, G.; Marona, H. Evaluation of mutagenic and antimutagenic properties of some bioactive xanthone derivatives using vibrio harveyi test. *Lett. Appli. Microbiol.* **2010**, *50*, 252–257. [CrossRef] [PubMed]
183. Sypniewski, D.; Szkaradek, N.; Loch, T.; Waszkielewicz, A.M.; Gunia-Krzyżak, A.; Matczyńska, D.; Sołtysik, D.; Marona, H.; Bednarek, I. Contribution of reactive oxygen species to the anticancer activity of aminoalkanol derivatives of xanthone. *Invest. New Drugs* **2018**, *36*, 355–369. [CrossRef] [PubMed]
184. Chen, X.; Leng, J.; Rakesh, K.P.; Darshini, N.; Shubhavathi, T.; Vivek, H.K.; Mallesha, N.; Qin, H.-L. Synthesis and molecular docking studies of xanthone attached amino acids as potential antimicrobial and anti-inflammatory agents. *Med. Chem. Commun.* **2017**, *8*, 1706–1719. [CrossRef] [PubMed]
185. Rakesh, K.P.; Darshini, N.; Manukumar, H.M.; Vivek, H.K.; Prasanna, D.S.; Mallesha, N. Xanthone conjugated amino acids as potential anticancer and DNA binding agents: Molecular docking, cytotoxicity and sar studies. *Anticancer Agents Med. Chem.* **2018**. [CrossRef] [PubMed]
186. Szkaradek, N.; Stachura, K.; Waszkielewicz, A.; Cegla, M.; Szneler, E.; Marona, H. Synthesis and antimycobacterial assay of some xanthone derivatives. *Acta Pol. Pharm.* **2008**, *65*, 21–28. [PubMed]
187. Marona, H.; Szkaradek, N.; Kubacka, M.; Bednarski, M.; Filipek, B.; Cegla, M.; Szneler, E. Synthesis and evaluation of some xanthone derivatives for anti-arrhythmic, hypotensive properties and their affinity for adrenergic receptors. *Archiv. der Pharmazie* **2008**, *341*, 90–98. [CrossRef] [PubMed]
188. Marona, H.; Librowski, T.; Cegła, M.; ERDOĞAN, C.; Sahin, N.O. Antiarrhythmic and antihypertensive activity of some xanthone derivatives. *Acta Pol. Pharm. Drug Res.* **2008**, *65*, 383–390.
189. Szkaradek, N.; Rapacz, A.; Pytka, K.; Filipek, B.; Siwek, A.; Cegła, M.; Marona, H. Synthesis and preliminary evaluation of pharmacological properties of some piperazine derivatives of xanthone. *Bioorg. Med. Chem.* **2013**, *21*, 514–522. [CrossRef] [PubMed]
190. Waszkielewicz, A.; Gunia, A.; Szkaradek, N.; Pytka, K.; Siwek, A.; Satała, G.; Bojarski, A.; Szneler, E.; Marona, H. Synthesis and evaluation of pharmacological properties of some new xanthone derivatives with piperazine moiety. *Bioorg. Med. Chem. Lett.* **2013**, *23*, 4419–4423. [CrossRef] [PubMed]
191. Rapacz, A.; Sapa, J.; Nowiński, L.; Mogilski, S.; Pytka, K.; Filipek, B.; Siwek, A.; Szkaradek, N.; Marona, H. Biofunctional studies of new 2-methoxyphenylpiperazine xanthone derivatives with α1-adrenolytic properties. *Phytochem. Rep.* **2015**, *67*, 267–274. [CrossRef] [PubMed]

192. Rapacz, A.; Sapa, J.; Pytka, K.; Dudek, M.; Filipek, B.; Szkaradek, N.; Marona, H. Antiarrhythmic activity of new 2-methoxyphenylpiperazine xanthone derivatives after ischemia/reperfusion in rats. *Phytochem. Rep.* **2015**, *67*, 1163–1167. [CrossRef] [PubMed]
193. Cheng, Y.-W.; Kang, J.-J. Mechanism of vasorelaxation of thoracic aorta caused by xanthone. *Eur. J. Pharmacol.* **1997**, *336*, 23–28. [CrossRef]
194. Cherkadu, V.; Kalavagunta, P.K.; Ningegowda, M.; Shivananju, N.S.; Madegowda, M.; Priya, B.S. Fecl3-catalyzed three-component one-pot synthesis of novel 4-[(benzo[d]thiazol-2-ylamino)(phenyl) methyl]-3-hydroxy-9h-xanthen-9-ones. *Synlett* **2016**, *27*, 1116–1120. [CrossRef]

molecules

MDPI

Article

Determination of Optical Purity of Lactic Acid-Based Chiral Liquid Crystals and Corresponding Building Blocks by Chiral High-Performance Liquid Chromatography and Supercritical Fluid Chromatography

Anna Poryvai [1], Terezia Vojtylová-Jurkovičová [2], Michal Šmahel [1], Natalie Kolderová [1,3], Petra Tomášková [2], David Sýkora [4] and Michal Kohout [1,*]

1 Department of Organic Chemistry, University of Chemistry and Technology Prague, Technická 5, 166 28 Prague, Czech Republic; poryvaia@vscht.cz (A.P.); smahelm@vscht.cz (M.Š.); natalie.kolderova@gmail.com (N.K.)

2 Institute of Physics, Academy of Sciences of the Czech Republic, Na Slovance 2, 182 21 Prague, Czech Republic; vojtyl@fzu.cz (T.V.-J.); tomasko@fzu.cz (P.T.)

3 Forensic Laboratory of Biologically Active Substances, University of Chemistry and Technology Prague, Technická 3, 166 28 Prague, Czech Republic

4 Department of Analytical Chemistry, University of Chemistry and Technology Prague, Technická 5, 166 28 Prague, Czech Republic; sykorad@vscht.cz

* Correspondence: michal.kohout@vscht.cz; Tel.: +420-22044-4118

Received: 30 January 2019; Accepted: 14 March 2019; Published: 20 March 2019

Abstract: Liquid crystals (LCs) are among the most prominent materials of the current information age, mainly due to their well-known application in liquid crystal displays (LCDs). Their unique electro-optical properties stem from their ability to form organised structures (mesophases) on the transition from solid state to isotropic liquid. Molecules of LCs in a mesophase still maintain the anisotropy of solid crystals, while simultaneously exhibiting the fluidity of liquids, which gives the system the ability to react immediately to external stimuli such as electric or magnetic fields, light, mechanical stress, pressure and, of course, temperature. For the proper function of LC-based devices, not only chemical, but also optical purity of materials is strongly desirable, since any impurity could be detrimental to the self-assembly of the molecules. Therefore, in this study we aimed to verify synthetic methods published in the literature, which are used nowadays to prepare chiral building blocks based on lactic acid, for their enantioselectivity. Moreover, we have focused on the development of an analytical chiral separation method for target liquid crystalline materials. Using a chiral polysaccharide-based column operated in liquid chromatography mode, we show that not all published methods of LC synthesis are enantioselective, which could lead to significant differences in the properties of the resulting materials. We show that high-performance liquid chromatography with UV detection and supercritical fluid chromatography with UV and mass spectrometry detection enable full control over the chemical and optical purity of the target LCs and the corresponding chiral building blocks. For the first time, we utilise supercritical fluid chromatography with mass detection for the direct chiral analysis of liquid crystalline materials and impurities formed during the synthesis.

Keywords: chiral liquid crystals; optical purity; chiral separation; supercritical fluid chromatography; enantioseparation of liquid crystals; mass spectrometry detection; mesomorphic properties

1. Introduction

Liquid crystals (LCs) represent one of the most prominent classes of materials of today's information age. The molecules of LCs can self-assemble into organised supramolecular structures (mesophases) that combine the fluidity of liquids with the long-range order of solids [1,2]. Molecules in a mesophase readily respond to external stimuli such as electric and magnetic field, light and mechanical stress. Thanks to these properties, LCs have found many applications, for example, in liquid crystal displays (LCDs), optical shutters, light beam steering and shaping [3,4].

Among LCs, chiral liquid crystals (CLCs) represent an important class of materials that can self-assemble into chiral mesophases [5]. Macroscopic chirality of the mesophases results in special properties, such as (anti)ferroelectricity, selective reflection of light, and heat sensitivity, which render CLCs ideal candidates for the fabrication of high-speed high-contrast LCDs, photonic devices, and contact thermography devices, respectively [1,6,7]. Since the self-assembly of LCs is very sensitive to any impurities present in the bulk material, control of chemical and optical purity during the synthesis of target CLCs is essential. Any impurity may result in the alteration of the mesophase structure or even complete loss of mesomorphic behaviour.

The optical purity of CLCs was in the past mainly controlled by optical rotation measurement and transformation of precursors to diastereoisomers [8]. However, such transformation does not usually provide information on trace amounts of other stereoisomers, which can be present in the final material. Even a trace of the opposite enantiomer may consequently lead to variations in the mesomorphic behaviour expected for the enantiomerically pure material. The source of such optical impurity may be the starting material, which frequently contains a trace amount of the opposite enantiomer, or it may occur during synthesis due to partial racemisation of an intermediate upon its synthetic modification. Although the optical purity of CLCs is a very important issue in the development and utilisation of novel chiral materials, only scarce information on chiral separation of CLCs can be found in the literature [9,10]. Most of the work thus far has been devoted to chiral liquid chromatography separation of photosensitive CLCs and the effect of light-induced *E*-to-*Z*-isomerisation on chiral recognition [11–13]. It has been found that the structure of CLCs plays an important role in the chiral resolution. It has also been documented that even materials bearing very small substituents at the chiral centre can be efficiently separated on polysaccharide-based chiral columns [9,10]. However, to the best of our knowledge, there is no systematic study dealing with the control of enantioselectivity of the synthetic protocol leading to CLCs using contemporary analytical chemistry instrumentation.

Therefore, we have focused on controlling enantioselectivity of the synthesis of a novel type of lactic acid-based CLCs. We used three different synthetic procedures reported in the literature, which have frequently been used for the preparation of the chiral intermediates required for the synthesis of target lactic acid-based CLCs. We developed a method for the chiral resolution of both enantiomers of the chiral precursors using the synthesised enantiomers and verified the enantioselectivity of the synthetic pathways. Moreover, we elaborated a method for the analytical enantioseparation of the target liquid crystalline materials, not only in HPLC, but also in supercritical fluid chromatography (SFC) with mass spectrometry (MS) detection. We document that the small percentage of the opposite enantiomer present in commercially available chiral starting materials increases during the synthesis, thereby affording materials that are not optically pure. The described chiral separation procedure enables very precise control of the chemical and optical purity of the target CLCs, which is required for practical applications.

2. Results and Discussion

2.1. Synthesis of Chiral Building Blocks

The studied chiral intermediates (Figure 1) were prepared using previously described synthetic procedures and characterised by nuclear magnetic resonance techniques to comply with the data published in the literature [8,14–18]. To clarify the differences among the three methods used here, the experimental procedures are described below in brief. The method (B) is a modified version of a synthetic procedure already described in the literature [17].

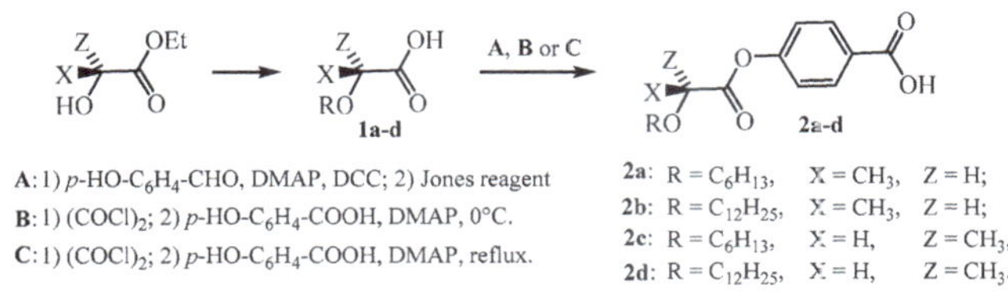

Figure 1. Synthesis and designation of the chiral precursors prepared by three different methods; (**A**) four step method using oxidation of an aldehyde as an intermediate step, (**B**) and (**C**) direct acylation of 4-hydroxybenzoic acid with the chiral acid under different conditions.

The first two steps [8,15,16] of the synthesis are common to all three procedures, and they were accomplished as follows: A suspension of silver(I) oxide, alkyl iodide and ethyl (*S*)-lactate (or methyl (*R*)-lactate) in diethyl ether was stirred for 90 h in the absence of UV light. Then, the reaction mixture was filtered, solvent was evaporated and the product (colourless liquid) was purified via distillation under reduced pressure (p = 0.6 Torr, t = 54–55 °C for *S*-C6; p = 2.4 Torr, t = 143–146 °C for *S*-C12). Subsequently, an aqueous solution of lithium hydroxide (1.3 M) was added dropwise to a stirred solution of ethyl *O*-alkyllactate in methanol (with THF in the case of dodecyl derivative) at 0 °C. The reaction mixture was stirred for 3 days at room temperature, then it was acidified (pH = 2) with 17% aq. hydrochloric acid and extracted with dichloromethane (3 × 20 mL). The combined organic solution was dried with anhydrous magnesium sulphate. Solvent was removed under reduced pressure and the corresponding *O*-alkyllactic acid was isolated as a colourless liquid. For analytical data of the derivatives, see ESI.

2.1.1. Method (A)

To the solution of **1a** in dry DCM, *p*-hydroxybenzaldehyde, *N,N'*-dicyclohexylcarbodiimide (DCC) and 4-*N,N*-dimethylaminopyridine (DMAP) were added, and the reaction mixture was stirred in an inert argon atmosphere overnight, and then it was decomposed with 17% aq. hydrochloric acid [14,18]. Layers were separated, and the aqueous layer was extracted with ethyl acetate (2 × 20 mL). The combined organic solution was washed with brine (10 mL) and dried with anhydrous magnesium sulphate. The solvent was removed, and the crude white product was purified by column chromatography (eluent toluene/*tert*-butyl methyl ether, 30/1, *v*/*v*). To a cold (0 °C) solution of aldehyde in acetone, Jones reagent was added dropwise under stirring. The reaction mixture was left to heat up to room temperature and stirred overnight. Then, the reaction mixture was poured onto crushed ice (400 mL). After all the ice had melted, the formed precipitate was filtered, washed with cold water and dried in air. The crude product was purified via crystallisation from hexane.

2.1.2. Method (B)

A mixture of acid **1a** with a catalytic amount of *N,N*-dimethylformamide (DMF) (0.05 mL) in oxalyl chloride (10–20 mL) was stirred at room temperature overnight. The unreacted oxalyl chloride was distilled off, and the residue was boiled for 1 min in hexane with a spatula-tip of active carbon. The suspension was filtered while hot and the solvent was evaporated. The crude acid chloride was dissolved in dry DCM (3 mL) and added dropwise to a cold (0 °C) solution of 4-hydroxybenzoic acid and DMAP in dry dichloromethane. The mixture was stirred at 0 °C in the argon atmosphere for 10 h. The reaction was then decomposed with 17% aq. hydrochloric acid. Layers were separated, and the aqueous layer was extracted with chloroform (2 × 20 mL). The combined organic solution was washed with brine (10 mL) and dried with anhydrous magnesium sulphate. The solvent was removed, and the crude white product was purified by column chromatography.

2.1.3. Method (C)

A mixture of acid **1a** with a catalytic amount of DMF (0.05 mL) in oxalyl chloride (10–20 mL) was stirred under reflux for 2 h [17]. Unreacted oxalyl chloride was distilled off, and the crude product was boiled for 1 min in hexane with a spatula-tip of active carbon. The suspension was filtered while hot and the solvent was evaporated. The crude acid chloride was dissolved in dry dichloromethane (3 mL) and added dropwise to a solution of 4-hydroxybenzoic acid and DMAP in dry DCM. The mixture was stirred and heated under reflux in the argon atmosphere for 5 h. Then, it was cooled to the ambient temperature and decomposed with 17% aq. hydrochloric acid. Layers were separated, and the aqueous layer was extracted with chloroform (2 × 20 mL). The combined organic solution was washed with brine (10 mL) and dried with anhydrous magnesium sulphate. The solvent was removed, and the crude white product was purified by column chromatography.

2.2. Synthesis of the Target Liquid Crystal

The synthesis of the target liquid crystalline material (Figure 2) started from chiral acids **2a** and **2c** prepared according to methods (A) and (B), respectively, which provided the best results. The acids were first transformed to appropriate acid chlorides, which were subsequently used for the acylation of hydroxy ester **3**, prepared according to a method described in the literature [19], in the presence of DMAP as a base.

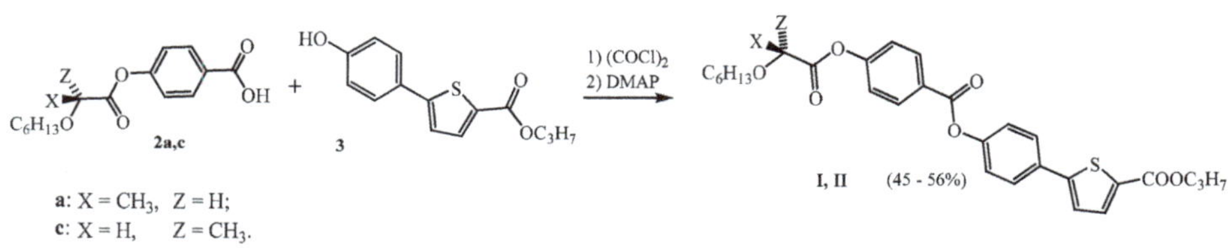

Figure 2. Synthesis of target LCs.

2.3. Chiral HPLC Separation of Chiral Building Blocks

The chiral separation method was based on a previously optimised methodology available in our laboratory for the enantioseparation of chiral photosensitive LCs [11–13]. Optimal separation conditions for precursors **2a–d** were heptane/propan-2-ol (9/1, *v*/*v*). Pure (*R*)- and (*S*)-enantiomers were screened individually, and the position of the enantiomeric impurity in the main substance was determined by comparing the retention times (Figure 3).

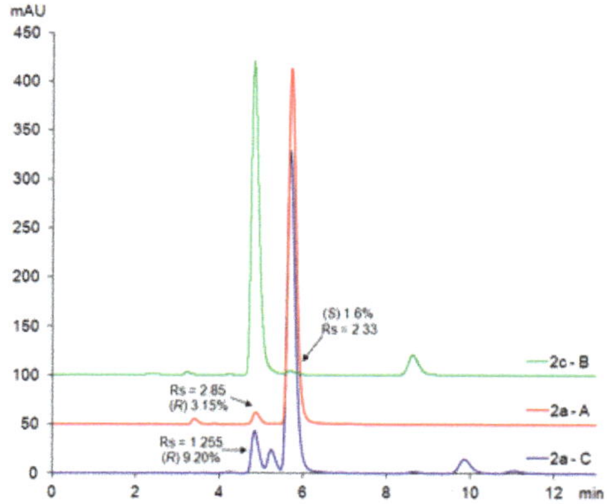

Figure 3. Enantioseparation of chiral acids **2a** and **2c** prepared according to experimental methods A–C in HPLC mode using ECOM HPLC system.

Under the given conditions, baseline resolution of the enantiomers was feasible. In addition to that, it was possible to study the effect of the particular synthetic strategy on enantiomeric purity of the chiral building blocks. The results (Table 1) show that synthetic method A provided the chiral building block **2a** with *ee* = 93.7%. The chemical purity of the substance was high. Synthetic method B, which was first used for the preparation of **2c**, gave rise to a product with higher optical purity (*ee* = 96.8%) than method A. However, several other chemical impurities were detected. An attempt to improve the reaction rate by heating up the reaction mixture (method C [17]) provided the target acid **2a** containing a broad range of chemical impurities, as well as the opposite enantiomer **2c**, resulting in *ee* = 81.6%. Similar results were obtained for acids **2b** and **2d**, possessing the C12 terminal alkyl chain (see Figure S1). Therefore, it is clear that method C should not be used for the synthesis of chiral building blocks **2a–d** or compounds possessing a similar structure with lactic acid as the chiral unit.

Table 1. Relative peak areas of chiral building blocks **2a** and **2c** obtained from chiral HPLC analysis and corresponding enantiomeric excess values of the respective enantiomers.

Peak Area (%)	Method A	Method B	Method C
2a	96.85	1.60	90.80
2c	3.15	98.40	9.20
% *ee*	93.70	96.80	81.60

Since the optical purity of starting esters, as stated by the manufacturer, is *ee* = 98% for (*S*)-ethyl lactate and *ee* = 99% for (*R*)-methyl lactate, partial racemisation also occurred when using the synthetic methods A and B. First, we decided to modify method A, because it is well known that DCC (and other carbodiimides) may induce partial or full racemisation of a chiral acid during its activation [20,21]. Therefore, O-(1H-6-chlorobenzotriazol-1-yl)-1,1,3,3-tetramethyluronium tetrafluoroborate (TCTU), a more selective coupling reagent, was employed for the synthesis of **2a** using a method described for analogous reagents [22]. However, an almost identical enantiomeric composition of **2a** as for DCC-mediated reaction was achieved (see Figure S2).

The results indicate that the partial racemisation observed for building blocks prepared according to methods A and B most probably takes place during the first two steps of the synthesis, although the methodology was previously described as racemisation-free [8,23]. Therefore, it is reasonable to assume that with the development of new analytical approaches, the precision of impurity determination has been significantly improved. The current technology of chiral CSPs obviously beats the former methods based on optical rotation measurements and determination of enantiomeric excess by transformation of the respective chiral building blocks to an amide [8].

2.4. Chiral HPLC Separation of Full-Length Liquid-Crystalline Materials

The developed separation method for the precursors was further successfully applied for the chiral separation of full-length liquid crystals (R_S = 2.5). Due to the shift in the absorption maximum of the target compound, the detection wavelength was set to 308 nm. To determine contamination of target CLCs with an opposite enantiomer, we prepared corresponding (*S*)- and (*R*)-enantiomers (**I** and **II**, respectively), and analysed them independently (Figure 4).

Analogously to the chiral precursors, (*R*)-enantiomer **II** eluted first. This documents that the enantiorecognition of the enantiomers is strictly governed by the substituents near the chiral centre, while the rest of the molecule only contributes to retention. Interestingly, the analysis of (*S*)-enantiomer **I** revealed *ee* = 98.8% and for target material **II** ((*R*)-enantiomer) *ee* = 98.4% was found. This finding contradicts the results obtained for the chiral precursors, which showed higher optical purity of the (*R*)-enantiomer. We can speculate that the target CLCs could be enantiomerically enriched via crystallisation, leading, in consequence, to target materials with higher optical purity than the starting compounds. However, intensive purification will be used, because the initial application of column chromatography and three consecutive crystallisation still provided a rather impure material (Figure 4).

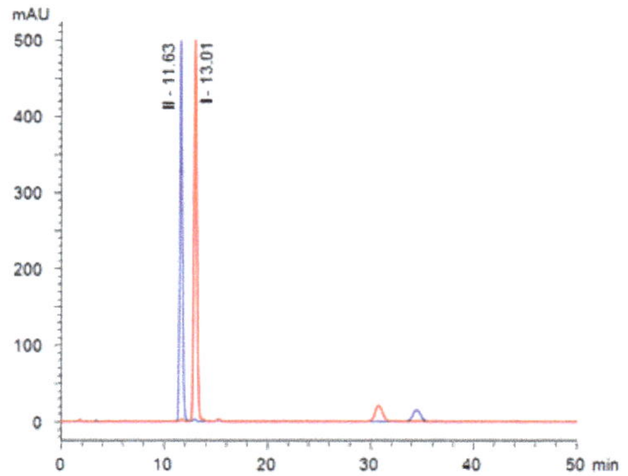

Figure 4. Chiral separation of target chiral liquid crystals. Overlaid chromatograms of **I** (red trace) and **II** (blue trace) acquired with Chiralpak AD-3 using Acquity UPC2 operated in HPLC mode; mobile phase heptane/IPA (9/1, *v*/*v*), flow rate 1 mL/min, temperature 25 °C, sample concentration 0.5 mg/mL, injection volume 5 μL.

2.5. Chiral SFC and SFC-MS Separation of Full-Length Liquid-Crystalline Materials

Recently, it has been shown that SFC offers striking selectivity in the enantioseparation of chiral liquid crystals [9,10]. Therefore, we prepared a spiked sample, which enabled us to determine the elution order of enantiomers and match the chiral impurities (observed in HPLC mode) to the corresponding materials. Indeed, the SFC analysis provided better resolution ($R_S = 3.1$), not only for the target materials, but also for chiral impurities present in the materials (Figure 5). The elution order of **I** and **II** remained unchanged; material **II** eluted in the first peak.

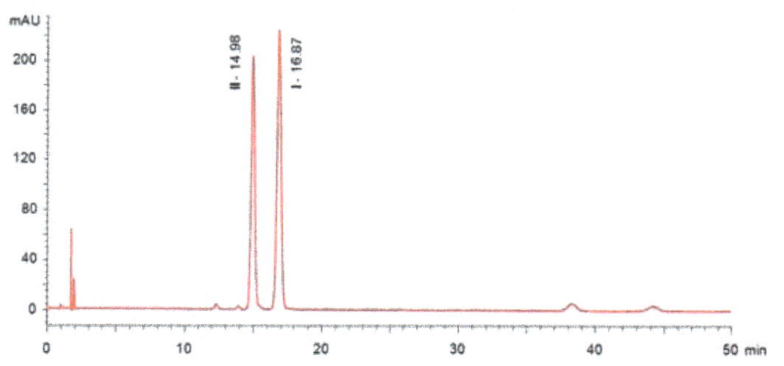

Figure 5. Mixture of **I** and **II** analysed in SFC mode on Chiralpak AD-3 to directly compare the performance of the column in HPLC and SFC mode. Mobile phase scCO_2/IPA (70/30), sample concentration 0.5 mg/mL, injection volume 5 μL, temperature 30 °C, backpressure 2000 psi and the flow rate of 1 mL/min were used.

It should be noted that the SFC separation was, in fact, performed outside the supercritical region of the fluid [24,25]. Yet, the term SFC is generally also accepted for the subcritical fluid region (also subFC), because the main advantages of lower viscosity and higher diffusivity of the mobile phase are fully preserved [26]. Unlike normal phase HPLC, direct coupling to a mass spectrometer (detector) is easily accessible in SFC. Therefore, we took this advantage to control the optical purity of the materials and also to shed light on their impurity profile. Due to specific SFC conditions, a dedicated SFC column (ChiralArt Amylose-C 250 × 4.6 mm, i.d., 5 μm) was used in all following SFC measurements. It should be noted that a dedicated SFC column should be used for each chromatographic mode, because a frequent change of chromatographic modes could reduce column lifetime. Although it is natural to use short columns with small particles in SFC (due to low viscosity and higher diffusivity of the fluid), in some cases it is economic to use older type of columns for SFC method development, in particular, when unusual conditions (e.g., outside supercritical region, highly acidic mobile phase) are expected to be used.

Analysis of the target material **I** (Figure 6) clearly demonstrates that the additional purification with gradient elution column chromatography and subsequent multiple crystallisations provide the chemically and enantiomerically pure material. No signals of impurities have been detected, vide infra. Furthermore, only a negligibly increased baseline was observed in the area where the peak of (*R*)-enantiomer (material **II**) should be present. Since the limit of quantification for SFC-MS is below 1 μg/mL (for details see ESI), the enantiomeric purity of the target material **I** analysed under the given conditions (Figure 6) is *ee* > 99.6%.

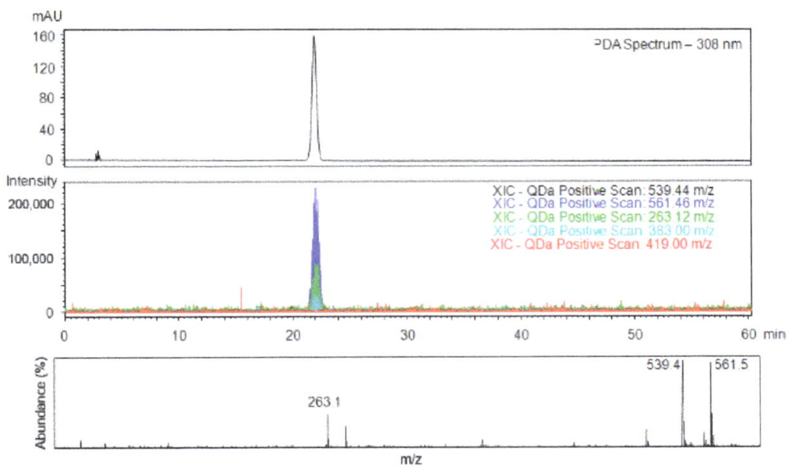

Figure 6. Analysis of purified target material **I** performed in SFC-MS mode. Upper part shows the UV trace, lower part shows reconstructed chromatograms (XIC) with the mass of the target material 539.4 $[M + H]^+$, 561.5 $[M + Na]^+$, fragments 263.1 $[M + H]^+$, 383.0 $[M + H]^+$, and the main impurity 419.0 $[M + H]^+$ and mass spectrum of the product. Conditions: ChiralArt Amylose-C (250 × 4.6 mm, i.d., 5 μm) column, mobile phase $scCO_2$/IPA (70/30), flow rate of 1 mL/min, sample concentration 0.5 mg/mL, injection volume 5 μL, temperature 30 °C, backpressure 2000 psi, ESI+.

For the analysis of impurities, impurity fractions of material **I** obtained from the column chromatography and mother liquor after re-crystallisation were collected and used. Since the mobile phase composed of $scCO_2$ and IPA (low molar mass alcohols in general) is acidic [27,28], ESI+ was the preferred mode also for the impurities analysis. Apart from the target material **I** and the enantiomeric impurity (material **II**), several other substances were identified (Figure 7). The SFC-MS analysis shows that molecular masses of impurities correspond to fragments observed for ionised target materials. This documents that all impurities present in the target material originate from the starting compounds and side reactions among them—mainly migration of *O*-alkyllactic acid from the target materials to phenol **3** (for more detailed analysis and structures of the impurities see ESI in the Supplementary Materials).

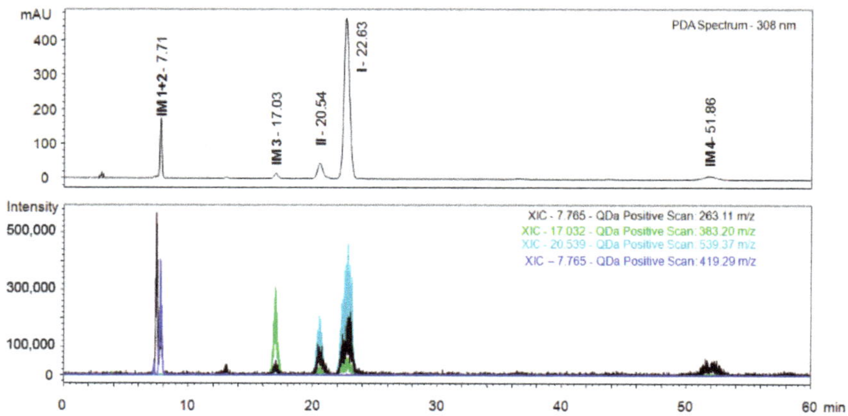

Figure 7. SFC-MS analysis of major impurities (IM) of the target compound **I**. Upper part shows the UV trace, lower part reconstructed chromatograms with masses of the target materials and impurities originating from the synthesis. Conditions: ChiralArt Amylose-C (250 × 4.6 mm, i.d., 5 µm) column, mobile phase $scCO_2$/IPA (70/30), flow rate of 1 mL/min, sample concentration 1.0 mg/mL, injection volume 10 µL, temperature 30 °C, backpressure 2000 psi, ESI+.

3. Materials and Methods

3.1. Chemicals

Heptane, dichloromethane (DCM), methanol (MeOH), and propan-2-ol (IPA) were purchased from LachNer s.r.o. (Neratovice, Czech Republic); all solvents were of HPLC grade. Carbon dioxide (SFC grade) was obtained from Linde Industrial Gasses (Prague, Czech Republic). Chemicals used for the synthesis of the chiral precursors and target LCs were commercial products from Sigma-Aldrich (Prague, Czech Republic), Fischer Scientific (Pardubice, Czech Republic) and they were used without further purification. Silica gel (Kieselgel 60) for the purification of the intermediates and target LCs was purchased from Merck (Darmstadt, Germany).

3.2. Instrumentation and Methods

The optical purity of chiral precursors was determined by HPLC using Chiralpak® AD-3 (150 × 4.6 mm ID, 3 µm) from Chiral Technologies Europe (Illkirch, France) column in a heptane/IPA (9/1) mixture, the flow rate was set to 1 mL/min, temperature 25 °C. Sample concentration was 0.2 mg/mL, injection volume 20 µL and detection wavelength was set to 235 nm. Optical purity of final liquid crystalline materials was verified under the same conditions, except for the detection wavelength, which was set to 305 nm. The ECOM HPLC system consisted of Alpha pump (ECOM, Prague, Czech Republic), CT050 controller (AZ Chrom, Bratislava, Slovakia) and ECDA2000 (ECOM, Prague, Czech Republic) detector. A part of HPLC measurements on Chiralpak AD-3 was carried out on an SFC system (operated in the HPLC mode), as specified below.

SFC measurements were performed on a SFC-dedicated column, namely ChiralArt Amylose-C (250 × 4.6 mm ID, 5 µm) from YMC Europe GmbH (Dinslaken, Germany) using an Acquity Ultra-Performance Convergence Chromatography (UPC2) system equipped with a binary solvent manager, sample manager, convergence manager, column manager 30S for eight 250 mm columns and PDA detector and single quadrupole mass detector (QDa) from Waters (Milford, CT, USA). The mobile phase consisted of supercritical carbon dioxide ($scCO_2$) with 30% of IPA. The flow rate was set to 1 mL/min, injection volume to 5 µL. For all measurements, the backpressure was set to 2000 psi (138 bar) and the column temperature to 30 °C. The sample concentration was 0.5 mg/mL in a mixture of heptane/IPA (9/1, v/v). The void volume ($t0$) was determined from the first negative peak observed after the injection. An equilibration window of 15 min was applied prior the first sample injection for each column. The PDA acquisition range was 210–400 nm and the detection wavelength was 308 nm. The mass detection was performed in positive ion mode (ESI+) with a mass range 200.00–600.00 Da, cone voltage 10 V. The ESI spray voltage was set to 0.8 kV. Empower 3 software was used for system control and data acquisition.

Nuclear magnetic resonance (NMR) spectra were acquired using an Agilent 400-MR DDR2 spectrometer (Santa Clara, CA, USA) operating at 400.13 MHz for ^{1}H and 100.62 MHz for ^{13}C cores. Elemental analysis was performed on a Perkin-Elmer 2400 Instrument (Waltham, MA, USA).

3.3. Experimental

(*S*)-Propyl 5-{4-(4-(2-hexyloxypropanoyloxy)benzoyloxy)phenyl}thiophene-2-carboxylate (**I**)

A mixture of acid **2a** (500 mg, 1.71 mmol) with a catalytic amount of DMF (0.05 mL) in oxalyl chloride (20 mL) was stirred at reflux for 2 h. The excess of the oxalyl chloride was distilled off and the residue was heated under reflux in hexane with a spatula-tip of active carbon for 1 min. The suspension was filtered, and the solvent evaporated; the crude acid chloride was dissolved in dry dichloromethane and added dropwise to a solution of hydroxyl ester **3** (374 mg, 1.43 mmol) with DMAP (174 mg, 1.43 mmol) in dry dichloromethane (25 mL). The reaction mixture was stirred in argon atmosphere at room temperature for 2 h. Then, the reaction mixture was decomposed with 17% aq. hydrochloric acid, layers were separated, and the aqueous layer was extracted with toluene. The combined organic solution was washed with brine and dried with anhydrous magnesium sulphate. The solvent was removed, and the crude product was purified by column chromatography (toluene/*tert*-butyl methyl ether 20:1, v/v) and multiple crystallisations from an ethyl acetate/ethanol mixture to obtain 430 mg (56%) of a white solid.

It should be noted that the purification procedure given above was found to be insufficient (vide infra), and therefore both target materials **I** and **II** were further purified. This additional purification step consisted of column chromatography using gradient elution with $CHCl_3$ – 0.8% MeOH in $CHCl_3$ and subsequent crystallisation from ethanol. The gradient elution column chromatography afforded a chemically pure substance while the subsequent crystallisation served as a tool for removal of the trace amount of the opposite enantiomer, which was confirmed by SFC-MS measurements.

^{1}H-NMR (400 MHz, $CDCl_3$): 0.89 (t, 3H, CH_3), 1.02 (t, 3H, CH_3), 1.23–1.45 (m, 8H, CH_2), 1 60 (d, 3H, CH_3), 1.79 (m, 2H, CH_2), 3.52 (m, 1H, OCH_2), 3.69 (m, 1H, OCH_2), 4.22 (q, 1H, OCH), 4.28 (t, 2H, OCH_2), 7.26–7.30 (m, 5H, H_{ar}), 7.70 (d, 2H, J = 8.7 Hz, H_{ar}), 7.77 (d, 1H, J = 3.9 Hz, H 21), 8.26 (d, 2H, J = 8.8 Hz, H_{ar}). ^{13}C-NMR (100 MHz, $CDCl_3$): 10.46 (CH_3), 14.04 (CH_3), 18.67 (CH_3), 22.13 (CH_2), 22.59 (CH_2), 25.72 (CH_2), 29.73 (CH_2), 31.62 (CH_2), 66.75 (OCH_2), 70.85 (OCH_2), 74.97 (OCH), 121.74 (CH_{ar}), 122.40 (CH_{ar}), 123.84 (CH_{ar}), 126.93 (C), 127.40 (CH_{ar}), 131.45 (C), 131.90 (CH_{ar}), 132.79 (C), 134.22 (CH_{ar}), 149.81 (C), 151.12 (C), 154.78 (C), 162.27 (C), 164.15 (C), 171.46 (C). Elemental analysis for $C_{30}H_{34}O_7S$ (538.67), calculated C 66.89, H 6.36, S 5.95%, found C 67.02, H 6.44, S 5.86%.

In the similar way, starting from acid **2c**, (*R*)-propyl 5-{4-(4-(2-hexyloxypropanoyloxy)benzoyloxy) phenyl}thiophene-2-carboxylate (**II**) was prepared, 300 mg (45%). ^{1}H-NMR: (400 MHz, $CDCl_3$): 0.89 (t, 3H, CH_3), 1.03 (t, 3H, CH_3), 1.24–1.45 (m, 8H, CH_2), 1.59 (d, 3H, CH_3), 1.79 (m, 2H, CH_2), 3.52 (m, 1H, OCH_2), 3.69 (m, 1H, OCH_2), 4.22 (q, 1H, OCH), 4.27 (t, 2H, OCH_2), 7.26–7.30 (m, 5H, H_{ar}), 7.69 (d, 2H, J = 8.7 Hz, H_{ar}), 7.77 (d, 1H, J = 3.9 Hz, H 21), 8.26 (d, 2H, J = 8.8 Hz, H_{ar}). ^{13}C-NMR (100 MHz, $CDCl_3$): 10.46 (CH_3), 14.04 (CH_3), 18.67 (CH_3), 22.13 (CH_2), 22.59 (CH_2), 25.72 (CH_2), 29.73 (CH_2), 31.62 (CH_2), 66.75 (OCH_2), 70.85 (OCH_2), 74.97 (OCH), 121.74 (CH_{ar}), 122.40 (CH_{ar}), 123.84 (CH_{ar}), 126.93 (C), 127.40 (CH_{ar}), 131.45 (C), 131.90 (CH_{ar}), 132.79 (C), 134.22 (CH_{ar}), 149.81 (C), 151.12 (C), 154.78 (C), 162.27 (C), 164.15 (C), 171.46 (C). Elemental analysis for $C_{30}H_{34}O_7S$ (538.67), calculated C 66.89, H 6.36, S 5.95%, found C 66.54, H 6.37, S 5.92%.

4. Conclusions

In this study, we focused on the evaluation of optical purity of lactic acid-based building blocks used for the synthesis of a broad range of CLCs (including the target LCs). We have shown that frequently utilised synthetic procedures do not afford enantiomerically pure building blocks. The use of such compounds does not inevitably result in impure target materials; however, very precise control of chemical and optical purity must be employed. Partial racemisation occurring during the synthesis of CLCs could result in modification of mesomorphic properties of the target materials. This may have caused the discrepancies in the mesomorphic behaviour of the same CLCs reported by different research groups. Most importantly, slight modification of enantiomeric composition of the chiral material could potentially lead to malfunctioning of a CLC-based device. Therefore, precise contemporary analytical methods with low limits of detection should be used to secure required quality of CLCs used in research, development and applications.

Supplementary Materials: The supplementary information is available online.

Author Contributions: A.P. and M.Š. synthesised the materials; T.V., P.T. and D.S. designed and performed HPLC separations, N.K. carried out SFC measurements; M.K. wrote the manuscript, M.K. and D.S. performed final formatting and revision of the manuscript.

Funding: The work was supported by Czech Science Foundation (project No. 16-17689Y) and Specific University Research (MSMT No. 21-SVV/2018).

Conflicts of Interest: The authors declare no conflicts of interest.

References

1. Goodby, J.W.; Collings, P.J.; Kato, T.; Tschierske, C.; Gleeson, H.F.; Raynes, P. *Handbook of Liquid Crystals*; Wiley-VCH Verlag: Weinheim, Germany, 2014.
2. Kato, T.; Uchida, J.; Ichikawa, T.; Sakamoto, T. Functional liquid crystals towards the next generation of materials. *Angew. Chem. Int. Ed.* **2018**, *57*, 4355–4371. [CrossRef]
3. Bisoyi, H.K.; Bunning, T.J.; Li, Q. Stimuli-driven control of the helical axis of self-organized soft helical superstructures. *Adv. Mater.* **2018**, *30*, 1706512. [CrossRef]
4. Jing, H.; Xu, M.; Xiang, Y.; Wang, E.; Liu, D.; Poryvai, A.; Kohout, M.; Éber, N.; Buka, Á. Light tunable gratings based on flexoelectric effect in photoresponsive bent-core nematics. *Adv. Opt. Mater.* **2019**. [CrossRef]
5. Collings, P.J.; Hird, M. *Introduction to Liquid Crystals: Chemistry and Physics*; Taylor & Francis: London, UK, 1997.
6. Gennes, P.D.; Prost, J.; Pelcovits, R. The physics of liquid crystals. *Phys. Today* **1995**, *48*, 67. [CrossRef]
7. Stephen, M.J.; Straley, J.P. Physics of liquid crystals. *Rev. Mod. Phys.* **1974**, *46*, 617–704. [CrossRef]
8. Černovská, K.; Košata, B.; Svoboda, J.; Novotná, V.; Glogarová, M. Novel ferroelectric liquid crystals based on fused thieno[3,2-b]furan and thieno[3,2-b]thiophene cores. *Liq. Cryst.* **2006**, *33*, 987–996. [CrossRef]
9. Vaňkátová, P.; Kalíková, K.; Kubíčková, A. Ultra-performance supercritical fluid chromatography: A powerful tool for the enantioseparation of thermotropic fluorinated liquid crystals. *Anal. Chim. Acta* **2018**, *1038*, 191–197. [CrossRef]
10. Vaňkátová, P.; Kubíčková, A.; Cigl, M.; Kalíková, K. Ultra-performance chromatographic methods for enantioseparation of liquid crystals based on lactic acid. *J. Supercrit. Fluids* **2019**, *146*, 217–225. [CrossRef]
11. Vojtylová, T.; Kašpar, M.; Hamplová, V.; Novotná, V.; Sýkora, D. Chiral HPLC for a study of the optical purity of new liquid crystalline materials derived from lactic acid. *Phase Trans.* **2014**, *87*, 758–769. [CrossRef]
12. Vojtylová, T.; Żurowska, M.; Milewska, K.; Hamplová, V.; Sýkora, D. Chiral HPLC and physical characterisation of orthoconic antiferroelectric liquid crystals. *Liq. Cryst.* **2016**, *43*, 1244–1250. [CrossRef]
13. Jurkovičová-Vojtylová, T.; Cigl, M.; Tomášková, P.; Hamplová, V.; Sýkora, D. Influence of photoinduced isomerization on the chiral separation of novel liquid crystalline materials with a diazene moiety. *J. Sep. Sci.* **2018**, *41*, 3034–3041. [CrossRef] [PubMed]
14. Bajzíková, K.; Kohout, M.; Tarábek, J.; Svoboda, J.; Novotná, V.; Vejpravová, J.; Pociecha, D.; Gorecka, E. All-organic liquid crystalline radicals with a spin unit in the outer position of a bent-core system. *J. Mater. Chem. C* **2016**, *4*, 11540–11547. [CrossRef]

15. Barberá, J.; Iglesias, R.; Serrano, J.L.; Sierra, T.; de la Fuente, M.R.; Palacios, B.; Pérez-Jubindo, M.A.; Vázquez, J.T. Switchable columnar metallomesogens. New helical self-assembling systems. *J. Am. Chem. Soc.* **1998**, *120*, 2908–2918. [CrossRef]
16. Chin, E.; Goodby, J.W.; Patel, J.S. The liquid-crystalline properties of some chiral derivatives of the 4-n-alkanoyloxbiphenyl-4′-carboxylic acids. *Mol. Cryst. Liq. Cryst. Inc. Nonlinear Opt.* **1988**, *157*, 163–191. [CrossRef]
17. Kobayashi, S.; Ishibashi, S. Ferroelectric liquid crystals with chiral groups on each side of the core. *Mol. Cryst. Liq. Cryst. Sci. Technol. Sect. A* **1992**, *220*, 1–17. [CrossRef]
18. Kohout, M.; Bubnov, A.; Šturala, J.; Novotná, V.; Svoboda, J. Effect of alkyl chain length in the terminal ester group on mesomorphic properties of new chiral lactic acid derivatives. *Liq. Cryst.* **2016**, *43*, 1472–1485. [CrossRef]
19. Kovářová, A.; Kohout, M.; Svoboda, J.; Novotná, V. New liquid crystal based on 2-phenylthiophene central core. *Liq. Cryst.* **2014**, *41*, 1703–1718. [CrossRef]
20. Joullie, M.M.; Lassen, K.M. Evolution of amide bond formation. *Arkivoc* **2010**, *2010*, 189–250.
21. Kohout, M.; Bielec, B.; Steindl, P.; Trettenhahn, G.; Lindner, W. Mechanistic aspects of the direct C-acylation of cyclic 1,3-diones with various unactivated carboxylic acids. *Tetrahedron* **2015**, *71*, 2698–2707. [CrossRef]
22. Twibanire, J.A.K.; Grindley, T.B. Efficient and controllably selective preparation of esters using uronium-based coupling agents. *Org. Lett.* **2011**, *13*, 2988–2991. [CrossRef]
23. Stevens, P.G. The configuration of methylisopropylcarbinol with a note on racemization. *J. Am. Chem. Soc.* **1932**, *54*, 3732–3738. [CrossRef]
24. Mangelings, D.; Vander Heyden, Y. Chiral separations in sub- and supercritical fluid chromatography. *J. Sep. Sci.* **2008**, *31*, 1252–1273. [CrossRef] [PubMed]
25. Tarafder, A.; Guiochon, G. Extended zones of operations in supercritical fluid chromatography. *J. Chromatogr. A* **2012**, *1265*, 165–175. [CrossRef] [PubMed]
26. Wolrab, D.; Frühauf, P.; Gerner, C.; Kohout, M.; Lindner, W. Consequences of transition from liquid chromatography to supercritical fluid chromatography on the overall performance of a chiral zwitterionic ion-exchanger. *J. Chromatogr. A* **2017**, *1517*, 165–175. [CrossRef] [PubMed]
27. Wolrab, D.; Kohout, M.; Boras, M.; Lindner, W. Strong cation exchange-type chiral stationary phase for enantioseparation of chiral amines in subcritical fluid chromatography. *J. Chromatogr. A* **2013**, *1289*, 94–104. [CrossRef] [PubMed]
28. Wolrab, D.; Macíková, P.; Boras, M.; Kohout, M.; Lindner, W. Strong cation exchange chiral stationary phase—A comparative study in high-performance liquid chromatography and subcritical fluid chromatography. *J. Chromatogr. A* **2013**, *1317*, 59–66. [CrossRef]

Sample Availability: Samples of all the compounds are available from the authors.

molecules

MDPI

Article

Enhanced Near-Field Chirality in Periodic Arrays of Si Nanowires for Chiral Sensing

Emilija Petronijevic * and Concita Sibilia

Department S.B.A.I., Sapienza Università di Roma, Via A. Scarpa 14, 00161 Rome, Italy; concita.sibilia@uniroma1.it

* Correspondence: emilija.petronijevic@uniroma1.it

Academic Editor: Derek J. McPhee

Received: 29 January 2019; Accepted: 27 February 2019; Published: 28 February 2019

Abstract: Nanomaterials can be specially designed to enhance optical chirality and their interaction with chiral molecules can lead to enhanced enantioselectivity. Here we propose periodic arrays of Si nanowires for the generation of enhanced near-field chirality. Such structures confine the incident electromagnetic field into specific resonant modes, which leads to an increase in local optical chirality. We investigate and optimize near-field chirality with respect to the geometric parameters and excitation scheme. Specially, we propose a simple experiment for the enhanced enantioselectivity, and optimize the average chirality depending on the possible position of the chiral molecule. We believe that such a simple achiral nanowire approach can be functionalized to give enhanced chirality in the spectral range of interest and thus lead to better discrimination of enantiomers.

Keywords: semiconductor nanowires; chirality; enantioselectivity; near-field optical chirality

1. Introduction

Chirality, a lack of mirror symmetry [1], is an important property of our world governing the behavior of many molecules, enzymes, DNA, and sugars. Optical isomers of the opposite handedness that are non-superimposable images of each other are called *enantiomers*. Specifically, two enantiomers of the same chiral drug have the same chemical structure and physical properties, but different spatial arrangement and optical activity [2], which leads to differences in biological activities such as toxicity [3,4] and enantioselective reactions [5]. Two enantiomers can therefore have extremely different effects on the human body where one can lead to serious side effects [6–8]. Conventionally, polarimetry and circular dichroism (CD) measurements are used to distinguish enantiomers since they differently interact with the circularly polarized (CP) light of the opposite handedness [9]. However, such experiments require high concentrations of enantiomers and long integration times as the intrinsic CD of the molecules is usually extremely low. Therefore, finding novel approaches to detect and recognize low concentrations of enantiomers, thus increasing the sensitivity and reducing the material waste, is of great interest for today's pharmaceutical industry.

In the past decade, the nanoscale photonics community has been dealing with chirality as well: artificial plasmonic nanostructures with broken symmetry were shown to provide a chiral response and CD [10–16]. We recently showed that semiconductor-based nanomaterials, asymmetrically covered by thin metallic layers, can also produce strong CD [17–19]. Merging of the chiral plasmonics with the field of chiral biomolecule enantioselectivity [20] has already given promising results in chiral sensing. For example, in [21], the authors reported on the several orders of magnitude enhancement of the enantiomer detection by means of twisted plasmonic unit cells periodically arranged to form a chiral *metamaterial*. Furthermore, investigation of the near-field effects has opened ways to enhance the interaction between the nanostructure and the chiral molecule via so called *superchiral* fields [22,23]. Namely, for a monochromatic electromagnetic field at frequency ω the optical chirality factor C is defined as:

$$C = -0.5\varepsilon_0 \omega \mathrm{Im}\{\vec{E}^*(\vec{r})\cdot\vec{B}(\vec{r})\}, \quad (1)$$

where $\vec{E}$ and ($\vec{B}$) are electric (magnetic) complex field amplitudes. For a plane wave, C is zero for linear polarization, maximum for right circular polarization (RCP), and minimum for left circular polarization (LCP), switching the sign between RCP and LCP. In the near-field of nanostructured materials at the resonance, the electric and magnetic fields can be parallel and out of phase, while their enhancement leads to a C greater than the one of CP, hence the term superchirality. The C factor directly influences the excitation rate of chiral molecules [24] as well as the absorption dissymmetry [25], and its switching of the sign between RCP and LCP is directly connected with the enantioselectivity. Namely, the difference in the absorption rate of two enantiomers can be significantly enhanced if the medium where the mixture exists has an enhanced C. If the medium is excited with CP so that it generates enhanced C of one sign, this will enhance the absorption rate of one enantiomer. Changing the handedness of CP will enhance the absorption rate of another enantiomer, leading to the enhanced difference between the two enantiomers. Finally, in this way the sensitivity of distinction for low enantiomer concentrations can be improved. In general, the medium should produce enhanced C in the volume of the enantiomers, and it must be of the same sign over the volume in order to contribute to the absorption rate of only one enantiomer. Plasmonic nanostructures can be designed to tailor C in the wavelength region of interest for a particular chiral molecule. We recently numerically investigated C enhancement in GaAs nanowires (NWs) asymmetrically covered by Au [26], while in [25,27,28], symmetric nanostructures were shown to give an enhanced near-field C as well. Moreover, in [25] the authors underlined the importance of having only near-field, local chiral effects, without the background chirality of the supporting medium itself. In this way, all the chiral enhancement and corresponding effects arise from the coupling of the chiral molecule with the electromagnetic field in the vicinity of the achiral structure.

In this work, we investigate intrinsically achiral 2D periodic arrays of Si NWs on a Si substrate, a device which is completely symmetric and easy to fabricate. Geometric parameters can be chosen so that these NWs support leaky waveguide modes [29–31] in the visible part of the spectrum. These modes are weakly guided along the NW borders and have a strong field leakage to the surrounding medium, where the evanescent field components can produce a near-field chirality. In [26] we investigated GaAs NW modes at ~800 nm, where we showed that for linearly polarized excitation, the near-field chirality cancels out and the additional symmetry breaking (e.g., the asymmetric layer of Au) must be introduced in order to have a C of prevalently one sign. In [32], NW dimers are used to induce the optical chirality and the hotspots of high C enhancement. Here, we show that under circularly polarized (CP) excitation, achiral NW structure can generate the enhanced chirality in the NW border vicinity, without the additional symmetry breaking which would complicate the processing of a perspective device. For Si NWs, we investigate the influence of geometric parameters on the mode wavelength, which in turn tailors the C distribution; we show that it is possible to spectrally place the enhanced C in the high frequency range of the visible spectrum, technologically relevant for chiral molecules. Further, we investigate the optimization of the average C so that it is distributed in the big part of the volume of the unit cell. For the opposite handedness of the CP excitation, average C changes sign, thus we propose these metasurfaces for enantioselectivity applications, with parameters that can be fixed in the wavelength range of interest. As these materials are easy to fabricate and implement in the existing technology, we believe that this approach can lead to efficient background-free tunable chiral recognition, with decreased waste of materials and increased efficiency.

2. Results

Structures under investigation are hexagonal Si NWs, that can be patterned on Si substrate, e.g., by means of conventional electron beam lithography [32,33]. In what follows, we use a commercial-grade simulator based on the 3D Finite Difference Time Domain (FDTD) method by Lumerical [34] to simulate complex electromagnetic fields in the Si NW array, and later extract C factor;

more details are given in Section 4. The scheme of the simulated device is given in Figure 1a: vertical hexagonal NWs of radius r and length L are periodically arranged with periodicity p in x and y direction. Such structures have been previously proposed for absorption enhancement, lasing and solar cells, as they can efficiently absorb the visible spectrum wavelengths. The structure is excited by a circularly polarized plane-wave under normal incidence. In Figure 1b we show the absorption spectra dependence on the NW radius, for L = 1000 nm, and p = 400 nm. As expected from theory [35], as well as experiments in [30,31], there is always a fundamental HE_{11} mode excited for these thin NWs, which leads to the resonant absorption; moreover, with the radius increase, the resonant wavelength of the mode red-shifts. For this periodicity, the mode closely resembles the leaky waveguide mode of the single Si NW. In Figure 1c, the electric field vector in xz cross-section of the unit cell is shown for the NW array with parameters r = 50 nm, L = 1000 nm, p = 400 nm, excited at its resonance λ = 525 nm with CP light. Clearly, the electric field keeps circularly polarized behavior, while it experiences enhancement, especially in two regions of the volume corresponding to the antinodes of the Fabry–Perot resonances (due to the boundary conditions at the top and the bottom of the NW). In Figure 1d, we show the electric field intensity enhancement at z = 800 nm xy cross-section of the unit cell; there is an obvious leakage of the mode to the surrounding medium, along with a 35-fold increase. In Figure 1e, for the same cross-section, the magnetic field enhancement is even higher, but it remains confined in the NW core. According to Equation (1), the enhanced C can be expected at the points where these field enhancements spatially overlap, while $\vec{E}$ and $\vec{H}$ have parallel components out of phase. In the close vicinity of the NW, molecules present in the surrounding medium will experience both electric and magnetic field enhancement, while for the near-field chirality calculations, their phase difference must be considered as well. Thus, in the following, we investigate C distribution dependence on the NW parameters at the resonant wavelengths; we report on the normalized $C^* = C/C_0$, where C_0 is extracted for the RCP excitation of the simulation domain without the Si NWs. For better visualization, we show only the points with $|C^*| > 1.2$.

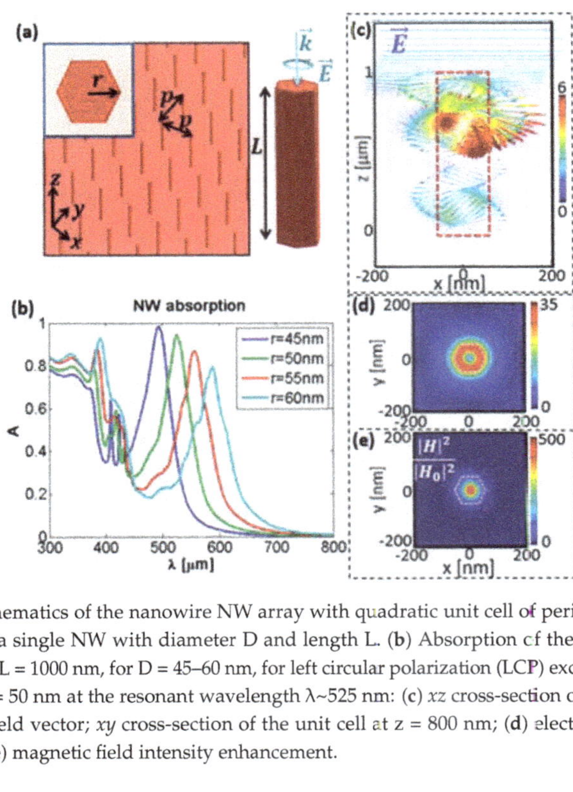

Figure 1. (**a**) Schematics of the nanowire NW array with quadratic unit cell of periodicity p x and y directions, and a single NW with diameter D and length L. (**b**) Absorption of the NW arrays with p = 400 nm, and L = 1000 nm, for D = 45–60 nm, for left circular polarization (LCP) excitation. (**c**–**e**) LCP excitation for r = 50 nm at the resonant wavelength λ~525 nm: (**c**) xz cross-section of the distribution of the electric field vector; xy cross-section of the unit cell at z = 800 nm; (**d**) electric field intensity enhancement; (**e**) magnetic field intensity enhancement.

2.1. *Influence of the NW Core Radius*

Firstly, C distribution in the unit cell is investigated for p = 400 nm and L = 1000 nm, for the radii investigated previously. In Figure 2, in all the cases, $|C^*|$ is higher than 5 at the NW borders and remains enhanced in the NW proximity, where the chiral molecules can be deposited for the experiment. More importantly, for RCP (LCP) excitation, C* keeps the positive (negative) sign of the excitation (normalization by RCP), while having increased values. As the whole domain is achiral, the resonant wavelengths for RCP and LCP are the same. The NW array with r = 45 nm has a resonant wavelength of λ = 493 nm, where Si has higher losses; therefore, the field is efficiently absorbed in the upper part of the NW, and leads to lower electromagnetic field enhancement on the borders closer to the bottom of the NW. For r = 50 nm, we can note that the z positions of the maximum C* correspond to the antinode enhancements in Figure 1c. With the radius increase leading to the resonance red-shift, the losses of the mode become lower, and C* spreads more in the unit cell volume; moreover, the second antinodal enhancement of C* (close to the substrate) also becomes more prominent. Therefore, tuning of the radius can effectively tailor the enhanced C* in terms of wavelength and 3D distribution. As an example, if the enhanced enantioselectivity experiment involved enantiomers with absorption dissymmetry around 590 nm, the last NW with r = 60 nm should be chosen from these four configurations (the enhanced $|C^*|$ has the highest spread in the unit cell volume).

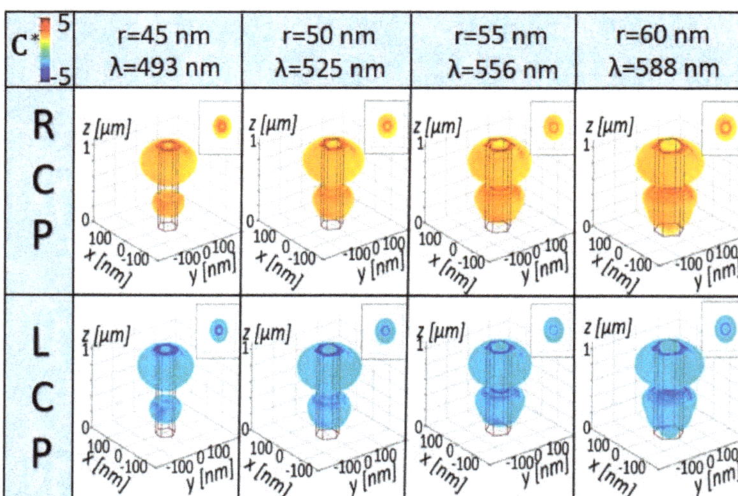

Figure 2. C* distribution in the unit cell for p = 400 nm, L = 1000 nm, and r = 45–60 nm, at the corresponding NW resonant wavelength for RCP (upper, positive) and LCP (bottom, negative) excitation. The insets show the *xy* top view of each C* distribution.

2.2. *Influence of the NW Array Periodicity*

As the coupling between the neighboring NWs influences the electromagnetic fields, we further investigated the absorption and C* dependence on the periodicity (Figure 3a) for r = 50 nm and L = 1000 nm. 2D periodic NW arrays are photonic crystals, where, for larger periods, the single NW modes are not considerably influenced by the array coupling [29,35]. However, for denser Si NW arrays, we note a blue-shift of the resonance with the decreasing period; this arises from the destructive coupling between the neighboring NWs due to enhanced near-field evanescent wave interaction. Unfortunately, in Figure 3b, for p = 200 nm, this leads to C* confinement inside the NW, which is detrimental for the applications where the enantiomers surround the NW in the unit cell. The situation can be optimized by increasing the periodicity, which confines C* also around the NW (e.g., p = 300 nm in the middle of Figure 3b, or p = 400 nm in the second graph in Figure 2). Finally, sparse arrays of p = 600 nm have stronger C* enhancement over the high part of the volume, as it will be shown later. This is due to the strong electromagnetic field enhancement in the near-field medium close to the NW, for the fundamental leaky waveguide modes of the single NW.

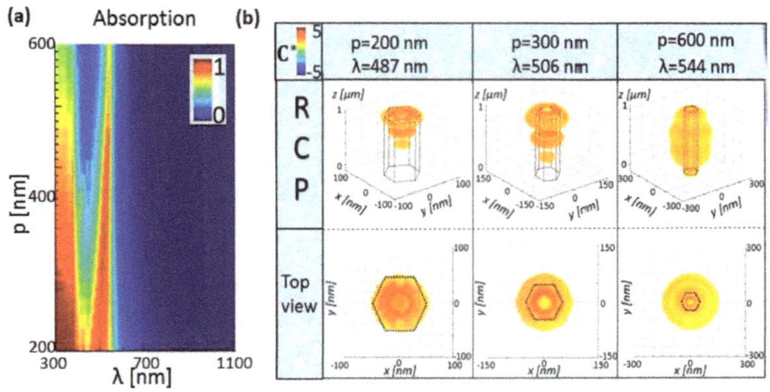

Figure 3. (**a**) Absorption dependence on the NW array periodicity for r = 50 nm and L = 1000 nm. (**b**) Upper part: C* distribution in the unit cell for p = 200 nm, p = 300 nm, and p = 600 nm at the corresponding NW resonant wavelengths for RCP excitation; bottom: *xy* top-view of the corresponding distribution from the upper part. Black hexagonal line shows the geometric position of the NW *xy* cross-section. Only RCP excitation is shown as LCP leads to the sign inversion.

2.3. *Influence of the NW Length*

For the enhanced enantioselectivity, as previously discussed, the nanostructures can be optimized in order to interact with the enantiomers in the most efficient way, which depends also on the enantiomer position. Therefore, we investigated the NW length influence on C* distribution, for r = 50 nm and p = 400 nm. In Figure 4a, the modes do not shift with increasing length, but the unitary absorption is reached only for longer lengths. This is rather expected for vertical high refractive index nanowires, where the absorption peak arises due to the mode, which is mainly defined by the radial boundary conditions of the nanowire core (as in dielectric waveguides). Therefore, the radius plays a major role in the spectral position of the absorption peak, while other geometric parameters have a minor influence, as we experimentally demonstrated in [17,18,30,31]. As expected, the shortest NW (L = 500 nm) has an enhanced C* only due to the first antinode of the resonant mode, while for L = 800 nm another antinode appears. Finally, for L = 1500 nm, three enhanced volume parts are distinguishable; however, the bottom one would not contribute to the enantioselectivity as its C* is more confined inside the NW. If we consider e.g., the enantiomers with absorption dissymmetry around 525 nm, the best Si NW substrate (with r = 50 nm and p = 400 nm) would be the one with L = 500 nm, and the matrix with chiral molecules should be deposited between z = 0 nm and z = 500 nm. Otherwise, if the NW length is fixed to L = 1500 nm, in order to use zones with the highest C*, one should deposit another non-absorbing and achiral medium (buffer layer) in the range 0 nm < z < 1000 nm, and then the chiral substance in the range 1000 nm < z < 1500 nm, which complicates the simple approach proposed here.

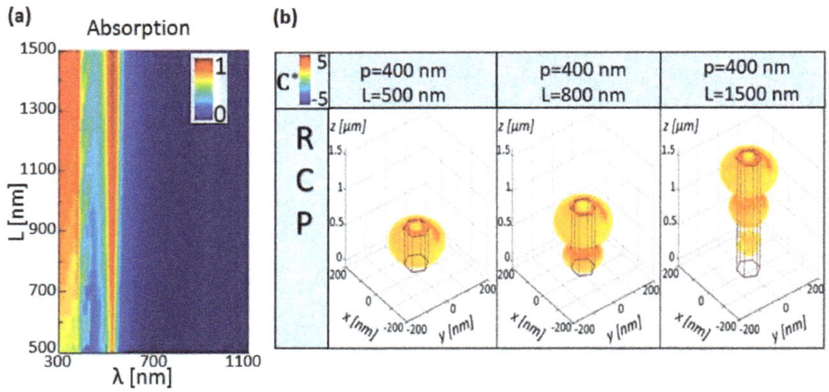

Figure 4. (**a**) Absorption dependence on the NW length for r = 50 nm and p = 400 nm. (**b**) C* distribution in the unit cell for L = 500 nm, L = 800 nm, and L = 1500 nm at the resonant wavelength of 525 nm. Only RCP excitation is shown as LCP leads to the sign inversion.

3. Discussion

The overall enantioselectivity enhancement will depend on the percentage of the chiral molecules that are positioned exactly in the part of the volume with enhanced C*, i.e., in the NW near-field. This is usually not the case for the periodicities that support leaky waveguide modes, so it is important to estimate the average chirality in the unit cell volume, with subtracted NW volume. In order to gain insight into the spectral behavior of the investigated chirality enhancement, we calculate the average normalized optical chirality as:

$$\widetilde{C}(\lambda) = \frac{1}{C_{RCP}(\lambda)} \frac{\iiint C(x,y,z,\lambda)dV}{\iiint dV}, \tag{2}$$

where we integrate C across the unit cell as follows: in the *xy* plane, only the parts of the unit-cell outside of the NW core are considered, while in the *z* direction, we investigate two possibilities. First, we integrate in the 0 < z < L range (upper sketch of Figure 5), i.e., total integration. Then, we take into account only the top C* antinode, integrating in the L/2 < z < L range (bottom sketch of Figure 5), i.e., half-integration, which corresponds to the case when the enantiomers are specially positioned on a buffer layer in order to experience higher C. We approximate that the buffer layer has negligible influence on the C* distribution, which is valid for low concentrations of chiral molecules, deposited in a solvent which subsequently evaporates. However, generally the buffer layer does change the modal confinement, and the NW array must be reoptimized in terms of r, L and p in order to give enhanced C* once the optical properties of the buffer layer are known. In Figure 5a, the total integration of L = 1000 nm and r = 50 nm, for the periodicities from 200 nm to 400 nm gives almost no enhancement at the resonant wavelengths, and it is vaguely present only for p = 500 nm (the lowest NW coupling, $|\widetilde{C}| \sim 1.2$). However, the half-integration in Figure 5b improves this difference; here, especially for p = 500 nm, $|\widetilde{C}|$ reaches almost 2, which is a considerable average enhancement for a $0.5 \times 0.5\ \mu m^2$ surface. For shorter NWs, L = 500 nm and r = 50 nm, the total integration in Figure 5c gives $\widetilde{C}$ improvement with respect to Figure 5a, due to inclusion of the part of the space where one antinode generates enhanced C* (Figure 4b-left). However, this case does not get improved with the half-integration, Figure 4d.

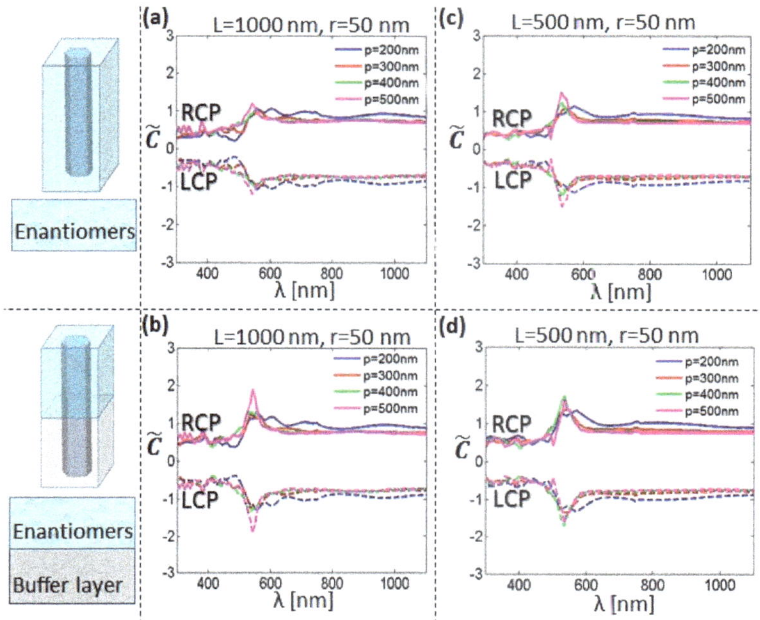

Figure 5. Normalized averaged C spectra for r = 50 nm, p = 200–425 nm, and RCP (full lines) and LCP (dashed lines) excitation. Upper and bottom sketches correspond to the integration from z = 0 nm to z = L, and z = L/2 to z = L, respectively. (**a**) L = 1000 nm, total integration; (**b**) L = 500 nm, total integration; (**c**) L = 1000 nm, half-integration; (**d**) L = 500 nm, half-integration.

It is worth noting that for all investigated cases in Figure 5, Si NW arrays give a rather modest average optical chirality. However, we underline the advantage of the use of Si in such nanostructures. Apart from the highly developed technology and perfectly known optical and electrical properties, Si NWs have already been proposed for bio and chemical sensing as the NW surface allows for the immobilization of the investigated substance in the NW near-field, thus affecting the sensor performance and enabling high sensitivity and selectivity [36,37]. For the enhanced enantioselectivity, one should position the chiral molecules in a very thin layer close to the surface which enhances the near-field chirality (a few tens of nm, according to [24]). Many approaches have been found to functionalize the Si nanoparticle surface with molecules [38]. Therefore, the ultimate optimization of the presented approach is the attaching of the chiral species to the NW surface, in the small part of the volume as in the sketch of Figure 6. When the solvent evaporates, a small concentration of chiral molecules which remain attached to the Si surface introduces a negligible change in the NW environment, so that the molecules experience the near-field chirality of the resonant modes presented above. Next, we focus on the best case from Figure 5b, i.e., r = 50 nm, L = 1000 nm, and p = 500 nm, which gave the $|\tilde{C}|$ peak at 542 nm. In Figure 6 we show the *xy* cross-section of the C* distribution at the resonance and at the z position of the C* antinode, for RCP and LCP excitation (we omit C* inside the NW as it is not important for the interaction with the enantiomers). There is a 36-fold C enhancement at the NW borders, and the chiral molecules in the dashed circles experience the average C at least on the order of 20, which will finally lead to significantly higher enhancement with respect to the approach in Figure 5c. Therefore, one can smartly functionalize Si NW sidewalls so that the enantiomers are positioned at points with maximum C*.

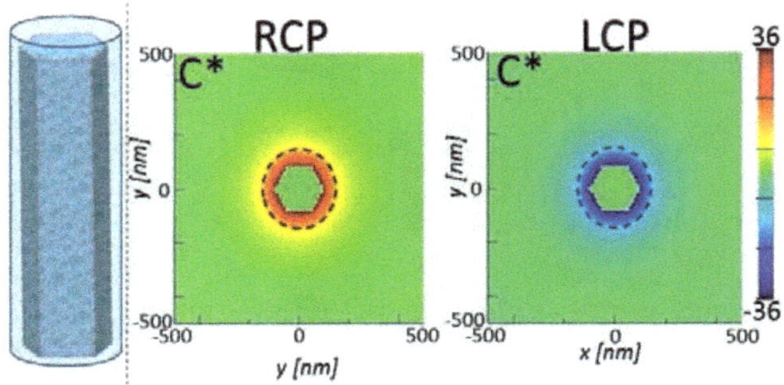

Figure 6. Left: sketch of the Si NW functionalization with bonded chiral substance for the improvement of enantioselectivity. Right: xy cross-section of C* distribution at the antinode of the NW with r = 50 nm, L = 1000 nm, and p = 500 nm (resonant wavelength at 542 nm) for RCP and LCP excitation. Dashed circles represent the volume of the enantiomers attached to the NW surface.

4. Materials and Methods

FDTD simulations solve the Maxwell's equations over discrete spatial and temporal grid, and give as a result complex electromagnetic fields in the unit cell consisting of one Si NW in the middle. Si NW lie on a Si substrate, which is considered semi-infinite for $z < 0$; for $z > 0$, the medium is air. Optical properties of Si were taken from the Lumerical database. In the *z* direction, perfectly matched layers (PML) were placed at least half the maximum wavelength from the top and bottom of the NW to ensure the numerical stability. For the normal incidence excitation, a plane-wave source was used from the top side (negative *z* direction in Figure 1a), and periodic boundary conditions were applied in the *xy* plane. Circular polarization was simulated as a combination of two orthogonal sources with a phase difference of $\pm 90^\circ$. Total absorption of the NW was simulated by integrating the absorption per unit volume σ_{abs} over the NW volume; σ_{abs} was calculated from $\sigma_{abs} = -0.5\omega^2 \,|\,E\,|^2 \mathrm{Im}\{\varepsilon\}$, from the electric field and complex refractive index monitors across the NW domain. Electric and magnetic field confinements were monitored by a cross-section field profile monitor in the *xz* and *xy* planes. C factor was extracted from a 3D field profile monitor encompassing the whole unit cell in the *xy* plane, and having a z span from 0 nm to the NW length. The calculated near-field chiral properties arise from electromagnetic field confinement of the NW modes, and are not an intrinsic property of molecules. Moreover, approximation of air as a medium surrounding the NWs is valid for low concentrations of chiral molecules deposited in the solution which evaporates, leaving the molecules in the NW vicinity. In future work, the experimental proof of principle will be done for chiral molecules that have intrinsic circular dichroism at the modal resonances of a chosen Si nanowire array.

5. Conclusions

In this work, we have proposed a path to enhanced near-field optical chirality by means of symmetric Si NW arrays, which support leaky waveguide modes that enhance the near-field optical chirality of CP excitation in the shorter wavelength part of the visible spectrum, which is of interest for many chiral molecules. The C enhancement can be optimized by choosing the wavelength range where enantiomers show CD, setting the radius, length and period of the NW array so that it gives resonances in that range, and optimizing the molecules position. Such an achiral approach does not suffer from the background chiral behavior present in intrinsically or extrinsically chiral plasmonic nanomaterials, and the absorption dissymmetry arises only for the near-field effects. Moreover, the use of conventional Si technology enables the functionalization of the NW surface which greatly enhances the overall chirality. We believe that this simple approach can lead to Si nanostructures-governed applications in enhanced enantioselectivity.

Author Contributions: Conceptualization, E.P. and C.S.; methodology, E.P.; software, E.P.; validation, C.S.; formal analysis, E.P.; investigation, E.P.; resources, C.S.; data curation, E.P.; writing—original draft preparation, E.P.; writing—review and editing, C.S.; visualization, E.P.; supervision, C.S.; project administration, C.S.; funding acquisition, C.S.

Funding: The authors have no funding to report.

Conflicts of Interest: The authors declare no conflict of interest.

References

1. Kelvin, L. *The Molecular Tactics of a Crystal;* Clarendon Press: Oxford, UK, 1894.
2. Nguyen, L.A.; He, H.; Pham-Huy, C. Chiral Drugs: An Overview. *Int J. Biomed. Sci.* **2006**, *2*, 85–100. [PubMed]
3. Hutt, A.J.; Tan, S.C. Drug chirality and its clinical significance. *Drugs* **1996**, *52*, 1–12. [CrossRef] [PubMed]
4. Smith, S.W. Chiral Toxicology: It's the same thing ... Only different. *Toxicol. Sci.* **2009**, *110*, 4–30. [CrossRef] [PubMed]
5. Nguyen, T.N.; Chen, P.-A.; Setthakarn, K.; May, J.A. Chiral Diol-Based Organocatalysts in Enantioselective Reactions. *Molecules* **2018**, *23*, 2317. [CrossRef] [PubMed]
6. Teo, S.K.; Colburn, W.A.; Tracewell, W.; Kook, K.; Stirling, D.; Jaworsky, M.; Scheffler, M.; Thomas, S.; Laskin, O. Clinical pharmacokinetics of thalidomide. *Clin. Pharmacokinet.* **2004**, *45*, 311–327. [CrossRef] [PubMed]
7. Ribeiro, C.; Santos, C.; Gonçalves, V.; Ramos, A.; Afonso, C.; Tiritan, M.E. Chiral Drug Analysis in Foresnic Chemistry: An Overview. *Molecules* **2018**, *23*, 262. [CrossRef] [PubMed]
8. Guo, J.; Li, M.; Liu, Y.; Wang, F.; Kong, Z.; Sun, Y.; Lu, J.; Jin, N.; Huang, Y.; Liu, J.; et al. Residue and Dietary Risk Assessment of Chiral Cyflumetofen in Apple. *Molecules* **2018**, *23*, 1060. [CrossRef] [PubMed]
9. Rodger, A.; Nordén, B. *Circular Dichroism and Linear Dichroism;* Oxford Univ. Press: Oxford, UK, 1997.
10. Valev, V.K.; Baumberg, J.J.; Sibilia, C.; Verbiest, T. Chirality and chiroptical effects in plasmonic nanostructures: Fundamentals, recent progress, and outlook. *Adv. Mater.* **2013**, *25*, 2517–2534. [CrossRef] [PubMed]
11. Wang, Z.; Cheng, F.; Winsor, T.; Liu, Y. Optical chiral metamaterials: A review of the fundamentals, fabrication methods and applications. *Nanotechnology* **2016**, *27*, 412001. [CrossRef] [PubMed]
12. Belardini, A.; Larciprete, M.C.; Centini, M.; Fazio, E.; Sibilia, C.; Chiappe, D.; Martella, C.; Toma, A.; Giordano, M.; Buatier de Mongeot, F. Circular dichroism in the optical second-harmonic emission of curved gold metal nanowires. *Phys. Rev. Lett.* **2011**, *107*, 257401. [CrossRef] [PubMed]
13. Belardini, A.; Centini, M.; Leahu, G.; Fazio, E.; Sibilia, C.; Haus, J.W.; Sarangan, A. Second harmonic generation on self-assembled tilted gold nanowires. *Faraday Discuss.* **2015**, *178*, 357–362. [CrossRef] [PubMed]
14. Belardini, A.; Centini, M.; Leahu, G.; Hooper, D.C.; Li Voti, R.; Fazio, E.; Haus, J.W.; Sarangan, A.; Valev, V.K.; Sibilia, C. Chiral light intrinsically couples to extrinsic/pseudo-chiral metasurfaces made of tilted gold nanowires. *Sci. Rep.* **2016**, *6*, 31796. [CrossRef] [PubMed]
15. Bertolotti, M.; Belardini, A.; Benedetti, A.; Sibilia, C. Second harmonic circular dichroism by self-assembled metasurfaces. *J. Opt. Soc. Am. B* **2015**, *32*, 1287–1293. [CrossRef]
16. Ma, X.; Pu, M.; Li, X.; Guo, Y.; Gao, P.; Luo, X. Meta-Chirality: Fundamentals, Construction and Applications. *Nanomaterials* **2017**, *7*, 116.
17. Leahu, G.; Petronijevic, E.; Belardini, A.; Centini, M.; Sibilia, C.; Hakkarainen, T.; Koivusalo, E.; Rizzo Piton, M.; Suomalainen, S.; Guina, M. Evidence of optical circular dichroism in GaAs-based nanowires partially covered with gold. *Adv. Opt. Mater.* **2017**, *5*, 1601063. [CrossRef]
18. Petronijevic, E.; Leahu, G.; Belardini, A.; Centini, M.; Li Voti, R.; Hakkarainen, T.; Koivusalo, E.; Rizzo Piton, M.; Suomalainen, S.; Guina, M.; et al. Photo-Acoustic Spectroscopy Reveals Extrinsic Optical Chirality in GaAs-Based Nanowires Partially Covered with Gold. *Int. J. Thermophys.* **2018**, *39*, 46. [CrossRef]
19. Petronijevic, E.; Leahu, G.; Li Voti, R.; Belardini, A.; Scian, C.; Michieli, N.; Cesca, T.; Mattei, G.; Sibilia, C. Photo-acoustic detection of chirality in metal-polystyrene metasurfaces. *Appl. Phys. Lett.* **2019**, *114*, 053101. [CrossRef]

20. Kong, X.-T.; Besteiro, L.V.; Wang, Z.; Govorov, A.O. Plasmonic Chirality and Circular Dichroism in Bioassembled and Nonbiological Systems: Theoretical Background and Recent Progress. *Adv. Mater.* **2018**, 1801790. [CrossRef] [PubMed]
21. Zhao, Y.; Askarpour, A.N.; Sun, L.; Shi, J.; Li, X.; Alù, A. Chirality detection of enantiomers using twisted optical metamaterials. *Nat. Commun.* **2017**, *8*, 14180. [CrossRef] [PubMed]
22. Hendry, E.; Carpy, T.; Johnston, J.; Popland, M.; Mikhaylovskiy, R.V.; Lapthorn, A.J.; Kelly, S.M.; Barron, L.D.; Gadegaard, N.; Kadodwala, M. Ultrasensitive detection and characterization of biomolecules using superchiral fields. *Nat. Nanotechnol.* **2010**, *5*, 783–787. [CrossRef] [PubMed]
23. Tang, Y.; Cohen, A.E. Enhanced enantioselectivity in excitation of chiral molecules by superchiral light. *Science* **2011**, *332*, 333–336. [CrossRef] [PubMed]
24. Tang, Y.; Cohen, A.E. Optical chirality and its interaction with matter. *Phys. Rev. Lett.* **2010**, *104*, 163901. [CrossRef] [PubMed]
25. Vázquez-Guardado, A.; Chanda, D. Superchiral Light Generation on Degenerate Achiral Surfaces. *Phys. Rev. Lett.* **2018**, *120*, 137601. [CrossRef] [PubMed]
26. Petronijevic, E.; Centini, M.; Belardini, A.; Leahu, G.; Hakkarainen, T.; Sibilia, C. Chiral near-field manipulation in Au-GaAs hybrid hexagonal nanowires. *Opt. Express* **2017**, *25*, 17841–17848. [CrossRef] [PubMed]
27. Schäferling, M.; Yin, X.; Giessen, H. Formation of chiral fields in a symmetric environment. *Opt. Express* **2012**, *20*, 26326–26336. [CrossRef] [PubMed]
28. Davis, T.J.; Hendry, E. Superchiral electromagnetic fields created by surface plasmons in nonchiral metallic nanostructures. *Phys. Rev. B* **2013**, *87*, 085405. [CrossRef]
29. Fountaine, K.T.; Whitney, W.S.; Atwater, H.A. Resonant absorption in semiconductor nanowires and nanowire arrays: Relating leaky waveguide modes to Bloch photonic crystal modes. *J. Appl. Phys.* **2014**, *116*, 153106. [CrossRef]
30. Leahu, G.; Petronijevic, E.; Belardini, A.; Centini, M.; Li Voti, R.; Hakkarainen, T.; Koivusalo, E.; Guina, M.; Sibilia, C. Photo-acoustic spectroscopy revealing resonant absorption of self-assembled GaAs-based nanowires. *Sci. Rep.* **2017**, *7*, 2833. [CrossRef] [PubMed]
31. Petronijevic, E.; Leahu, G.; Belardini, A.; Centini, M.; Li Voti, R.; Hakkarainen, T.; Koivusalo, E.; Guina, M.; Sibilia, C. Resonant Absorption in GaAs-Based Nanowires by Means of Photo-Acoustic Spectroscopy. *Int. J. Thermophys.* **2018**, *39*, 45. [CrossRef]
32. Zhao, X.; Alizadeh, M.H.; Reinhard, B.M. Generating Optical Birefringence and Chirality in Silicon Nanowire Dimers. *ACS Photonics* **2017**, *4*, 2265–2273. [CrossRef]
33. Seo, K.; Wober, M.; Steinvurzel, P.; Schonbrun, E.; Dan, Y.; Ellenbogen, T.; Crozier, K.B. Multicolored Vertical Silicon Nanowires. *Nano Lett.* **2011**, *11*, 1851–1856. [CrossRef] [PubMed]
34. Lumerical Solutions, Inc. Available online: http://www.lumerical.com/tcad-products/fdtd/ (accessed on 20 February 2019).
35. Azizur-Rahman, K.M.; LaPierre, R.R. Wavelength-selective absorptance in GaAs, InP and InAs nanowire arrays. *Nanotechnology* **2015**, *26*, 295202. [CrossRef] [PubMed]
36. Cao, A.; Sudhölter, E.J.R.; de Smet, L.C.P.M. Silicon Nanowire-Based Devices for Gas-Phase Sensing. *Sensors* **2014**, *14*, 245–271. [CrossRef] [PubMed]
37. Su, S.; Wei, X.; Zhong, Y.; Guo, Y.; Su, Y.; Huang, Q.; Lee, S.-T.; Fan, C.; He, Y. Silicon nanowire-based molecular beacons for high-sensitivity and sequence-specific DNA multiplexed analysis. *ACS Nano* **2012**, *6*, 2582–2590. [CrossRef] [PubMed]
38. Ruizendaal, L.; Pujari, S.P.; Gevaerts, V.; Paulusse, J.M.J.; Zuilhof, H. Biofunctional Silicon Nanoparticles by Means of Thiol-Ene Click Chemistry. *Chem. Asian J.* **2011**, *6*, 2776–2786. [CrossRef] [PubMed]

Sample Availability: Samples of the compounds are not available from the authors.

molecules

MDPI

Article

An Enantioselective Potentiometric Sensor for 2-Amino-1-Butanol Based on Chiral Porous Organic Cage CC3-R

Bang-Jin Wang [1,2], Ai-Hong Duan [2], Jun-Hui Zhang [2], Sheng-Ming Xie [2], Qiu-E Cao [1,*] and Li-Ming Yuan [2,*]

1 Key Laboratory of Medicinal Chemistry for Natural Resource, Ministry of Education, School of Chemical Science and Technology, Yunnan University, Kunming 650091, China; wangbangjin711@163.com

2 Department of Chemistry, Yunnan Normal University, Kunming 650500, China; duanaihong215@aliyun.com (A.-H.D.); zjh19861202@126.com (J.-H.Z.); xieshengming_2006@163.com (S.-M.X.)

* Correspondence: qecao@ynu.edu.cn (Q.-E.C.); yuan_limingpd@126.com (L.-M.Y.); Tel./Fax: +86-871-6503-3679 (Q.-E.C.); +86-871-6594-1088 (L.-M.Y.)

Received: 23 December 2018; Accepted: 23 January 2019; Published: 24 January 2019

Abstract: Porous organic cages (POCs) have attracted extensive attention due to their unique structures and tremendous application potential in numerous areas. In this study, an enantioselective potentiometric sensor composed of a polyvinyl chloride (PVC) membrane electrode modified with CC3-R POC material was used for the recognition of enantiomers of 2-amino-1-butanol. After optimisation, the developed sensor exhibited enantioselectivity toward *S*-2-amino-1-butanol ($\log K_{S,R}^{Pot} = -0.98$) with acceptable sensitivity, and a near-Nernstian response of 25.8 ± 0.3 mV/decade within a pH range of 6.0–9.0.

Keywords: enantioselective potentiometric sensor; 2-amino-1-butanol; chiral porous organic cage; CC3-R; PVC membrane electrode

1. Introduction

Chirality is a general phenomenon and an important characteristic in naturally occurring molecules. For instance, most amino acids are levorotatory and sugars are dextrorotatory in biological systems. Consequently, chiral discrimination has attracted tremendous attention on account of its significance in pharmaceutical, biomedicine and chemical fields. Currently, chiral discrimination can be precisely achieved in many ways including gas chromatography (GC), high-performance liquid chromatography (HPLC) and high-performance capillary electrophoresis (HPCE). Although these methods have different advantages in terms of sensitivity or applicability, they suffer similar drawbacks including complicated operation and the need for expensive equipment. By contrast, ion-selective electrodes are simple, rapid and affordable, and have been widely applied to the enantioselective recognition and detection of chiral compounds in recent years [1–9].

As versatile functional material platforms, porous organic cages (POCs) have attracted much attention [10–13], and have been widely applied in various areas such as gas-selective adsorption and separation [14–17], molecular recognition [18–27], catalysis [28], water treatment [29] and sensing [30]. As shown in Figure 1, the *R*-type chiral POC CC3-R has an interlinked chiral pore channel structure with adjacent tetrahedral cages packed together [31,32]. The chiral pore channel structures and cyclohexyl, imino and phenyl groups of cage molecules constitute a chiral microenvironment comprising a variety of enantioselective factors including dispersion forces, dipolar interactions and hydrogen bonds [33]. These properties combine to make CC3-R an excellent chiral selector for use in chiral recognition methods. Recently, a CC3-R-modified GC stationary phase was developed for the separation of racemates, and exhibited excellent enantioselectivity [34].

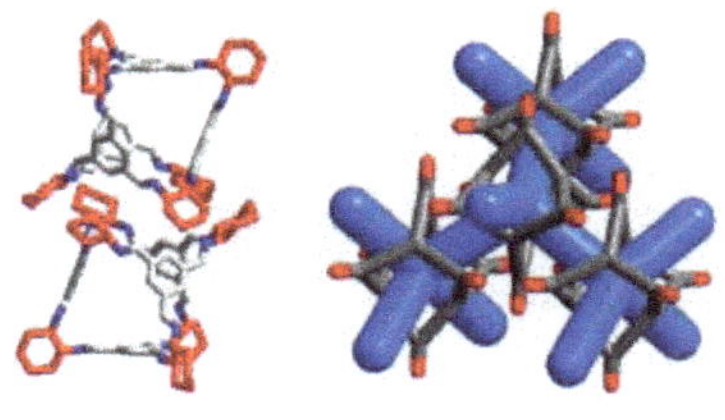

Figure 1. Schematic diagram of the structure of CC3-R.

2-Amino-1-butanol (Figure 2) is generally used as an intermediate in the synthesis of pharmaceuticals such as the bacteriostatic antituberculosis agent (*S*,*S*)-ethambutol [35–38]. In the present work, CC3-R was applied as a chiral selector in PVC membrane electrodes, resulting in impressive enantioselectivity for *S*-2-amino-1-butanol. Factors influencing the enantioselectivity of the CC3-R-based membrane electrode, such as the content of CC3-R, the category of plasticiser and the pH value of analyte solutions, were systematically investigated.

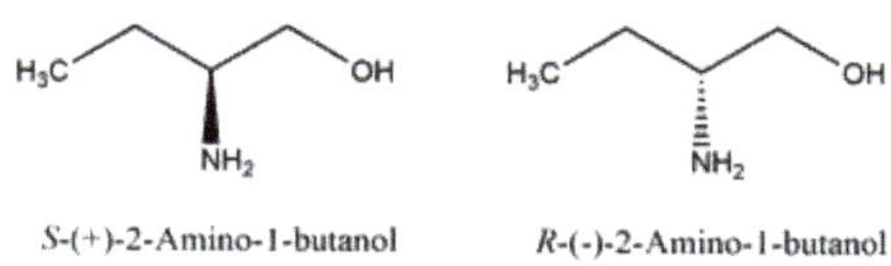

Figure 2. Molecular structure of 2-amino-1-butanol.

2. Results and Discussion

2.1. Characterisation of the Synthesised CC3-R

The synthesised CC3-R crystals were characterised by Nuclear Magnetic Resonance (NMR), Powder X-ray diffraction (PXRD) and elemental analysis. As can be seen in Figure 3, the PXRD pattern of synthesised CC3-R crystals was consistent with the Singlecrystal simulation. Furthermore, CC3-R retained the same crystallinity and structure whether recrystallised from tetrahydrofuran or rinsed with water for 48 h, demonstrating excellent stability as chiral selector in the membrane electrode.

Elemental analysis was performed on CC3-R ($C_{72}H_{85}N_{12}$); calculated = C 77.31, H 7.66, N 15.03; detected = C 77.08, H 7.76, N 14.88. ^{1}H-NMR ($CDCl_3$) δ = 8.18 (s, 12H, -C*H*=N-), 7.92 (s, 12H, -Ar-*H*), 3.36 (m, 12H, -CHN-), 1.86–1.54 (m, 48H, -CH_2-) ppm. ^{13}C-NMR ($CDCl_3$) δ = 159.15, 136.64, 129.53, 74.65, 33.02, 24.39 ppm. All characterisation data confirmed that CC3-R was successfully synthesised.

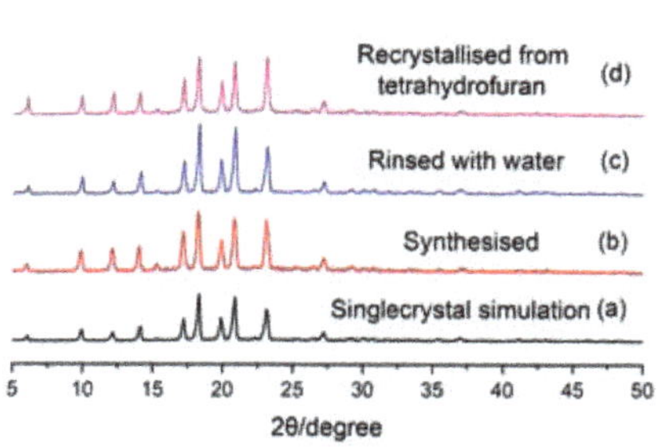

Figure 3. PXRD patterns of CC3-R: (**a**) Singlecrystal simulation, (**b**) synthesised, (**c**) rinsed with water for 48 h, (**d**) recrystallised from tetrahydrofuran.

2.2. Optimisation of Membrane Components

The nature and amount of chiral selector and plasticiser contained in the membrane can strongly influence the selectivity and sensitivity of the membrane electrode. Consequently, the potential response characteristics of multiple electrodes with different quantities of CC3-R and three types of plasticiser (*o*-NPOE, DOS and DBP) were evaluated.

Figure 4 shows the potential response characteristics of membrane electrodes with different CC3-R mass percentages. The performance of the membrane electrode improved with increasing CC3-R content, and the best enantioselectivity toward *S*-2-amino-1-butanol was achieved with 3% CC3-R (by weight). However, the enantioselectivity and sensitivity decreased slightly when the amount of CC3-R reached 4%. It is possible that the PVC membrane becomes saturated, hence the number of recognition sites does not increase proportionately with the chiral selector. Moreover, excess CC3-R could affect the ion-exchange capacity of the membrane electrode.

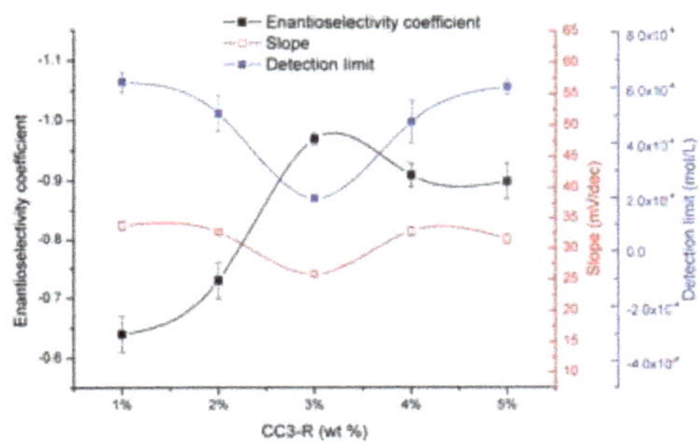

Figure 4. Potential response characteristics of membrane electrodes with different CC3-R contents.

The influence of the type of plasticiser is shown in Figure 5. DOS and DBP were clearly inferior to *o*-NPOE in terms of detection limit and enantioselectivity coefficient for *S*-2-amino-1-butanol.

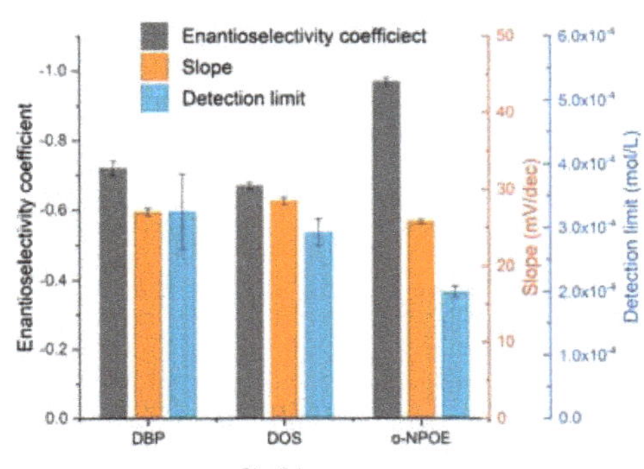

Figure 5. Potential response characteristics of membrane electrodes with different types of plasticiser.

2.3. Effect of pH on the Electrode

In order to investigate the effect of pH on the response performance of the optimised membrane electrode, the potential response value of the 2-amino-1-butanol solution (1.0×10^{-3} mol·L^{-1}) was measured at different pH values (pH 2.0–12.0). As shown in Figure 6, the potential response value was stable within a pH range of 5.0–9.0. Furthermore, a large difference between the two enantiomers was observed at pH 9.0. Therefore, pH 9.0 was adopted for measurements using the optimised membrane electrode.

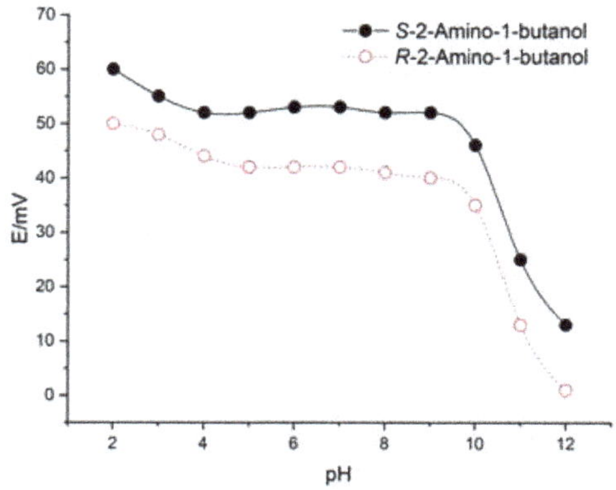

Figure 6. The influence of pH on the response performance of the membrane electrode to *S*/*R*-2-amino-1-butanol.

2.4. Enantioselectivity Coefficient of the Electrode

Figure 4 shows $\log K_{S,R}^{Pot}$ values for membrane electrodes of varying composition. The optimised membrane electrode containing 3 wt% CC3-R displayed impressive enantioselectivity toward *S*-2-amino-1-butanol over *R*-2-amino-1-butanol ($\log K_{S,R}^{Pot} = -0.98$). By comparison, the CC3-S (3 wt%) modified membrane electrode showed similar enantioselectivity toward *R*-2-amino-1-butanol ($\log K_{S,R}^{Pot} = -0.94$).

Furthermore, the $\log K_{S,int}^{Pot}$ value was used to evaluate the enantioselectivity of the optimised electrode in the presence of interfering ions with a similar configuration to 2-amino-1-butanol [39]. Specifically, the potential values of *R*/*S*-2-amino-3-phenyl-1-propanol, *R*/*S*-2-amino-3-methyl-1-butanol and *R*/*S*-3-amino-1,2-propanediol (0.1 mol·L^{-1}) were measured, and $\log K_{S,int}^{Pot}$ values are shown in Table 1.

Table 1. $\text{Log}K_{S,int}^{Pot}$ values of ions potentially interfering with *S*-2-amino-1-butanol.

Interference Ion	$\text{Log}K_{S,int}^{Pot}$
R-2-Amino-1-butanol	−0.98
S-2-Amino-3-phenyl-1-propanol	−0.59
R-2-Amino-3-phenyl-1-propanol	−0.59
S-2-Amino-3-methyl-1-butanol	0.31
R-2-Amino-3-methyl-1-butanol	0.26
S-3-Amino-1,2-propanediol	−0.41
R-3-Amino-1,2-propanediol	−0.41

As shown in Table 1, the membrane electrode exhibited comparable responses to other alkamines with similar configurations to 2-amino-1-butanol. Steric hindrance caused by additional organic groups of other alkamines could impair the recognition performance during ion exchange. The electrode displayed slight enantioselective recognition of enantiomers of 2-amino-3-methyl-1-butanol, which have the most similar configuration. However, 2-amino-3-methyl-1-butanol yielded similar potential response values, and caused significant interference.

2.5. Recognition of Mixing Samples

To further explore the enantioselectivity of the developed membrane electrode, a mixing sample test was conducted using different molar ratios of *S*- and *R*-enantiomers of 2-amino-1-butanol (Figure 7). The results showed that the potential response values of the mixing solution increased with increasing proportion of *S*-2-amino-1-butanol, revealing a clear positive linear correlation between the proportion of *S*-2-amino-1-butanol and potential response values of mixing solutions (0.1 mol·L^{-1}). These results demonstrate the selective recognition of *S*-2-amino-1-butanol in the presence of *R*-2-amino-1-butanol.

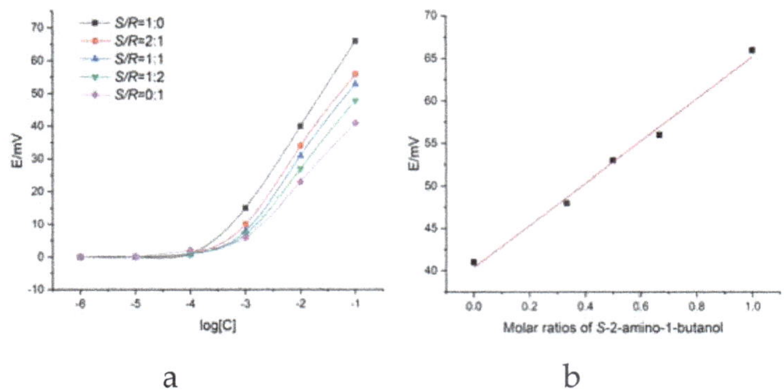

Figure 7. (**a**) Potential responses of *S*- and *R*-enantiomers of 2-amino-1-butanol mixing solutions. (**b**) Positive linear correlation between the proportion of *S*-2-amino-1-butanol and potential response values of mixing solutions (0.1 mol·L^{-1}).

3. Materials and Methods

3.1. Materials

Enantiomers of 2-amino-1-butanol, 2-amino-3-phenyl-1-propanol, 2-amino-3-methyl-1-butanol, and 3-amino-1,2-propanediol were obtained from Aladdin (Shanghai, China). (*R,R*)-1,2-Diaminocyclohexane and 1,3,5-triformylbenzene were purchased from Acros (Geel, Belgium). *o*-Nitrophenyl Octyl Ether (*o*-NPOE), dioctyl sebacate (DOS) and dibutyl phthalate (DBP) were obtained from TCI (Tokyo, Japan). Polyvinyl chloride (PVC) powder and trifluoroacetic acid were purchased from Sigma-Aldrich (St. Louis, MO, USA). All other reagents were of analytical grade. Deionised water was used to prepare and dilute all buffer and analyte solutions.

3.2. Synthesis of CC3-R

CC3-R was synthesised using a previously reported method [29]. Briefly, 20 mL dichloromethane was added dropwise onto 1.0 g 1,3,5-triformylbenzene in a two-necked flask without stirring at room temperature, and 20 μL trifluoroacetic acid was added as a catalyst. Within minutes, 20 mL dichloromethane containing 1.0 g (*R,R*)-1,2-diaminocyclohexane was dripped slowly into the mixture. After reaction for 72 h at room temperature, white crystals were present on the wall of the flask, which were filtered and rinsed with ethanol/dichloromethane (95:5 v/v).

3.3. Preparation of Enantioselective Membrane Electrodes

To prepare the PVC membranes, PVC powder, plasticiser (*o*-NPOE), and CC3-R were added to 3 mL tetrahydrofuran and stirred to form a transparent solution [40]. This was poured onto a glass sheet and volatilised for 24 h to form a semitransparent film ~0.5 mm thick. The obtained film was incised into an appropriately sized disc and assembled using a PVC tube, which was subsequently filled with 0.1 mol·L^{-1} KCl as an internal reference solution. A silver chloride electrode was applied as an internal reference electrode, and a saturated calomel electrode was utilised as a reference electrode. For comparison, a CC3-*S*-modified membrane electrode was prepared in the same way. The overall strategy for enantioselective potentiometric sensor fabrication is depicted in Scheme 1.

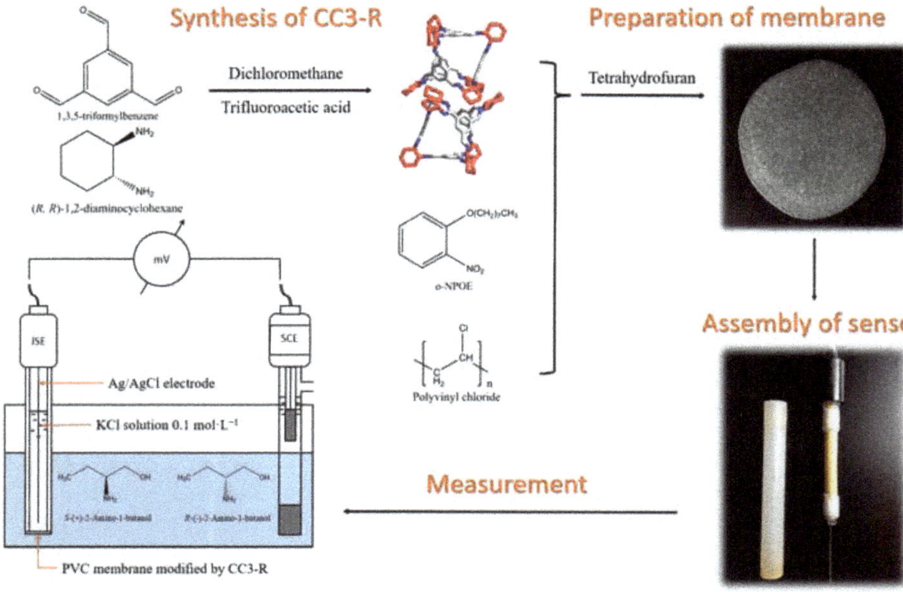

Scheme 1. Schematic illustration of enantioselective potentiometric sensor fabrication.

3.4. Potentiometric Measurement

The direct potentiometric method was applied to measure the potential value of each S/R-2-amino-1-butanol solution at different molar concentrations (1.0×10^{-6} to 1.0×10^{-1} mol·L^{-1}) and mixing solution with different molar ratios (S/R = 1:0, 2:1, 1:1, 1:2, and 0:1). A Model PHS-3C pH meter (Leici, Shanghai, China) was used for potentiometric and pH measurements, and all potentiometric measurements were performed during stirring at room temperature. Before measurement, membrane electrodes were soaked in S-2-amino-1-butanol solution (1.0×10^{-3} mol·L^{-1}) for 24 h.

A revised separate solution method was used to calculate the enantioselectivity coefficient ($\log K_{S,R}^{Pot}$) with the following formula:

$$\log K_{S,R}^{Pot} = \frac{E_R - E_S}{D}$$

where E_R and E_S are the potentials of 0.1 mol·L^{-1} R- and S-2-amino-1-butanol solutions, respectively, and D is the slope of the response curve of S-2-amino-1-butanol.

4. Conclusions

The chiral porous organic cage CC3-R proved to be a useful chiral selector for the modification of PVC membrane electrodes to generate enantioselective potentiometric sensors. The optimised membrane electrode containing 3 wt% CC3-R exhibited enantiomeric recognition toward S-2-amino-1-butanol over R-2-amino-1-butanol ($\log K_{S,R}^{Pot} = -0.98$) with acceptable sensitivity, and a near-Nernst response of 25.8 ± 0.3 mV/decade toward S-2-amino-1-butanol at pH 9.0.

Author Contributions: Conceptualization, Q.-E.C. and L.-M.Y.; Methodology, A.-H.D.; Writing—Original draft preparation, B.-J.W.; Writing—Review and editing, J.-H.Z. and S.-M.X.

Funding: This research was funded by the National Nature Science Foundation of China (No. 21665028, No. 21265026, and No. 21465025).

Conflicts of Interest: The authors declare no conflict of interest.

References

1. Duan, A.H.; Wang, B.J.; Xie, S.M.; Zhang, J.H.; Yuan, L.M. A chiral, porous, organic cage-based, enantioselective potentiometric sensor for 2-aminobutanol. *Chirality* **2017**, *29*, 172–177. [CrossRef] [PubMed]
2. Xu, L.; Yang, Y.Y.; Wang, Y.Q.; Gao, J.Z. Chiral salen Mn(III) complex-based enantioselective potentiometric sensor for L-mandelic acid. *Anal. Chim. Acta* **2009**, *653*, 217–221. [CrossRef] [PubMed]
3. Chen, Y.; Chen, L.; Bi, R.L.; Xu, L.; Liu, Y. A potentiometric chiral sensor for l-Phenylalanine based on crosslinked polymethylacrylic acid–polycarbazole hybrid molecularly imprinted polymer. *Anal. Chim. Acta* **2012**, *754*, 83–90. [CrossRef] [PubMed]
4. Stefan-van, S.R.I.; Nhlapo, N.S.; Staden-van, J.F.; Aboul-Enein, H.Y. Enantioanalysis of (-)butaclamol using vancomycin and teicoplanin as chiral Selectors. *Comb. Chem. High Throughput Screen.* **2010**, *13*, 690–693. [CrossRef]
5. Zhou, Y.X.; Nagaoka, T.; Yu, B.; Levon, K. Chiral ligand exchange potentiometric aspartic acid sensors with polysiloxane films containing a chiral ligand N-carbobenzoxy-aspartic acid. *Anal. Chem.* **2009**, *81*, 1888–1892. [CrossRef]
6. Stefan-van, S.R.I.; Staden-van, J.F.; Aboul-Enein, H.Y. Macrocyclic antibiotics as chiral selectors in the design of enantioselective, potentiometric membrane electrodes for the determination of S-flurbiprofen. *Anal. Bioanal. Chem.* **2009**, *394*, 821–826. [CrossRef]
7. Yin, X.L.; Ding, J.J.; Zhang, S.; Kong, J.L. Enantioselective sensing of chiral amino acids by potentiometric sensors based on optical active polyaniline films. *Biosens. Bioelectron.* **2006**, *21*, 2184–2187. [CrossRef]
8. Ozoemena, K.I.; Stefan, R.I. Enantioselective potentiometric membrane electrodes based on α-, β- and γ-cyclodextrins as chiral selectors for the assay of l-proline. *Talanta* **2005**, *66*, 501–504. [CrossRef]
9. Rat'ko, A.A.; Stefan, R.I.; Staden-van, J.F.; Aboul-Enein, H.Y. Macrocyclic antibiotics as chiral selectors in the design of enantioselective, potentiometric membrane electrodes for the determination of L- and D-enantiomers of methotrexate. *Talanta* **2004**, *64*, 145–150. [CrossRef]
10. Hui, M.D.; Yang, Y.H.; Li, B.J.; Pan, F.; Zhu, G.Z.; Zeller, M.; Yuan, D.Q.; Wang, C. Targeted synthesis of a large tria zine-based [4+6] organic molecular cage: Structure, porosity and gas separation. *Chem. Commun.* **2015**, *51*, 1976–1979.
11. Zhang, G.; Presly, O.; White, F.; Oppel, I.M.; Mastalerz, M. A permanent mesoporous organic cage with an exceptionally high surface area. *Angew. Chem. Int. Ed.* **2014**, *53*, 1516–1520. [CrossRef] [PubMed]
12. Zhang, G.; Mastalerz, M. Organic cage compounds—From shape-persistency to function. *Chem. Soc. Rev.* **2014**, *43*, 1934–1947. [CrossRef] [PubMed]
13. Liu, M.; Little, M.A.; Jelfs, K.E.; Jones, J.T.A.; Schmidtmann, M.; Chong, S.Y.; Hasell, T.; Cooper, A.I. Acid- and base-stable porous organic cages: Shape persistence and pH stability via post-synthetic "tying" of a flexible amine cage. *J. Am. Chem. Soc.* **2014**, *136*, 7583–7586. [CrossRef] [PubMed]
14. Hasell, T.; Miklitz, M.; Stephenson, A.; Little, M.A.; Chong, S.Y.; Clowes, R.; Chen, L.J.; Holden, D.; Tribello, G.A.; Jelfs, K.E.; et al. Porous organic cages for sulfur hexafluoride separation. *J. Am. Chem. Soc.* **2016**, *138*, 1653–1659. [CrossRef] [PubMed]
15. Mastalerz, M.; Schneider, M.W.; Oppel, I.M.; Presly, O. A salicylbisimine cage compound with high surface area and selective CO_2/CH_4 adsorption. *Angew. Chem. Int. Ed.* **2011**, *50*, 1046–1051. [CrossRef] [PubMed]
16. Jin, Y.H.; Voss, B.A.; Jin, A.; Long, H.; Noble, R.D.; Zhang, W. Highly CO_2-selective organic molecular cages: What determines the CO_2 selectivity. *J. Am. Chem. Soc.* **2011**, *133*, 6650–6658. [CrossRef] [PubMed]
17. Mitra, T.; Wu, X.F.; Clowes, R.; Jones, J.T.A.; Jelfs, K.E.; Adams, D.J.; Trewin, A.; Bacsa, J.; Steiner, A.; Cooper, A.I. A soft porous organic cage crystal with complex gas sorption behavior. *Chem. Eur. J.* **2011**, *17*, 10235–10240. [CrossRef] [PubMed]

18. Zhang, J.H.; Xie, S.M.; Wang, B.J.; He, P.G.; Yuan, L.M. A homochiral porous organic cage with large cavity and pore windows for the efficient gas chromatography separation of enantiomers and positional isomers. *J. Sep. Sci.* **2018**, *41*, 1385–1394. [CrossRef]
19. Zhang, J.H.; Zhu, P.J.; Xie, S.M.; Zi, M.; Yuan, L.M. Homochiral porous organic cage used as stationary phase for open tubular capillary electrochromatography. *Anal. Chim. Acta* **2018**, *999*, 169–175. [CrossRef]
20. Song, Q.L.; Jiang, S.; Hasell, T.; Liu, M.; Sun, S.J.; Cheetham, A.K.; Sivaniah, E.; Cooper, A.I. Porous organic cage thin films and molecular-sieving membranes. *Adv. Mater.* **2016**, *28*, 2629–2637. [CrossRef]
21. Xie, S.M.; Zhang, J.H.; Fu, N.; Wang, B.J.; Hu, C.; Yuan, L.M. Application of homochiral alkylated organic cages as chiral stationary phases for molecular separations by capillary gas chromatography. *Molecules* **2016**, *21*, 1466. [CrossRef] [PubMed]
22. Xie, S.M.; Zhang, J.H.; Fu, N.; Wang, B.J.; Chen, L.; Yuan, L.M. A chiral porous organic cage for molecular recognition using gas chromatography. *Anal. Chim. Acta* **2016**, *903*, 156–163. [CrossRef] [PubMed]
23. Zhang, J.H.; Xie, S.M.; Wang, B.J.; He, P.G.; Yuan, L.M. Highly selective separation of enantiomers using a chiral porous organic cage. *J. Chromatogr. A* **2015**, *1426*, 174–182. [CrossRef] [PubMed]
24. Kewley, A.; Stephenson, A.; Chen, L.J.; Briggs, M.E.; Hasell, T.; Cooper, A.I. Porous organic cages for gas chromatography separations. *Chem. Mater.* **2015**, *27*, 3207–3210. [CrossRef]
25. Mitra, T.; Jelfs, K.E.; Schmidtmann, M.; Ahmed, A.; Chong, S.Y.; Adams, D.J.; Cooper, A.I. Molecular shape sorting using molecular organic cages. *Nat. Chem.* **2013**, *5*, 276–281. [CrossRef] [PubMed]
26. Chen, L.J.; Reiss, P.S.; Chong, S.Y.; Holden, D.; Jelfs, K.E.; Hasell, T.; Little, M.A.; Kewley, A.; Briggs, M.E.; Stephenson, A.; et al. Separation of rare gases and chiral molecules by selective binding in porous organic cages. *Nat. Mater.* **2014**, *13*, 954–960. [CrossRef] [PubMed]
27. Xie, S.M.; Yuan, L.M. Recent progress of chiral stationary phases for separation of enantiomers in gas chromatography. *J. Sep. Sci.* **2017**, *40*, 124–137. [CrossRef]
28. Zhang, Y.; Xiong, Y.; Ge, J.; Lin, R.; Chen, C.; Peng, Q.; Wang, D.S.; Li, Y.D. Porous organic cage stabilised palladium nanoparticles: Efficient heterogeneous catalysts for carbonylation reaction of aryl halides. *Chem. Commun.* **2018**, *54*, 2796–2799. [CrossRef]
29. Kong, X.; Jiang, J.W. Amorphous porous organic cage membranes for water desalination. *J. Phys. Chem. C* **2018**, *122*, 1732–1740. [CrossRef]
30. Brutschy, M.; Schneider, M.W.; Mastalerz, M.; Waldvogel, S.R. Porous organic cage compounds as highly potent affinity materials for sensing by quartz crystal microbalances. *Adv. Mater.* **2012**, *24*, 6049–6052. [CrossRef]
31. Bojdys, M.J.; Briggs, M.; Jones, J.T.A.; Adams, D.J.; Chong, S.Y.; Schmidtmann, M.; Cooper, A.I. Supramolecular engineering of intrinsic and extrinsic porosity in covalent organic cages. *J. Am. Chem. Soc.* **2011**, *133*, 16566–16571. [CrossRef] [PubMed]
32. Tozawa, T.; Jones, J.T.A.; Swamy, S.I.; Jiang, S.; Adams, D.J.; Shakespeare, S.; Clowes, R.; Bradshaw, D.; Hasell, T.; Chong, S.Y.; et al. Porous organic cages. *Nat. Mater.* **2009**, *8*, 973–978. [CrossRef] [PubMed]
33. Yamamoto, C.; Yashima, E.; Okamoto, Y. Structural analysis of amylose tris(3,5-dimethylphenylcarbamate) by NMR relevant to its chiral recognition mechanism in HPLC. *J. Am. Chem. Soc.* **2002**, *124*, 12583–12589. [CrossRef] [PubMed]
34. Zhang, J.H.; Xie, S.M.; Chen, L.; Wang, B.J.; He, P.G.; Yuan, L.M. Homochiral porous organic cage with high selectivity for the separation of racemates in gas chromatography. *Anal. Chem.* **2015**, *87*, 7817–7824. [CrossRef] [PubMed]
35. Goncalves, R.S.B.; Da-Silva, E.T.; De-Souza, M.V.N. An environmentally friendly, scalable and highly efficient synthesis of (*S,S*)-Ethambutol, a first line drug against tuberculosis. *Lett. Org. Chem.* **2015**, *12*, 478–481. [CrossRef]
36. Dobrikov, G.M.; Valcheva, V.; Stoilova-Disheva, M.; Momekov, G.; Tzvetkova, P.; Chimov, A.; Dimitrov, V. Synthesis and in vitro antimycobacterial activity of compounds derived from (R)- and (S)-2-amino-1-butanol—The crucial role of the configuration. *Eur. J. Med. Chem.* **2012**, *48*, 45–56. [CrossRef] [PubMed]
37. Dobrikov, G.M.; Valcheva, V.; Nikolova, Y.; Ugrinova, I.; Pasheva, E.; Dimitrov, V. Efficient synthesis of new (R)-2-amino-1-butanol derived ureas, thioureas and acylthioureas and in vitro evaluation of their antimycobacterial activity. *Eur. J. Med. Chem.* **2013**, *63*, 468–473. [CrossRef]

38. Dobrikov, G.M.; Valcheva, V.; Nikolova, Y.; Ugrinova, I.; Pasheva, E.; Dimitrov, V. Enantiopure antituberculosis candidates synthesized from (-)-fenchone. *Eur. J. Med. Chem.* **2014**, *77*, 243–247. [CrossRef]
39. Horváth, V.; Takécs, T.; Horvai, G.; Huszthy, P.; Bradshaw, J.S.; Izatt, R.M. Enantiomer-Selectivity of ion-selective electrodes based on a chiral crown-ether ionophore. *Anal. Lett.* **1997**, *9*, 1591–1609. [CrossRef]
40. Moody, G.J.; Oke, R.B.; Thomas, J.D.R. A calcium-sensitive electrode based on a liquid ion exchanger in a poly(vinyl chloride) matrix. *Analyst* **1970**, *95*, 910–918. [CrossRef]

Sample Availability: Not available.

MDPI
St. Alban-Anlage 66
4052 Basel
Switzerland
Tel. +41 61 683 77 34
Fax +41 61 302 89 18
www.mdpi.com

Molecules Editorial Office
E-mail: molecules@mdpi.com
www.mdpi.com/journal/molecules

www.ingramcontent.com/pod-product-compliance
Lightning Source LLC
Chambersburg PA
CBHW061607220926
51850CB00005B/447
9783039362387